Pareto Distributions

Second Edition

MONOGRAPHS ON STATISTICS AND APPLIED PROBABILITY

General Editors

F. Bunea, V. Isham, N. Keiding, T. Louis, R. L. Smith, and H. Tong

1. Stochastic Population Models in Ecology and Epidemiology *M.S. Barlett* (1960)
2. Queues *D.R. Cox and W.L. Smith* (1961)
3. Monte Carlo Methods *J.M. Hammersley and D.C. Handscomb* (1964)
4. The Statistical Analysis of Series of Events *D.R. Cox and P.A.W. Lewis* (1966)
5. Population Genetics *W.J. Ewens* (1969)
6. Probability, Statistics and Time *M.S. Barlett* (1975)
7. Statistical Inference *S.D. Silvey* (1975)
8. The Analysis of Contingency Tables *B.S. Everitt* (1977)
9. Multivariate Analysis in Behavioural Research *A.E. Maxwell* (1977)
10. Stochastic Abundance Models *S. Engen* (1978)
11. Some Basic Theory for Statistical Inference *E.J.G. Pitman* (1979)
12. Point Processes *D.R. Cox and V. Isham* (1980)
13. Identification of Outliers *D.M. Hawkins* (1980)
14. Optimal Design *S.D. Silvey* (1980)
15. Finite Mixture Distributions *B.S. Everitt and D.J. Hand* (1981)
16. Classification *A.D. Gordon* (1981)
17. Distribution-Free Statistical Methods, 2nd edition *J.S. Maritz* (1995)
18. Residuals and Influence in Regression *R.D. Cook and S. Weisberg* (1982)
19. Applications of Queueing Theory, 2nd edition *G.F. Newell* (1982)
20. Risk Theory, 3rd edition *R.E. Beard, T. Pentikäinen and E. Pesonen* (1984)
21. Analysis of Survival Data *D.R. Cox and D. Oakes* (1984)
22. An Introduction to Latent Variable Models *B.S. Everitt* (1984)
23. Bandit Problems *D.A. Berry and B. Fristedt* (1985)
24. Stochastic Modelling and Control *M.H.A. Davis and R. Vinter* (1985)
25. The Statistical Analysis of Composition Data *J. Aitchison* (1986)
26. Density Estimation for Statistics and Data Analysis *B.W. Silverman* (1986)
27. Regression Analysis with Applications *G.B. Wetherill* (1986)
28. Sequential Methods in Statistics, 3rd edition *G.B. Wetherill and K.D. Glazebrook* (1986)
29. Tensor Methods in Statistics *P. McCullagh* (1987)
30. Transformation and Weighting in Regression *R.J. Carroll and D. Ruppert* (1988)
31. Asymptotic Techniques for Use in Statistics *O.E. Bandorff-Nielsen and D.R. Cox* (1989)
32. Analysis of Binary Data, 2nd edition *D.R. Cox and E.J. Snell* (1989)
33. Analysis of Infectious Disease Data *N.G. Becker* (1989)
34. Design and Analysis of Cross-Over Trials *B. Jones and M.G. Kenward* (1989)
35. Empirical Bayes Methods, 2nd edition *J.S. Maritz and T. Lwin* (1989)
36. Symmetric Multivariate and Related Distributions *K.T. Fang, S. Kotz and K.W. Ng* (1990)
37. Generalized Linear Models, 2nd edition *P. McCullagh and J.A. Nelder* (1989)
38. Cyclic and Computer Generated Designs, 2nd edition *J.A. John and E.R. Williams* (1995)
39. Analog Estimation Methods in Econometrics *C.F. Manski* (1988)
40. Subset Selection in Regression *A.J. Miller* (1990)
41. Analysis of Repeated Measures *M.J. Crowder and D.J. Hand* (1990)
42. Statistical Reasoning with Imprecise Probabilities *P. Walley* (1991)
43. Generalized Additive Models *T.J. Hastie and R.J. Tibshirani* (1990)
44. Inspection Errors for Attributes in Quality Control *N.L. Johnson, S. Kotz and X. Wu* (1991)
45. The Analysis of Contingency Tables, 2nd edition *B.S. Everitt* (1992)
46. The Analysis of Quantal Response Data *B.J.T. Morgan* (1992)
47. Longitudinal Data with Serial Correlation—A State-Space Approach *R.H. Jones* (1993)

48. Differential Geometry and Statistics *M.K. Murray and J.W. Rice* (1993)
49. Markov Models and Optimization *M.H.A. Davis* (1993)
50. Networks and Chaos—Statistical and Probabilistic Aspects
O.E. Barndorff-Nielsen, J.L. Jensen and W.S. Kendall (1993)
51. Number-Theoretic Methods in Statistics *K.-T. Fang and Y. Wang* (1994)
52. Inference and Asymptotics *O.E. Barndorff-Nielsen and D.R. Cox* (1994)
53. Practical Risk Theory for Actuaries *C.D. Daykin, T. Pentikäinen and M. Pesonen* (1994)
54. Biplots *J.C. Gower and D.J. Hand* (1996)
55. Predictive Inference—An Introduction *S. Geisser* (1993)
56. Model-Free Curve Estimation *M.E. Tarter and M.D. Lock* (1993)
57. An Introduction to the Bootstrap *B. Efron and R.J. Tibshirani* (1993)
58. Nonparametric Regression and Generalized Linear Models *P.J. Green and B.W. Silverman* (1994)
59. Multidimensional Scaling *T.F. Cox and M.A.A. Cox* (1994)
60. Kernel Smoothing *M.P. Wand and M.C. Jones* (1995)
61. Statistics for Long Memory Processes *J. Beran* (1995)
62. Nonlinear Models for Repeated Measurement Data *M. Davidian and D.M. Giltinan* (1995)
63. Measurement Error in Nonlinear Models *R.J. Carroll, D. Rupert and L.A. Stefanski* (1995)
64. Analyzing and Modeling Rank Data *J.J. Marden* (1995)
65. Time Series Models—In Econometrics, Finance and Other Fields
D.R. Cox, D.V. Hinkley and O.E. Barndorff-Nielsen (1996)
66. Local Polynomial Modeling and its Applications *J. Fan and I. Gijbels* (1996)
67. Multivariate Dependencies—Models, Analysis and Interpretation *D.R. Cox and N. Wermuth* (1996)
68. Statistical Inference—Based on the Likelihood *A. Azzalini* (1996)
69. Bayes and Empirical Bayes Methods for Data Analysis *B.P. Carlin and T.A Louis* (1996)
70. Hidden Markov and Other Models for Discrete-Valued Time Series *I.L. MacDonald and W. Zucchini* (1997)
71. Statistical Evidence—A Likelihood Paradigm *R. Royall* (1997)
72. Analysis of Incomplete Multivariate Data *J.L. Schafer* (1997)
73. Multivariate Models and Dependence Concepts *H. Joe* (1997)
74. Theory of Sample Surveys *M.E. Thompson* (1997)
75. Retrial Queues *G. Falin and J.G.C. Templeton* (1997)
76. Theory of Dispersion Models *B. Jørgensen* (1997)
77. Mixed Poisson Processes *J. Grandell* (1997)
78. Variance Components Estimation—Mixed Models, Methodologies and Applications *P.S.R.S. Rao* (1997)
79. Bayesian Methods for Finite Population Sampling *G. Meeden and M. Ghosh* (1997)
80. Stochastic Geometry—Likelihood and computation
O.E. Barndorff-Nielsen, W.S. Kendall and M.N.M. van Lieshout (1998)
81. Computer-Assisted Analysis of Mixtures and Applications—Meta-Analysis, Disease Mapping and Others
D. Böhning (1999)
82. Classification, 2nd edition *A.D. Gordon* (1999)
83. Semimartingales and their Statistical Inference *B.L.S. Prakasa Rao* (1999)
84. Statistical Aspects of BSE and vCJD—Models for Epidemics *C.A. Donnelly and N.M. Ferguson* (1999)
85. Set-Indexed Martingales *G. Ivanoff and E. Merzbach* (2000)
86. The Theory of the Design of Experiments *D.R. Cox and N. Reid* (2000)
87. Complex Stochastic Systems *O.E. Barndorff-Nielsen, D.R. Cox and C. Klüppelberg* (2001)
88. Multidimensional Scaling, 2nd edition *T.F. Cox and M.A.A. Cox* (2001)
89. Algebraic Statistics—Computational Commutative Algebra in Statistics
G. Pistone, E. Riccomagno and H.P. Wynn (2001)
90. Analysis of Time Series Structure—SSA and Related Techniques
N. Golyandina, V. Nekrutkin and A.A. Zhigljavsky (2001)
91. Subjective Probability Models for Lifetimes *Fabio Spizzichino* (2001)
92. Empirical Likelihood *Art B. Owen* (2001)
93. Statistics in the 21st Century *Adrian E. Raftery, Martin A. Tanner, and Martin T. Wells* (2001)
94. Accelerated Life Models: Modeling and Statistical Analysis
Vilijandas Bagdonavicius and Mikhail Nikulin (2001)

95. Subset Selection in Regression, Second Edition *Alan Miller* (2002)
96. Topics in Modelling of Clustered Data *Marc Aerts, Helena Geys, Geert Molenberghs, and Louise M. Ryan* (2002)
97. Components of Variance *D.R. Cox and P.J. Solomon* (2002)
98. Design and Analysis of Cross-Over Trials, 2nd Edition *Byron Jones and Michael G. Kenward* (2003)
99. Extreme Values in Finance, Telecommunications, and the Environment
Bärbel Finkenstädt and Holger Rootzén (2003)
100. Statistical Inference and Simulation for Spatial Point Processes
Jesper Møller and Rasmus Plenge Waagepetersen (2004)
101. Hierarchical Modeling and Analysis for Spatial Data
Sudipto Banerjee, Bradley P. Carlin, and Alan E. Gelfand (2004)
102. Diagnostic Checks in Time Series *Wai Keung Li* (2004)
103. Stereology for Statisticians *Adrian Baddeley and Eva B. Vedel Jensen* (2004)
104. Gaussian Markov Random Fields: Theory and Applications *Håvard Rue and Leonhard Held* (2005)
105. Measurement Error in Nonlinear Models: A Modern Perspective, Second Edition
Raymond J. Carroll, David Ruppert, Leonard A. Stefanski, and Ciprian M. Crainiceanu (2006)
106. Generalized Linear Models with Random Effects: Unified Analysis via H-likelihood
Youngjo Lee, John A. Nelder, and Yudi Pawitan (2006)
107. Statistical Methods for Spatio-Temporal Systems
Bärbel Finkenstädt, Leonhard Held, and Valerie Isham (2007)
108. Nonlinear Time Series: Semiparametric and Nonparametric Methods *Jiti Gao* (2007)
109. Missing Data in Longitudinal Studies: Strategies for Bayesian Modeling and Sensitivity Analysis
Michael J. Daniels and Joseph W. Hogan (2008)
110. Hidden Markov Models for Time Series: An Introduction Using R
Walter Zucchini and Iain L. MacDonald (2009)
111. ROC Curves for Continuous Data *Wojtek J. Krzanowski and David J. Hand* (2009)
112. Antedependence Models for Longitudinal Data *Dale L. Zimmerman and Vicente A. Núñez-Antón* (2009)
113. Mixed Effects Models for Complex Data *Lang Wu* (2010)
114. Intoduction to Time Series Modeling *Genshiro Kitagawa* (2010)
115. Expansions and Asymptotics for Statistics *Christopher G. Small* (2010)
116. Statistical Inference: An Integrated Bayesian/Likelihood Approach *Murray Aitkin* (2010)
117. Circular and Linear Regression: Fitting Circles and Lines by Least Squares *Nikolai Chernov* (2010)
118. Simultaneous Inference in Regression *Wei Liu* (2010)
119. Robust Nonparametric Statistical Methods, Second Edition
Thomas P. Hettmansperger and Joseph W. McKean (2011)
120. Statistical Inference: The Minimum Distance Approach
Ayanendranath Basu, Hiroyuki Shioya, and Chanseok Park (2011)
121. Smoothing Splines: Methods and Applications *Yuedong Wang* (2011)
122. Extreme Value Methods with Applications to Finance *Serguei Y. Novak* (2012)
123. Dynamic Prediction in Clinical Survival Analysis *Hans C. van Houwelingen and Hein Putter* (2012)
124. Statistical Methods for Stochastic Differential Equations
Mathieu Kessler, Alexander Lindner, and Michael Sørensen (2012)
125. Maximum Likelihood Estimation for Sample Surveys
R. L. Chambers, D. G. Steel, Suojin Wang, and A. H. Welsh (2012)
126. Mean Field Simulation for Monte Carlo Integration *Pierre Del Moral* (2013)
127. Analysis of Variance for Functional Data *Jin-Ting Zhang* (2013)
128. Statistical Analysis of Spatial and Spatio-Temporal Point Patterns, Third Edition *Peter J. Diggle* (2013)
129. Constrained Principal Component Analysis and Related Techniques *Yoshio Takane* (2014)
130. Randomised Response-Adaptive Designs in Clinical Trials *Anthony C. Atkinson and Atanu Biswas* (2014)
131. Theory of Factorial Design: Single- and Multi-Stratum Experiments *Ching-Shui Cheng* (2014)
132. Quasi-Least Squares Regression *Justine Shults and Joseph M. Hilbe* (2014)
133. Data Analysis and Approximate Models: Model Choice, Location-Scale, Analysis of Variance, Nonparametric Regression and Image Analysis *Laurie Davies* (2014)
134. Dependence Modeling with Copulas *Harry Joe* (2014)
135. Hierarchical Modeling and Analysis for Spatial Data, Second Edition *Sudipto Banerjee, Bradley P. Carlin, and Alan E. Gelfand* (2014)

136. Sequential Analysis: Hypothesis Testing and Changepoint Detection *Alexander Tartakovsky, Igor Nikiforov, and Michèle Basseville* (2015)
137. Robust Cluster Analysis and Variable Selection *Gunter Ritter* (2015)
138. Design and Analysis of Cross-Over Trials, Third Edition *Byron Jones and Michael G. Kenward* (2015)
139. Introduction to High-Dimensional Statistics *Christophe Giraud* (2015)
140. Pareto Distributions: Second Edition *Barry C. Arnold* (2015)

Monographs on Statistics and Applied Probability 140

Pareto Distributions

Second Edition

Barry C. Arnold
University of California, Riverside
Riverside, CA, USA

CRC Press is an imprint of the
Taylor & Francis Group, an **informa** business
A CHAPMAN & HALL BOOK

CRC Press
Taylor & Francis Group
6000 Broken Sound Parkway NW, Suite 300
Boca Raton, FL 33487-2742

First issued in paperback 2020

ISBN-13: 978-1-4665-8484-6 (hbk)
ISBN-13: 978-0-367-73847-1 (pbk)

Visit the Taylor & Francis Web site at
http://www.taylorandfrancis.com

and the CRC Press Web site at
http://www.crcpress.com

To my best friend,
my wife, Carole,
and to my youngest granddaughter,
Kaelyn

Contents

List of Figures xvii

List of Tables xix

Preface to the First Edition xxi

Preface to the Second Edition xxiii

1 Historical sketch with emphasis on income modeling 1
1.1 Introduction 1
1.2 The First Steps 2
1.3 The Modern Era 7

2 Models for income distributions 19
2.1 What Is a Model? 19
2.2 The Law of Proportional Effect (Gibrat) 20
2.3 A Markov Chain Model (Champernowne) 21
2.4 The Coin Shower (Ericson) 24
2.5 An Open Population Model (Rutherford) 25
2.6 The Yule Distribution (Simon) 27
2.7 Income Determined by Inherited Wealth (Wold-Whittle) 30
2.8 The Pyramid (Lydall) 30
2.9 Competitive Bidding for Employment (Arnold and Laguna) 31
2.10 Other Models 33
2.11 Parametric Families for Fitting Income Distributions 35

3 Pareto and related heavy-tailed distributions 41
3.1 Introduction 41
3.2 The Generalized Pareto Distributions 41
3.3 Distributional Properties 47
3.3.1 Modes 47
3.3.2 Moments 47
3.3.3 Transforms 49
3.3.4 Standard Pareto Distribution 50
3.3.5 Infinite Divisibility 50
3.3.6 Reliability, $P(X_1 < X_2)$ 50

3.3.7 Convolutions 51
3.3.8 Products of Pareto Variables 53
3.3.9 Mixtures, Random Sums and Random Extrema 54
3.4 Order Statistics 55
3.4.1 Ratios of Order Statistics 57
3.4.2 Moments 59
3.4.3 Moments in the Presence of Truncation 63
3.5 Record Values 64
3.6 Generalized Order Statistics 67
3.7 Residual Life 69
3.8 Asymptotic Results 71
3.8.1 Order Statistics 71
3.8.2 Convolutions 73
3.8.3 Record Values 74
3.8.4 Generalized Order Statistics 74
3.8.5 Residual Life 75
3.8.6 Geometric Minimization and Maximization 75
3.8.7 Record Values Once More 77
3.9 Characterizations 78
3.9.1 Mean Residual Life 78
3.9.2 Truncation Equivalent to Rescaling 82
3.9.3 Inequality Measures 83
3.9.4 Under-reported Income 84
3.9.5 Functions of Order Statistics 86
3.9.6 Record Values 93
3.9.7 Generalized Order Statistics 94
3.9.8 Entropy Maximization 97
3.9.9 Pareto (III) Characterizations 98
3.9.10 Two More Characterizations 99
3.10 Related Distributions 100
3.11 The Discrete Pareto (Zipf) Distribution 110
3.11.1 Zeta Distribution 110
3.11.2 Zipf Distributions 111
3.11.3 Simon Distributions 112
3.11.4 Characterizations 114
3.12 Remarks 115

4 Measures of inequality 117
4.1 Apologia for Prolixity 117
4.2 Common Measures of Inequality of Distributions 118
4.2.1 The Lorenz Curve 121
4.2.2 Inequality Measures Derived from the Lorenz Curve 123
4.2.3 The Effect of Grouping 135
4.2.4 Multivariate Lorenz Curves 139
4.2.5 Moment Distributions 145

4.2.6 Related Reliability Concepts 148
4.2.7 Relations Between Inequality Measures 149
4.2.8 Inequality Measures for Specific Distributions 149
4.2.9 Families of Lorenz Curves 157
4.2.10 Some Alternative Inequality Curves 166
4.3 Inequality Statistics 170
4.3.1 Graphical Techniques 171
4.3.2 Analytic Measures of Inequality 175
4.3.3 The Sample Gini Index 183
4.3.4 Sample Lorenz Curve 185
4.3.5 Further Sample Measures of Inequality 189
4.3.6 Relations Between Sample Inequality Measures 193
4.4 Inequality Principles and Utility 194
4.4.1 Inequality Principles 195
4.4.2 Transfers, Majorization and the Lorenz Order 196
4.4.3 How Transformations Affect Inequality 203
4.4.4 Weighting and Mixing 209
4.4.5 Lorenz Order within Parametric Families 211
4.4.6 The Lorenz Order and Order Statistics 212
4.4.7 Related Orderings 214
4.4.8 Multivariate Extensions of the Lorenz Order 216
4.5 Optimal Income Distributions 221

5 Inference for Pareto distributions 223
5.1 Introduction 223
5.2 Parameter Estimation 224
5.2.1 Maximum Likelihood 224
5.2.2 Best Unbiased and Related Estimates 227
5.2.3 Moment and Quantile Estimates 233
5.2.4 A Graphical Technique 236
5.2.5 Bayes Estimates 236
5.2.6 Bayes Estimates Based on Other Data Configurations 243
5.2.7 Bayes Prediction 246
5.2.8 Empirical Bayes Estimation 248
5.2.9 Miscellaneous Bayesian Contributions 249
5.2.10 Maximum Likelihood for Generalized Pareto Distributions 249
5.2.11 Estimates Using the Method of Moments and Estimating Equations for Generalized Pareto Distributions 254
5.2.12 Order Statistic Estimates for Generalized Pareto Distributions 259
5.2.13 Bayes Estimates for Generalized Pareto Distributions 263
5.3 Interval Estimates 264
5.4 Parametric Hypotheses 268
5.5 Tests to Aid in Model Selection 269
5.6 Specialized Techniques for Various Data Configurations 275

5.7 Grouped Data 283
5.8 Inference for Related Distributions 288
5.8.1 Zeta Distribution 289
5.8.2 Simon Distributions 289
5.8.3 Waring Distribution 292
5.8.4 Under-reported Income Distributions 293
5.8.5 Inference for Flexible Extensions of Pareto Models 296
5.8.6 Back to Pareto 298

6 Multivariate Pareto distributions 299
6.0 Introduction 299
6.1 A Hierarchy of Multivariate Pareto Models 299
6.1.1 Mardia's First Multivariate Pareto Model 299
6.1.2 A Hierarchy of Generalizations 300
6.1.3 Distributional Properties of the Generalized Multivariate Pareto Models 304
6.1.4 Some Characterizations of Multivariate Pareto Models 309
6.2 Alternative Multivariate Pareto Distributions 310
6.2.1 Mixtures of Weibull Variables 310
6.2.2 Transformed Exponential Variables 312
6.2.3 Trivariate Reduction 313
6.2.4 Geometric Minimization and Maximization 315
6.2.5 Building Multivariate Pareto Models Using Independent Gamma Distributed Components 319
6.2.6 Other Bivariate and Multivariate Pareto Models 321
6.2.7 General Classes of Bivariate Pareto Distributions 323
6.2.8 A Flexible Multivariate Pareto Model 325
6.2.9 Matrix-variate Pareto Distributions 327
6.3 Related Multivariate Models 328
6.3.1 Conditionally Specified Models 328
6.3.2 Multivariate Hidden Truncation Models 332
6.3.3 Beta Extensions 334
6.3.4 Kumaraswamy Extensions 334
6.3.5 Multivariate Semi-Pareto Distributions 335
6.4 Pareto and Semi-Pareto Processes 336
6.5 Inference for Multivariate Pareto Distributions 340
6.5.1 Estimation for Mardia's Multivariate Pareto Families 340
6.5.2 Estimation for More General Multivariate Pareto Families 342
6.5.3 A Confidence Interval Based on a Multivariate Pareto Sample 347
6.5.4 Remarks 349
6.6 Multivariate Discrete Pareto (Zipf) Distributions 350

A Historical income data sources 353

B Two representative data sets 359

C A quarterly household income data set 369

References 371

Subject Index 409

Author Index 421

List of Figures

4.2.1 Minor concentration ratio. 133
4.2.2 The Lorenz zonoid for a bivariate Pareto (II) distribution with $\alpha = 9$ and parameters $(\mu_1, \mu_2, \sigma_1, \sigma_2) = (0, 0, 1, 1)$. 145
4.3.1 A Pareto chart for Indian income data (based on super tax records) from Shirras (1935). 172

B.1 Pareto (II) model fitted to golfer data. 363
B.2 Standard Pareto model (left) and classical Pareto model (right), each fitted to Texas counties data. 365

C.1 Pareto (II) model fitted to Mexican income data. 370

List of Figures

List of Tables

4.3.1 Asymptotic variance (σ_1^2) of the mean deviation in the case of a $P(I)(\sigma,\alpha)$ distribution (from Gastwirth, 1974). 176

5.2.1 Changes in the hyperparameters of the prior (5.2.76) when combined with the likelihood (5.2.83) 242

A.1 Data in the form of a distribution function, Lorenz curve or probit diagram (usually grouped data) 353
A.2 Bivariate distributional data 356
A.3 Data in the form of inequality indices only 356

B.1 Golfer data (lifetime earnings in thousands of dollars). (Source: Golf magazine, 1981 yearbook) 360
B.2 Texas county data (total personal income in 1969 in millions of dollars). (Source: County and City Data Book, 1972; Bureau of the Census) 360
B.3 Maximum likelihood estimates for generalized Pareto models for golfer data 362
B.4 Goodness-of-fit statistics for fitted maximum likelihood models 362
B.5 Alternative estimates for the golfer data 363
B.6 Maximum likelihood estimates for generalized Pareto models for Texas counties data 364
B.7 Goodness-of-fit statistics for fitted maximum likelihood models 366
B.8 Alternative estimates for the Texas counties data 367

C.1 Maximum likelihood estimates for generalized Pareto models for the Mexican income data 369
C.2 Goodness-of-fit statistics for fitted maximum likelihood models 370

List of Tables

Preface to the First Edition

This monograph is the end result of a self-help program. In 1974 my colleague Leonor Laguna asked me what I knew about Pareto distributions with relation to income modeling. I did not know much. I still have much to learn, but I am much more aware of the considerable body of literature which has developed in this context. Many would dismiss Pareto distributions as "just transformed exponential variables" or "just random variables with regularly varying survival functions" and thus not worthy of separate treatment. In fact, many statisticians view the Pareto distribution as only being valuable in the sense that it can be used to construct classroom exercises which will not overtax their students' meager calculus skills. There is an element of truth in all these viewpoints. But, as I believe the monograph will show, appropriate specialized techniques related to Pareto distributions have been developed. I have endeavored to gather together material on the statistical and distributional aspects of Pareto distributions. No economic expertise is claimed, and I have, in general, eschewed comments regarding how econometric theory might tie in with the prevalence of Pareto-like distributions in real world income data sets. Occasionally, I have succumbed to the temptation to gratuitously put forward naive hypotheses about underlying processes. I am sure I will be informed of how far I have put my foot in my mouth in these instances.

Many friends have encouraged me to begin, get on with and/or finish the project. The aforementioned Leonor Laguna really got me started. Together we wrote a brief report on Pareto distributions to provide technical background for certain analyses of Peruvian income data that she was undertaking at the time. H. A. David encouraged me to carry through my project for extending that report by augmenting the breadth of coverage and attempting to put matters in historical perspective. G. P. Patil has been a patient, helpful and supportive editor. J. K. Ord read a version of the manuscript and made many excellent suggestions for improvement. In addition to providing essential support and encouragement throughout the project, my wife, Carole, worked indefatigably in correcting my grammar, logic and punctuation. Computational assistance was provided by S. Ganeshalingam and Robert Houchens. Typing the manuscript involved not only interpreting my justly famous quasi-legible scrawl, but also numerous revisions, renumberings, resectionings, reinterpretations and the like. Peggy Franklin not only achieved this in a thoroughly competent fashion, but also did it in a manner cheerful enough to keep both our spirits up. I am grateful. I know we are both glad to be finished.

Finally, I wish to thank the Committee on Statistical Distributions in Scientific Work for its International Summer School at Trieste, Italy held in 1980, and for the support and encouragement given for the preparation and presentation of this monograph during its proceedings.

July 1981

Barry C. Arnold
Riverside, California

Preface to the Second Edition

After a lapse of 33 years, it is inevitable that any volume will begin to show its age. The literature related to Pareto distributions has blossomed in the interval since the first edition of this book was completed. If the book were to retain relevance, an updated version seemed essential. Several of the more historical chapters are little changed in the revision. Chapters 4, 5 and 6 are considerably expanded to accommodate more recent results. Inference procedures are currently more computer intensive than previously. The revised Chapter 5, together with classical inference material, includes discussion of a spectrum of more recent proposals to handle the problems of inference for Pareto models and their several generalizations and extensions. In Chapter 4, new material on multivariate inequality has been added. Chapter 6, dealing with multivariate Pareto models, has grown to reflect an increasing interest shown in recent decades in bivariate and multivariate income and survival models.

Complete coverage of the Pareto literature is not claimed. It is hoped that, at least, a sufficiently broad coverage has been achieved to make the volume useful to most researchers interested in facets of the life and times of the Pareto model and its offspring.

As usual, the author is ultimately responsible for the material, including any errors, in the book. Equally as usual, the author owes debts of gratitude to numerous co-workers and colleagues. Bill Hanley graciously provided Figure 4.2.2 for inclusion in the book. Appendix C owes much to Humberto Vaquera. My wife Carole, as in the first edition, tried to rein in my sometimes florid prose and to remind me that short sentences can often be preferable to some of the long rambling ones that I often propose.

My editor, John Kimmel, demonstrated patience, forbearance and encouragement throughout this project. To him and to all who have helped me, I offer my thanks.

September 2014

Barry C. Arnold
Riverside, California

Chapter 1

Historical sketch with emphasis on income modeling

1.1 Introduction

A logical starting point for a discussion of Pareto and Pareto-like distributions is Vilfredo Pareto's Economics textbook published in Rome in 1897. In it he observed that the number of persons in a population whose incomes exceed x is often well approximated by $Cx^{-\alpha}$ for some real C and some positive α.

Accumulating experience rapidly pointed out the fact that it is only in the upper tail of the income distributions that Pareto-like behavior can be expected. In fact, Pareto's "law" soon became less constraining (and, simultaneously, more believable) as it shaded into statements like: "Income distributions have heavy tails." Pareto's distributions and their close relations and generalizations do indeed provide a very flexible family of heavy-tailed distributions which may be used to model income distributions as well as a wide variety of other social and economic distributions. It is for this reason that Pareto's law and Pareto's distribution remain evergreen topics.

The central position played by heavy-tailed distributions in this context may be compared with the central role played by the normal distribution in many experimental sciences. There the unifying element is the central limit theorem. The experimental errors are surely sums of many not very dependent components, none of which is likely to play a dominant role. Invoking names like De Moivre and Lindeberg, the assumption of normality becomes plausible.

Is there an analogous principle which will justify the prevalence of heavy tails in socio-economic distributions? Many of the models proposed for income distributions seek to enunciate such a principle. Adherents of log-normal models would allude to multiplicative effects and call on the central limit theorem applied to the logarithms of the observed values. Stable distribution enthusiasts will point out that, if we delete the assumption of finite second moments, the arguments which point towards a normal distribution actually point toward stable distributions. These stable distributions have heavy Pareto-like tails. From this viewpoint the only difference between the kinds of distributions encountered in experimental data and those prevalent in socio-economic data is a fattening of the tails due to lack of moments.

Defenders of Pareto's distribution and its close relatives frequently adopt an empirical pose. They have seen many data sets, and the Pareto distribution undeniably fits the upper tails remarkably well. To further buttress their position they argue that the shapes of income and wealth distributions appear to be invariant under changes of definition of income, changes due to taxation, etc., and to be insensitive to the choice of measuring individual or family income, or income per unit household member. Variants of the Pareto (or Pareto-like distributions) can be shown to exhibit such invariance properties. Parenthetically, we remark that many of these invariance arguments point to Pareto-like tails by way of stable distributions. Finally, several stochastic models of economic systems may be shown to lead to Pareto-like wealth and/or income distributions. Chapters 2 and 3 of this monograph will survey income models and properties of Pareto distributions, respectively.

The development and refinement of Pareto's model was influential in centering attention on desirable properties of summary measures of income inequality. Some of the measures tacitly assumed Paretian behavior. Others were more general measures of dispersion. Again a dichotomy appears. In experimental sciences the sample distribution function and the standard deviation hold unquestioned sway. In the income literature the Lorenz curve and relatives of the Gini index are omnipresent. We will summarize some of the development of measures of inequality in Chapter 4. Several generations of Italian statisticians following Gini's leadership have dedicated time to this topic. Attention here will be restricted to measures of inequality found useful in the study of income distributions.

The fifth chapter will survey inference procedures which have been developed for Pareto distributions. The last chapter treats the various multivariate Pareto distributions extant in the literature. Certain stochastic processes, with stationary and/or long run distributions of the Pareto form, are also discussed in this chapter.

Appendix A provides a guide to the location of income inequality data in the literature. Two small data sets are analyzed for illustrative purposes in Appendix B. Appendix C contains an analysis of a representative income data set. The remaining sections of the present chapter are devoted to a brief historical survey of our topic.

1.2 The First Steps

We return to Vilfredo Pareto's influential economics text. Based on observation of many income distributions, Professor Pareto suggested two models. The simplest model asserts that if we let $\overline{F}(x)$ be the proportion of individuals in a population whose income exceeds x, then for large values of x, we have approximately

$$\overline{F}(x) = Cx^{-\alpha}. \tag{1.2.1}$$

This is a distribution which we will call the classical Pareto distribution. We will later introduce several related families of distributions. According to Pareto's observations the parameter α in (1.2.1) was usually not much different from 1.5. He asserted that there was some kind of underlying law which determined the form of income distributions. On occasion he even claimed that the value of α was invariant or almost invariant under changes of population and of the precise definition of income or wealth used. He was not always adamant about this. In his book he proposed a more complicated model to account for the distribution of wealth, namely

$$\overline{F}(x) = C(x+b)^{-\alpha}e^{-\beta x}. \tag{1.2.2}$$

A very readable concise discussion of Pareto's life and theories is to be found in Cirillo (1979). A valuable feature of Cirillo's book is its inclusion of a smooth translation of Pareto's original introduction of what was to become Pareto's law in his "Cours." Cirillo also includes a chapter on the life and times of the Pareto law. His book appeared after the first draft of the present chapter was written. It is gratifying that the two accounts are in essential agreement, but reference to Cirillo's Chapter V may still be recommended since the occasional interpretational divergences are of interest.

The classical Pareto distribution (1.2.1) with its heavy tail soon became an accepted model for income. That is not to say that competitors did not abound. Nevertheless, it became quite socially acceptable to go ahead and estimate the "Pareto index" α without bothering to check whether the data were in agreement with a Pareto distribution. One should not be too critical, however. Whenever they did fit data by plotting $\log \overline{F}(x)$ against $\log x$, an approximately linear relation was verified. In fact as George Zipf (1949) amply demonstrates in his work, it is amazing how many economic and social phenomena seem to obey a Pareto-like law.

This might have been the end of the story except for two major lacunae. First, there remained the nagging question of why income (and other phenomena) obeyed Vilfredo Pareto's law. Second, the law seemed to hold only for the upper tail of the income distribution. What kind of model would account for income distribution throughout its entire range? How might one determine the cutoff point, above which the Pareto law could be expected to hold sway?

In 1905 M. O. Lorenz suggested a novel way of graphically presenting data on income distributions. His chief concern was to provide a more flexible tool for reporting and measuring income inequality. Several summary measures of income inequality had been presented. Measures associated with the names of Holmes and Bowley were prominent. Lorenz proposed plotting a curve with points $(L(u),u)$ where $L(u)$ represents the percentage of the total income of the population accruing to the poorest u percent of the population. This curve with axes interchanged has come down to us as the Lorenz curve or, sometimes, as the concentration curve. Actually Lorenz may have been scooped. Chatelain, Gini, Seailles and Monay might qualify as prior discussants of the device. Nevertheless, Lorenz popularized this scale-free graphical measure of inequality. As we will see in Chapter 4, the Lorenz curve actually determines the parent distribution up to a scale factor. Consequently it is more than just a measure of inequality. The line with slope 1 would represent a completely egalitarian distribution of income. Roughly speaking, the more extreme is the gap between the Lorenz curve of the population and the egalitarian line, the more "unequal" is the income distribution.

The idea that the measure of income inequality should be scale invariant was not universally accepted. Holmes (1905), although impressed by the potential of Lorenz curves, defended his own non-scale-invariant summary measure of inequality quite vigorously. The question is not open and shut. If one thinks of the possibility of measuring income in pennies instead of dollars, then surely we want a scale invariant measure. If, however, we think of the effect of doubling everyone's income or halving it so that many fall below a poverty line, then the argument for scale invariance is no longer compelling.

Watkins (1905) argued against summary measures of inequality, preferring the continuous and more informative Lorenz curve. However, he later (Watkins, 1908) pointed out that in practice you need a good pair of eyes to distinguish between typical Lorenz curves. This was undoubtedly a consequence of the prevalence of rather extreme instances of income inequality at that time. A general trend of diminution of income inequality can be observed in the twentieth century (see Hagstroem, 1960). Watkins proposed instead that one plot $\log u$ against $\log L(u)$. These plots in his experience were approximately linear, and the corresponding slope could be used as a summary measure of concentration.

Gini (1909) proposed a new summary index of inequality or concentration. It was not however the one which is now associated with his name. Perhaps tacitly assuming an underlying Pareto distribution, his index was obtained by plotting $\overline{F}(x)$, the proportion of people whose incomes exceed x, against $\overline{F}_1(x)$, the proportion of the income of the population obtained by those individuals whose incomes exceed x. If this is done on log-log paper, the plot will be approximately linear (exactly so, if the underlying population is of the Pareto form). The slope of this line, denoted by δ, was Gini's suggested index of income concentration. If the population is indeed Paretian, then one may readily verify that $\delta = \alpha/(\alpha-1)$. Furlan (1911) demonstrated that for certain German income data sets, if one fitted a Pareto distribution, thus effectively estimating α, and if one plotted $\log \overline{F}(x)$ against $\log \overline{F}_1(x)$ in order to estimate Gini's index δ, the equation $\delta = \alpha/(\alpha-1)$ was not satisfied. As Harter (1977) suggested, this is most likely due to the fact that the actual income distribution is not Paretian. We observe that Gini's index δ is a monotonically decreasing function of the Pareto index α. Again this tacitly assumes a Paretian parent distribution. Porru (1912) observed that with some real data sets δ does not seem to decrease when α increases.

Both δ and α are questionable summary measures of inequality since they both are only meaningful when the parent distribution is Paretian. Nevertheless, in the early years of the twentieth century considerable discussion centered around the relative merits of these indices. Porru (1912) for example came out in favor of Gini's δ.

In 1914 Corrado Gini introduced the ratio of concentration. This has come down to us as the Gini index. As we shall see in Chapter 4, there are many equivalent definitions of this Gini index (which we will denote by G). It may be expressed as the ratio of the mean difference (introduced by Gini in 1912) to twice the mean. More picturesquely it can be shown to be equivalent to twice the area between the Lorenz curve and the egalitarian Lorenz curve (the 45° line). Gini suggested several other indices, but the ratio of concentration, G, was soon accepted as the most suitable. It has generally been accepted as the best summary index of inequality although its limitations have been recognized. Considerable subsequent effort has been directed to determining what other aspects of income distributions should be reported in addition to the Gini index. That G is worth computing and reporting soon become unquestioned. The analogous situation with experimental data leads to an almost instinctive computation of means.

Pietra in 1915 derived some relationships among the various competing measures of variability. These relationships do not really aid in selecting which measure to use. It was left to Dalton (1920) to break the ground for discussions regarding exactly what are the desirable properties for a measure of income inequality or variability. For example, a Robin Hood axiom could be adopted: A redistribution which takes from the rich and gives to the poor should decrease inequality. He also proposed that adding a constant to everyone's income should decrease inequality. He came out in favor of relative measures of inequality rather than absolute measures. The relative mean difference (Gini's index) and the relative standard deviation seemed to fare best in the competition. Nevertheless Dalton could not resist proposing yet another candidate. He was particularly fond of the ratio of the logarithms of the mean and the geometric mean. The importance of Dalton's work is its groundbreaking character in trying to determine optimal properties of measures of income inequality which do not overtly or implicitly assume a Paretian model.

Winkler (1924) returned to the question of comparing measures of inequality, while Gumbel (1928) provided an early example of computation of the Gini index for definitely non-Paretian distributions (e.g., exponential and half normal) (see also von Bortkiewicz, 1931, and Castellano, 1933). Amoroso (1925) entered the lists with a more complicated generalized Pareto distribution for which, in special cases, he could compute the Gini index. See d'Addario (1936) for the form of the corresponding Lorenz curves. To Gibrat (1931) we owe the law of proportional effect which was used to justify a log-normal distribution for incomes. d'Addario (1931) questioned Gibrat on technical details but nevertheless decided that in many cases a log-normal model provided better fits than a Pareto model.

During the early thirties, Pietra (1932, 1935a,b), d'Addario (1934a,b) and Castellano (1935) wrestled with problems of assignment of priority regarding the introduction of several competing measures of inequality. The relationship $\delta = \alpha/(\alpha - 1)$ between the Gini and the Pareto index was repeatedly discovered to hold when the underlying distribution is overtly or tacitly assumed to be Paretian. Pietra (1935b) indicated that the relation holds if and only if the underlying distribution is Paretian, providing an early example in the list of characterizations of that distribution. A much earlier one, Hagstroem (1925), had lain relatively unnoticed. Hagstroem's result dealt with invariance of the mean residual life function, a concept which in the context of Pareto distributions did not attract much attention until reconsidered by Bhattacharya (1963) and thought of in terms of under-reported incomes by Krishnaji (1970). An often overlooked reference in this context is d'Addario (1939).

Yntema (1933) undertook an empirical comparison of eight competing summary measures of inequality. Included were Gini's δ and G, Pareto's α, the relative mean deviation, coefficient of variation, and the mean deviation and standard deviation of the logarithm of income. He gave graphical interpretations of the relative mean deviation and the relative mean difference ($= 2G$) in terms of the Lorenz curve. The suggestion of the relative mean deviation as a measure of inequality has been attributed to Bresciani-Turroni (1939), but see Holmes (1905). It was certainly studied by von Bortkiewicz (1931), and its possible interpretations in terms of the Lorenz curve were noted by Pietra (1932). Schutz (1951) rediscovered and popularized this interpretation. See also Elteto and Frigyes (1968). An interesting brief historical survey of this topic may be found in Kondor (1971). Yntema, on the basis of his empirical study, favored the relative mean deviation and the relative mean difference ($2G$) above the competing measures. Remarkably, in the light of the continuing popularity of G, Yntema proposed that the relative mean deviation was preferable because of its lack of sensitivity to grouping. There was however a notable lack of uniformity in the rankings of the 17 empirical populations provided by the 8 measures studied.

Shirras (1935) decided to test the plausibility of the Pareto model for Indian data. He plotted $\log x$ against $\log \overline{F}(x)$ and, following Pareto, fitted straight lines by least squares.[1] Bowley had earlier noticed that such curves were slightly parabolic. Shirras was harsh: "The points of the income tax data do not lie even roughly on a straight line," and decisive:

> "There is indeed no Pareto law. It is time that it should be entirely discarded in studies of the distribution of income."

Such cautionary notes regarding uncritical use of Pareto's law are a recurring theme in the literature. For example Hayakawa (1951) found that Japanese data are not well fitted by the Pareto model. Shirras' blast is perhaps the most strident. Macgregor (1936) rose to defend Pareto's law. He adduced that British income data for the years 1918–19 were reasonably well described by the Pareto law. The problem of deciding what constitutes an acceptable fit remained. Shirras condemned the Pareto law after "eyeballing" a linear fit to a slightly parabolic configuration of points. Macgregor resuscitated the model after visual inspection of actual and fitted tables. To complicate matters, it appears to me, to introduce a third subjective impression, that the Shirras data is, if anything, better fitted by the Pareto model than is Macgregor's! Johnson (1937) joined Macgregor in defense of the Pareto law using U.S. income data (1914–1933). See Figure 4.3.1 for an example of Shirras' plots.

Two survey articles merit attention at this point. Gini (1936) surveyed work on competing measures of inequality with special emphasis on Italian contributions. He argued the superiority of his δ over Pareto's α for incomes and rents. For general distributions he advocated the relative mean deviation and the relative mean difference.

Bowman (1945) provided a convenient compendium of graphical methods. The five described were:

(i) Pareto chart: $\log x$ against $\log \overline{F}(x)$

(ii) Gini chart: $\log \overline{F}(x)$ against $\log \overline{F}_1(x)$

(iii) Reversed Gini chart: $\log F(x)$ against $\log F_1(x)$

(iv) Lorenz curve: $F(x)$ against $F_1(x)$

[1] Shirras alluded to earlier use of such plots by Josiah Stamp to identify tax evaders!

(v) Semi-log graph; $\log x$ against $\overline{F}(x)$.

Charts (i) and (ii) will be linear if the population is Paretian (with respective slopes given by Pareto's α and Gini's δ). Bowman criticized (i) as insensitive. The Gini chart focuses attention on high incomes; the reversed Gini chart is more sensitive to changes in the lower levels of income. The Lorenz curve gives information about the full range of incomes, but difficulties of interpretation occur when Lorenz curves cross. Bowman proposed (v) as a modification of the Pareto chart designed to "bring out some characteristics of the income distribution in the modal ranges." Bowman included graphical representation of several interesting data sets. Ultimately she proposed that (iv) and (v) may prove to be the most useful for graphical analysis.

It is appropriate at this point to recommend Harter's (1977) annotated bibliography on order statistics. It is an invaluable source of information on early and sometimes obscure publications dealing with income modeling and inequality measurement.

1.3 The Modern Era

Identification of the onset of what we call the modern era is necessarily somewhat arbitrary. Roughly speaking we have used 1940 as a demarcation line. The second world war caused a hiatus in scientific publication which serves as a plausible breaking point. A few pre-war papers are much in the spirit of things to come and have been tabbed modern. The post-war period brought a diminution of the dichotomy between discrete distributions and continuous distributions. Wold's 1935 paper had paved the way. Using Stieltjes integrals he provided a uniform treatment pointing out a slight inconsistency between the generally accepted discrete definition of the Gini mean difference (due to Gini) and the natural analog of the definition used in the continuous case. He also discussed convergence of Lorenz curves assuming weak convergence of the corresponding distributions.

We may discern two other main currents in post-war work. One concerns the introduction of flexible families of distributions as potential income models; often just focusing on whether a good fit is obtainable with few parameters and rarely with description of a stochastic mechanism to account for the distribution. The second theme was just such a search for plausible models. Some papers contribute to both areas by proposing flexible families of putative income distributions together with a plausible stochastic mechanism to account for them. Gibrat's earlier cited 1931 law of proportional effect leading to a log-normal distribution is perhaps the earliest example of this genre. The log-normal remains a strong competitor in any effort to fit income curves. The obvious advantage of the log-normal distribution is that, following a simple transformation, the enormous armature of inference for normal distributions is readily available. Kalecki (1945) provided a convenient review of Gibrat's multiplicative central limit theorem argument and suggested a modification of the Gibrat model. In the modified structure the variance of log-income is constant over time. This is more plausible than the linearly increasing variance which is characteristic of Gibrat's model. Kalecki also proposed somewhat arbitrary transformations of the data before taking logarithms to improve the fit.

Champernowne (1937) introduced the name income power for the logarithm of income. He argued that in the absence of inheritance, the equilibrium distribution can be expected to be normal. However he pointed out plausible circumstances under which no equilibrium can be expected. Dissatisfied with Gibrat's model and even with a more complicated modified Gibrat distribution, he turned to other possible models. He discarded a simple Pareto model in order to fit incomes throughout the entire range, rather than concentrate on the upper tail. The family of densities for *income power* proposed by Champernowne may be written in the form

$$f(x) = A/[\cosh(Bx - C) - D]. \tag{1.3.1}$$

In Champernowne (1952) he provided interpretations for the parameters appearing in (1.3.1). There are really 3 parameters in (1.3.1) since A is determined by the requirement that the density should integrate to 1. To account for skewness an additional parameter was introduced (see Champernowne's equation (5.1)). He illustrated the flexibility of the proposed family by providing good fits to Bohemian and United Kingdom income data sets. Incidentally, no less than seven possible fitting techniques were proposed. Although he did provide interpretations for the parameters, the formula (1.3.1) cannot be construed as having arisen from a stochastic model for the generation of the income distribution.

Two special cases of (1.3.1) merit attention as providing particularly simple models. Written in terms of the upper tail of the distribution of income [not income power] they take the forms

$$\overline{F}(x) = \left[1 + \left(\frac{x}{x_0}\right)^{\alpha}\right]^{-1}, \quad x > 0 \tag{1.3.2}$$

and

$$\overline{F}(x) = \frac{2}{\pi}\tan^{-1}\left(\frac{x_0}{x}\right)^{\alpha}, \quad x > 0. \tag{1.3.3}$$

Fisk (1961a,b) focused on the family (1.3.2) which he called the sech^2 distribution, although it might well have been named log-logistic. Fisk also noted that if we find log income to be approximately logistic, then perhaps we might try to fit other functions of income by the logistic distribution. This somewhat ad hoc procedure met with some success with the Bohemian data earlier studied by Champernowne.

The distribution (1.3.2) is actually a special case of Burr's (1942) twentieth family of distributions. The Burr family, after the addition of location and scale parameters as suggested by Johnson, Kotz and Balakrishnan (1994, p. 54), assumes the form

$$\overline{F}(x) = \left[1 + \left(\frac{x-\mu}{\sigma}\right)^{1/\gamma}\right]^{-\alpha}, \quad x > \mu. \tag{1.3.4}$$

The distribution (1.3.4) is sometimes called the Burr distribution. In the present monograph, to highlight its position in a hierarchy of increasingly complicated (and flexible) Pareto-like distributions, it will be called the Pareto (IV) distribution. In this hierarchy Fisk's sech^2 (or log-logistic) distribution is endowed with yet another name, Pareto (III). Hatke (1949) gave moment charts for the Burr (Pareto IV) family of distributions enabling one to compare it with the Pearson family of curves. More recent work in this area was provided by Burr and Cislak (1968), Rodriguez (1977) and Pearson, Johnson and Burr (1978). The latter authors pointed out the wide difference between higher moments of the Burr distribution and corresponding moments of matched log-normal and Pearson (IV) distributions.

In Chapter 2, an even more flexible model, dubbed the Feller-Pareto distribution, will be introduced, It includes as special cases the Burr and Pareto (IV) models and provides a unified framework for developing related distributional theory.

The quantity log-income or income power was called moral fortune by Frechet (1958), a name he attributed to Bernoulli. He noted that Gibrat's model is accounted for by normally distributed moral fortunes, while a Laplace distribution for moral fortunes would lead to the Pareto distribution for incomes above the median. Adoption of the sech^2 model is, as observed above, a tacit assumption that moral fortunes have a logistic distribution. We could of course continue the exercise of assuming a particular form for the distribution of income power and deriving the implied income distribution. The log-Cauchy model could be obtained by assuming Cauchy distributed moral incomes, and so on.

Are there any plausible models besides Gibrat's to suggest sensible distributions for income or moral income? Perhaps the first stochastic model to lead to non-normally distributed moral income was Ericson's (1945) coin shower. By using Bose-Einstein statistics for a random partition of a finite fortune among n individuals, he arrived at the exponential distribution as an income model. He had mixed success when he applied it to actual data. Hagstroem (1960) refined the argument somewhat, but also arrived at the exponential model.

Champernowne (1953) suggested a Markov chain model for income distribution. As usual he treated log-income, now discretized, and he prescribed a transition matrix which led to a geometric long run distribution of income power, i.e., a discrete Pareto distribution for income. A key assumption is that the percentage change in income is independent of level of income. If this is violated, as it might well be under the influence of progressive income taxes, for example, the discrete Paretian limit will not obtain. Stiglitz (1969) later provided alternative economic interpretations for the transition model assumed by Champernowne. Inter alia he pointed out that many deterministic models lead to uniform long run distributions. Evidently such models are not generally acceptable.

Diffusion models related to Champernowne's were naturally considered. Sargan (1957) however modified Champernowne's assumption about the special role of minimum income and arrived at a long run income distribution of log-normal type. Mandelbrot (1961) suggested diffusion explanations for both Pareto and log-normal distributions.

Lydall (1959) pointed out that mixtures of Pareto distributions are not Pareto though the tail behavior may still be Paretian. This is important since income comes from many sources. It is often possible to postulate a plausible Markovian model for a particular kind of income or a particular method of generation of wealth. But when all sources are lumped together, we are probably expecting too much if we ask for a single distribution to fit more than the upper tail. Here we are clearly getting back to the early modest assertions regarding Pareto's law. Lydall (1959) postulated a pyramid model for employment income, assuming that the income of a supervisor is proportional to the sum of the incomes of his supervisees. This suggests a discrete Pareto income distribution. Earlier, Wold and Whittle (1957) derived a Paretian long run distribution for wealth using a diffusion model involving estate growth and equal division of inherited estates. Even earlier Castellani (1950) had noted that Pearson (III) (including Paretian) distributions could arise in the long run in certain diffusion models. Ord (1975) provided a convenient summary of the Champernowne model. He highlighted the critical role played by the assumption of a minimum income. With it we are led to a Paretian model. Without it, the log-normal model is encountered.

Simon (1955) supported Champernowne's model to some extent. He proposes a stream of dollars model, analogous to models proposed to explain biological species diversity. This model leads to the Yule distribution which exhibits Paretian tail behavior. The discrete Pareto distribution had been frequently used to model the size distribution of business firms and other socio-economic variables. Simon and Bonini (1958) suggested the more flexible Yule distribution and its relatives for such applications.

Mandelbrot (1960, 1963) echoed Lydall's reservations regarding a Paretian model for the complete income distribution. He argued for the weak Pareto law (i.e., a Paretian upper tail) on the basis of closure under aggregation, mixture and maximization. He also pointed out that Pareto-like distributions of many physical features might well trigger similar distributions in related socio-economic phenomena.

Rutherford (1955) moved from Gibrat in another direction. He described a stochastic process to account for income distributions, but it is not easy to give it an economic interpretation. He assumed a stream of newcomers with normally distributed income power subject to random normal shocks. The resulting family of distributions is somewhat richer than the log-normal which is included as a limiting case. He naturally got a better fit (using Champernowne's data) than he would have with the log-normal. It is interesting that the fit remains unsatisfactory for the upper tail.

The decade of the 1960's saw the introduction of multivariate Pareto distributions (by Mardia, 1962, 1964) and multivariate Pareto (IV) (or Burr) distributions (by Takahasi, 1965). Efforts were also made to consider multivariate concentration curves; Blitz and Brittain (1964) and, later, Taguchi (1972a,b, 1973). The multivariate theory remained shallow. Little effort had been given to the problem of deriving models leading to multivariate Paretian distributions.

Pareto characterizations came into vogue once more. Many were spinoffs from concerted effort in the field of characterizing the exponential distribution. Several authors dealt with the truncated mean or, equivalently, mean residual life. Most of the other characterizations are in terms of properties of Pareto order statistics. Few of the characterizations have what may be termed economic interpretations. The exceptions are the under-reported income characterizations of Krishnaji (1970) and Revankar, Hartley and Pagano (1974). Krishnaji characterized the standard Pareto distribution under assumptions relating true and reported income by an independent multiplicative factor. Revankar, Hartley and Pagano postulated an additive relation between true and reported income. They are led to what we call the Pareto (II) distribution of the form

$$\overline{F}(x) = \left[1 + \left(\frac{x-\mu}{\sigma}\right)\right]^{-\alpha}, \quad x > \mu. \tag{1.3.5}$$

Distributions of the form (1.3.5) have been used for modeling purposes by Maguire, Pearson and Wynn (1952), Lomax (1954), Silcock (1954) and Harris (1968). Most of these authors arrived at the distribution via a mixture of exponential distributions using a gamma mixing distribution. This genesis suggests that the Pareto (II) might be well adapted to modeling reliability problems, and many of its properties are interpretable in this context. It does have Pareto-like tails and might be expected to be a viable competitor for fitting income distributions. Balkema and de Haan (1974) showed that it arises as a limit distribution of residual lifetime at great age. Bryson (1974) advocated use of the Pareto (II) solely because it is a simple, quite flexible family of heavy-tailed alternatives to the exponential.

What if you mix Weibull random variables rather than exponentials? Dubey (1968) and Harris and Singpurwalla (1969) did just this and arrived at the previously described Pareto (IV) distributions (1.3.4).

In the context of extreme value theory, especially in the study of peaks over thresholds, Pickands (1975) introduced what he called a generalized Pareto distribution. The corresponding density is

$$f(x;\sigma,k) = \frac{1}{\sigma}\left(1 - \frac{kx}{\sigma}\right)^{(1-k)/k} I(x > 0, (kx)/\sigma < 1), \tag{1.3.6}$$

where $\sigma > 0$ and $-\infty < k < \infty$. The density corresponding to $k = 0$ is obtained by taking the limit as $k \uparrow 0$ in (1.3.6). In fact (1.3.6) includes three kind of densities. When $k < 0$, it yields a Pareto (II) density (with $\mu = 0$), when $k = 0$ it yields an exponential density, while for $k > 0$, it corresponds to a scaled Beta distribution (of the first or standard kind). The density (1.3.6) thus unifies three models that are of interest in the study of peaks over thresholds. However, for income modeling it will generally be true that only the case $k < 0$ (and perhaps $k = 0$) will be of interest. Nevertheless, it bears remarking that the literature dealing with Pickands' generalized Pareto distribution is quite extensive and can provide useful information about distributional properties and inference for the Pareto (II) distribution.

Arnold and Laguna (1976) provided a characterization of the Pareto (III) distribution (1.3.2). It arose as the limiting distribution under repeated geometric minimization. An economic interpretation is possible in terms of inflation and a hiring policy involving selection of the cheapest of a random number of available employees.

The standard battery of inference techniques (minimum variance unbiased estimates, maximum likelihood estimates, best linear estimates, uniformly most powerful tests, unbiased tests, etc.) have been developed for the classical Pareto distribution. Most could be obtained by transforming to the exponential distribution, but they were derived independently in the literature. Muniruzzaman (1957) provided important early impetus in this direction. Quandt (1966) surveyed the available techniques of estimation and suggested a new one based on spacings. Malik (1966, 1970a,b,c) discussed distributional and estimation problems for the classical Pareto distribution. Kabe (1972) and Huang (1975) provide further discussion of order statistics (leading back into characterizations). Aigner (1970) and Aigner and Goldberger (1970) described suitable estimation techniques when the data are grouped (as is typically the case). Lwin (1972) developed estimates for the tail of the Pareto distribution. W-T Huang (1974) pointed out that selection problems for the classical Pareto distribution may, by a logarithmic transformation, be related to selection problems for gamma populations. Hartley and Revankar (1974) and Hinkley and Revankar (1977) described inference techniques particularly appropriate for the Revankar, Hartley and Pagano (1974) under-reported income model.

Blum (1970) returned to the problem of convolutions of Pareto variables earlier studied by Hagstroem (1960) and Taguchi (1968). Blum also pointed out that the standard Pareto is in the domain of attraction of a stable law and gave a series expansion for the limiting distribution of sample sums. Mandelbrot had earlier used such observations in the development of his models for income distribution. The question of infinite divisibility of the classical Pareto distribution was settled in the affirmative. See Goldie (1967), Steutel (1969) and Thorin (1977).

Inference techniques for the Pareto (II) distribution are not uncommon in the engineering literature. Moore and Harter (1967) were among the early contributors. Linear systematic estimates were considered by Chan and Cheng (1973), Kulldorff and Vannman (1973), Kaminsky and Nelson (1975) (who consider prediction rather than estimation) and Vannman (1976). Saksena (1978) compared several competing estimates in his doctoral thesis. Some numerical comparisons had earlier been made by Lucke, Myhre and Williams (1977). Inference techniques for Pareto (III) and Pareto (IV) populations have typically assumed several of the parameters to be known. An exception is encountered in Harris and Singpurwalla (1969) where the Pareto (IV) likelihood equations were derived.

In 1974 Salem and Mount suggested that a simple gamma distribution might be adequate to fit income distributions. Cramer (1978) pointed out that it fares quite well when compared to the log-logistic (i.e., Pareto (III) with $\mu = 0$). Singh and Maddala (1976) had proposed the Pareto (IV) model using an argument involving decreasing failure rates. Singh and Maddala (1975, 1978) proposed a rich family of Lorenz curves for fitting purposes. This family included the Weibull, logistic and Pareto IV curves. These suggestions further augmented the list of possible distributions one might consider in fitting income data. Ord, Patil and Taillie (1981b) pointed out that the Pareto, gamma and log-normal distributions might be selected if we used a criterion of maximum entropy (different measures of entropy of course lead to different maximizing distributions). McDonald studied the properties of the beta distribution as a candidate income model (see McDonald and Ransom, 1979).

Two late additions to the field were the Bradford distribution and the prize competition distribution. The Bradford distribution had as its Lorenz curve

$$L(u) = \log(1+\beta u)/\log(1+\beta). \tag{1.3.7}$$

Leimkuhler (1967) described some of its properties and a fitting technique involving the method of moments. Bomsdorf (1977) introduced the prize competition distribution. Its density is of the form

$$f(x) = c/x, \quad a_1 < x < a_2. \tag{1.3.8}$$

Ramified by the addition of location, scale and shape parameters, both (1.3.7) and (1.3.8) might provide flexible families for income modeling.

Kloek and Van Dijk (1978) provided an attempt to empirically select a suitable distribution from the wide variety available. It is interesting that they still failed to fit the data well. Perhaps we must fall back to Mandelbrot's observation that, since for low incomes the typical distribution is erratic, it is unlikely that a single theory can account for all features of income distribution. Rather than throw up our hands in dismay, we may well go right back to Pareto and restrict efforts to modeling only the upper tail of the income distribution. This is probably challenging enough.

In the 1960's and 1970's there was a resurgence of interest in measures of income inequality. Recall that Dalton (1920) had initiated discussion of inequality principles (including the Robin Hood axiom) with regard to the measurement of income inequality. When we left the story, the Gini index and the relative mean deviation appeared to be the best summary measures of income inequality, while Lorenz curves seemed most useful for comparing populations (provided they did not intersect).

Morgan (1962) cited the Gini index as being the best single index of inequality. He observed that it does not seem to matter whether one uses income before or after taxes, per individual or per family, nor is it sensitive to whether it is based on single year or multi-year data. He did present a list of factors which apparently affect the Gini index. While age appears to be a factor, he singled out three major influences: female employment patterns, male unemployment levels and rural-urban migration. Verway (1966) ranked the states (of the U.S.) according to their Gini indices of income inequality and found this was related to the Gini indices of inequality of owner-occupied housing and education level, among other factors. Conlisk (1967) accounted for 91% of the variability in state Gini indices by multiple regression on seven variables including education, unemployment level, racial structure, age structure and percentage nonfarm rural population. Metcalf (1969) fitted a log-normal model to time series income data and studied the behavior of the parameters in certain subpopulations. Long, Rasmussen and Haworth (1977) considered Gini indices for standard metropolitan areas and fit a regression on various socio-economic variables. These studies, although they may have predictive utility, address neither the problem of modeling income distributions nor the question of how should we measure inequality.

Before turning to more recent work in this direction, one final empirical study merits mention. Ranadive (1965), using Reserve Bank of India figures, made the startling observation that the Indian Lorenz curve lay wholly within those of all other countries considered, including several "developed" countries. At face value this suggests that income inequality is markedly less in India than elsewhere. Ranadive pointed out that by excluding, for example, dividend income (which the rich typically get) much inequality is masked. He was really pointing out the dangers of unthinking comparison of Lorenz curves (and other measures of inequality) without checking to see that at least approximately the same definition of income has been used in both cases.

Following Dalton's footsteps 50 years later, Atkinson (1970) returned to the question of how one should measure inequality. A welfare measure which represents expected utility was suggested, i.e.,

$$W_U(f) = \int_0^{\bar{y}} U(y)f(y)dy \tag{1.3.9}$$

where $\bar{y}$ is the upper level of income, f is the density of income in the population and $U(y)$ is the utility of an income of level y. This opened a Pandora's box. A whole literature on utility theory could be tapped and related to income inequality. Atkinson restricted attention to increasing concave utility functions. He showed that if two income densities f_1 and f_2 have the same mean, then $W_U(f_1) > W_U(f_2)$ for every increasing concave U, if and only if the Lorenz curve of f_1 lies wholly within that of f_2. In terms of inequality principles this occurs if and only if we can get f_2 from f_1 simply by transfers from poor individuals to richer ones (reverse Robin Hood). These observations confirm the old belief that if Lorenz curves are nested, there is no difficulty in ordering populations with regard to income inequality. The problems arise, of course, because Lorenz curves are rarely nested. Nevertheless, Atkinson did provide helpful insights into the nature of the partial ordering of income distributions provided by nested Lorenz curves. He observed that the Gini index is sensitive to transfers among middle income individuals, while the standard deviation of log income gives more weight to lower income level transfers. He proposed a family of measures of income inequality, assuming scale invariance and constant relative inequality aversion, of the form

$$I_\varepsilon(f) = 1 - \left[\int_0^{\bar{y}} \left(\frac{y}{\mu} \right)^{1-\varepsilon} f(y)dy \right]^{1/(1-\varepsilon)} \tag{1.3.10}$$

where $\varepsilon \neq 1$ and where μ is the mean of the density f. By varying ε one can emphasize sensitivity to transfers at different levels of income.

Newbery (1970) observed that for no choice of U in (1.3.9) would ranking in terms of W_U coincide with the rankings provided by the Gini index. Sheshinski (1972) pointed out that it is not at all clear that welfare should be integrated utility. He gave examples of non-additive utility functions for which the welfare is a function of the mean and the Gini index.

Rothschild and Stiglitz (1973) argued for additivity. In their article and a companion article by Dasgupta, Sen and Starrett (1973) the ideas of Atkinson and Newbery were related back to classical utility theory and, further back, to Hardy, Littlewood and Polya's (1929) work on majorization. They considered distributions of a fixed total income among n individuals (let $X = (X_1, X_2, \ldots, X_n)$ be such a generic distribution). Several partial orders may be defined on this space of distributions. Thus we may define: $X \geq Y$ if

(i) the Lorenz curve of X is inside that of Y,

(ii) for every symmetric quasi-concave function W, $W(X) \geq W(Y)$,

(iii) Y is obtained from X by transfers from the poor to the rich, or

(iv) $\sum_i U(X_i) \leq \sum_i U(Y_i)$ for every concave U.

These are in fact equivalent. Rothschild and Stiglitz (1973) had an even more extensive list of equivalent partial orders. Dasgupta, Sen and Starrett (1973) gave a simple proof of Newbery's observation regarding the Gini index (actually they proved a technically stronger result).

So the Gini index cannot be expressed in the form (1.3.9) for any U. Even if that were possible, would we agree that the corresponding utility function was necessarily reasonable? How should one decide what is a reasonable choice for U? Perhaps introduction of the utility function has only confused the issue or at best replaced one difficult problem (choice of a measure of inequality or a ranking of inequality) by a different difficult problem (choice of a utility function). Of course, if the Lorenz curves are nested, no one will argue.

Gastwirth (1971) used the inverse distribution function (or quantile function) to give a convenient definition of the Lorenz curve (valid for any distribution):

$$L(u) = [E(X)]^{-1} \int_0^u F^{-1}(t)dt, \quad 0 \leq u \leq 1. \tag{1.3.11}$$

Much earlier, Pietra (1915) provided a somewhat less general version of this representation using different notation. Gastwirth, in 1972, used this representation to get bounds on the actual Gini index for grouped data. Gastwirth and Smith (1972) exploited these bounds to test whether U.S. income data could be fitted by log-normal or sech^2 distributions. Neither distribution appeared to be satisfactory.

Gastwirth (1974) worked out the large-sample normal distribution of several empirical measures of income inequality. In particular he treated the relative mean deviation and three other related measures introduced by Elteto and Frigyes (1968). The small sample distribution of Gini's mean difference in the exponential and uniform case was discussed by Kabe (1975).

The empirical Lorenz curve can be shown to converge almost surely uniformly to the population Lorenz curve (Goldie, 1977). Since many measures of income inequality are merely measures of distance between the Lorenz curve and the egalitarian line, we are not surprised to find that sample measures of inequality converge almost surely to the corresponding population measures. Under regularity conditions, Goldie also showed that the sample measures of inequality are asymptotically normal. In particular, he treated the sample Gini index and the sample relative mean deviation.

Mehran (1976) had earlier recognized that these were linear measures of income inequality, i.e., of the form

$$\frac{1}{m} \int_0^1 [F^{-1}(u) - m] W(u) du \tag{1.3.12}$$

where $m = E(X)$ or equivalently

$$\int_0^1 [u - L(u)] dW(u) \tag{1.3.13}$$

for a suitable choice of weight function W. Under regularity conditions, the corresponding sample versions are shown to be asymptotically normal. How should one choose the weight function? By appealing to inequality principles one may argue that W should be strictly increasing (a property held by the weight function corresponding to the Gini index but not by the function which corresponds to the relative mean deviation). Many commonly used measures of inequality are not linear in Mehran's sense, i.e., are not of the form (1.3.12). An example is the coefficient of variation.

Gail (1977) used the sample Lorenz curve to develop a relatively simple test for the exponential distribution. Osborne and Severini (2002) discuss an analogous analysis in the context of a truncated exponential model. Gastwirth (1972) noted that a decreasing hazard rate distribution has a Lorenz curve which uniformly encloses the Lorenz curve of an exponential distribution with the same mean. This suggests a tie-in with the reliability literature. Chandra and Singpurwalla (1978) related Lorenz curves to scaled time on test transforms and pointed out that star-ordering, a concept of considerable interest in reliability, under the assumption of equal means, implies nested Lorenz curves. In this light Gastwirth's observation regarding exponential and decreasing hazard rate distributions is less surprising. Chandra and Singpurwalla also discussed large sample behavior of the sample Lorenz curve.

Additional contributions to Lorenz curve technology included Kakwani and Podder's (1973, 1976) suggestions of a new coordinate system and weighted least squares estimates for grouped data. Fellman (1976) studied the effects of transformations on the Lorenz curve, e.g., if we use a transformation $g(x)$ with $g(x)/x$ decreasing, then inequality will be reduced. Ord, Patil and Taillie (1981b) identified conditions under which truncation will reduce inequality. Truncation might thus cause underestimation of inequality. In a similar vein, Peterson (1979) pointed out that if Gini indices are computed using grouped data, then uniform growth of incomes (inflation) can lead to a shift in the Lorenz curve and Gini index even though the actual income distribution is unchanged in shape. This casts considerable doubt on intertemporal comparisons of income inequality. Taguchi (1968) had earlier discussed the effects of truncation on Lorenz curves. He also considered the characterization of symmetric Lorenz curves. Hart (1975) reminded us that reporting the Lorenz curve does not constitute much condensation of data with respect to the distribution function. Almost all concentration measures are moments or functions of moments of the distribution function $F(x)$ and the first moment distribution $F_1(x)$. He noted that the log-normal distribution is unusual in that both its distribution and its first moment distribution are of the same form.

Whereas the full Lorenz curve is not enough condensation of the data, reporting a single index, say the Gini index, or the relative mean deviation probably represents too much condensation. Hagerbaumer (1977) proposed that what he called the minor concentration ratio be reported in addition to the Gini index. Levine and Singer (1970) concentrated on the area beneath the Lorenz curve between two fixed levels of incomes. They studied how it is affected by linear and lump sum taxes.

What was missing for many years was a generally agreed upon extension of the Lorenz curve concept and Lorenz ordering to the multivariate case. Koshevoy (1995), with the introduction of the Lorenz zonoid, filled this gap. The key element in the development of the Lorenz zonoid was a method of defining the Lorenz curve without reference to ordering of the data. The inability to envision suitable multivariate ordering concepts had been the stumbling block to the development of appropriate multivariate inequality orderings. Further details on the zonoid approach will be found in Chapter 4.

A remarkable resurgence of interest in Pareto distributions has been in evidence in the Mathematics and Statistics literature in the period 1980-2013. As an imperfect gauge of this phenomenon, it can be noted that a search in Mathematics Reviews under the keyword "Pareto distribution" yields 58 references prior to 1980, as compared with 613 references in the period 1980–2013. Most of these papers deal with inferential issues rather than distributional properties of Pareto distributions, since the latter had been subject to detailed discussion prior to 1980.

Two distributional advances can be mentioned. Expressions for the characteristic function and the Laplace transform of the Pareto (II) distribution appeared (Seal, 1980; Takano, 1992). More recently, Nadarajah (2007) discussed the distribution of YZ and YZ/W where Y, Z and W are independent Pareto (I) variables. Manas (2011) considered $Y/(Y+Z)$ in this context.

Several authors suggested new multivariate Pareto models with potential for modeling incomes of related individuals, or income from multiple sources or in multiple currencies. Further discussion of this topic may be found in Chapter 6.

Numerous alternatives to the Pareto distributions have been proposed for modeling non-negative variables, often, but not always, with income modeling in mind. Airth (1984) proposed a hybrid log-logistic log-normal model. Schwartz (1985) suggested a cube root model. Introduction of a change point in the classical Pareto model was suggested (Picard and Tribouley, 2002). Double Pareto models (Reed, 2003), product Pareto models (Nadarajah and Gupta, 2008) and even truncated Cauchy models have been discussed (Nadarajah, 2011). Perusal of Chapter 2 will yield a considerably more extensive and detailed list.

The variety of potential fields of application for Pareto and/or Pareto related models has grown in recent decades. The distributions were found to be useful for description and inference in the study of: forest fires (Robertson, 1972); wire length distributions (Donath, 1981); receiver operating characteristics (Campbell and Ratnaparkhi, 1993); modeling claim premiums (Brazauskas, 2000; Gay, 2004; Mert and Saykan, 2005); library data (Amati and van Rijsbergen, 2002); risk pricing and ruin computation in insurance (Ramsay, 2003; Wei and Yang, 2004); file sizes (Mitzenmacher, 2004); queueing models (Shortle, Gross, Fischer and Masi, 2006); medical insurance claims (Zisheng and Chi, 2006); financial durations (de Luca and Zuccolotto, 2006); software reliability (Tsokos and Nadarajah, 2003) and current duration data in gestation studies (Keiding, Kvist, Hartvig, Tvede and Juul, 2002; Keiding, Hansen, Sorensen and Slama, 2012). In short, the flexible generalized Pareto models (Pareto (IV), Feller-Pareto and further generalizations) are viable candidates for model fitting and prediction for almost any kind of non-negative (or bounded below) data sets.

With this in mind, it is appropriate to remark that although the lion's share of the papers referred to in this book have treated the Pareto-like distributions and measures of inequality in an income distribution context, the body of results will be applicable in any context where heavy-tailed distributions and related measures of spread or concentration are considered. Zipf (1949) and Mandelbrot (1963) would have us believe that there are inherent relationships and general laws which tie together many of the areas in which heavy-tailed distributions are encountered. Whether we believe this or not, it would be foolish to neglect the commonality of distributional features that many disparate socio-economic, geographic, scientific and engineering phenomena seem to share.

Chapter 2

Models for income distributions

2.1 What Is a Model?

How can one account for the observed shape of income distributions? In practice, a very complicated stochastic process is involved. The set of, say, all U.S. incomes in 1979 will be a random perturbation of the set of incomes in 1978. Some new individuals will have entered the market, some will have left. Those present in both years will typically have different incomes. The myriad contributory causes of these changes in the income-earning population and in the incomes of individuals defy enumeration. If a tractable model is to be obtained, it must necessarily ignore most of the possible causative agents and (over-) simplify the manner in which the remaining ones affect the income distribution. Similar observations apply to wealth distributions. We will chiefly speak of income distributions, but many of the models obtained apply, not surprisingly, almost as well to wealth distributions. The invariably heavy-tailed nature of income and wealth distributions was a matter of observational common knowledge in the nineteenth century. Pareto's two models (1.2.1, 1.2.2) constitute, essentially, selection of parametric families of distributions which seem to fit the upper tails of the observed distribution in a satisfactory manner. Their law-like statement and their remarkable resilience to matters such as how we measure wealth, what population we use, etc., have certainly affected subsequent modeling efforts. Models which lead to Paretian tail behavior have been popular and reasonable in the light of observational studies. Pareto's suggested laws will not be included in our list of models. In the final section of this chapter we will list several families of distributions which have been proposed as candidates for fitting or graduating income distributions, usually with little indication as to why the fit should be good. The two Pareto laws will appear in this list. The title, "model," in this chapter, will be reserved for stochastic mechanisms which mirror, to some extent, socio-economic forces which affect income and which lead to delineation of at least some properties of resulting income distributions. The line is not always easy to draw. Authors frequently supply post-hoc economic interpretations of their parameters. If they had written their paper the other way around, maybe they could have qualified as "models." Gibrat will be included in the list of modelers, although his law of proportional effect is, perhaps, more of an axiom than a model. He does provide a convenient starting point.

A convenient survey of pre-1975 income models was provided by Ord (1975).

2.2 The Law of Proportional Effect (Gibrat)

Gibrat (1931) ignored fluctuations in the earning population and merely postulated that the income of an individual viewed as a function of time, say $X(t)$, is a stochastic process subject to small random fluctuations. The fluctuations were assumed to be multiplicative in nature. Or, equivalently, at the end of a period of time the logarithm of the income was assumed to be given by

$$\log X(t) = \log X(0) + \sum_{i=1}^{N(t)} Z_i \tag{2.2.1}$$

where $X(0)$ is the initial income to which has been added a random number $N(t)$ of independent increments $Z_1, Z_2, \ldots$. For large values of t, provided $\lim_{t\to\infty} N(t) = \infty$ and that the Z_i's are not too dependent on $N(t)$ and more or less homogeneous in distribution, the asymptotic normality of $\log X(t)$ can be justified by various versions of the Central Limit Theorem. Thus $X(t)$ may be expected to have a log-normal distribution. As Ord (1975) pointed out, this model is not much different in character from a model which assumes $\log X(t)$ is an unrestricted Wiener process (i.e., a process with stationary independent increments with $\log X(t)$ normally distributed with mean mt and variance $\sigma^2 t$). The extreme nature of the sample path fluctuations encountered in the Wiener process is well known. This may already cause us to question the plausibility of the model. Related to this and perhaps easier to check empirically is the fact that under such a model the variance of $\log X(t)$ can be expected to grow linearly with t. Kalecki (1945) was aware of this anomalous behavior of the Gibrat model. Kalecki felt that such growth was not characteristic of actual income distributions. In periods of rampant inflation the Gibrat model might fare better. It is not difficult to adjust the Gibrat model to cut down its variability. We merely need to postulate that increments in $\log X(t)$ will be negatively correlated with $X(t)$. This can be viewed as a consequence of vaguely specified economic forces which hold the process in check (Kalecki), or it can be viewed as a replacement of the Wiener process by the less variable Ornstein-Uhlenbeck process (Ord, 1975). The limiting distribution of income will be log-normal in these modified Gibrat models. The persistence of the log-normal model in the literature is probably not to be attributed to its being a faithful mirror of the income determining process, but instead to the flexibility of the log-normal distribution as a tool for fitting non-negative variables (if we don't pay too much attention to the fit in the tails). In middle income ranges an assumption of log-normality of income seems compatible with much of the evidence.

As mentioned in the introductory chapter, Mandelbrot emphasized the somewhat arbitrary nature of the assumption of normally attracted increments inherent in Gibrat's analysis. If we dispense with the variance assumptions (or whatever we used to invoke the Central Limit Theorem), then we are led to a model in which log income will have a (probably non-normal) stable distribution. The upper tail of such a distribution will be Paretian with $\alpha < 2$ (in equation (1.2.1)).

Sargan (1957) presented a continuous time model for wealth distributions which are assumed to be subject to changes due to savings and gifts and to foundation and disappearance of households. Under certain homogeneity assumptions he arrived at a log-normal model. A more recent proposal of a log-normal wealth distribution was contained in Pestieau and Possen (1979).

An interesting alternative argument for the prevalence of income distributions well fitted by log-normal models was presented by Brown and Sanders (1981). They argue that if individuals in a large population are sequentially classified according to roughly independent cross-classifications, then after a large number of classifications the variable representing the numbers of individuals in the resulting categories has approximately a log-normal distribution.

A variation on the Gibrat log-normal theme was proposed by Reed (2003). He assumed that the time since entering the workforce of a randomly chosen individual could be approximated by an exponential random variable with intensity λ. It then was argued that, assuming a geometric Brownian motion for the growth of an individual's income (a la Gibrat), the logarithm of the income of a randomly selected individual will be a random variable expressible as a sum of a normal random variable and an independent two-tailed exponential random variable with density

$$f(u) = \frac{\alpha\beta}{\alpha+\beta}\left[e^{\beta u}I(u<0)+e^{-\alpha u}I(u\geq 0)\right], \tag{2.2.2}$$

where α and β are positive parameters. Eventually this leads to density for the income X of a randomly chosen individual of the form

$$\begin{aligned} f_X(x) &= \frac{\alpha\beta}{\alpha+\beta}\Bigg[x^{-\alpha-1}exp[\alpha\mu+\alpha^2\sigma^2/2]\Phi(\frac{logx-\mu-\alpha\sigma^2}{\sigma}) \\ &\quad +x^{\beta-1}exp[\beta\mu+\beta^2\sigma^2/2]\overline{\Phi}(\frac{logx-\mu-\beta\sigma^2}{\sigma})\Bigg], \end{aligned} \tag{2.2.3}$$

where μ and σ^2 are the parameters of the normal distribution involved in the development of the model. Reed calls this a double-Pareto-log-normal model.This distribution exhibits Paretian behavior in both tails, thus

$$P(X\geq x)\sim c_1x^{-\alpha} \qquad [x\to\infty]$$

and

$$P(X\leq x)\sim c_2x^{\beta} \qquad [x\to 0].$$

Applications of this model in a variety of subject areas are discussed in Reed and Jorgensen (2004).

A simplified version of this model had been earlier proposed by Colombi (1990). Colombi suggested a model for income X as a product of two independent variables, one with a classical Pareto distribution and the other with a log-normal distribution. Reed notes that this can be viewed as a limiting form of the model (2.2.3) as $\beta\to\infty$.

2.3 A Markov Chain Model (Champernowne)

Champernowne developed his model for income distribution in the late 1930's, but publication of the detailed derivation was considerably delayed (see Champernowne,

1937, 1953, 1973). The original formulation was in discrete time with income restricted to discrete levels. That is the form we will present here. An analogous diffusion model in continuous time with continuous state space is readily defined, but the qualitative predictions regarding income distribution are essentially unchanged.

An individual income is assumed to develop through time according to a Markov process with state space $0, 1, 2, \ldots$. State j corresponds to an income in the interval $(10^{jh}y_{\min}, 10^{(j+1)h}y_{\min})$ where $y_{\min}$ is the minimum income in the population and h is a positive real number. If one prefers, one can think of the states as corresponding to the different income ranges for reporting purposes rather than the actual levels of income. It is assumed that income cannot move up by more than one income range in a time period, which we conveniently speak of as being a year. In addition, downward mobility is assumed to be constrained. It is postulated that an income cannot move down more than N income ranges for some positive integer N. A homogeneity assumption is also invoked. Thus, except for edge effects due to the existence of a minimum income, the probability of moving k levels down (or one level up) is independent of the level presently occupied. Thus, if we let p_{ij} be the probability of going from state i to state j and let $P = (p_{ij})_{i=0,j=0}^{\infty,\infty}$ be the (infinite) transition matrix we have, for $k = 1, 0, -1, -2, \ldots, -N$:

$$\begin{aligned} p_{ij} &= c_k > 0 \\ &\quad \text{if } i \geq 0, j > 0 \text{ and } j - i = k \\ p_{i0} &= 1 - \sum_{j=1}^{\infty} p_{ij} \end{aligned} \tag{2.3.1}$$

and

$$p_{ij} = 0 \quad \text{otherwise.}$$

For example if $N = 4$, P takes the form

$$\begin{pmatrix} * & c_1 & 0 & 0 & 0 & 0 & 0 & 0\ldots \\ * & c_0 & c_1 & 0 & 0 & 0 & 0 & 0\ldots \\ * & c_{-1} & c_0 & c_1 & 0 & 0 & 0 & 0\ldots \\ * & c_{-2} & c_{-1} & c_0 & c_1 & 0 & 0 & 0\ldots \\ c_{-4} & c_{-3} & c_{-2} & c_{-1} & c_0 & c_1 & 0 & 0\ldots \\ 0 & c_{-4} & c_{-3} & c_{-2} & c_{-1} & c_0 & c_1 & 0\ldots \\ \vdots & \vdots & & & & & & \end{pmatrix} \tag{2.3.2}$$

where the starred entries are chosen so that the rows sum to 1. A convenient reference dealing with the analysis of Markov chains is Isaacson and Madsen (1976). Since all the c_i's are assumed to be strictly positive, the chain is aperiodic and irreducible. Consequently, if the equation

$$\pi P = \pi \tag{2.3.3}$$

has a solution $\pi = (\pi_0, \pi_1, \pi_2, \ldots)$ such that $\pi_i > 0$, $(i = 0, 1, 2, \ldots)$ and $\sum_{i=0}^{\infty} \pi_i = 1$, this solution will represent the long run distribution of the (necessarily) ergodic Markov chain. The solution to (2.3.3) will satisfy the difference equation

$$\pi_j = \sum_{i=-N}^{1} c_i \pi_{j-1}, \quad j = 1, 2, \ldots.$$

If we try for a solution of the form $(1-b)b^i$, $(i = 0, 1, 2, \ldots)$, we find that b must be a non-zero root of the polynomial

$$g(z) = z - \sum_{i=-N}^{1} c_i z^{1-i}. \tag{2.3.4}$$

Observe that $g(0) = c_1 > 0$ while $g(1) = 0$. By Descartes' rule of signs, g has no more than two real roots. Thus, (a sketch of g may help here) a necessary and sufficient condition for the existence of a root b in the interval $(0, 1)$ is that $g'(1)$ should be strictly positive. Note that, since $g'(1) = -\sum_{i=-N}^{1} i c_i$, the condition invoked is that for higher levels of income $(i \geq N)$, the expected change in income is negative. This is Champernowne's stability assumption. It prevents the process from drifting out to infinity. If we let b be the unique root of g in the interval $(0, 1)$ (whose existence is guaranteed by the stability assumption) then a geometric long run distribution exists. Thus, in the long run, denoting income by X we have: for $y_k = 10^{kh} y_{\min}$

$$\begin{aligned} P(X > y_k) &= b^k \\ &= (y_k / y_{\min})^{\frac{1}{h}\log_{10} b}, \end{aligned} \tag{2.3.5}$$

i.e., a discrete Pareto distribution.

The above represents, with minor notational changes, Champernowne's (1953) original presentation. He observed at that time that many of the assumptions in the model could be relaxed without disturbing the Paretian character of the stationary distribution for large incomes (i.e., the geometric tail of the stationary log income distribution). It is even possible to allow a structured population with migration between subpopulations, without destroying the Pareto-like tail behavior.

The basic assumptions which must be approximately true in order to get Champernowne's result are:

(i) The process should be Markovian with state space corresponding to log incomes;

(ii) At least for large incomes the distribution of the change should be approximately independent of the present state of the process;

(iii) A minimum income level should exist; and

(iv) The process should be non-dissipative.

Ord (1975) pointed out that a simple continuous time, continuous state space process with these characteristics is a Wiener process for log income with reflecting barrier and negative drift.

Champernowne also considered a Markovian model without a lower bound for income, obtained by splicing together an upward and a downward process of the type outlined above. Paretian behavior in both tails is, then, encountered. He also pointed out, by an example, the crucial nature of assumption (ii) of the last paragraph in determining the Paretian limiting behavior.

Pakes (1981) suggests that a wide variety of Markov models will yield Paretian tail behavior. Geometric ergodicity of the transition matrix is almost sufficient. Several alternative Markovian models were discussed in Shorrocks (1975).

2.4 The Coin Shower (Ericson)

The coin shower model presumes that N individuals in a given population are to be identified with cells into which a stream of S dollars (the coins) are to be directed according to some random mechanism. The S dollars are thus partitioned into N cells. If we assume all partitions are equally likely (i.e., if we use Bose-Einstein statistics), then the probability that a particular cell receives x dollars is given by

$$\frac{S!(N-1)!}{(S+N-1)!}\frac{(S-x+N-2)!}{(S-x)!(N-2)!} \tag{2.4.1}$$

(cf. Feller, 1968, p. 61, problem 14). If we use Stirling's approximation in conjunction with (2.4.1), a good continuous approximation to the distribution of the number of dollars in a particular cell is provided by the exponential distribution. In proposing this model for income distribution, Ericson (1945) did not advance any rationale for assuming that Bose-Einstein statistics should be used to describe the random allocation of coins to individuals. If, instead, it had been assumed that each coin was equally likely to fall in any one of the cells, then a binomial distribution of income would result (or a normal distribution, if we use a continuous approximation). Hagstroem (1960) mentioned both possibilities. Ericson had mixed success in his efforts to fit observed income distributions by his exponential model. Hagstroem (1960) flatly stated that the fit is unsatisfactory and discarded the model in favor of a Pareto model.

The model can survive in a modified sense. Observed distributions might well be mixtures of distributions, which do arise by a coin shower mechanism. The resulting mixed exponential can have Pareto-like tail behavior. As mentioned in Chapter 1, use of a gamma mixing distribution leads to a Lomax or Pareto (II) distribution. The success of Sahin and Hendrick (1978) in fitting strike duration data by a mixture of just two exponentials, rather than by a Lomax distribution, suggested the advisability of checking whether income distributions can be modeled by mixtures of two (or a few) exponential distributions.

The stochastic mechanism involved in Ericson's coin shower is distressingly unrelated to economic influences. However, in its defense, if we restrict attention to small subpopulations, it may well be that income is basically spread around completely at random. It would appear that the coin shower has its chief utility as a building block to lead us to either a Lomax (Pareto II) distribution or a finite mixture of exponentials as candidate income distributions.

There is another possible use for the shower paradigm as a model-building device. Keith Ord (personal communication) observed that if Ericson's model is modified to represent an income-power shower rather than an income shower, the resulting distribution for income is log-exponential, i.e., a classical Pareto distribution. Presumably one could shower other functions of income to obtain other models. Such exercises will be of interest if one can argue plausibly that the particular function of income in question is "spread around at random" in the population.

2.5 An Open Population Model (Rutherford)

Rutherford's (1955) starting point was the observation that when one tries to fit income data with a log-normal model using normal probability paper, the resulting curves are often only piecewise linear. Alternatively, if one pays less attention to their angular character (due to grouping), they can be likened to polynomials of order 1, 2 or 3. Rutherford supported the view that the variance of income power (log-income) is relatively constant over time and, hence, regarded the Kalecki modification of Gibrat's model as a step in the right direction. One key feature of the Gibrat-Kalecki and the Champernowne models is their closed nature. The set of individuals in the population is fixed from the beginning, and each of their incomes develops according to a random-walk mechanism. In practice, individuals are constantly entering and leaving the population under study. The addition of some birth and death aspect to the model would appear appropriate. Rutherford sought a relatively simple model incorporating random shocks, a la Gibrat, but allowing new income earners to occasionally enter and allowing for mortality of income earners. If the resulting plots of log-income on normal paper are low degree polynomial in nature, so much the better.

Rutherford's basic model was described in terms of income power (log-income). Time was measured in discrete units (years). In the t'th year, a random number α_t of new income earners enter the population. Each income earner is assumed to have a normally distributed income power. The income earners are assumed to have independent initial income powers, and it is assumed that their income power is then subjected to year by year independent normally distributed zero-mean shocks. Finally, we assume a mortality rate independent of income-power. Thus, from demographic tables we will have available estimates of d_T (the probability that an individual born at time $t-T$ will not be alive at time t, given that the individual was alive at times $t-T, t-T+1, \ldots, t-1$), $T = 1, 2, \ldots$. In particular Rutherford assumed a geometric life table which is probabilistically equivalent to assuming that there is some number d, $(0 < d < 1)$, which represents the probability of an individual dying in any year independent of the age of the individual (and independent of the income power of the individual). After a period of time the population will stabilize. The long run distribution will depend on d (the death rate) and the distribution of α_t (the number of new entrants in the t'th year). Any individual alive after k years will have an income power which is normally distributed with mean m (the mean of the initial income power) and variance $\sigma^2 + k\sigma_s^2$ (where σ_s^2 is the variance of an annual shock). The resulting long run income-power distribution can be expected to be a mixture of these normal distributions with rather complicated weights.

One special case can be handled in some detail. If we assume that the number of individuals entering in a given year is Poisson distributed with mean λ (assumed large), in the long run the total population size will be Poisson distributed with mean λ/d divided into independent Poisson distributed cohorts. The number of individuals in the cohort of age i will have a Poisson distribution with mean $\lambda(1-d)^i$, $i = 0, 1, 2, \ldots$ (new entrants are accorded age zero). Ignoring the possibility that the population size will be zero (as we can if λ is large), the distribution of income power of a randomly selected individual (after the process has reached equilibrium) will have its moment generating function given by

$$\begin{aligned}\varphi(t) &= e^{tm}\left\{\sum_{i=0}^{\infty} d(1-d)^i \exp[(\sigma^2 + i\sigma_s^2)t^2/2]\right\} \\ &= de^{tm+(\sigma^2 t^2/2)}/[1-(1-d)e^{\sigma_s^2 t^2/2}]. \end{aligned} \tag{2.5.1}$$

From (2.5.1) or directly from the representation of the limiting distribution as a geometric mixture of normal distributions, we find the second and fourth central moments to be

$$\begin{aligned}\mu_2 &= \sigma^2 + \left(\frac{1-d}{d}\right)\sigma_s^2 \\ \mu_4 &= 3\left[\sigma^2 + \sigma_s^2\left(\frac{1-d}{d}\right)\right]^2 + 3\frac{\sigma_s^2(1-d)}{d^2}. \end{aligned} \tag{2.5.2}$$

These results are not markedly different from Rutherford's equation 6. Analogous expressions are obtainable for higher moments.

Rutherford apparently obtained his results by ignoring the distribution of the number of entrants in a year. His results could be duplicated by assuming a constant number of entrants per year and then using a continuous approximation for the required summations.

It is possible to use the moments (either those given in (2.5.2) or Rutherford's moments) to develop a Gram-Charlier expansion for the corresponding density. In this manner Rutherford obtains the following density for standardized income power

$$\frac{e^{-z^2/2}}{\sqrt{2\pi}}[1+\gamma(z^4-6z^2+3)] \tag{2.5.3}$$

where γ is a non-negative parameter. The corresponding probit diagram (i.e., the normal probability plot) obtained by Rutherford (1955, p. 283) may be viewed as being roughly piecewise linear or a polynomial of low degree.

Rutherford also presented an analogous model which incorporated a skewness parameter (η) to account for possibly variable means of the entrants' income power and possible nonzero means of subsequent shocks. The resulting four parameter model (location, scale, shape (γ) and skewness (η)) was then fitted to some British and American income data. The results were encouraging except for the upper tails. They were, of course, better than one would obtain with a log-normal model (which corresponds to the choice $\gamma = \eta = 0$). Rutherford suggested that improvement could be obtained by including more terms in the Gram-Charlier expansion used. An alternative avenue in the search for improvement would return to Rutherford's shock model and try to incorporate a more realistic birth and death process. The Poisson stream with linear death rate assumed to derive (2.5.2) is probably not realistic. Although no alternative to the Gram-Charlier expansion comes to mind, this technique is, undoubtedly, suspect, since it generally yields poor approximations in the tails (the precise region we wish to emphasize). A small Monte-Carlo study of the model (2.5.1), a geometric mixture of normals, indicated that samples from such distributions when plotted on normal probability paper do generally exhibit the low degree-polynomial features encountered in Rutherford's several empirical examples.

2.6 The Yule Distribution (Simon)

Simon (1955) endeavored to present a unified derivation of J-shaped frequency curves encountered in a variety of sociological, biological and economic settings. In particular he considered the distribution of words in prose samples by frequency of occurrence, the distribution of cities by population, the distribution of biological genera by numbers of species, the distribution of scientists by number of papers published and the distribution of income by size. The distributions in question are concentrated on the positive integers ($i = 1, 2, \ldots$), and empirical observation suggests densities which for large i take the form

$$f(i) \simeq ab^i / i^k \tag{2.6.1}$$

where a, b and k are positive constants. Typically, b is close enough to 1 to be ignored, except very far out in the tail. With $b = 1$, (2.6.1) would be known as Pareto's law in the income setting. In the other settings described by Simon it would be associated with the name of Zipf. Zipf (1949) is replete with examples in which (2.6.1) with ($b = 1$) is apparently a plausible model. What kind of stochastic process for assigning dollars to individuals, species to genera, people to cities, etc., will lead to distributions of the form (2.6.1)? Simon focused on the J-shaped tail of the distribution. Such tails are encountered in the study of contagious distributions. Yule (1924) proposed a model with the density

$$f(i) = C \cdot B(i, \rho + 1) \tag{2.6.2}$$

where C and ρ are positive constants and $B(.,.)$ is the usual beta function. Such distributions are called Yule distributions. It is a single parameter family of distributions. The constant C is determined by the condition that $\sum_{i=1}^{\infty} f(i) = 1$. It was proposed as a model for the species-genera problem. Using Stirling's approximation, one may verify that for large i the density may be written approximately as

$$f(i) \simeq \Gamma(\rho + 1) / i^{\rho+1}. \tag{2.6.3}$$

Simon described a stochastic mechanism to explain the appearance of a Yule distribution. His description was in the context of prose writing. We describe it here in terms of a stream of dollars (still using Simon's notation). We assume a potentially unlimited field of income recipients and a stream of dollars showering over them according to the following rules. Let $f(i,k)$ be the number of individuals in the population who have received exactly i of the first k dollars in the stream. Assume that the probability that the next (the $(k+1)$'th) dollar goes to a person who already has i dollars is proportional to $if(i,k)$. In addition, assume that there is a constant probability α that the $(k+1)$'th dollar will go to an individual who up until then had received no money. Letting $m(i,k) = E[f(i,k)]$, it follows by conditioning that

$$\begin{aligned} m(i,k+1) - m(i,k) &= \frac{1-\alpha}{k}[(i-1)m(i-1,k) - im(i,k)], \quad i = 2,\ldots,k+1 \\ m(1,k+1) - m(1,k) &= \alpha - \frac{1-\alpha}{k}m(1,k). \end{aligned} \tag{2.6.4}$$

By assuming

$$\frac{f(i,k+1)}{f(i,k)} = \frac{k+1}{k} \text{ for all } i \text{ and } k, \tag{2.6.5}$$

Simon arrived at the steady-state solution (as $k \to \infty$) satisfying

$$\frac{f(i)}{f(i-1)} = \frac{(1-\alpha)(i-1)}{1+(1-\alpha)i}. \tag{2.6.6}$$

If we set $\rho = (1-\alpha)^{-1}$, this may be seen to be the Yule distribution (2.6.2). Since (2.6.6) does satisfy (2.6.4), the questionable assumption (2.6.5) could have been dispensed with at the price of solving (2.6.4) directly.

Simon presented a more complicated model involving (in our setting) deaths or losses. It is assumed that the probability of the death of an individual with i dollars (after k dollars have been distributed) is proportional to $f(i,k)$. This leads to a distribution satisfying

$$\frac{f(i)}{f(i-1)} = \frac{(1-\alpha)(i-1)}{i}, \tag{2.6.7}$$

i.e., a log-series distribution. By a yet more complicated model he arrived at a distribution satisfying

$$\frac{f(i)}{f(i-1)} = \frac{i-1+c}{i+c+d} \tag{2.6.8}$$

for positive constants c and d. In a later paper (Simon, 1960), still more complicated birth and death rules led to the density

$$f(i) = A\lambda^i B(i+c, d-c+1). \tag{2.6.9}$$

Simon (1955, 1960) and Simon and Bonini (1958) argue that Champernowne's

(1953) model involved assumptions leading to a stochastic process not dissimilar to the stream of dollars which leads to the Yule distribution (2.6.2) and its relatives (2.6.8), (2.6.9).

The Yule distributions can be thought of as mixtures of geometric distributions. Specifically, if we assume X, given p, has a geometric distribution on $1, 2, \ldots$ and that p has a beta distribution with parameters α and β, then the unconditional density of X is given by

$$\begin{aligned} P(X = i) &= \int_0^1 p^{i-1}(1-p)p^{\alpha-1}(1-p)^{\beta-1}dp/B(\alpha, B) \\ &= B(\alpha + k - 1, \beta + 1)/B(\alpha, \beta), \quad i = 1, 2, \ldots. \end{aligned} \tag{2.6.10}$$

If we set $\alpha = 1$ and $\beta = \rho$, this yields the Yule distribution (2.6.2). If we set $\alpha - 1 = c$ and $\beta = d$, we get the more general (2.6.8). This distribution (2.6.10) is sometimes called the Waring distribution. It will be discussed further in Sections 3.8 and 5.8.

If income is measured on a continuous scale, the above development suggests a mixture of exponentials as a model. The continuous analog to the Waring distribution would be obtained by assuming that X given λ is exponentially distributed and that $e^{-\lambda}$ has a beta distribution. This leads to a survival function of the form

$$\overline{F}(x) = B(\alpha + x, \beta)/B(\alpha, \beta), \quad x > 0. \tag{2.6.11}$$

The special case $\alpha = 1$, $\beta + 1 = \rho$ (the continuous analog to the Yule distribution) has the approximate form, when x is large, given by

$$\overline{F}(x) \simeq (1 + x)^{\rho - 1}, \tag{2.6.12}$$

i.e., the tail is like that of the Pareto (II) distribution (see (3.2.2)).

Finally, it should be remarked that in the context of word frequencies, Efron and Thisted (1976) found that fits to certain data sets could be improved by allowing improper priors in the development of the Waring distribution. Specifically, the assumption that c and d be positive in (2.6.8) may profitably be relaxed. In conjunction with Simon's model (2.6.9), this would suggest consideration of the general family of densities satisfying

$$\frac{f(i)}{f(i-1)} = \frac{i+b}{ci+d}, \quad i = 2, 3, \ldots, \tag{2.6.13}$$

where $b > -2$, $d > -2c$ and $c \geq 1$. If $c = 1$, we need to assume that $d > b$.

Remark. *The appearance of Simon's (1955) paper was not without criticism. Mandelbrot (1959) was particularly unhappy with the fact that Simon argued against information theory based development of word frequency models, specifically those developed by Mandelbrot (1953). A lively and protracted interchange between the two authors, Simon and Mandelbrot, can be followed in subsequent issues of the journal Information and Control during the years 1960 and 1961. The titles of these volleys include: Notes, Further Notes, Final Note, Post Scriptum, and Reply to Post Scriptum. The two protagonists and the editor of the journal finally terminated the discussion without a clear resolution of the differing viewpoints.*

2.7 Income Determined by Inherited Wealth (Wold-Whittle)

Wold and Whittle (1957) set out to develop a model to explain the observed Paretian nature of wealth distributions. If we are willing to assume that incomes are more or less proportional to wealth, the resulting model can be used to justify the assumption of a Pareto model for income distributions.

In its simplest form, the model assumes that an individual's wealth grows at a fixed compound interest rate β during his life. Upon death, which can occur with probability $\gamma\Delta t$ during any time interval of length Δt, his fortune is equally divided among n inheritors. If we let $N(x,t)$ be the expected number of individuals whose income at time t is less than x and let $f(x,t) = \partial N(x,t)/\partial x$, then we readily verify that f should satisfy the partial differential equation

$$\frac{\partial f(x,t)}{\partial t} = -(\beta+\gamma)f(x,t) - \beta x\frac{\partial f(x,t)}{\partial x} + \gamma n^2 f(nx,t) \qquad (2.7.1)$$

assuming differentiability as required. Equation (2.7.1) is to hold for every $x > 0$ and $t > 0$. Solutions of (2.7.1) of the form $f(x,t) = x^p e^{qt}$ exist, but these cannot be densities, since $\int_0^\infty x^p dx = \infty$. This suggests that the given model will not lead to the desired Pareto distribution. The absence of a lower bound for income might be the cause. Such a bound was present in Champernowne's Pareto model but absent in the Kalecki-Gibrat log-normal model.

Wold and Whittle revived the model by postulating that (2.7.1) should hold only for values of x greater than a value k. The members of the population with wealth less than k are assumed to be an undifferentiated lower pool, growing in numbers as a result of a linear birth process and as a result of the entry of persons who inherit from a parent whose wealth is less than nk. Finally, it is assumed that individuals can emigrate from the lower pool into the propertied class (entering with minimal income k). With these revisions, assuming equal rates of population growth for the lower pool and the wealthier class, it may be shown that a Pareto solution to (2.7.1) is obtainable where p satisfies

$$(\beta/\gamma)(p+1) = n(n^{p+1} - 1). \qquad (2.7.2)$$

Eichhorn and Gleissner (1985) discuss the general solution of a variant form of equation (2.7.1) and identify certain non-Paretian solutions.

If the number of inheritors is allowed to be a random variable, the equation (2.7.1) becomes more complicated, but it is usually still possible to obtain a Pareto solution. The question of existence of other non-Pareto stable distributions remains open.

2.8 The Pyramid (Lydall)

The pyramid model proposed by Lydall (1959) is somewhat difficult to classify. Effectively, it states a few relatively plausible assumptions about the wage structure of large organizations which deterministically lead directly to a Pareto income distribution. A large corporation may be thought of as a pyramid involving several grades of employees. Let the number of individuals employed in the grades be $y_1, y_2, \ldots$

(where y_1 corresponds to the lowest grade). Now, we assume that each member of grades $2,3,\ldots$ serves as the supervisor of n individuals in the next lower grade, i.e., $y_i = ny_{i+1}$. Finally, assume that the income of a supervisor is directly proportional to the aggregate income of his immediate supervisees. Denoting the income of grade i employees by x_i, we thus have

$$x_{i+1} = pnx_i. \tag{2.8.1}$$

It is an immediate consequence of these assumptions that income power (log income) will be geometrically distributed within the organization, and thus income will have a discrete Pareto distribution (cf. the earlier discussion of Champernowne's model in Section 2.3). Lydall argued that in practice the number of grades will be large, and incomes within grades will be spread out from the average. This, in his view, will lead to a continuous Pareto distribution for income of the form

$$\overline{F}(x) = \left(\frac{x}{x_0}\right)^{-\alpha}, \quad x > x_0 \tag{2.8.2}$$

where x_0 is the minimum income in the population. It is possible to relate the parameter α in (2.8.2) to the supervisory ratio n and the salary factor p. In fact, we have

$$\alpha = -\log p / \log np. \tag{2.8.3}$$

Lydall's two assumptions (a uniform supervisory ratio and income proportional to the total income of one's supervisees) seem unlikely to be even approximately valid in complex populations. Nevertheless, the result may be robust with respect to violations of the assumptions. The pyramid model does provide a very simple route to the Pareto distribution.

2.9 Competitive Bidding for Employment (Arnold and Laguna)

This model assumes that the pool of employees is divided into non-overlapping generations. Let $F_n(x)$ be the income distribution associated with the n'th generation. Members of the $(n+1)$'st generation compete for jobs. The salaries they request are assumed to have the same distribution as those of the n'th generation, except that they are inflated by a factor c. Each job in the $(n+1)$'st generation is filled after a random number of applicants have been interviewed, and the job is awarded to the applicant requesting the lowest salary. Thus, an $(n+1)$'st generation salary is the minimum of a random number of independent identically distributed random variables with common distribution function $F_n(x/c)$. Let N be the random number of individuals interviewed for a job, and let P_N be the generating function of N. It follows that

$$\overline{F}_{n+1}(x) = E([\overline{F}_n(x/c)]^N) = P_N(\overline{F}_N(x/c)). \tag{2.9.1}$$

Stable long run distributions in such a model will be encountered if the distributions F_{n+1} and F_n in (2.9.1) differ by, at most, a scale factor. A particularly tractable case occurs when N is a geometric random variable, i.e., if $P_N(s) = ps/(1-qs)$ for some $p \in (0,1)$. In this case, if we define

$$\gamma_n(x) = F_n(x)/[1-F_n(x)], \tag{2.9.2}$$

we may rewrite (2.9.1) as

$$\gamma_{n+1}(x) = \gamma_n(x/c)/p, \tag{2.9.3}$$

whence

$$\gamma_n(x) = \gamma_1(x/c^{n-1})/p^{n-1}. \tag{2.9.4}$$

Now the limiting behavior of (2.9.4) depends on the rate at which $F_1(x)$ (the initial income distribution) converges to 0 as $x \to 0$.

If $\lim_{x\to 0} F(x)/x^a = b^a$ (finite, non-zero), then, provided $c = p^{-1/a}$, it may be verified that

$$\lim_{n\to\infty} \gamma_n(x) = (bx)^a,$$

and consequently,

$$\lim_{n\to\infty} \overline{F}_n(x) = (1+(bx)^a)^{-1}, \quad x > 0 \tag{2.9.5}$$

which is a Pareto (III) distribution (see (3.2.5)).

Thus, repeated geometric minimization for a wide variety of initial distributions will lead to a limiting distribution of Paretian form. In Arnold and Laguna (1976, 1977) the c's (the inflation factors) were allowed to differ from generation to generation. Arnold and Laguna considered only the case where N is geometric. What happens if N has a different distribution? If a stable long run distribution F is to exist, it will have to satisfy

$$\overline{F}(x) = P_N[\overline{F}(x/d)] \tag{2.9.6}$$

for some d. It is readily verified that if F is to satisfy (2.9.6) for some non-trivial N, then its support must be $(0,\infty)$. Non-Pareto solutions are possible. For example, if $N = 2$ with probability one, then an exponential solution is obtainable. A distribution function F can appear as a stable distribution in (2.9.6), provided that the function

$$\tilde{P}_N(s) = \overline{F}(d\overline{F}^{-1}(s)) \tag{2.9.7}$$

is indeed a generating function of some positive integer valued random variable. Not every choice of $\overline{F}$ (with support $(0,\infty)$) will yield a generating function via (2.9.7) (for example: $\overline{F}(x) = e^{-(x+x^2)}$). A. G. Pakes (personal communication) observed that the functional equation (2.9.6) is familiar in the theory of branching processes (see for example Athreya and Ney, 1972, p. 29). It may be verified that solutions to (2.9.6) must be scale mixtures of Weibull distributions. It would be interesting to characterize the class of distribution functions which can satisfy (2.9.6) and thus could arise as limiting distributions under repeated minimization. Some work by Baringhaus (1980) on an analogous problem suggested the possibility that, under certain regularity conditions, only Pareto (IV) distributions may arise in this way.

Pakes (1983) discusses a much more general version of this model in which the number of individuals interviewed from the n'th generation is a positive integer valued random variable denoted by N_n, with probability generating function $f_n(s)$.In addition the inflation factor, c, in the model is permitted to vary from generation to generation.

2.10 Other Models

It would be difficult to construct an exhaustive list of the many stochastic models which have been proposed to explain observed configurations of inequality, concentration, diversity and the like. Any one of them could be invoked as a possible model for income distributions and, with a little ingenuity, could be cloaked in wealth, income and welfare trappings to make it a, perhaps, plausible model. We will be content to mention only a few of these models, whose genesis is really outside the income sphere, but which can be, and sometimes are, suggested as income models.

Cohen's (1966) book provided an in-depth study of two models: a broken-stick model and a balls-in-boxes model. The first envisioned selecting $n-1$ independent uniform random variables to divide up a stick of unit length into n pieces. The second model involved a complicated ritual of assigning balls to boxes with ground rules which are of similar character to those which were used to generate Simon's distribution (2.6.6). For both models, Cohen found that the i'th largest object (interval or number of balls in a box) can be expected to be approximately proportional to

$$\frac{1}{n}\sum_{j=1}^{i}(n+1-j)^{-1}. \tag{2.10.1}$$

If we then consider the proportion of objects whose lengths fall in an interval $(x, x+\delta x)$ and denote this by $f(x)\delta x$, we find that $f(x)$ is approximately an exponential density. Perhaps, this could be phrased in Ericson's coin shower language (Section 2.4) to provide another model suggesting an exponential model for income distributions. Cohen (1966, Chapter 5) actually used the broken-stick or balls-in-boxes model to arrive at a Pareto distribution for sizes of firms. The Pareto distribution arises, however, by means of an exponential transformation, which seems to be justified chiefly because it does the trick of transforming an exponential distribution into a Pareto distribution.

Hill (1970) gave an interesting derivation of Zipf's law (the discrete Pareto distribution) in the species-genera context. His work has subsequently been refined, extended and generalized (see Hill, 1974; Hill and Woodroofe, 1975). The original paper gave an adequate picture of the general nature of the available results. Suppose that N species are to be distributed among M (non-empty) genera. Let L_i be the number of species belonging to the i'th genus. Hill considered two probabilistic models for the distribution of the vector $\underline{L}$ given M and N: Bose-Einstein allocation and Maxwell-Boltzmann allocation (as defined for example in Feller, 1968, pp. 39–41). Now we let $G(s)$ be the number of genera with exactly s species. Finally, assume M is a random variable, and assume that as $N \to \infty$, the random variable M/N has a well-defined limiting distribution, i.e., assume that

$$\lim_{N\to\infty} P(M/N \leq x|N) = F(x). \tag{2.10.2}$$

Under these assumptions Hill determined the asymptotic distribution of $G(s)/M$ under both a Bose-Einstein and a Maxwell-Boltzmann model. The simpler result is obtained using Bose-Einstein statistics. One finds

$$\lim_{N\to\infty} P(G(s)/M \leq x|N) = P(\theta(1-\theta)^{s-1} \leq x) \tag{2.10.3}$$

where θ has F (from (2.10.2)) as its distribution. In fact, one can get more than convergence in distribution using an extended form of the model (see Hill and Woodroofe, 1975). The limiting distribution depends on the choice of F. What we get from (2.10.3) are all possible mixtures of geometric distributions. This includes the Yule and Waring distributions (2.6.2) and (2.6.8). How are we to select which mixture of geometric distributions is appropriate? The simplest possible choice $(F(x) = x)$ leads to a limiting distribution of the form

$$P(Z = s) = [s(s+1)]^{-1}, \quad s = 1, 2, \ldots. \tag{2.10.4}$$

This is, except for a translation of one unit, the Standard Zipf distribution to be discussed in Section 3.11 (see (3.11.7)). Moving back to the realm of continuous variables, the Hill development prompts one to consider mixtures of exponential distributions as income distribution models. The choice of the mixing distribution would, hopefully, be based on some reasonable model of the true stochastic mechanism which governs the distribution of income. If one uses Maxwell-Boltzmann statistics, the Hill model leads to a mixture of (translated) Poisson variables. The form of the mixing distribution is determined by F (from (2.10.2)). Not all the distributions, which arise in the Hill framework, exhibit Paretian tail behavior. Perhaps, it would be possible to agree on a few general properties that F should possess. Presumably, if we are modeling income, these properties should be sufficient to guarantee Paretian tail behavior of the resulting mixture of distributions.

The following simple "explanation" of Paretian income distributions was communicated to me by R. F. Green. We assume that incomes of economically active individuals grow at an exponential rate (governed essentially by inflation and increasing ability and experience). Now, assume an exponential distribution for the amount of time T that a typical individual has been economically active and assume all individuals have the same initial income $X_0 (> 1)$. The income of a typical individual will then be

$$X = x_0^{\gamma T+1}, \tag{2.10.5}$$

where $T \sim \Gamma(1, \beta)$ and γ reflects the income growth rate. It is easy to verify that X has a classical Pareto distribution (2.11.1) with parameters $\sigma = x_0$ and $\alpha = \gamma\beta \log x_0$.

To finish this Section, we present a sampling of recent arguments that have been advanced to justify the frequency with which Paretian behavior is observed in income and/or wealth distributions.

Midlarsky (1989) presents an intriguing argument based on sequential acquisition of resources which ultimately leads to a log-exponential (i.e., Pareto) distribution of income.

Dutta and Michel (1998) argue that the presence of imperfect altruism, that is to say the random occurrence of individuals who selfishly spend all their wealth leaving none for their descendents, is another scenario which, under regularity conditions, leads to a Pareto distribution of wealth.

Levy (2003) observed that if one considers wealth accumulation via capital investment, it is only under an assumption of homogeneous investment talent that a Pareto wealth distribution will evolve. If investment talent is differentiably spread in the population, non-Pareto distributions will result. He then argues that the observed prevalence of Pareto wealth distributions suggests that investment success is actually attributable more to chance than to skill.

Benhabib, Bisin and Zhu (2011) consider a model with overlapping generations, finitely lived individuals and intergenerational transmission of wealth. Paretian behavior in the right tail of the stationary distribution is a consequence.

2.11 Parametric Families for Fitting Income Distributions

We turn now to several "models" for income distribution which have been proposed, usually with little more justification than: "this family of distributions has properties similar to those exhibited by many observed income distributions" or: "one can improve the fit of the X model by introducing new parameters or by assuming the model actually applies to some function of income rather than directly to income." For predictive purposes, such contrived "models" may well be more useful than true models obtained by attempting to mirror observed behavior by considering well-defined stochastic phenomena.

The first family of distributions suggested for fitting income data was the classical Pareto distribution

$$\overline{F}(x) = (x/\sigma)^{-\alpha}, \quad x > \sigma. \tag{2.11.1}$$

As we have seen, several true models suggest that Vilfredo Pareto's distribution should be, at least approximately, correct for large values of x. Pareto also proposed the density

$$f(x) = kx^{-(\alpha+1)}e^{-\beta x}, \quad x > \sigma. \tag{2.11.2}$$

Castellani (1950) postulated the existence of certain potentials in order to arrive at Pearson curves of several types including as a special case Pareto's curve (2.11.1). Ferreri (1964, 1967, 1975) suggested several variations of the classical Pareto distribution some of which involve truncation above and below.

We have already considered Gibrat's model (Section 2.1) which leads to a lognormal income distribution. Several ad hoc modifications of the Gibrat model have been suggested. Metcalf (1969) suggested that log(income + constant) might be normally distributed. Kalecki (1945) had earlier suggested that a log-normal model might apply to some function of income, rather than income itself. No theoretical basis for the choice of the function was provided. Rutherford (1955) made similar suggestions before developing his mixture of log-normals model (Section 2.5).

Champernowne (1952) proposed the density for income given by

$$f(x) = \frac{c}{x\left[\left(\frac{x}{x_0}\right)^{-\alpha} + \left(\frac{x}{x_0}\right)^{\alpha} + 2\lambda\right]} \tag{2.11.3}$$

(this is equivalent to (1.3.1) for income power). Although he gave no stochastic backing for this distribution, it is possible (see Ord, 1975) to describe a diffusion model for log-income which leads to (2.11.3). Although the formulation proposed by Ord is of interest, his choice of infinitesimal generators must be considered somewhat arbitrary. Effectively, they are chosen to lead to (2.11.3). As mentioned in Chapter 1, Champernowne's equation (2.11.3) contains interesting and potentially useful special cases. Two of these are Fisk's log-logistic or sech^2 distribution (1.3.2) and the powered Cauchy distribution (1.3.3). The powered Cauchy distribution has not been used much. It will be recalled that the Arnold-Laguna model (Section 2.9) led to the sech^2 distribution. It is, then, a short step to consider the richer Pareto (IV) family or Burr family (equation (1.3.4)). This family in full generality was suggested by Arnold and Laguna (1977) as potentially useful for fitting income data. The special case when $\mu = 0$ was proposed by Singh and Maddala (1976). This case was also investigated in some detail by Cronin (1977, 1979). The representation of Pareto (IV) distributions as mixtures of Weibull distributions suggests a possible stochastic genesis via a generalized coin shower (cf. Ericson's model, Section 2.4).

The still more general Feller-Pareto distribution (see (3.2.11)) was described in Arnold and Laguna (1977). Dagum (1980) suggested a mixed distribution with positive probability attached to the minimum income value and the remaining probability distributed according to a Feller-Pareto distribution (with $\gamma_1 = 1$).

The list of candidate distributions that have been proposed for modeling income data continues to grow. It even includes unlikely proposals such as the uniform distribution. Stiglitz (1969), however, described a variety of deterministic models for which the stable distribution is uniform.

Lomax (1954) introduced the Pareto (II) distribution, (1.3.5), often called the Lomax distribution, together with a second Lomax distribution with survival function of the form

$$\overline{F}(x) = exp[a/(be^{-bx} - 1], \;\; x > 0. \tag{2.11.4}$$

The Bradford (1948) distribution, (1.3.7), and Bomsdorf's prize competition distribution must also be included in this list.

Several standard distributions have, on occasion, been advocated for income modeling. Salem and Mount (1974) suggested use of the gamma distribution. McDonald (1981) proposed the scaled beta distribution, while Vartia and Vartia (1980) recommended the scaled F-distribution for fitting Finnish income data. Schmittlein

(1983) argued in favor of use of the Burr distribution (i.e., a Pareto (IV) distribution, (1.3.4), with the location parameter $\mu = 0$).

An intriguing alternative model was introduced by Schwartz (1985). He considered power transformations of income data (instead of the commonly used logarithmic transformation). He concluded that a good fit to a normal distribution could often be obtained by applying a cube-root transformation to the data. Thus the model is of the form $X = (\mu + \sigma Z)^3$ where Z is a standard normal variable.

Arnold and Villaseñor noted that the Lorenz curve for Mexican income in 1980 strongly resembled a segment of a circle. This led them to investigate general quadratic Lorenz curves and their corresponding distributions. These will be discussed in more detail in Chapter 4. An example of a family of densities with hyperbolic Lorenz curves is

$$f(x) \propto (x - \lambda_2 + \lambda_1)^{-3/2} I(\lambda_2 < x < \lambda_3), \tag{2.11.5}$$

where $\lambda_1 > 0$ and $0 < \lambda_2 < \lambda_3 < \infty$ (Arnold, 1986).

Instead of quadratic Lorenz curves, several other parametric families of Lorenz curves have been proposed for fitting empirical Lorenz curves. Each such curve of course corresponds to a scale parameter family of distributions, but many of them do not admit a convenient closed form expression for the corresponding distribution or density.

If the income data is discrete, e.g., rounded or grouped, then the Geeta distribution, discussed by Consul (1990), together with other discrete models (to be described in Chapter 3) are worthy of consideration.

Krishnan, Ng and Shihadeh (1991) proposed consideration of modified Pareto models of the form

$$\overline{F}(x) = Cx^{-\alpha} exp[-\sum_{i=1}^{k} \beta_i x^i] \, I(x > 0). \tag{2.11.6}$$

This includes the Pareto model (1.2.1) when $k = 1$, and essentially coincides with Pareto's more general model (1.2.2) when $k = 2$. Even more general models can be obtained by replacing $\sum_{i=1}^{k} \beta_i x^i$ by $Q(x; \underline{\theta})$, a parametric family of functions corresponding to situations with a more general form of elasticity. Kagan and Schoenberg (2001) refer to the model (1.2.2) as a tapered Pareto distribution and remark that it has been used to model the size of earthquakes. Presumably Krishnan's more general model might find applications in that area also.

The exponentiated Pareto distribution, discussed by Nadarajah (2005) with density

$$F(x) = \left[1 - \left(\frac{x - \mu}{\sigma}\right)^{-\alpha}\right]^{\theta}, \quad x > \mu, \tag{2.11.7}$$

represents a suggested alternative generalization of the Pareto distribution for added modeling flexibility. See Section 3.10 for further discussion of this distribution.

Efforts to model the entire income distribution and not just its Paretian upper tail have led to the proposal of composite densities. In these, two densities are spliced together smoothly so that, while the upper tail of the spliced density is Paretian, the lower part of the density has a shape which can be more adequately adjusted to model lower and intermediate incomes. Three representative versions of this construction will suffice as illustrations.

Teodorescu and Vernic (2006) propose splicing an exponential model with a Pareto model. Carreau and Bengio (2009) suggest splicing a normal density with a Pareto density. Picard and Tribouley (2002) splice two Pareto (I) densities with differing inequality parameters (differing α's).

Numerous other researchers have considered such hybrid spliced models. The combination of a log-normal density (for lower values) and a Pareto density (for large values) has quite naturally been the most popular. For examples, see Mitzenmacher (2004) and Teodorescu (2010).

Colombi (1990) suggested consideration of a product model of the form $X = YW$ in which Y and W are independent random variables where $Y \sim P(I)(1,\alpha)$ and W has a log-normal distribution. Subsequently, Nadarajah (2007) discussed properties of the densities corresponding to thirteen variants of the Colombi model in which W has various densities distinct from the log-normal density assumed by Colombi. Nadarajah also provides results for models of the form $X = (Y_1 W)/Y_2$ where all three variables W, Y_1 and Y_2 are independent, Y_1 and Y_2 have Pareto distributions and W has a density belonging to one of those thirteen different families of densities.

Other ratio models have been proposed. For example, Nadarajah (2012) suggests the model $X = Y/Z$ in which Y and Z are independent gamma and beta random variables. Manas (2011) considers $X = Y_1/(Y_2+Y_3)$ where the Y_i's are independent Pareto (I) variables.

Muralidharan and Khabia (2011) instead suggest a distribution that is a convex combination of a Pareto (I) distribution and a uniform distribution with a small mean and a small variance. This might be useful to model income distributions when a proportion of the population members have relatively negligible income.

The Lindley distribution, corresponding to a one parameter family of densities introduced by Lindley (1958), has been suggested as a possible competitor for the exponential distribution by Ghitany, Atieh and Nadarajah (2008). Its survival function is of the form

$$\overline{F}(x) = \frac{\theta+1+\theta x}{\theta+1} e^{-\theta x}, \quad x > 0, \tag{2.11.8}$$

where $\theta \in (0,\infty)$. This density has an increasing failure rate. As such, it may well not be suitable for modeling income distributions, for which heavier tails would be more appropriate. Bakouch, Al-Zahrani, Al-Shomrani, Marchi and Louzada (2012) introduced a more flexible version of the Lindley distribution which has enhanced potential for income modeling. The corresponding survival function is of the form

$$\overline{F}(x) = \left(\frac{\theta+1+\theta x}{\theta+1}\right)^{\alpha} e^{-(\theta x)^{\beta}}, \quad x > 0, \tag{2.11.9}$$

where $\alpha \in (-\infty,0) \cup \{0,1\}$, $\theta > 0$ and $\beta \geq 0$. For example, this model includes both of Pareto's models, (1.2.1) and (1.2.2), as special cases.

Nadarajah and Gupta (2007), motivated albeit by ideas outside of the realm of income modeling, proposed consideration of what they called a product Pareto distribution. Its density is of the form

$$f(x) \propto \left(1+\frac{x}{\sigma_1}\right)^{-(\alpha_1+1)}\left(1+\frac{x}{\sigma_2}\right)^{-(\alpha_2+1)}, \quad x>0, \tag{2.11.10}$$

where $\alpha_1, \alpha_2, \sigma_1$ and σ_2 are positive parameters. Though lacking in any obvious economic motivation, this density clearly exhibits the Paretian tail behavior that is typical of income distributions and the additional parameters enhance the flexibility of the model. Multicomponent versions of this density, i.e.,

$$f(x) \propto \prod_{j=1}^{k}\left(1+\frac{x}{\sigma_j}\right)^{-(\alpha_j+1)}, \quad x>0, \tag{2.11.11}$$

might also be considered.

Akinsete, Famoye and Lee (2008) discuss the beta-Pareto distribution. Its density is of the form

$$f(x) \propto \left[1-\left(\frac{x}{\sigma}\right)^{-\gamma}\right]^{\alpha-1}\left(\frac{x}{\sigma}\right)^{-\gamma\beta-1}, \quad x>\sigma, \tag{2.11.12}$$

where $\alpha>0, \beta>0, \gamma>0$ and $\sigma>0$. It is a technical generalization of the Pareto distribution with additional flexibility induced by the introduction of two additional parameters. Though not motivated by income considerations, the model might be useful for fitting and prediction in income settings. The density (2.11.12) represents a beta generalization of the classical Pareto (I) distribution. Akinsete, Famoye and Lee also describe analogous generalizations of more general Pareto distributions, such as the Pareto (IV) distribution. Further discussion of these models may be found in Section 3.10.

Cormann and Reiss (2009) consider the log-Pareto distribution with survival function

$$\overline{F}(x) = \left[1+\frac{1}{\beta}log\left(1+\frac{x}{\sigma}\right)\right]^{-\delta}, \quad x>0, \tag{2.11.13}$$

where β, δ and σ are positive parameters. Such distributions can be characterized as ones having super-heavy tails. Higher order log-Pareto models are also described by Cormann and Reiss. It is, however, not clear whether such distributions will find applications in income modeling; perhaps they might in situations involving extreme inequality.

Cai (2010) suggested that, instead of combining distributions, survival functions or densities, we might consider the possibility of modeling income by building an appropriate quantile function that is a combination of a Pareto quantile function and a power quantile function. A drawback of such an approach is that, typically, the resulting quantile functions do not have closed form expressions for their densities or distribution functions.

The phase-type (PH) distributions, introduced by Neuts (1975), are natural generalizations of the exponential distribution. They correspond to times till absorption in a single absorbing state in a continuous time Markov chain. Since the classical Pareto distribution can be described as a log-exponential distribution, i.e., it is representable as $X = e^Y$ where Y has an exponential distribution, it is natural to consider the more general model $X = e^Y$ where Y has a PH distribution. Such log-PH distributions were introduced by Ahn, Kim and Ramaswami (2012). The potential of such models for studying income and wealth distributions is evident.

As a final entry in this extensive list of possible income distribution models, consider a density which can be viewed as having arisen via a "hidden truncation" route. Assume that we begin with a two-dimensional random variable (X,Y) having some version of a bivariate Pareto distribution (many such distributions will be discussed in Chapter 6). Now assume that X will only be observed if Y does not exceed a specific threshold c. An example of a density for the observed X's obtained in such a fashion is

$$f(x) \propto \left[\left(1+\frac{x}{\sigma}\right)^{-(\alpha+1)} - \left(1+\theta+\frac{x}{\sigma}\right)^{-(\alpha+1)} \right], \quad x>0, \tag{2.11.14}$$

where $\alpha > 0$, $\sigma > 0$ and $\theta > 0$ (Arnold and Ghosh (2011). More general models involving linear combinations of Pareto distributions might also be considered, such as

$$\overline{F}(x) = \sum_{j=1}^{k} \delta_j \left[1+\frac{x}{\sigma_j}\right]^{-\alpha_j}, \quad x>0, \tag{2.11.15}$$

where the δ_j's are not necessarily non-negative but do sum to 1. Undoubtedly, models corresponding to only relatively small values of k would merit consideration. Further discussion of distributions of the forms (2.11.14) and (2.11.15) may be found in Section 3.10.

Chapter 3

Pareto and related heavy-tailed distributions

3.1 Introduction

In this chapter we will introduce a hierarchy of successively more complicated Pareto distributions. We begin with the classical Pareto distribution and, by the introduction of parameters which relate to location, scale, Gini index and shape, progress to a family called the Pareto IV family (essentially the Burr distributions). It turns out, that for purposes of deriving distributional results, it is sometimes convenient to consider an even more general family, which is here called the Feller-Pareto family. The full five-parameter Feller-Pareto distribution has, to the author's knowledge, not much been used for fitting income data. (It should do well, when it is used, since with 5 parameters you could fit just about any unimodal distribution!) This chapter draws heavily on the material in Arnold and Laguna (1977).

3.2 The Generalized Pareto Distributions

For the benefit of readers who have not read Chapter 1, we recall that the name "generalized Pareto distribution" has also been applied to the Pickands class of distributions (1.3.6), useful in the study of peaks over thresholds. That class includes the exponential distribution together with scaled beta distributions and certain Pareto distributions. The relationship between the Pickands generalized Pareto model and the models which are called generalized Pareto in this book was indicated in Chapter 1 and will be revisited later in this Section.

The starting point in our hierarchy of generalized Pareto models is the classical Pareto distribution, called here the Pareto (I) distribution. Its survival function is of the form

$$\overline{F}(x) = (x/\sigma)^{-\alpha}, \quad x > \sigma. \tag{3.2.1}$$

Here σ, the scale parameter, is positive and α (Pareto's index of inequality) is also positive. In practice, it is frequently assumed that $\alpha > 1$, so that the distribution has a finite mean. If a random variable X has (3.2.1) as its survival function, then we write $X \sim P(I)(\sigma, \alpha)$.

In much of the literature the standard Pareto distribution is permitted to have an additional location parameter. We will call the resulting distribution the Pareto (II) distribution. A convenient parameterization is provided by

$$\overline{F}(x) = \left[1 + \left(\frac{x-\mu}{\sigma}\right)\right]^{-\alpha}, \quad x > \mu \tag{3.2.2}$$

where μ, the location parameter, is real, σ is positive and α is positive. In most applications μ will be non-negative, but it does not complicate matters if we allow negative values. In this light, Schutz (1951) has pointed out that negative incomes are real world phenomena and cannot be simply swept under the table. If a random variable X has (3.2.2) as its survival function, we will write $X \sim P(\mathrm{II})(\mu, \sigma, \alpha)$. Here too, the parameter α is often assumed to be greater than 1, in order to have a finite first moment. We will simply call α the shape parameter. It is sometimes interpreted as an index of inequality. We will later see that, with $\mu = 0$, it is indeed a monotone function of the Gini index.

It was noted in Chapter 1 that there is an intimate relation between the Pareto (II) distribution and the Pickands generalized Pareto model. Repeating the comments in Chapter 1, we recall that the density of the Pickands generalized Pareto model is

$$f(x;\sigma,k) = \frac{1}{\sigma}\left(1 - \frac{kx}{\sigma}\right)^{(1-k)/k} I(x > 0, (kx)/\sigma < 1), \tag{3.2.3}$$

where $\sigma > 0$ and $-\infty < k < \infty$. The density corresponding to $k = 0$ is obtained by taking the limit as $k \uparrow 0$ in (3.2.3). This model, (3.2.3), includes three kinds of densities. When $k < 0$, it yields a Pareto (II) density (with $\mu = 0$), when $k = 0$ it yields an exponential density, while for $k > 0$, it corresponds to a scaled Beta distribution (of the first or standard kind). As will be seen later in this chapter, several results for the Pareto (II) distribution can be proved to remain valid in the more general context of the Pickands generalized Pareto model. However, since our focus is on heavy-tailed distributions, it will be the Pareto (II) submodel to which our attention will usually be directed.

Truncated versions of the Pareto (II) distribution have found application in simulation contexts, and have been proposed as plausible income distribution models (e.g., Arnold and Austin (1987)). Such distributions are of the form

$$\begin{aligned} F(x) &= 0, & x \leq \mu, \\ &= \frac{1-\left(1+\frac{x-\mu}{\sigma}\right)^{-\alpha}}{1-\left(1+\frac{\tau-\mu}{\sigma}\right)^{-\alpha}}, & \mu < x < \tau, \\ &= 1, & x \geq \tau, \end{aligned} \tag{3.2.4}$$

where $-\infty < \mu < \tau < \infty,\ \sigma > 0$, and $\alpha > 0$.

An alternative variation on the Pareto theme, which provides tail behavior similar to (3.2.2), is provided by the Pareto (III) family with survival function

$$\overline{F}(x) = \left[1 + \left(\frac{x-\mu}{\sigma}\right)^{1/\gamma}\right]^{-1}, \quad x > \mu \tag{3.2.5}$$

where μ is real, σ is positive and γ is positive. We will call γ the inequality parameter. If $\mu = 0$ and $\gamma \leq 1$, then γ turns out to be precisely the Gini index of inequality. If X has (3.2.5) as its survival function, we will write $X \sim P(\text{III})(\mu, \sigma, \gamma)$.

Pillai (1991) introduced a distribution that is closely related to the Pareto (III) distribution, which he called a semi-Pareto distribution. It has a survival function of the form

$$\overline{F}(x) = [1 + \psi(x)]^{-1}, \quad x > 0, \tag{3.2.6}$$

where the function ψ satisfies the functional equation

$$\psi(x) = p^{-1}\psi(p^{1/\alpha}x), \quad x > 0, \tag{3.2.7}$$

for some $p \in (0,1)$ and some $\alpha > 0$. The choice $\psi(x) = x^{1/\gamma}$ with $\alpha = 1/\gamma$ yields the $P(\text{III})(0,1,\gamma)$ distribution. Further discussion of the semi-Pareto distribution may be found in Section 3.10.

We may generalize further. By introducing both a shape and an inequality parameter, we arrive at the Pareto (IV) family:

$$\overline{F}(x) = \left[1 + \left(\frac{x-\mu}{\sigma}\right)^{1/\gamma}\right]^{-\alpha}, \quad x > \mu \tag{3.2.8}$$

where μ (location) is real, σ (scale) is positive, γ (inequality) is positive and α (shape) is positive. No longer will γ be identifiable with the Gini index. This will only occur when $\alpha = 1$ and $\mu = 0$. If a random variable X has (3.2.8) as its survival function, we will write $X \sim P(\text{IV})(\mu, \sigma, \gamma, \alpha)$.

The Pareto (IV) family provides a convenient vehicle for computing distributional results for the three more specialized families. They may be identified as special cases of the Pareto (IV) family as follows:

$$\begin{aligned} P(\text{I})(\sigma, \alpha) &= P(\text{IV})(\sigma, \sigma, 1, \alpha), \\ P(\text{II})(\mu, \sigma, \alpha) &= P(\text{IV})(\mu, \sigma, 1, \alpha), \\ P(\text{III})(\mu, \sigma, \gamma) &= P(\text{IV})(\mu, \sigma, \gamma, 1). \end{aligned} \tag{3.2.9}$$

Feller (1971, p. 49) defined a Pareto distribution in a somewhat different manner. Let Y have a Beta distribution with parameters γ_1 and γ_2, i.e.,

$$f_Y(y) = y^{\gamma_1 - 1}(1-y)^{\gamma_2 - 1}/B(\gamma_1, \gamma_2), \quad 0 < y < 1, \tag{3.2.10}$$

and define $U = Y^{-1} - 1$. Then U has what Feller called a Pareto distribution. By considering a linear function of a power of such a random variable, we arrive at a very general family, which we call the Feller-Pareto family. More precisely if $Y \sim \text{Beta}(\gamma_1, \gamma_2)$ and if for μ real, $\sigma > 0$ and $\gamma > 0$ we define

$$W = \mu + \sigma(Y^{-1} - 1)^{\gamma}, \tag{3.2.11}$$

then W has a Feller-Pareto distribution, and we write $W \sim FP(\mu, \sigma, \gamma, \gamma_1, \gamma_2)$. To see how this family represents a generalization of the Pareto (IV) family (and thus of the other Pareto families), we may begin by trying to obtain its survival function in closed form. To this end, note that if $Y \sim \text{Beta}(\gamma_1, \gamma_2)$ and if $U = Y^{-1} - 1$, then U has a Beta distribution of the second kind, equivalently a scaled F distribution with density

$$f_U(u) = u^{\gamma_2 - 1}(1+u)^{-\gamma_1 - \gamma_2}/B(\gamma_1, \gamma_2), \quad u > 0. \tag{3.2.12}$$

If W is as defined in (3.2.11), then

$$P(W > w) = P(\mu + \sigma U^\gamma > w) = P\left(U > \left(\frac{w-\mu}{\sigma}\right)^{1/\gamma}\right). \tag{3.2.13}$$

Thus we need a closed form for the survival function of U. This will be simplest when either γ_1 or γ_2 is 1. Let us consider the case when $\gamma_2 = 1$, so that

$$\overline{F}_U(u) = (1+u)^{-\gamma_1}, \quad u > 0. \tag{3.2.14}$$

Combining (3.2.13) and (3.2.14), we see that, with $\gamma_2 = 1$,

$$P(W > w) = \left[1 + \left(\frac{w-\mu}{\sigma}\right)^{1/\gamma}\right]^{-\gamma_1}. \tag{3.2.15}$$

Referring to (3.2.8), we see that the Pareto (IV) distributions are identifiable with the Feller-Pareto distributions with $\gamma_2 = 1$, i.e.,

$$P(\text{IV})(\mu, \sigma, \gamma, \alpha) = FP(\mu, \sigma, \gamma, \alpha, 1). \tag{3.2.16}$$

In the case when $\gamma_1 = 1$, one may verify that

$$P(W > w) = 1 - \left[1 + \left(\frac{w-\mu}{\sigma}\right)^{-1/\gamma}\right]^{-\gamma_2}. \tag{3.2.17}$$

It may be seen that the Feller-Pareto family includes distributions (such as (3.2.17)) whose behavior is distinctly non-Paretian in the upper tail.

Hurlimann (2009) develops certain generalized Pareto models using a general affine transformation schema. In one dimension, such models are of the form

$$X = h(\mu + \sigma g(Y)), \tag{3.2.18}$$

where g and h are well-behaved functions and Y has a specified distribution. In particular, the choice of an exponential distribution for Y, $h(u) = u$, and $g(u) = e^{cu}$ leads to a Pareto (II) model; compare equation (3.2.26) below in which the notation is slightly changed. Alternative choices for g, h and the distribution of Y yield interesting competing models. The generalized Pareto distributions that Hurlimann derives in this manner can be viewed as specialized cases of the general Feller-Pareto model (3.2.11). Care must be taken when reading the Hurlimann (2009) paper because of notational differences. His generalized Pareto model (his equation (4.1)) is recognizable as a Pareto (II) model, while his Pareto Type IV model (his equation (4.8)) is not the same as the Pareto (IV) model of the present book; it is instead identifiable as a Feller-Pareto distribution with $\gamma = 1$. He does present, in his equation (5.2), a formula which represents a recent rediscovery of the Pareto (IV) distribution.

It is not difficult to obtain the density of the general Feller-Pareto distribution defined by (3.2.11). It is of the form

$$f_W(w) = \left(\frac{w-\mu}{\sigma}\right)^{(\gamma_2/\gamma)-1}\left[1+\left(\frac{w-\mu}{\sigma}\right)^{1/\gamma}\right]^{-\gamma_1-\gamma_2} \Big/ \left[\gamma\sigma B(\gamma_1,\gamma_2)\right],$$
$$w > \mu. \tag{3.2.19}$$

Kalbfleisch and Prentice (1980) called this density (with $\mu = 0$) the generalized F density. The corresponding survival function is obtainable from tables of the incomplete beta function, using the fact that

$$P(W > w) = P\left(Y < \left[1+\left(\frac{w-\mu}{\sigma}\right)^{1/\gamma}\right]^{-1}\right] \tag{3.2.20}$$

where $Y \sim \text{Beta}(\gamma_1, \gamma_2)$. In many computations it is simpler to work directly with the representation (3.2.11). As we have seen, the Pareto (IV) distributions correspond to the case when $\gamma_2 = 1$. The Pareto (III) distributions are obtained when $\gamma_1 = \gamma_2 = 1$, i.e. when Y is uniformly distributed.

An alternative representation of the Feller-Pareto distribution is possible. A random variable X has the $FP(\mu, \sigma, \gamma, \gamma_1, \gamma_2)$ distribution, if it is of the form

$$X = \mu + \sigma(Z_2/Z_1)^{\gamma} \tag{3.2.21}$$

where the Z_i's are independent gamma random variables with unit scale parameter, $Z_i \sim \Gamma(\gamma_i, 1), i = 1,2$. The representation (3.2.21) leads to the following observation regarding the closure of the Feller-Pareto class under reciprocation when $\mu = 0$:

$$X \sim FP(0, \sigma, \gamma, \gamma_1, \gamma_2) = X^{-1} \sim FP(0, \sigma^{-1}, \gamma, \gamma_2, \gamma_1). \tag{3.2.22}$$

A generalized Feller-Pareto distribution was proposed by Zandonatti (2001). It involves the introduction of one additional parameter in the model (3.2.11). We begin with $Y \sim \text{Beta}(\gamma_1, \gamma_2)$ and for μ real, $\sigma > 0$, $\delta > 0$ and $\gamma > 0$ we define

$$V = \mu + \sigma(Y^{-\delta} - 1)^{\gamma}. \tag{3.2.23}$$

Such a random variable V is then said to have a generalized Feller-Pareto distribution, and we write $V \sim GFP(\mu, \sigma, \delta, \gamma, \gamma_1, \gamma_2)$. The corresponding density function may be found in Subsection 5.2.10 in equation (5.2.103). An alternative source for the density, with slightly different parameterization, is Kleiber and Kotz (2003, p.197).

At the risk of being accused of getting carried away in the quest for yet richer families of generalized Pareto families, one might consider (motivated by (3.2.21)) random variables of the form

$$X = (\mu_1 + \sigma_1 Z_1^{\gamma_1})/(\mu_2 + \sigma_2 Z_2^{\gamma_2}) \tag{3.2.24}$$

where the Z_i's are independent gamma variables $(Z_1 \sim \Gamma(\lambda_i, 1), i = 1,2)$. Such an eight-parameter (!) family, though eminently flexible, would find few friends; unless the parameters could be endowed with economic interpretations, which appears unlikely.

We have now met the full array of generalized Pareto distributions to be considered in this chapter. Almost all are subsumed in the Feller-Pareto family and a unified derivation of distributional results is, thus, often possible. The generalized versions of the Feller-Pareto distribution defined in equations (3.2.23) and (3.2.24) will not be considered further. To conclude this section, we review again (cf. Chapter 1) the origins of the subfamilies of the Feller-Pareto distributions. Inter alia, some alternative representations of the distributions will be pointed out.

Pareto, of course, is the name to be rightfully attached to the Pareto (I) and, undoubtedly, the Pareto (II) family. Much of the statistical theory relating to the Pareto (I) distribution has been (or could have been) derived from the following representation. A random variable X has a $P(\mathrm{I})(\sigma, \alpha)$ distribution if it is of the form

$$X = \sigma e^{V/\alpha} \tag{3.2.25}$$

where $V \sim \Gamma(1,1)$ (a standard exponential random variable). An analogous representation of the $P(\mathrm{II})$ in terms of an exponential random variable is possible, i.e.,

$$X = \mu + \sigma(e^{V/\alpha} - 1). \tag{3.2.26}$$

Both (3.2.25) and (3.2.26) are simple consequences of the representation (3.2.11).

A second important representation of the Pareto (II) distribution, known to Maguire, Pearson and Wynn (1952), is as a mixture of exponentials. We may describe it in terms of the conditional survival function, given an auxiliary gamma distributed random variable Z. Thus, if

$$P(X > x \mid Z = z) = e^{-z(x-\mu)/\sigma}, \quad x > \mu,$$

i.e., a (translated) exponential distribution and if $Z \sim \Gamma(\alpha, 1)$, then it follows that unconditionally $X \sim P(\mathrm{II})(\mu, \sigma, \alpha)$.

This representation of the Pareto (II) distribution as a gamma mixture of exponential distributions arises naturally in reliability and survival contexts, see, e.g., Keiding, Kvist, Hartvig, Tvede and Juul (2002). It is also familiar in Bayesian analysis of exponential data, where the gamma density enters as a convenient prior.

The Pareto (III) distribution was apparently first considered by Fisk (1961a,b) under the label5sech^2 distribution. It is closely related to the logistic distribution. We say that a random variable X has a logistic (μ, σ) distribution, if its distribution function assumes the form

$$F_X(x) = [1 + e^{-(x-\mu)/\sigma}]^{-1}, \quad -\infty, < x < \infty,$$

and we write $X \sim L(\mu, \sigma)$. It is not difficult to verify that

$$X \sim L(\mu, \sigma) \Leftrightarrow e^X \sim P(\mathrm{III})(0, e^{\mu}, \sigma). \tag{3.2.27}$$

It is, as a consequence of the relation (3.2.27), that the Pareto (III) distribution is sometimes called the log-logistic distribution.

The Pareto (IV) family was suggested by Arnold and Laguna (1976), as well as Ord (1975) and Cronin (1977, 1979). As mentioned earlier, much of the distribution theory regarding the Pareto (IV) distribution can be obtained by using available results for the Burr distribution (see, e.g., Johnson, Kotz and Balakrishnan, 1994, p. 54). The Pareto (IV) distribution has a representation as a mixture of Weibull distributions. If

$$P(X > x|Z = z) = e^{-z[(x-\mu)/\sigma]^{1/\gamma}}, \quad x > \mu,$$

i.e., a translated Weibull distribution, and if $Z \sim \Gamma(\alpha, 1)$ then $X \sim P(\text{IV})(\mu, \sigma, \gamma, \alpha)$. In this guise the Pareto (IV) distribution appeared early in the engineering literature (although it does not seem to have been christened), see, e.g., Dubey (1968).

The Feller-Pareto family, based on Feller's idiosyncratic definition of the Pareto distribution, was suggested by Arnold and Laguna (1976, 1977). Its importance rests in the representation (3.2.11) and the consequent possibility of relating most distributional questions regarding Pareto distributions of varying levels of generality to equivalent distributional questions relating to beta random variables. The incomplete Beta function is not pretty, but it is tabulated and no new distributional tables are needed for the generalized Pareto distributions.

3.3 Distributional Properties

3.3.1 Modes

The Feller-Pareto distributions are unimodal. By differentiating the density (3.2.19), we may verify that the mode is at μ if $\gamma > \gamma_2$, while if $\gamma \leq \gamma_2$, we find (here $W \sim FP(\mu, \sigma, \gamma, \gamma_1, \gamma_2)$)

$$\text{mode}(W) = \mu + \sigma[(\gamma_2 - \gamma)/(\gamma_1 + \gamma)]^{\gamma}. \tag{3.3.1}$$

As special cases of (3.3.1) we find:

$$\text{(Pareto I)} \quad \text{mode}(X) = \sigma, \tag{3.3.2}$$

$$\text{(Pareto II)} \quad \text{mode}(X) = \mu, \tag{3.3.3}$$

$$\text{(Pareto III)} \quad \text{mode}(X) = \mu + \sigma[(1-\gamma)/(1+\gamma)]^{\gamma}, \gamma < 1, \tag{3.3.4}$$

$$\text{(Pareto IV)} \quad \text{mode}(X) = \mu + \sigma[(1-\gamma)/(\alpha+\gamma)]^{\gamma}, \gamma < 1. \tag{3.3.5}$$

3.3.2 Moments

The intimate relationship between the Feller-Pareto and Beta distributions is profitably exploited to obtain distributional properties of the various Pareto families. Since the Beta distribution has no closed form moment generating function, but readily computed moments, we are not surprised to encounter an analogous situation among the Pareto distributions.

In order to compute moments of the Pareto distributions, it is convenient to work with the representation (3.2.11). An alternative derivation involving the gamma representation (3.2.21) is, of course, possible. Suppose that $W \sim FP(\mu,\sigma,\gamma,\gamma_1,\gamma_2)$. Define $W^* = (W-\mu)/\sigma$, then $W^* \sim FP(0,1,\gamma,\gamma_1,\gamma_2)$, and, in fact, $W^* = (Y^{-1}-1)^{\gamma}$ where $Y \sim \text{Beta}(\gamma_1,\gamma_2)$.

For which real numbers δ does the δ'th moment of W^* exist? Operating formally we have:

$$\begin{aligned} E(W^{*\delta}) &= E[(Y^{-1}-1)^{\gamma\delta}] \\ &= \int_0^1 \left(\frac{1-y}{y}\right)^{\gamma\delta} y^{\gamma_1-1}(1-y)^{\gamma_2-1}dy/B(\gamma_1,\gamma_2) \\ &= B(\gamma_1-\gamma\delta,\gamma_2+\gamma\delta)/B(\gamma_1,\gamma_2). \end{aligned}$$

Clearly, the required integral will converge, and, thus, the δ'th moment of W^* will exist, provided $\gamma_1-\gamma\delta$ and $\gamma_2+\gamma\delta$ are both positive. Therefore, we may write, cancelling $\Gamma(\gamma_1+\gamma_2)$ from the numerator and denominator,

$$\begin{aligned} E(W^{*\delta}) &= \Gamma(\gamma_1-\gamma\delta)\Gamma(\gamma_2+\gamma\delta)/\Gamma(\gamma_1)\Gamma(\gamma_2), \\ &-(\gamma_2/\gamma) < \delta < (\gamma_1/\gamma). \end{aligned} \tag{3.3.6}$$

For the general Feller-Pareto distribution it follows that for integral values of $k < (\gamma_1/\gamma)$, we have

$$E(W^k) = \sum_{j=0}^{k} \binom{k}{j} \mu^{k-j}\sigma^j\Gamma(\gamma_1-j\gamma)\Gamma(\gamma_2+j\gamma)/\Gamma(\gamma_1)\Gamma(\gamma_2). \tag{3.3.7}$$

This expression simplifies to a single term when $\mu=0$, and from it we can get expressions for the δ'th moments of the various Pareto families. Thus, when $\mu=0$

$$\text{(Pareto II)}\quad E(X^\delta) = \sigma^\delta\Gamma(\alpha-\delta)\Gamma(1+\delta)/\Gamma(\alpha),\ -1<\delta<\alpha, \tag{3.3.8}$$

$$\text{in particular:}\quad E(X) = \sigma/(\alpha-1),\ \alpha>1, \tag{3.3.9}$$

$$\begin{aligned} \text{(Pareto III)}\quad E(X^\delta) &= \sigma^\delta\Gamma(1-\gamma\delta)\Gamma(1+\gamma\delta),\ -\gamma^{-1}<\delta<\gamma^{-1} \\ &= \sigma^\delta\pi\gamma\delta/\sin(\pi\gamma\delta), \end{aligned} \tag{3.3.10}$$

$$\text{(Pareto IV)}\quad E(X^\delta) = \sigma^\delta\Gamma(\alpha-\gamma\delta)\Gamma(1+\gamma\delta)/\Gamma(\alpha),\ -\gamma^{-1}<\delta<\alpha/\gamma. \tag{3.3.11}$$

Moments of the Pareto (I) distribution cannot be obtained in this way since, for it, $\mu=\sigma\neq 0$. They are obtainable by direct integration:

$$\text{(Pareto I)}\quad E(X^\delta) = \sigma^\delta\left(1-\frac{\delta}{\alpha}\right)^{-1},\quad \delta<\alpha. \tag{3.3.12}$$

Cohen and Whitten (1988) provide tables for the mean, variance, skewness and kurtosis of the Pareto (I) distribution for a range of values of α and σ.

The expressions (3.3.8)–(3.3.11) have been rediscovered recurrently in the literature in varying contexts. See for example Burr (1942), Dubey (1968), Johnson,

Kotz and Balakrishnan (1994) and Rodriguez (1977). In most of these references the distributions are labeled Burr distributions. The moments of the classical Pareto distribution (3.3.12) were known in the nineteenth century.

It is only in the Pareto (I) and (II) cases that simple expressions are obtainable for the variance. Thus

$$\text{(Pareto I and II)} \quad \text{var}(X) = \sigma^2\alpha(\alpha-1)^{-2}(\alpha-2)^{-1}, \quad \alpha > 2. \tag{3.3.13}$$

The unpleasant form of the expressions for the variances of the various distributions should not be cause for alarm. In the context of modeling income distributions, the variance is not often of importance. It is much more important that other measures of inequality be related in as simple a manner as possible to the parameters in the distribution. In Chapter 4, we will see that the Pareto families fare quite well in this regard.

Tadikamalla (1980) provides discussion and a diagram displaying the range of possible values of the skewness and kurtosis measures for the Pareto (IV) distribution.

3.3.3 Transforms

Instead of directly computing the moments of Pareto distributions, it is natural to investigate the possibility of evaluating them by using a suitable transform.

For the Pareto (I)(σ,α) distribution, Lorah and Stark (1978) argue in favor of using the Mellin transform defined by

$$M_X(v) = \int_0^\infty x^{v-1} f_X(x)\, dx.$$

In the Pareto (I) case, this assumes the form

$$M_X(v) = \sigma^{v-1}\frac{\alpha}{1+\alpha-v}, \quad v < 1+\alpha.$$

The k-th moment can be obtained by setting $v = k+1$. The Mellin transform is particularly convenient for identifying the distribution of powers of Pareto (I) variables and the distribution of products and quotients of independent Pareto (I) variables.

If we set the lower bound μ equal to zero in the Pareto (II) model, then we will be dealing with non-negative variables and the natural tool for moment investigation is the Laplace transform of the distribution. Unfortunately, the corresponding Laplace transform is only expressible in terms of special functions. If we denote the Laplace transform by $L(s)$, $s > 0$, it can be expressed as

$$L(s) = E(e^{-sX}) = \int_0^\infty e^{-sx}\frac{\alpha}{\sigma}\left(1+\frac{x}{\sigma}\right)^{-(\alpha+1)} dx. \tag{3.3.14}$$

Seal (1980) provides an expression for (3.3.14) involving the exponential integral function. Nadarajah and Kotz (2006a) give an expression in terms of the Whittaker function, while Takano (2007) presents an expression involving confluent hypergeometric functions.

3.3.4 *Standard Pareto Distribution*

It is convenient to introduce, what we will call, the standard Pareto distribution. A random variable Z will be said to have a standard Pareto distribution, if its survival function is of the form

$$\overline{F}_Z(z) = (1+z)^{-1}, \quad z > 0 \tag{3.3.15}$$

(it is Feller-Pareto with parameters (0,1,1,1,1)). Although the standard Pareto distribution is a special case of each of the families Pareto II, III and IV and of the Feller-Pareto family, it does not play as central a role as does the standard normal distribution among normal distributions. This is due to the lack of a representation theorem. For example, it is not possible to express a general Feller-Pareto random variable as a simple function of a standard Pareto variable. Such a representation theorem does exist for the Pareto (III) family. Thus, if Z is standard Pareto, then

$$\mu + \sigma Z^{\gamma} \sim P(\text{III})(\mu, \sigma, \gamma). \tag{3.3.16}$$

Following a time-honored tradition, the letter Z will be reserved in this book to indicate a standard random variable; here it is always standard Pareto (3.3.15) (rather than normal). The representation (3.3.16) will prove to be particularly useful in the study of the order statistics of a Pareto (III) distribution.

The standard Pareto distribution is self-reciprocal, i.e., if Z has a standard Pareto distribution, then Z and Z^{-1} are identically distributed. This same property is shared by a $P(\text{III})(0,1,\gamma)$ distribution. These observations can be verified directly using (3.3.15) and (3.3.16), or they may be considered to be consequences of the representation (3.2.27) of the Pareto (III) (with $\mu = 0$) as a log-logistic distribution.

3.3.5 *Infinite Divisibility*

Thorin (1977) showed that the Pareto $(\text{II})(0,\sigma,\alpha)$ distribution is a generalized Γ-transform and thus is infinitely divisible. A shorter alternative proof was provided by Berg (1981).

3.3.6 *Reliability, $P(X_1 < X_2)$*

In a stress-strength context, interest focuses on the evaluation of $P(X_1 < X_2)$ in which X_1 and X_2 are independent random variables. If $X_i \sim P(\text{I})(\sigma_i, \alpha_i)$, $i = 1,2$ then it is possible to evaluate this reliability quantity. The relevant expression was provided by Nadarajah (2003). Thus

$$P(X_1 < X_2) = \frac{\alpha_1}{\alpha_1 + \alpha_2}\left(\frac{\sigma_1}{max\{\sigma_1,\sigma_2\}}\right)^{\alpha_1}\left(\frac{\sigma_2}{max\{\sigma_1,\sigma_2\}}\right)^{\alpha_2}. \tag{3.3.17}$$

The corresponding expression in the case in which the X_i's have independent Pareto (II) distributions involves confluent hypergeometric functions (Nadarajah, 2003).

3.3.7 Convolutions

What if we add independent Pareto variables? Specifically, consider $X_1 \sim P(\text{IV})(\mu_1, \sigma_1, \gamma_1, \alpha_1)$ and $X_2 \sim P(\text{IV})(\mu_2, \sigma_2, \gamma_2, \alpha_2)$ where X_1 and X_2 are independent. Let $Y = X_1 + X_2$. The exact distribution seems quite intractable, but it is possible to quickly identify the tail behavior of the distribution of Y using the concept of regular variation (see Feller, 1971, p. 268–272). Since as $x \to \infty$

$$\overline{F}_{X_1}(x) \sim \left(\frac{x}{\sigma_1}\right)^{-\alpha_1/\gamma_1} L_1(x)$$

and

$$\overline{F}_{X_2}(x) \sim \left(\frac{x}{\sigma_2}\right)^{-\alpha_2/\gamma_2} L_2(x),$$

where L_1 and L_2 are slowly varying, it follows from Feller's equations (8.15) and (8.17) that as $x \to \infty$,

$$\overline{F}_Y(x) \sim x^{-\min(\alpha_1/\gamma_1, \alpha_2/\gamma_2)} L(x) \tag{3.3.18}$$

where L is slowly varying. Thus, the tail of the convolution will exhibit Paretian behavior. Roehner and Winiwarter (1985) consider $S(n) = \sum_{i=1}^{n} X_i$ where the X_i's are i.i.d. P(I)(σ, α_i), $i = 1, 2, ..., n$, random variables. They conclude that the density of $S(n)$ satisfies, as $x \to \infty$,

$$f_{S(n)}(x) \sim \frac{\alpha_{1:n}}{\sigma} \left(\frac{x}{\sigma}\right)^{-(\alpha_{1:n}+1)} L(x),$$

where $\alpha_{1:n} = \min\{\alpha_1, ..., \alpha_n\}$.

Similarly, if $X_1, X_2, \ldots, X_n$ are independent identically distributed $P(\text{IV})(\mu, \sigma, \gamma, \alpha)$ variables and if $Y = \sum_{i=1}^{n} X_i$, then as $x \to \infty$

$$\overline{F}_{X_i}(x) \sim \left(\frac{x}{\sigma}\right)^{-\alpha/\gamma} L(x) \tag{3.3.19}$$

and

$$\overline{F}_Y(y) \sim n \left(\frac{x}{\sigma}\right)^{-\alpha/\gamma} L(x)$$

where L is slowly varying. These results not only give insight into the persistence of Paretian tail behavior, but also are useful in identifying the limiting distributions to be expected when we consider normalized sums of Pareto variables in Section 3.6 of this chapter.

Can we expect more detailed results when we consider less complicated Pareto variables? To visualize the problem, it is convenient to introduce uniform representations of Pareto II and III random variables. If U has a uniform (0, 1) distribution, then

$$(\mu - \sigma) + \sigma U^{-1/\alpha} \sim P(\text{II})(\mu, \sigma, \alpha) \tag{3.3.20}$$

and

$$\mu+\sigma(U^{-1}-1)^{\gamma}\sim P(\mathrm{III})(\mu,\sigma,\gamma). \tag{3.3.21}$$

These are, really, the results of inverse distribution function transforms. The corresponding representation for a Pareto (IV) is sufficiently complicated to deter all but the most hardy from trying to consider the exact form of convolutions.

If we focus on Pareto (II) variables, the basic problem (using (3.3.20)) may be stated as follows. Let U_1 and U_2 be independent uniform $(0,1)$ random variables. Determine the distribution of $V=U_1^{-1/\alpha_1}+cU_2^{-1/\alpha_2}$ where α_1, α_2 and c are positive constants. This is an elementary, though tiresome, exercise. A series expansion for the density of V may be found as follows, for $v>1+c$,

$$\begin{aligned} f_V(v) &= \int_{(v-c)^{-\alpha_1}}^{1}\left(\frac{\alpha_2}{c}\right)\left(\frac{v-z^{-1/\alpha_1}}{c}\right)^{-\alpha_2-1} dz \\ &= \left(\frac{\alpha_2}{c}\right)\left(\frac{v}{c}\right)^{-\alpha_2-1}\int_{(v-c)^{-\alpha_1}}^{1}\left[\sum_{j=0}^{\infty}\binom{-\alpha_2-1}{j}\frac{z^{-j/\alpha_1}}{(-v)^j}\right]dz. \end{aligned} \tag{3.3.22}$$

The integration may be performed term by term, and we obtain, provided α_1 is not an integer,

$$f_V(v)=\frac{\alpha_1\alpha_2}{c}\left(\frac{v}{c}\right)^{-\alpha_2-1}\sum_{j=0}^{\infty}\binom{-\alpha_2-1}{j}\frac{1-(v-c)^{j-\alpha_1}}{(\alpha_1-j)(-v)^j},$$
$$1+c<v<\infty. \tag{3.3.23}$$

If α_1 is an integer, a logarithmic term is involved. If α_1 and α_2 are small integers, simpler expressions can be obtained.

Using the representation (3.3.20) for the Pareto (II) family (and the Pareto (I) family, which corresponds to setting $\mu=\sigma$ in (3.3.20)), equation (3.3.23) allows one to, in principle, write down a series expansion for the density of a convolution of two Pareto (II) distributions with possibly different parameters.

It will be recalled that (3.3.19) gave information about the tail of the distribution of a sum of n independent identically distributed Pareto (IV) variables. In the Pareto I and II case more detailed information is available. Hagstroem (1960) obtained (3.3.19) from a series expansion and, in addition, worked out the exact survival functions for sums of 2 and 3 independent Pareto (I) variables in the special case when $\alpha=1$. The result for 2 variates could be obtained from (3.2.27). Thus, if X_1,X_2 are independent identically distributed Pareto $I(1,1)$ random variables, then

$$P(X_1+X_2>x)=\frac{2}{x}+\frac{2\log(x-1)}{x^2},\quad x>2. \tag{3.3.24}$$

The corresponding expression for the sum of three such variables is considerably more cumbersome. Blum (1970) considered the case of a sum of n independent Pareto I $(\alpha,1)$ random variables. He showed that an asymptotic expansion obtained earlier by Brennan, Reed and Sollfrey (1968) was, in fact, an exact convergent series expression for the density of the n-fold convolution. The density $f_n(x)$ is of the form

$$f_n(x) = \frac{-1}{\pi}\sum_{j=1}^{n}\binom{n}{j}[-\Gamma(1-\alpha)]^j \sin(\pi\,\alpha\,j)\sum_{m=0}^{\infty}\frac{C_{N-j,m}\Gamma(m+\alpha j+1)}{x^{m+\alpha j+1}}$$

$$x > n, \tag{3.3.25}$$

where $C_{p,m}$ is the mth coefficient in the series expansion of the pth power of the confluent hypergeometric function:

$$_jF_1(-\alpha, 1-\alpha; t) = \sum_{j=0}^{\infty}\left(\begin{array}{c}-\alpha\\ j-\alpha\end{array}\right)\frac{1}{j!}t^j. \tag{3.3.26}$$

Blum devoted considerable attention to the form of these coefficients. The expression (3.3.25) is probably only of technical interest. It, obviously, has potential as a tool for identifying the limiting distribution for sums of Pareto variables. Blum made use of it in that context. Ramsay (2003) provides an alternative expression for $f_n(x)$ in terms of exponential integral functions for the case in which α is a positive integer.

What can be said about sums of independent Pareto (III) variables? Referring to either of the representations (3.3.16) or (3.3.21), the basic question may be phrased as follows. Let X_1 and X_2 be independent standard Pareto variables and define $Y = X_1^{\gamma_1} + cX_2^{\gamma_2}$. What is the distribution of Y? By conditioning on X_2, one readily finds

$$P(Y > y) = \left[1+\left(\frac{y}{c}\right)^{1/\gamma_2}\right]^{-1} + \int_0^{(y/c)^{1/\gamma_2}}\frac{[1+(y-cx^{\gamma_2})^{1/\gamma_1}]^{-1}}{(1+x)^2}dx. \tag{3.3.27}$$

From this expression one could obtain a series expansion for the survival function of Y in a manner analogous to that used to derive (3.3.23).

3.3.8 Products of Pareto Variables

If we multiply independent Pareto variables rather than adding them, it is sometimes possible to get simple expressions for the density of the resulting product. In the case of the Pareto I distribution the key lies in utilization of representation (3.2.25). Thus, if $X_1, X_2, \ldots, X_n$ are independent Pareto I variables with $X_i \sim P(\mathrm{I})(\sigma_i, \alpha_i)$, then their product W has the representation

$$W = \left(\prod_{i=1}^{n}\sigma_i\right)\exp\left(\sum_{i=1}^{n}(V_i/\alpha_i)\right) \tag{3.3.28}$$

where the V_i's are independent standard exponential variables. Expressions are available for the distribution of $\sum_{i=1}^{n} V_i/\alpha_i$.

In particular, if $\alpha_i = \alpha, (i = 1, 2, \ldots, n)$, then $\sum_{i=1}^{n} V_i/\alpha \sim \Gamma(n, 1/\alpha)$, and we may readily obtain the density of W:

$$f_W(w) = \frac{\left(\sigma \log\frac{w}{\sigma}\right)^{n-1}\left(\frac{w}{\sigma}\right)^{-\alpha}\frac{\alpha}{w}}{\Gamma(n)}, \quad w > \sigma \tag{3.3.29}$$

where $\sigma = \prod_{i=1}^{n} \sigma_1$ (cf. Malik, 1970a). The other case in which simple closed form expressions are available is when all the α_i's are distinct. In this situation using a result for weighted sums of exponentials given in, for example, Feller (1971, p. 40), we may write the survival function of the product in the form:

$$P(W > w) = \sum_{i=1}^{n} \left(\frac{w}{\sigma}\right)^{-\alpha_i} \prod_{\substack{k=1 \\ k \neq i}}^{n} \left(\frac{\alpha_k}{\alpha_i - \alpha_k}\right), \quad w > \sigma \tag{3.3.30}$$

where $\sigma = \prod_{i=1}^{n} \sigma_i$ and $\alpha_i \neq \alpha_j$ if $i \neq j$. Equation (3.3.30) may be proved inductively, after checking it directly for $n = 2$. The distribution of products of independent Pareto (IV) variables can, via the representation (3.2.21), be reduced to a problem involving the distribution of products of powers of independent gamma random variables. Unlike the Pareto (I) case, closed form expressions for the resulting density are apparently not obtainable. See also remark (i) in Section 3.12.

The Pareto (IV) family is closed under minimization. Thus, if X_1 and X_2 are independent random variables with $X_i \sim P(\mathrm{IV})(\mu, \sigma, \gamma, \alpha_i), i = 1, 2$, then

$$\min(X_1, X_2) \sim P(\mathrm{IV})(\mu, \sigma, \gamma, \alpha_1 + \alpha_2). \tag{3.3.31}$$

It is possible to combine this observation with the earlier result regarding products of Pareto (I) variables and obtain results for products of minima of Pareto (I) variables (for details see Rider, 1964, and Malik, 1970a).

3.3.9 *Mixtures, Random Sums and Random Extrema*

We now turn to the distributions which arise when the parameters of the Pareto (IV) distribution are themselves random variables. In most cases the resulting distributions are intractable, but in some cases simple results are obtainable. As an example, consider $X|\alpha \sim P(\mathrm{IV})(\mu, \sigma, \gamma, \alpha)$ where $\alpha \sim \text{geometric}(p)$. In this case the unconditional survival function is of the form

$$\begin{aligned} P(X > x) &= \sum_{j=1}^{\infty} pq^{j-1} \left[1 + \left(\frac{x-\mu}{\sigma}\right)^{1/\alpha}\right]^{-j} \\ &= \left[1 + \left(\frac{x-\mu}{\sigma p^{\gamma}}\right)^{1/\gamma}\right]^{-1}. \end{aligned} \tag{3.3.32}$$

Thus, $X \sim P(\mathrm{III})(\mu, \sigma p^{\gamma}, \gamma)$. This is merely a mixture formulation of the geometric minima of Pareto (III) variables described in Section 2.9.

If α is assumed to have an exponential distribution with mean γ^{-1} instead of a geometric distribution, a closed form survival function for the resulting mixture is

$$\overline{F}(x) = \left[1 + \gamma^{-1} \log\left(1 + \left(\frac{x-\mu}{\sigma}\right)^{1/\gamma}\right)\right]^{-1}. \tag{3.3.33}$$

If μ is assumed to be a random variable, then the resulting mixture is a convolution of the distribution of μ and a $P(\mathrm{IV})(0,\sigma,\gamma,\alpha)$ distribution. Only in very special circumstances will this be obtainable in closed form. If μ is zero and σ is a random variable, then, after taking logarithms, evaluating the mixture can be reduced to a convolution. But, again, closed form results are not to be expected save in contrived situations. No tractable examples are known in which γ is used as a mixing parameter.

Next, we consider random sums of Pareto variables. The initial work in this direction was by Hagstroem (1960), who considered Poisson sums of Pareto (I) variables. It is not surprising that his results were only approximate, when one recalls the complex form of the n-fold convolution (3.3.25). Hagstroem argued that if $Y = \sum_{i=1}^{N} X_i$ where the X_i's are independent $P(\mathrm{I})(1,\alpha)$ random variables and where N is Poisson(γ) independent of the X_i's, then

$$P(Y > y) \sim \gamma y^{-\alpha}. \tag{3.3.34}$$

This is a consequence of the fact that $P(Y > y|N = n) \sim ny^{-\alpha}$ (refer to (3.3.19)). In fact, Hagstroem's result is not dependent on the fact that N has a Poisson distribution. Equation (3.3.34) will be valid for any non-negative integer-valued random variable N with expectation γ.

Random extrema results arise in the following scenario (for motivation, recall the competitive bidding for employment model discussed in Chapter 2). Consider a sequence $X_1, X_2, \ldots$ of i.i.d. Pareto (III)(μ,σ,γ) random variables. Suppose that for some $p \in (0,1)$, N_p is independent of the X_i's and has a geometric(p) distribution, i.e., $P(N = n) = p(1-p)^{n-1}, n = 1,2,\ldots$. Define the corresponding random extrema by

$$U_p = \min\{X_1, X_2, \ldots, X_{N_p}\}, \tag{3.3.35}$$

and

$$V_p = \max\{X_1, X_2, \ldots, X_{N_p}\}. \tag{3.3.36}$$

It is readily verified that U_p and V_p each have Pareto (III) distributions. Specifically it is the case that

$$U_p \sim P(III)(\mu, \sigma p^{\gamma}, \gamma), \tag{3.3.37}$$

and

$$V_p \sim P(III)(\mu, \sigma p^{-\gamma}, \gamma). \tag{3.3.38}$$

Observe that, if $\mu = 0$, then

$$p^{-\gamma} U_p \stackrel{d}{=} p^{\gamma} V_p \stackrel{d}{=} X_1.$$

Some characterizations based on this observation will be discussed in Section 3.9.

3.4 Order Statistics

Consider a sample $X_1, X_2, \ldots, X_n$ from the Feller-Pareto $(\mu,\sigma,\gamma,\gamma_1,\gamma_2)$ distribution and denote the corresponding order statistics by $X_{1:n} < X_{2:n} < \ldots < X_{n:n}$ (ties may be ignored since we are dealing with a continuous distribution). If we let f and F denote the common density and distribution functions of the X_i's, then we may write down well-known formulas for the joint and marginal densities of the order statistics. Thus

$$f_{X_{1:n},X_{2:n},\ldots,X_{n:n}}(x_{1:n},x_{2:n},\ldots,x_{n:n}) = n!\prod_{i=1}^{n} f(x_{i:n}),$$
$$\mu < x_{1:n} < x_{2:n} < \ldots < x_{n:n}, \tag{3.4.1}$$

and for each i:

$$f_{X_{i:n}}(x_{i:n}) = n\binom{n-1}{i-1}[F(x_{i:n})]^{i-1}[1-F(x_{i:n})]^{n-1}f(x_{i:n}). \tag{3.4.2}$$

Such general results are conveniently cataloged in David and Nagaraja (2003). David and Nagaraja's book will be referred to repeatedly in this section. For the general Feller-Pareto distribution it does not seem possible to obtain useful expressions from (3.4.1) and (3.4.2). We do always have available a representation in terms of the inverse distribution function. Thus, in general, if $X_{i:n}$ is the ith order statistic from a sample of size n with common distribution function F, then

$$X_{i:n} \stackrel{d}{=} F^{-1}(U_{i:n}) \tag{3.4.3}$$

where $\stackrel{d}{=}$ means that the two random variables are identically distributed and where $U_{i:n}$ is the ith order statistic of a sample of size n from a uniform (0,1) distribution. It is well known (see, e.g., David and Nagaraja, 2003) that $U_{i:n} \sim \text{Beta}(i, n-i+1)$.

The study of the joint distribution of several order statistics is facilitated by the following general result.

Theorem 3.4.1. *Let $X_1, X_2, \ldots, X_n$ be a random sample of size n from a continuous distribution F and let $X_{1:n} < X_{2:n} < \ldots < X_{n:n}$ be the corresponding order statistics. For any $k_1 < k_2$, the conditional distribution of $X_{k_2:n}$, given that $X_{k_1:n} = x_0$, is the same as the distribution of the (k_2-k_1)st order statistic in a sample of size $n-k_1$ drawn from F truncated at the left at x_0.*

A convenient reference for this result is David and Nagaraja (2003, p. 17) It is proved there, as a consequence of the following well-known result regarding exponential order statistics (see, e.g., Sukhatme, 1937).

Theorem 3.4.2. *Let $X_1, X_2, \ldots, X_n$ be sample of size n from the exponential distribution $(\overline{F}(x) = e^{-\mu x}, x > 0)$ with corresponding order statistics $X_{i:n}, i = 1,2,\ldots n$, and spacings $Y_{i:n} = X_{i:n} - X_{i-1:n}, i = 1,2,\ldots,n$ ($X_{0:n} = 0$, by definition). The spacings are independent exponential random variables. In fact, the random variables $\{(n-i+1)Y_{i:n} : i = 1,2,\ldots,n\}$ are independent identically distributed random variables with common exponential survival function $\overline{F}(x) = e^{-\mu x}, x > 0$.*

The intimate relationship between the classical Pareto distribution and the exponential distribution (3.2.25) renders Theorem 3.4.2 an important tool in discussing the distribution of classical Pareto order statistics. Huang (1975) explicitly used Theorem 3.4.2 in this way to rederive many results obtained earlier in the literature by direct computation. Thanks to the memoryless property of the exponential distribution, Theorems 3.4.1 and 3.4.2 are equivalent. In what follows, reference will usually be made to Theorem 3.4.1. The choice is a matter of taste.

In certain special cases the density of the ith order statistic (3.4.2) assumes a recognizable form. For example, if the parent distribution is standard Pareto, then using the inverse distribution function representation (3.4.3), we find that the order statistics themselves have Feller-Pareto distributions, specifically $X_{i:n} \sim FP(0,1,1,n-i+1,i)$. A more general result is available:

$$X_i\text{'s} \sim P(\text{III})(\mu,\sigma,\gamma) \Rightarrow X_{i:n} \sim FP(\mu,\sigma,\gamma,n-i+1,i). \qquad (3.4.4)$$

It is possible to write down the density of the ith order statistic from P(II) and P(IV) families, but the corresponding distributions are not Feller-Paretian. The exception is the smallest order statistic. One finds, since $X_{1:n} > x$ if and only if $X_i > x$ for every i,

$$X_i\text{'s} \sim P(\text{IV})(\mu,\sigma,\gamma,\alpha) \Rightarrow X_{1:n} \sim P(\text{IV})(\mu,\sigma,\gamma,n\alpha). \qquad (3.4.5)$$

The particular case of the sample median of a Pareto (IV) distribution has been studied in detail by Burr and Cislak (1968). They presented tables of moments and compared the variances of the sample mean and median in such populations. They considered the case $\mu = 0$, $\sigma = 1$ and, naturally, referred to the distributions as a family introduced by Burr.

Expressions for the joint density of two or more order statistics from Pareto (IV) samples can be written down easily using the inverse probability transformation, applied to the corresponding joint density for uniform order statistics.

3.4.1 Ratios of Order Statistics

In general, the distributions of systematic statistics (functions of order statistics) do not simplify to any useful degree. An exception is provided by ratios of order statistics from the classical Pareto distribution. The simplification encountered here is a consequence of the exponential representation of the Classical Pareto (3.2.25) and the Sukhatme (1937) results on exponential spacings. Consider $X_{i:n}, (i = 1,2,\ldots,n)$, the order statistics of a sample of size n from a $P(\text{I})(\sigma,\alpha)$ distribution. For $k_1 < k_2$, define

$$R_{k_1,k_2:n} = X_{k_2:n}/X_{k_1:n}. \qquad (3.4.6)$$

In order to determine the distribution of $R_{k_1,k_2:n}$, we may argue as follows. From (3.2.25) (the exponential representation) $\log R_{k_1,k_2:n}$ has the same distribution as $\alpha^{-1}(U_{k_2:n} - U_{k_1:n})$ where the $U_{i:n}$'s are order statistics from a standard exponential distribution. Thus, using the Sukhatme result for exponential spacings,

$$\log R_{k_1,k_2:n} = \frac{1}{\alpha} \sum_{j=k_1+1}^{k_2} Y_{j:n}$$

where the $Y_{j:n}$'s are independent random variables with $P(Y_j > y) = e^{-(n-j+1)y}$. It follows that

$$R_{k_1,k_2:n} \stackrel{d}{=} \prod_{j=k_1+1}^{k_2} W_j \tag{3.4.7}$$

where the W_j's are independent random variables with $W_j \sim P(\mathrm{I})(1, \alpha(n-j+1))$. We have earlier discussed the distribution of the product of independent Pareto (I) random variables with differing α's. Using (3.3.30), it is possible to immediately write down the survival function of $R_{k_1,k_2:n}$, in the form

$$P(R_{k_1,k_2:n} > x) = \sum_{j=k_1+1}^{k_2} x^{-\alpha(n-j+1)} \prod_{\substack{\ell=k_1+1 \\ \ell \neq j}}^{k_2} \left(\frac{n-\ell+1}{\ell-j}\right), \quad x > 1. \tag{3.4.8}$$

The foregoing analysis requires only minor modification (let $k_1 = 0$ and define $X_{0:n} = 1$) to yield a computing form for the survival function of the ith order statistics from a $P(\mathrm{I})(\sigma, \alpha)$ sample:

$$\begin{aligned} P(X_{i:n} > x) &= \sum_{j=1}^{i} \left(\frac{x}{\sigma}\right)^{-\alpha(n-j+1)} \prod_{\substack{\ell=1 \\ \ell \neq j}}^{i} \left(\frac{n-\ell+1}{\ell-j}\right) \\ &= \binom{n}{i} \sum_{j=1}^{i} \binom{i}{j} \frac{j(-1)^{j-1}}{(n-j+1)} \left(\frac{x}{\sigma}\right)^{-\alpha(n-j+1)}, \quad x > \sigma. \end{aligned} \tag{3.4.9}$$

An alternative derivation of (3.4.9) involves determination of the survival function of the ith order statistic of a sample from a $P(\mathrm{I})(1,1) = P(\mathrm{III})(1,1,1)$ distribution using (3.4.4), and then using the transformation $x \to x^{1/\alpha}$, it could also be obtained using the inclusion-exclusion formula (see, e.g., Feller, 1968, p. 106), since $X_{i:n} > x$ occurs, if and only if, no more than $i-1$ of the observations are less than or equal to x.

If, instead of considering $R_{k_1,k_2:n}$ with $k_1 < k_2$, we focus on its reciprocal $\widetilde{R}_{k_1,k_2:n} = X_{k_1:n}/X_{k_2:n}$, which is always bounded above by 1, it is possible to derive simple expression for the corresponding density (Malik and Trudel, 1982). Thus

$$f_{\widetilde{R}_{k_1,k_2:n}}(x) = \frac{\alpha x^{(n-k_2+1)\alpha-1}(1-x^{\alpha})^{k_2-k_1-1}}{B(k_2-k_1, n-k_2+1)} I(0 < x < 1). \tag{3.4.10}$$

Equivalently we may observe that $(\widetilde{R}_{k_1,k_2:n})^{\alpha} \sim B(k_2-k_1, n-k_2+1)$, a result which can be derived in a straightforward manner using the following two facts:

(1) If $X \sim P(\mathrm{I})(\sigma, \alpha)$, then $(X/\sigma)^{-\alpha} \sim Uniform(0,1)$.

(2) The spacings corresponding to a sample from a $Uniform(0,1)$ distribution have a $Dirichlet(1,1,...,1)$ distribution.

Since the Pareto (II) distribution results from a translation of $\mu - \sigma$ applied to the Pareto (I), we may immediately modify (3.4.9) to yield an expression for the survival function of the ith order statistic of a sample of size n from a $P(\text{II})(\mu, \sigma, \alpha)$ distribution:

$$P(X_{i:n} > x) = \sum_{j=1}^{i} \left(1 + \frac{x - \mu}{\sigma}\right)^{-\alpha(n-j+1)} \prod_{\substack{\ell=1 \\ \ell \neq j}}^{i} \left(\frac{n - \ell + 1}{\ell - j}\right), \quad x > \mu. \tag{3.4.11}$$

It should be noted that in the expressions (3.4.8)–(3.4.11), which are linear combinations of Pareto survival functions, some of the coefficients are negative.

If one attempts a similar analysis for ratios of order statistics from a Pareto III or IV family, the analysis is more cumbersome. For example, if $X_1, X_2, \ldots, X_n$ is a sample from a $P(\text{III})(0, \sigma, \gamma)$ population with order statistics $X_{i:n}$, then for $k_1 < k_2$ we find, using the uniform representation (3.3.21), that

$$X_{k_2:n}/X_{k_1:n} \stackrel{d}{=} [(U_{n-k_2+1:n}^{-1} - 1)/(U_{n-k_1+1:n}^{-1} - 1)]^{\gamma}. \tag{3.4.12}$$

The expression in square brackets in (3.4.12) is a relatively simple function of the random vector

$$(U_{n-k_2+1:n}, U_{n-k_1+1:n} - U_{n-k_2+1:n}, 1 - U_{n-k_1+1:n}),$$

which has a Dirichlet$(n - k_2 + 1, k_2 - k_1, k_1)$ distribution, but this observation does not seem to lead to any useful computing formula for the survival function of $X_{k_2:n}/X_{k_1:n}$. See, however, remark (ii) in Section 3.9.

3.4.2 Moments

Although the actual distributions of Paretian order statistics are not often simple, we may reasonably hope that their means and covariances will be more accessible. Several strategies are available to us in our efforts to evaluate these quantities. We may try to exploit the exponential representation of the classical Pareto via Sukhatme's result (Theorem 3.4.2). We may, in some cases, know that the order statistics are themselves Feller-Paretian variables whose moments are known (see equations (3.4.4) and (3.4.5)). The approach we will follow is to consider first the case of the Standard Pareto. Theorem 3.4.1 and a truncation property of the Pareto (II) distribution will be useful in the analysis. Along the way we will point out available results for the means and covariance structure of the order statistics of more general Paretian distributions. The level of generality achieved includes the Pareto (IV) family, for which tables can be constructed, but falls short of the Feller-Pareto generalization.

Recall that a random variable, Z, is said to have a standard Pareto distribution, if its survival function is of the form

$$\overline{F}_Z(z) = (1 + z)^{-1}, \quad z > 0, \tag{3.4.13}$$

with corresponding inverse distribution function

$$F_Z^{-1}(z) = z/(1 - z), \quad 0 < z < 1. \tag{3.4.14}$$

If we let $Z_{k:n}$ denote the kth order statistic in a sample of size n from a standard Pareto distribution, we find (cf. David and Nagaraja, 2003, p. 34)

$$\begin{aligned} E(Z_{k:n}) &= n\binom{n-1}{k-1}\int_0^1 [z/(1-z)]z^{k-1}(1-z)^{n-k}dz \\ &= \frac{k}{n-k}, \quad k=1,2,\ldots,n-1. \end{aligned} \tag{3.4.15}$$

The largest order statistic, $Z_{n:n}$, has infinite expectation. More generally, the δth moment of $Z_{k:n}$ may be computed, provided $k+\delta<n+1$. One finds

$$\begin{aligned} E(Z_{k:n}^{\delta}) &= \int_0^1 [z/(1-z)]^{\delta} z^{k-1}(1-z)^{n-k}dz/B(k,n-k+1) \\ &= \frac{\Gamma(k+\delta)\Gamma(n-k-\delta+1)}{\Gamma(k)\Gamma(n-k+1)}. \end{aligned} \tag{3.4.16}$$

From (3.4.16) the expectations of order statistics for a Pareto (III) distribution follow readily, since a Pareto (III) random variable is of the form $\mu+\sigma Z^{\gamma}$ where Z is standard Pareto. Thus:

$$\text{(Pareto III)}: \quad E(X_{k:n}) = \mu+\sigma\frac{\Gamma(k+\gamma)\Gamma(n-k-\gamma+1)}{\Gamma(k)\Gamma(n-k+1)}, \tag{3.4.17}$$

provided $k+\gamma<n+1$. Closed form expressions for the expectations of order statistics from a Feller-Pareto distribution do not seem to be obtainable. The same problem occurs with the Pareto (IV) distribution. In this case, however, it is possible to generate the expectations in a straightforward manner.

If we let $\mu_{k:n}$ denote the expectation of the kth order statistic from a sample of size n from the distribution F, then since

$$\mu_{k:n} = \int_0^1 F^{-1}(z)z^{k-1}(1-z)^{n-k}/B(k,n-k+1),$$

it is easy to verify the following recurrence relation

$$(n-k)\mu_{k:n} = n\mu_{k:n-1} - k\mu_{k+1:n}. \tag{3.4.18}$$

Relation (3.4.18) may be used to generate all the $\mu_{k:n}$'s, if we know $\mu_{1:n}, (n=1,2,\ldots)$. In the case of the Pareto (IV) distribution we know from (3.4.5) that $X_{1:n}$ itself has a Pareto (IV) distribution, and so, using (3.3.11), we find

$$\text{(Pareto IV)} \quad E(X_{1:n}) = \mu+\sigma\Gamma(n\alpha-\gamma)\Gamma(1+\gamma)/\Gamma(n\alpha), \quad \gamma<n\alpha. \tag{3.4.19}$$

A table of expected values of order statistics from a Pareto (IV) distribution generated by means of (3.4.18) and (3.4.19) may be found in Arnold and Laguna (1977). An earlier table for the Pareto (I) distribution was provided by Malik (1966).

The recurrence relation (3.4.18) is valid for any parent distribution for the X_i's. More specialized recurrence relations which take into account the nature of the distribution of the X_i's are sometimes available. For example, Adeyemi (2002) and

Adeyemi and Ojo (2004) provide some recurrence relations for moments and product moments of order statistics in the Pareto (II) case (with $\mu = 0$). Analogous recurrences for the Pareto (III) distribution (with $\mu = 0$) were presented by Ali and Khan (1987). Ahmad (2001) presents some relations in the Pareto (IV) case (also with $\mu = 0$).

The computation of the covariances between Pareto order statistics is more complicated. No difficulty is encountered when dealing with the standard Pareto distribution, but results for more general Pareto families are not readily obtainable. Theorem 3.4.1 proves to be useful in this context. It is used in conjunction with the observation that a truncated Pareto (II) distribution is itself a Pareto (II) distribution. More specifically, if $X \sim P(\mathrm{II})(\mu,\sigma,\alpha)$, then the distribution of X truncated at the left at x_0 is $P(\mathrm{II})(x_0,\sigma+x_0-\mu,\alpha)$, since, for $x > x_0$,

$$\begin{aligned}\overline{F}(x)/\overline{F}(x_0) &= \left(1+\frac{x-\mu}{\sigma}\right)^{-\alpha} \Big/ \left(1+\frac{x_0-\mu}{\sigma}\right)^{-\alpha} \\ &= \left(1+\frac{x-x_0}{\sigma+x_0-\mu}\right)^{-\alpha}. \end{aligned} \tag{3.4.20}$$

It follows that if $Z_{1:n}, Z_{2:n}, \ldots, Z_{n:n}$ are standard Pareto order statistics (i.e. from a P(II) (0,1,1) distribution) then, given $Z_{k_1:n} = z_0$, $Z_{k_2:n}$ is distributed as the $(k_2 - k_1)$st order statistic from a sample of size $n-k_1$ from a $P(\mathrm{II})(z_0,1+z_0,1)$ distribution (which is the same as a $P(\mathrm{III})(z_0,1+z_0,1)$ distribution). From (3.4.17) it follows that

$$E(Z_{k_2:n} \mid Z_{k_1:n} = z_0) = z_0 + (1+z_0)\frac{k_2-k_1}{n-k_2}, \tag{3.4.21}$$

so that, for $k_1 < k_2$,

$$\begin{aligned} E(Z_{k_1:n}Z_{k_2:n}) &= E[Z_{k_1:n}E(Z_{k_2:n} \mid Z_{k_1:n})] \\ &= E[Z_{k_1:n}^2 + (Z_{k_1:n}+Z_{k_1:n}^2)(k_2-k_1)/(n-k_2)]. \end{aligned}$$

The quantities $E(Z_{k_1:n}), E(Z_{k_2:n})$ and $E(Z_{k_1:n}^2)$ can be evaluated using (3.4.15) and (3.4.16), and after algebraic manipulation, one obtains

$$\operatorname{cov}(Z_{k_1:n},Z_{k_2:n}) = nk_1/[(n-k_1)(n-k_2)(n-k_1-1)]. \tag{3.4.22}$$

Introduction of location and scale parameters presents no problems, so that one may conclude for samples from a $P(\mathrm{IV})(\mu,\sigma,1,1)$ distribution that

$$\operatorname{cov}(X_{k_1:n},X_{k_2:n}) = \sigma^2\, nk_1/[(n-k_1)(n-k_2)(n-k_1-1)]. \tag{3.4.23}$$

The truncation property displayed in (3.4.20) assumes a simple form in the case of the classical Pareto distribution (when $\mu = \sigma$). Thus, if $X \sim P(\mathrm{I})(\sigma,\alpha)$, then X truncated at x_0 has a $P(\mathrm{I})(x_0,\alpha)$ distribution. It follows that for $P(\mathrm{I})(\sigma,\alpha)$ order statistics

$$\begin{aligned} E(X_{k_1:n}X_{k_2:n}) &= E[X_{k_1:n}E(X_{k_2:n} \mid X_{k_1:n})] \\ &= E[X_{k_1:n}^2 E(X'_{k_2-k_1:n-k_1})] \\ &= \sigma^2 E(X'^2_{k_1:n})E(X'_{k_2-k_1:n-k_1}) \end{aligned} \tag{3.4.24}$$

where the $X'_{k:n}$'s are order statistics from a $P(\mathrm{I})(1,\alpha)$ distribution. Equation (3.4.24) can be used to generate covariances between classical Pareto order statistics. To accomplish this, one uses recurrence relation (3.4.18) and the observation that

$$X \sim P(I)(\sigma,\alpha) \Rightarrow X^2 \sim P(\mathrm{I})(\sigma^2,\alpha/2). \tag{3.4.25}$$

The corresponding covariances can alternatively be obtained by direct integration using the joint density of the order statistics (as was done by Malik, 1966).

A much simpler approach is, however, available (only) in the case of the classical Pareto distribution. Huang (1975) first pointed out the utility of the Sukhatme representation of exponential order statistics (Theorem 3.4.2) in this regard. Using that theorem, it follows that if $X_{k:n}$ is an order statistic from a classical Pareto distribution ($P(\mathrm{I})(\sigma,\alpha)$), then

$$X_{k:n} \stackrel{d}{=} \sigma \prod_{j=1}^{k} W_{j:n} \tag{3.4.26}$$

where the $W_{j:n}$'s are independent random variables with $W_{j:n} \sim P(\mathrm{I})(1,\alpha(n-j+1))$ (cf. equation (3.4.7)). Observation (3.4.25) is valid for arbitrary powers, not just the square, so that

$$X \sim P(\mathrm{I})(\sigma,\alpha) \Rightarrow X^\gamma \sim P(\mathrm{I})(\sigma^\gamma,\alpha/\gamma), \tag{3.4.27}$$

and as a consequence, or by direct integration (cf. (3.3.12)),

$$E(W_{j:n}^\gamma) = \left(1 - \frac{\gamma}{\alpha(n-j+1)}\right)^{-1}, \quad \alpha > \gamma/(n-j+1). \tag{3.4.28}$$

It is, then, easy to write down arbitrary product moments of classical Pareto order statistics. In this manner one may verify that, for the Pareto (I) distribution,

$$E(X_{k:n}) = \sigma \frac{n!}{(n-k)!} \frac{\Gamma(n-k+1-\alpha^{-1})}{\Gamma(n+1-\alpha^{-1})} \tag{3.4.29}$$

$$\begin{aligned}\operatorname{cov}(X_{k_1:n}, X_{k_2:n}) = \sigma^2 &\frac{n!}{(n-k_2)!} \frac{\Gamma(n-k_1+1-2\alpha^{-1})}{\Gamma(n+1-2\alpha^{-1})} \frac{\Gamma(n-k_2+1-\alpha^{-1})}{\Gamma(n-k_1+1-\alpha^{-1})} \\ &- E(X_{k_1:n})E(X_{k_2:n})\end{aligned} \tag{3.4.30}$$

where $k_1 < k_2$ and, for convergence, $\alpha > (n-k+1)^{-1}$ in (3.4.29) and $\alpha > \max\{2/(n-k_1+1),(n-k_2+1)^{-1}\}$ in (3.4.30). Kabe (1972) had earlier obtained product moments of classical Pareto order statistics, but since he did not directly utilize the Sukhatme representation, the development was more complicated. Malik (1966) had equation (3.4.28) together with certain recurrence formulae such as

$$E(X_{k_1:n}X_{k_2:n}) = \frac{n-k_2+1}{n-k_2-\alpha^{-1}+1} E(X_{k_1:n}X_{k_2-1:n}). \tag{3.4.31}$$

Here again, use of the Sukhatme representation renders the result transparent. Malik

(1970a) also considered products of minima from independent classical Pareto samples. Since such minima are themselves classical Pareto variables, the corresponding results can be obtained from (3.3.29) and (3.3.30). Earlier Rider (1964) had described analogous results for samples from power function populations (i.e. reciprocals of classical Pareto variables).

If one attempts to investigate the covariance structure of order statistics from a Pareto (III) sample, one is led to consider expressions of the form $E(Z_{k_1:n}^{\gamma} Z_{k_2:n}^{\gamma})$ where the $Z_{k:n}$'s are standard Pareto order statistics. Applying Theorem 3.4.1 brings us up against expressions of the form $E[(a + bZ_{k:n})^{\gamma}]$. These are not troublesome if γ is an integer. In cases of interest (when $0 < \gamma < 1$) they appear to be intractable.

3.4.3 Moments in the Presence of Truncation

It is not uncommon to encounter data sets which have been subject to one or two sided truncation. Balakrishnan and Joshi (1982) investigated the behavior of moments of order statistics from a doubly truncated Pareto (I) distribution. Their results can be somewhat simplified if we take advantage of the fact that truncation from below in a Pareto (I) distribution is equivalent to rescaling. As a consequence it is only necessary to consider one-sided truncation from above.

The truncated Pareto (I) density is of the form

$$f_c(x) = \frac{\alpha x^{-(\alpha+1)}}{1 - c^{-\alpha}}, \quad 1 < x < c. \tag{3.4.32}$$

The corresponding quantile function is

$$F_c^{-1}(u) = [1 - (1 - c^{-\alpha})u]^{-1/\alpha}, \quad 0 < u < 1. \tag{3.4.33}$$

The δ-moment of the corresponding k'th order statistic can thus be expressed as

$$E(X_{k:n}^{\delta}) = \frac{n!}{(k-1)!(n-k)!} \int_0^1 [1 - (1 - c^{-\alpha})u]^{-\delta/\alpha} u^{k-1} (1-u)^{n-k} \, du. \tag{3.4.34}$$

Upon replacing the factor $[1 - (1 - c^{-\alpha})u]^{-\delta/\alpha}$ by its Taylor series expansion (Newton's binomial formula) and integrating, it follows that

$$\begin{aligned} E(X_{k:n}^{\delta}) &= \tfrac{n!}{(k-1)!} \textstyle\sum_{j=0}^{\infty} \binom{-\delta/\alpha}{j} [-(1-c^{-\alpha})]^j \tfrac{(j+k-1)!}{(n+j)!} \\ &= \textstyle\sum_{j=0}^{\infty} \tfrac{(\delta/\alpha)_j (k)_j}{(n+1)_j} \tfrac{(1-c^{-\alpha})^j}{j!}, \end{aligned} \tag{3.4.35}$$

where $(b)_j = b(b+1)...(b+j-1)$. Khurana and Jha (1991) list 5 recurrence relations between the coefficients of a hypergeometric series such as (3.4.35) which yield corresponding recurrence relations between moments of the order statistics of the truncated Pareto (I) distribution.

In a subsequent paper Khurana and Jha (1995) consider expressions for $E(X_{k_1:n}^{\delta_1} X_{k-2:n}^{\delta_2})$ in the truncated Pareto (I) case. These are in terms of Kampe de Feriet's functions whose coefficients also satisfy several recurrence relations which yield potentially useful relations among the product moments. Khurana and Jha (1995) also present the corresponding results for one sided truncations and in the absence of truncation. These recurrence formulas in the untruncated case include and extend many of those mentioned in Section 3.4.2.

Ahmad (2001) provides some recurrence relations for moments and product moments of order statistics from a doubly truncated Pareto (IV) distribution (with $\mu = 0$).

3.5 Record Values

Let $X_1, X_2, \ldots$ be a sequence of i.i.d. random variables with common distribution function F_X. Assume that F_X is continuous (as it is for our generalized Pareto models). An observation X_j is said to be a record value, or simply a record, if it is larger than all preceding observations, i.e., if $X_j > X_i$ for every $i < j$.The first observation, X_1, is viewed as a record value which will be called the zero'th record, denoted by R_0. Subsequent record values are sequentially denoted by $R_1, R_2, \ldots$. In the study of record values, the exponential case is most easily resolved.

Using the lack of memory property of the exponential distribution, and using the notation R_n^* to denote the n'th record value corresponding to a sequence X_i's that are i.i.d. $exp(1)$, it can be verified that

$$R_n^* \sim \Gamma(n+1, 1), \quad n = 0, 1, 2, \ldots . \tag{3.5.1}$$

A convenient reference for discussion of general properties of record values is Arnold, Balakrishnan and Nagaraja (1998).

Since, in general, the X_i's are assumed to have a continuous distribution function F_X and corresponding quantile function F_X^{-1}, it follows that

$$-\log(1 - F_X(X)) \sim exp(1)$$

and so $X \stackrel{d}{=} F_X^{-1}(1 - e^{-X^*})$ where $X^* \sim exp(1)$. Since X is then a monotone increasing function of X^*, it is possible to express the n'th record value, R_n, of the $\{X_i\}$ sequence as a function of the n'th record value, R_n^*, of a standard exponential sequence $\{X_i^*\}$. Thus

$$R_n \stackrel{d}{=} F_X^{-1}(1 - e^{-R_n^*}), \quad n = 0, 1, 2, \ldots, \tag{3.5.2}$$

where, as remarked before, $R_n^* \sim \Gamma(n+1, 1)$.

Introducing the notation

$$\Psi_{F_X}(u) = F_X^{-1}(1 - e^{-u}), \tag{3.5.3}$$

we may write

$$R_n \stackrel{d}{=} \Psi_{F_X}\left(\sum_{i=0}^{n} X_i^*\right), \tag{3.5.4}$$

where the X_i^*'s are i.i.d. $exp(1)$ random variables. Since the density of $\sum_{i=0}^{n} X_i^* \sim$

$\Gamma(n+1,1)$ is readily available, it is then possible to write the density of R_n in the form

$$f_{R_n}(r) = f_X(r)[-\log(1-F_X(r))]^n/n!. \tag{3.5.5}$$

In the case in which the X_i's have a Pareto (IV)$(\mu,\sigma,\gamma,\alpha)$ distribution, we have

$$\text{(Pareto IV)} \quad F_X^{-1}(u) = \mu + \sigma[(1-u)^{-1/\alpha} - 1]^\gamma. \tag{3.5.6}$$

The corresponding quantile functions for the Pareto (I), (II) and (III) submodels are

$$\text{(Pareto I)} \quad F_X^{-1}(u) = \sigma(1-u)^{-1/\alpha}, \tag{3.5.7}$$

$$\text{(Pareto II)} \quad F_X^{-1}(u) = \mu - \sigma + \sigma(1-u)^{-1/\alpha}, \tag{3.5.8}$$

$$\text{(Pareto III)} \quad F_X^{-1}(u) = \mu + \sigma\left(\frac{u}{1-u}\right)^\gamma. \tag{3.5.9}$$

Consequently we have the following representations for the n'th record, R_n,

$$\text{(Pareto I)} \quad R_n \stackrel{d}{=} \sigma e^{R_n^*/\alpha}, \tag{3.5.10}$$

$$\text{(Pareto II)} \quad R_n \stackrel{d}{=} (\mu - \sigma) + \sigma e^{R_n^*/\alpha}, \tag{3.5.11}$$

$$\text{(Pareto III)} \quad R_n \stackrel{d}{=} \mu + \sigma(e^{R_n^*} - 1)^\gamma. \tag{3.5.12}$$

$$\text{(Pareto IV)} \quad R_n \stackrel{d}{=} \mu + \sigma(e^{R_n^*/\alpha} - 1)^\gamma \tag{3.5.13}$$

where, as usual, $R_n^* \sim \Gamma(n+1,1)$.

Not surprisingly, the classical Pareto case (Pareto I) is the simplest to deal with. From (3.5.10) we may conclude that

$$\text{(Pareto I)} \quad R_n \stackrel{d}{=} \sigma\left(\prod_{j=0}^{n} U_j\right)^{-1/\alpha}, \quad n = 0,1,2,\dots, \tag{3.5.14}$$

where the U_j's are i.i.d. $Uniform(0,1)$ random variables. From (3.5.14) it is evident that R_n can be represented in terms of independent Pareto I$(1,\alpha)$ random variables, i.e.,

$$\text{(Pareto I)} \quad R_n \stackrel{d}{=} \sigma \prod_{j=0}^{n} X_j, \quad n = 0,1,2,\dots, \tag{3.5.15}$$

where the X_j's are i.i.d. Pareto I$(1,\alpha)$ random variables.

In this case it is evident that quotients of successive records are independent Pareto (I) random variables. Thus if we define $Q_n = R_n/R_{n-1}$, $n = 1,2,\dots$, then the Q_j's are i.i.d. P(I)$(1,\alpha)$ random variables. Several characteristic properties of the classical Pareto distribution are associated with this property of the corresponding record quotients (see Section 3.9.6 below).

For examples we have the following properties:

$$R_{n+k}/R_n \text{ and } R_n \text{ are independent,} \tag{3.5.16}$$

$$E(R_n|R_{n-1}=r)=\alpha r, \text{ for } \alpha>1, \tag{3.5.17}$$

$$\frac{R_n}{R_{n-1}} \stackrel{d}{=} cR_0. \tag{3.5.18}$$

The mean and variance of classical Pareto records are readily obtained from (3.5.14) or (3.5.15), thus

$$E(R_n)=\sigma\left(1-\frac{1}{\alpha}\right)^{-(n+1)}, \quad \alpha>1, \tag{3.5.19}$$

Pareto (I)

$$Var(R_n)=\sigma^2\left[\left(1-\frac{2}{\alpha}\right)^{-(n+1)}-\left(1-\frac{1}{\alpha}\right)^{-2(n+1)}\right], \quad \alpha>2. \tag{3.5.20}$$

Other moments are of the form

$$\text{Pareto (I)} \quad E(R_n^\delta)=\sigma^\delta\left(1-\frac{\delta}{\alpha}\right)^{-(n+1)}, \quad \delta<\alpha. \tag{3.5.21}$$

The covariance between two records is also obtainable using (3.5.14) or (3.5.15). Thus, for $m<n$ and $\alpha>2$, the covariance between Pareto (I) records is given by

$$cov(R_m,R_n)=\sigma^2\left(\frac{\alpha}{\alpha-1}\right)^{n+1}\left[\left(\frac{\alpha-1}{\alpha-2}\right)^{m+1}-\left(\frac{\alpha}{\alpha-1}\right)^{m+1}\right]>0. \tag{3.5.22}$$

Analogous properties of Pareto (II) records require only minor algebraic manipulation since the n'th Pareto (II) record is obtainable by adding $\mu-\sigma$ to a corresponding n'th Pareto (I) record. Turning to Pareto (III) and (IV) records, the plot thickens somewhat. In principle, it is not difficult to write an expression for the density of the n'th record in such cases, but we no longer have nice relationships between successive records. Consequently, we are not able to develop simple expressions for the means, variances and covariances of the records. Series expansions are possible but numerical integration may well be the simplest route to follow in these cases.

It is possible to derive recurrence relations for the moments and product moments of Pareto (I) and (II) records. In the Pareto (I) case, the recurrence is a trivial consequence of the representation, (3.5.15), of R_n as σ multiplied by a product of i.i.d. Pareto (I) variables. Thus we have, for $k<\alpha$,

$$\text{Pareto (I)} \quad E(R_n^k)=E(R_{n-1}^k)E(X_1^k)=\frac{\alpha}{\alpha-k}E(R_{n-1}^k). \tag{3.5.23}$$

The recurrences for the Pareto (II) distribution are more interesting. For simplicity, assume that $\mu = 0$ and $\sigma = 1$. The Pareto (II)$(0, 1, \alpha)$ survival function and density are readily determined to be related as follows

$$\overline{F}(x) = \frac{1}{\alpha}(1+x)f(x). \tag{3.5.24}$$

For $k+1 < \alpha$, consider

$$\begin{aligned} E(R_n^k) + E(R_n^{k+1}) &= \int_0^\infty x^k(1+x)\frac{[-\log\overline{F}(x)]^n}{n!}f(x)\,dx \\ &= \alpha\int_0^\infty x^k \frac{[-\log\overline{F}(x)]^n}{n!}\overline{F}(x)\,dx, \end{aligned}$$

upon using (3.5.24). Now integrate by parts, integrating x^k and differentiating the other factor of the integrand. Upon rearrangement this yields

$$E(R_n^{k+1}) = \frac{1}{\alpha-(k+1)}[(k+1)E(R_n^k) + \alpha E(R_{n-1}^{k+1})]. \tag{3.5.25}$$

This recurrence can be used to generate values of $E(R_n^k)$ for all n and k with $k < \alpha$. For further details and related recurrence formulas, see Balakrishnan and Ahsanullah (1994), who derive the results in the context of the Pickands generalized Pareto distribution.

We remark that many results are available for what are called k-records or kth record values in the sense defined by Dziubdziela and Kopocinski (1976). However in the Pareto (I)-(IV) cases, the k-records behave like ordinary records from a corresponding Pareto distribution with α replaced by $k\alpha$, so that no new results are obtained by this extension.

3.6 Generalized Order Statistics

Kamps (1995) introduced the concept of generalized order statistics. Initially they were viewed as a convenient vehicle to allow simultaneous treatment of order statistics and record values. Subsequently, several experimental designs used in quality evaluation have been recognized as ones leading to data configurations that fall into the category of generalized order statistics. In the income distribution arena, it is not obvious that data of the generalized order statistics form will be encountered, but it is worthwhile reviewing properties of generalized order statistics for Pareto type distributions for their intrinsic merit and because such Pareto distributions do sometimes find application in survival analysis in biological and engineering contexts. In such settings, generalized order statistics data configurations may well be encountered.

Following Kamps (1995) we begin by defining a set of n uniform generalized order statistics, $U_1 < U_2, < \ldots < U_n$, to be random variables with joint density

$$f_{U_1,U_2,\ldots,U_n}(u_1,u_2,\ldots,u_n)$$

$$\propto \left(\prod_{i=1}^{n-1}(1-u_i)^{m_i}\right)(1-u_n)^{k-1}I(0 < u_1 < u_2 < \ldots < u_n < 1) \tag{3.6.1}$$

where $k > 0$ and $\underline{m} = \{m_k\}_{k=1}^{n-1}$ is a set of $n-1$ real parameters satisfying $\gamma_r = k + n - r + \sum_{j=r}^{n-1} m_j > 0$ for every $r \in \{1,2,...,n-1\}$. From this, for an arbitrary continuous distribution function, F, with quantile function F^{-1}, a corresponding set of n generalized order statistics, $X_1 < X_2 < ... < X_n$, is defined by setting $X_i = F^{-1}(U_i),\ i = 1,2,...,n$.

An alternative parameterization of the uniform generalized order statistic joint density will be used here. The joint density will be written in the form

$$f_{U_1,U_2,....,U_n}(u_1,u_2,....,u_n) =$$

$$\left(\prod_{i=1}^{n} \tau_i\right)\left(\prod_{i=1}^{n-1}(1-u_i)^{\tau_i-\tau_{i+1}-1}\right)(1-u_n)^{\tau_n-1}$$

$$\times\, I(0 < u_1 < u_2 < ... < u_n < 1). \qquad (3.6.2)$$

In this density the parameters $\underline{\tau} = \{\tau_i\}_{i=1}^{n}$ that are involved are constrained to be positive.

Using this parameterization, Arnold and Villaseñor (2012) verify that a set, $X_1 < X_2 < ... < X_n$, of generalized order statistics with base distribution F can be represented in the convenient form

$$X_i = \Psi_F(\sum_{j=1}^{i} X_j^*/\tau_j),\ \ i = 1,2,...,n, \qquad (3.6.3)$$

where $\Psi_F(u) = F^{-1}(1-e^{-u})$, the X_j^*'s are i.i.d. standard exponential random variables and the τ_j's are positive parameters.

Comparison of this representation of generalized order statistics with the corresponding representation (3.5.4) of ordinary records highlights the observation that ordinary record values correspond the special case in which the τ_j's are all equal to 1. Note also that the representation (3.6.3) can be readily extended to define a sequence of generalized order statistics $\{X_j\}_{j=0}^{\infty}$ with parameters $\{\tau_j\}_{j=0}^{\infty}$. The n variables defined in (3.6.3) can also be viewed as corresponding to the first n Pfeifer records, the initial segment of an infinite sequence of Pfeifer records (Pfeifer, 1984).

As was the case for ordinary records, when the base distribution F involved in the definition of the generalized order statistics is a classical Pareto distribution, the resulting distribution theory is particularly simple, since in this case $\Psi_F(u) = \sigma e^{u/\alpha}$. Thus the i'th generalized order statistic, X_i, can be expressed as

$$\text{Pareto (I)} \quad X_i = \sigma \prod_{j=1}^{i} Y_j, \qquad (3.6.4)$$

where the Y_j's are independent random variables with $Y_j \sim P(I)(1, \alpha\tau_j),\ \ j = 1,2,...,i$. An alternative representation is

$$\text{Pareto (I)} \quad X_i = \sigma \prod_{j=1}^{i} Z_j^{1/\tau_j}, \qquad (3.6.5)$$

where the Z_j's are i.i.d. Pareto(I)$(1,\alpha)$ variables. With these representations at hand, we can enunciate certain distributional characteristics of Pareto (I) generalized order statistics which parallel the corresponding properties of ordinary records. Thus:

$$X_n/X_{n-1} \text{ and } X_{n-1} \text{ are independent,} \tag{3.6.6}$$

$$E(X_n|X_{n-1}=x)=c_n x, \text{ for } \alpha>1, \tag{3.6.7}$$

$$\frac{X_n}{X_{n-1}} \stackrel{d}{=} X_1^{\tau_1/\tau_n}. \tag{3.6.8}$$

To what extent these properties of generalized order statistics are characteristic of the classical Pareto model will be discussed in Section 3.9.7.

Expressions for moments and joint moments of generalized order statistics in the Pareto (I) case are readily obtained from the representation (3.6.4). Kamps (1995) derived general recurrence relations for moments of generalized order statistics which can be applied when the base distribution is Paretian. Ahmad and Fawzy (2003) provide recurrence relations for doubly truncated distributions which also can be applied in Paretian cases. As usual, with such recurrence relations, there will be some concern with error propagation if the recurrence relations are used to construct extensive tables. Mahmoud, Rostom and Yhiea (2008) provide recurrence relations which are appropriate for the case of progressive Type-II right censored order statistics (a special case of generalized order statistics) in the Pickands generalized Pareto setting. The key ingredient in the development of these recurrence relations is the fact that, for the Pickands distributions we have

$$(1-\beta x)f(x)=\overline{F}(x).$$

Saran and Pandey (2004) have similar recurrences for the moment generating functions of generalized order statistics in the Pareto (II) case.

3.7 Residual Life

It was pointed out in Chapter 1 that the Pareto families have potential for modeling in both economic and reliability contexts. The study of residual life immediately conjures up images of online testing of light bulbs and the like, but if it is described as the study of the mean when the distribution is truncated from the left, its potential in the study of income distributions (which are frequently so truncated) becomes apparent.

It will be observed, that within our hierarchy of generalized Pareto distributions, it is the Pareto (II) family which is most suited to the present analysis. This observation, together with some observations regarding constraints on the Gini index of Pareto (II) families (to be discussed in Chapter 4), suggest that use of the Pareto (II) family might be most appropriate in reliability contexts.

For a general distribution function, F, we define its residual life distribution at time x, denoted by F^x, by

$$F^x(y) = P(x < X \leq x+y \mid X > x), \quad y \geq 0. \tag{3.7.1}$$

The residual life distribution at time x is only defined for x's for which $P(X > x) > 0$. Although in most applications F is assumed to correspond to a non-negative random variable, this assumption is not necessary for the development of the theory.

For the Pareto (IV) distribution we find, for $x > \mu$,

$$\overline{F}^x(y) = \left[\frac{\sigma^{1/\gamma} + (x+y-\mu)^{1/\gamma}}{\sigma^{1/\gamma} + (x-\mu)^{1/\gamma}}\right]^{-\alpha}, \quad y > 0 \tag{3.7.2}$$

where, as usual, the bar is used to denote a survival function. It is only when $\gamma = 1$ (i.e. in the Pareto II case) that the residual life distribution (3.7.2) becomes tractable. The resulting expression

$$\overline{F}^x(y) = \left(1 + \frac{y}{\sigma + x - \mu}\right)^{-\alpha}, \quad y > 0 \tag{3.7.3}$$

is, for each x, again a Pareto (II) survival function. This observation is, of course, equivalent to the fact that the $P(\text{II})(\mu, \sigma, \alpha)$ distribution truncated at x is $P(\text{II})(x, \sigma + x - \mu, \alpha)$, as observed in Section 3.4.

Provided that $\alpha > 1$, one may readily compute the mean residual life at time x denoted by $\mu(x)$, corresponding to a Pareto (II) distribution,

$$\mu(x) = \int_0^\infty \overline{F}^x(y)dy = (\sigma + x - \mu)/(\alpha - 1), \quad x \geq \mu. \tag{3.7.4}$$

It is striking that the Pareto (II) distribution has a linear mean residual life function $\mu(x)$ as a function of x. It is well known that the mean residual life function characterizes the distribution (see, e.g., Laurent, 1974). Thus, if we assume a linearly increasing mean residual life function, we are tacitly assuming a Pareto (II) distribution. See Morrison (1978) for an alternative derivation of this fact and see, for example, Vartak (1974) for the case of a linearly decreasing mean residual life function. The problem was also discussed by Sullo and Rutherford (1977). Constant mean residual life functions are associated with, possibly, translated exponential distributions. It is readily verified that exponential distributions are limits of Pareto distributions. Specifically, if in a $P(\text{II})(\mu, \sigma, \alpha)$ distribution we let $\alpha \to \infty$ in such a manner that $\alpha/\sigma \to \gamma > 0$, then

$$\lim \overline{F}(x) = \lim \left(1 + \frac{x-\mu}{\sigma}\right)^{-\alpha} = e^{-\gamma(x-\mu)}. \tag{3.7.5}$$

The mean residual life function of the classical Pareto distribution is particularly simple:

$$(\text{Pareto I}) \quad \mu(x) = x/(\alpha - 1), \quad x > \sigma. \tag{3.7.6}$$

The failure rate function ($r = F'/\overline{F}$) corresponding to a Pareto (II) distribution may be obtained directly or via the relation

$$r(x) = \left[1 + \frac{d}{dx}\mu(x)\right] / \mu(x). \tag{3.7.7}$$

One finds

$$\text{(Pareto I)} \quad r(x) = \alpha/x, \quad x > \sigma \tag{3.7.8}$$

$$\text{(Pareto II)} \quad r(x) = \alpha/(\sigma + x - \mu), \quad x > \mu. \tag{3.7.9}$$

Equations (3.7.8) and (3.7.9) are valid for any $\alpha > 0$. A derivation using (3.7.7) is, of course, only possible if $\alpha > 1$, but the direct route using the definition of r remains open, when $0 < \alpha \leq 1$.

The Pareto (II) family, having a linear rather than a constant mean residual life function, has and will continue to appear in the reliability literature as a convenient family of alternatives to the exponential distribution. Its place in the income modeling sphere, where the exponential distribution has less appeal, is secure, but for other reasons.

3.8 Asymptotic Results

In this section we will catalog some large sample distributional results for generalized Pareto distributions. The list will be far from exhaustive, but will treat the obvious questions such as attraction of sums and extremes and asymptotic behavior of order statistics. Several important asymptotic results for Pareto distributions will not be found here. It has seemed convenient to defer them until they arise naturally in subsequent chapters on measures of inequality and inference.

3.8.1 Order Statistics

Suppose we have a large sample $X_1, \ldots, X_n$ from a Pareto (IV) population with corresponding order statistics $X_{i:n}, (i = 1, 2, \ldots, n)$. Since the corresponding density has a convex support set, the non-extreme order statistics will have approximately a normal distribution (cf. e.g., Wilks, 1962, p. 273). Thus,

$$X_{k:n} \dot{\sim} N\left(F^{-1}\left(\frac{k}{n}\right), \left[\frac{k}{n}\left(1 - \frac{k}{n}\right) / nf^2\left(F^{-1}\left(\frac{k}{n}\right)\right)\right]\right) \tag{3.8.1}$$

where f and F are, respectively, the Pareto (IV) density and distribution functions. The symbol $\dot{\sim}$ in (3.8.1) is to be read as "is approximately distributed as." The inverse distribution function is relatively simple for the Pareto (IV), i.e.,

$$F^{-1}(y) = \mu + \sigma[(1-y)^{-1/\alpha} - 1]^{\gamma}. \tag{3.8.2}$$

However, when this is substituted in (3.8.1), it is only in the classical Pareto (I) case that any significant simplification results. We find

$$\text{(Pareto I)} \quad X_{k:n} \dot{\sim} N\left(\sigma\left(1 - \frac{k}{n}\right)^{-1/\alpha}, \frac{k\sigma^2}{n^2\alpha^2}\left(1 - \frac{k}{n}\right)^{-(1+2\alpha^{-1})}\right) \tag{3.8.3}$$

Corresponding approximately normal joint distributions for several order statistics can be readily computed using standard results summarized conveniently in Wilks (1962, p. 273-4).

The asymptotic behavior of the extreme order statistics may be verified directly in the Pareto (IV) case. For the minimum, one finds

$$\begin{aligned}\lim_{n\to\infty} P((\alpha n)^{\gamma}(X_{1:n}-\mu)/\sigma > z) &= \lim_{n\to\infty}\{1+[z/(\alpha n)^{\gamma}]^{1/\gamma}\}^{-\alpha n}\\ &= \lim_{n\to\infty}\{1+[z^{1/\gamma}/(\alpha n)]\}^{-\alpha n}\\ &= e^{-z^{1/\gamma}},\end{aligned} \tag{3.8.4}$$

for $z > 0$. Thus, the Pareto (IV) minimum, suitably normalized, is asymptotically a Weibull variable. The maximum can also be treated directly. After some experimentation with candidate normalizations one finds:

$$\begin{aligned}\lim_{n\to\infty} P(n^{-\gamma/\alpha}(X_{n:n}-\mu)/\sigma \le z) &= \lim_{n\to\infty}[1-(1+(n^{\gamma/\alpha}z)^{1/\gamma})^{-\alpha}]^{n}\\ &= \lim_{n\to\infty}[1-z^{-\alpha/\gamma}(1+z^{-1/\gamma}n^{-1/\alpha})/n]^{n}\\ &= e^{-z^{(-\alpha/\gamma)}}.\end{aligned} \tag{3.8.5}$$

Alternatively (3.8.4) and (3.8.5) can be obtained by appeal to Gnedenko's general theorem on the asymptotic distribution of maxima (see, e.g., Lamperti, 1966, p. 57). Results for the asymptotic joint distribution of the k largest (or smallest) order statistics (k fixed) can be deduced from Dwass' (1964) work on extremal processes (an alternative source is Weissman, 1975).

Adler (2003) considers an infinite triangular array of i.i.d. Pareto (I) random variables, $\{X_{n,i} : i = 1,2,...,m; n = 1,2,...\}$. For a fixed value $k \le m$, he discusses the question of whether there exist sequences $\{a_j\}$ and $\{b_n\}$ such that $b_n^{-1}\sum_{j=1}^{n} a_j X_{j(k)}$ converges to a non-zero constant in probability or almost surely. Here $X_{j(k)}$ denotes the k'th order statistic corresponding to the j'th row of the array. He also considers (Adler, 2005a, 2006a) analogous questions in which $X_{j(k)}$ is replaced by $X_{j(k_1)}/X_{j(k_2)}$ where $k_1 < k_2$.

In Adler (2004) he discusses the convergence of $b_n^{-1}\sum_{j=1}^{n} a_j X_{j(k)}$ in the case in which the $X_{n,i}$'s have a common generalized Pareto distribution of the following form:

$$\overline{F}(x;\lambda,p) = \left(\frac{pe}{\lambda}\right)^{\lambda}\frac{(\lg x)^{j}}{x^{p}},\quad x \ge e^{\lambda/p},$$

where $\lg x = \log(\max\{x,e\})$. When $\lambda = 0$ this corresponds to a classical Pareto (I) distribution.

In addition such convergence questions are addressed in the case in which, in each row of the array, the ratio of two randomly selected adjacent order statistics is considered instead of $X_{j(k_1)}/X_{j(k_2)}$ (Adler, 2005b). The case involving the ratio of two randomly selected order statistics from each row is treated in Adler (2006b).

3.8.2 *Convolutions*

Now suppose $X_1, X_2, \ldots$ is a sequence of independent identically distributed $P(\text{IV})(\mu, \sigma, \gamma, \alpha)$ random variables. Let $S_n = \sum_{i=1}^{n} X_i$. What is the asymptotic distribution of S_n (suitably normalized)? If $\alpha/\gamma > 2$, then variances exist, and the classical central limit theorem predicts a limiting normal distribution. In this case the normalizing constants are obvious, i.e., $[S_n - E(S_n)]/\sigma(S_n) \dot{\sim} N(0,1)$. If $\alpha/\gamma = 2$, then the central limit still applies, since

$$\lim_{y \to \infty} \left[y^2 \int_{|x|>y} dF(x) \right] \Big/ \left[\int_{|x| \le y} x^2 dF(x) \right] = 0,$$

(cf. Chung, 2001, Theorem 7.2.3). In this instance it is not transparent how one should choose the sequence of constants $\{b_n\}_{n=1}^{\infty}$ such that $[S_n - E(X_n)]/b_n \dot{\sim} N(0,1)$. A satisfactory choice is $b_n = n \log n$. When $\alpha/\gamma < 2$, the Pareto (IV) distribution is in the domain of normal attraction of a stable law with characteristic exponent α/γ (see Gnedenko and Kolmogorov, 1954, p. 181ff). If $\alpha/\gamma \in (1,2)$, then $E(X_i)$ exists and $[S_n - E(S_n)]/n^{\alpha/\gamma}$ will have an asymptotic distribution which is stable and has log-characteristic function of the form

$$\log \varphi(t) = -c \mid t|^{\alpha/\gamma} \left[1 + i \frac{t}{|t|} \tan \frac{\pi\alpha}{2\gamma} \right] \tag{3.8.6}$$

for some $c > 0$. If $\alpha/\gamma \in (0,1)$, then a somewhat simpler normalization, namely $S_n/n^{\alpha/\gamma}$, will have an asymptotically stable log-characteristic function of the form (3.8.6). Feller (1971, p. 548ff) provided a series expansion for the stable densities which correspond to (3.8.6), both for $\alpha/\gamma \in (0,1)$ and for $\alpha/\gamma \in (1,2)$. We remark that if $\alpha/\gamma < 1$, the limiting distribution is concentrated on $(0,\infty)$ and has a particularly simple Laplace transform, namely $\exp(-s^{\alpha/\gamma})$. The remaining case $\alpha/\gamma = 1$ requires a normalization of the form $(S_n - a_n)/n$ where $a_n = \int_{-\infty}^{\infty} \sin(x/n) dF(x)$ and leads to a limiting stable distribution whose log-characteristic function is given by

$$\log \varphi(t) = -c|t| \left[1 + \frac{2i}{\pi} \frac{t}{|t|} \log |t| \right] \tag{3.8.7}$$

for some $c > 0$. This asymptotic theory for $\alpha/\gamma < 2$ is perhaps of only technical interest. The normalized sums are attracted to stable laws, but these laws are not easy to deal with, Feller's asymptotic series for their densities notwithstanding. For example, the limiting survival function when $\alpha/\gamma < 1$ (cf. Blum, 1970) is given by

$$\overline{F}(x) = \frac{1}{\pi} \sum_{j=1}^{\infty} (-1)^{j+1} \left[\frac{\Gamma(1-\alpha/\gamma)}{x^{\alpha/\gamma}} \right]^j \frac{\sin(j\pi\alpha/\gamma)}{j!} \Gamma(j\alpha/\gamma),$$

$$x > 0. \tag{3.8.8}$$

3.8.3 *Record Values*

What can we say about the asymptotic behavior of Pareto records? Specifically, we may ask whether there exist sequences $\{a_n\}$ and $\{b_n\}$ such that

$$\frac{R_n - a_n}{b_n} \xrightarrow{d} U,$$

for some non-degenerate random variable U.

Resnick (1973) showed that there are only three kinds of possible limit laws for record values. He identified conditions on the behavior of the function Ψ_F that are necessary and sufficient for convergence of the record sequence to each of the three types. For Pareto (I)–(IV) records the discussion of the asymptotic distribution of R_n can be short. In each case, we have for large values of n,

$$R_n \sim e^{cR_n^*},$$

indicating rapid growth as $n \to \infty$. Indeed this growth is so rapid as to preclude the existence of suitable centering and scaling sequences $\{a_n\}$ and $\{b_n\}$ to ensure convergence to a non-degenerate limit law. Obviously non-linear transformations could lead to a non-degenerate limit law (e.g., we could transform back to R_n^* which is asymptotically normal), but the untransformed R_n's are not attracted to any of the three limit distributions (normal, log-normal and negative-log-normal) described by Resnick (1973).

3.8.4 *Generalized Order Statistics*

The study of the asymptotic distribution of Pareto generalized order statistics is more complicated than the parallel investigation for ordinary records. What happens depends heavily on the nature of the sequence $\{\tau_n\}_{n=1}^{\infty}$. For example, if the series $\sum_{i=1}^{\infty} 1/\tau_i$ is convergent, then X_n (the n'th generalized order statistic) will converge to a non-degenerate random variable X_∞ almost surely and thus in distribution, without any centering or scaling. If $\sum_{i=1}^{\infty} 1/\tau_i = \infty$ but $\sum_{i=1}^{\infty} 1/\tau_i^2 < \infty$, then a Weibull type extremal distribution will be encountered for suitably centered and scaled X_n's. If both $\sum_{i=1}^{\infty} 1/\tau_i$ and $\sum_{i=1}^{\infty} 1/\tau_i^2$ are divergent, then it is unlikely that a non-degenerate limit distribution will be encountered, but the issue is not completely resolved. A good reference for discussion of convergence questions for generalized order statistics is Cramer (2003). See also the discussion in Arnold and Villaseñor (2012) where the notation (involving τ's) used is that used in this book rather than the original Kamps (1995) notation used by Cramer. Finally we remark that Nasri-Roudsari and Cramer (1999) provide discussion of the limiting distribution of X_n for certain configurations of the parameters. Their discussion of the Pareto case for those parametric configurations includes results on the rate of convergence.

We now turn from asymptotic distribution results based on large Paretian samples to the study of situations in which Paretian distributions arise themselves as limit distributions.

3.8.5 Residual Life

Balkema and de Haan (1974) studied the problem of residual life at great age. From Section 3.7, recall that the residual life distribution at time x corresponding to a distribution function F is given by

$$F^x(y) = 1 - \overline{F}(x+y)/\overline{F}(x).$$

It turns out that the class of possible non-degenerate limits laws for normalized residual life (i.e., $\lim_{x\to\infty} F^x(a(x)y)$) is very restricted. In fact, its only members are

$$\Gamma_\alpha(x) = 1 - (1+x)^{-\alpha}, \quad x > 0 \tag{3.8.9}$$

and

$$\Pi(x) = 1 - e^{-x}, \quad x > 0, \tag{3.8.10}$$

that is to say either Pareto (II) (for some $\alpha > 0$) or exponential. The class of distributions attracted to $\Gamma_\alpha(x)$ in the asymptotic residual life sense is particularly easy to describe. In fact, F belongs to the domain of attraction of $\Gamma_\alpha(x)$, if and only if, $\overline{F}$ is a regularly varying function with exponent $-\alpha$. That is to say, if and only if $\overline{F}(x)$ can be expressed in the form

$$\overline{F}(x) = x^{-\alpha}L(x) \tag{3.8.11}$$

where $L(x)$ satisfies, for each $x > 0$,

$$\lim_{t\to\infty} L(tx)/L(t) = 1. \tag{3.8.12}$$

We are following the notation used in Feller (1971, p. 268ff).

In one direction this result is trivial. Suppose that $\overline{F}$ is regularly varying with exponent $-\alpha$ and consider the choice $a(x) = x$. One finds

$$\begin{aligned}\lim_{x\to\infty} \overline{F}^x(xy) &= \lim_{x\to\infty} \frac{\overline{F}((1+y)x)}{\overline{F}(x)} = \lim_{x\to\infty} \frac{[x(1+y)]^{-\alpha}L((1+y)x)}{x^{-\alpha}L(x)} \\ &= (1+y)^{-\alpha}\end{aligned} \tag{3.8.13}$$

using (3.8.11) and (3.8.12). The converse, that is the statement that only distributions with $(-\alpha)$ regularly varying tails are attracted to $\Gamma_\alpha(x)$, is not as easy. See Balkema and de Haan (1974) for a proof which draws heavily on Gnedenko's (1943) characterization of domains of maximal attraction. As examples of distributions attracted to $\Gamma_\delta(x)$ in the asymptotic mean residual life sense, one may consider $P(\text{IV})(\mu, \sigma, \gamma, \alpha)$ distributions with $\alpha/\gamma = \delta$.

3.8.6 Geometric Minimization and Maximization

The Pareto (III) distribution can arise as the limit distribution under repeated geometric minimization. This was the essential ingredient in the Arnold-Laguna income model described in Section 2.9. The result may be put in slightly more general form as follows. Let $X_1^{(1)}, X_2^{(1)}, \ldots$ be a sequence of i.i.d. random variables with common distribution function F_1. Define F_2 to be the distribution function of the minimum of a geometrically distributed number of i.i.d. random variables with common distribution function F_1. Denote the parameter of the geometric distribution by p_1. Then define F_3 to correspond to a geometric (p_2) minimum of random variables with distribution function F_2, etc.

Theorem 3.8.1. *Suppose that F_1 is a distribution function satisfying*

$$\lim_{x \to 0} F_1(x)/x^{1/\gamma} = (1/\sigma)^{1/\gamma} \tag{3.8.14}$$

for some real $\sigma > 0$ and $\gamma > 0$. Define $F_2, F_3, \ldots$ sequentially in such a manner that F_k corresponds to a geometric (p_{k-1}) minimum of random variables with common distribution function F_{k-1}. For large k, F_k is approximately a Pareto (III) distribution in the sense that, for $x > 0$,

$$\lim_{k \to \infty} F_k\left(x\left(\prod_{i=1}^{k-1} p_i\right)^{\gamma}\right) = 1 - \left(1 + \left(\frac{x}{\sigma}\right)^{1/\gamma}\right)^{-1}, \tag{3.8.15}$$

provided that $\lim_{k \to \infty} \prod_{i=1}^{k-1} p_i = 0$.

Proof. From the definition of F_k as the distribution of a geometric (p_{k-1}) minimum of random variables with common distribution F_{k-1}, it follows that

$$\overline{F}^k(x) = p_{k-1}\overline{F}_{k-1}(x)/[1 - (1 - p_{k-1})\overline{F}_{k-1}(x)].$$

So that, if we define $\varphi_k(x) = F_k(x)/\overline{F}_k(x)$, we can write $\varphi_k(x) = \varphi_{k-1}(x)/p_{k-1}$. Consequently, $\varphi_k(x) = (\prod_{i=1}^{k-1} p_i)^{-1}\varphi_1(x)$ and

$$\lim_{k \to \infty} \varphi_k(x(\prod_{i=1}^{k-1} p_i)^{\gamma}) = x^{1/\gamma} \lim_{k \to \infty} \varphi_1(x(\prod_{i=1}^{k-1} p_i)^{\gamma})/x^{1/\gamma}(\prod_{i=1}^{k-1} p_i) = (x/\sigma)^{1/\gamma}$$

using (3.8.14) and the fact that $\prod_{i=1}^{k-1} p_i \to 0$. Equation (3.8.15) follows immediately. □

If, in Theorem 3.8.1, F_1 was chosen to have support (μ, ∞) rather than $(0, \infty)$, only minor modifications are required. One merely replaces x by $x - \mu$ in (3.8.14) and (3.8.15) and throughout the proof. As an example of application of the theorem, one might consider F_1 corresponding to a $P(\text{IV})(0, \sigma, \gamma, \alpha)$ distribution. With this choice of F_1

$$\lim_{x \to 0} F_1(x)/x^{1/\gamma} = (\alpha^{\gamma}/\sigma)^{1/\gamma},$$

and so the asymptotic distribution of suitably normalized geometric minima will be $P(\text{III})(0, \sigma\alpha^{-\gamma}, \gamma)$. If F_1 were standard exponential, then the corresponding asymptotic distribution would be standard Pareto. We observe that the form of the limiting distribution obtained from Theorem 3.8.1 depends on the local behavior of F_1 near its lower bound. This is remarkable when we recall the great emphasis usually placed on upper tails in the literature on Pareto distributions. A parallel version of Theorem 3.8.1 involving geometric maxima can be proved in a fashion similar to that used in the geometric minima version.

Pakes (1983) considers more general versions of the random minima model in which the geometric random variables are replaced by integer valued random variables with other distributions. For example, if the geometric random variables are replaced by negative-binomial variables, a Pareto (IV) limit distribution can be encountered. See also Janjic (1986) for related discussion of random extrema.

3.8.7 *Record Values Once More*

The Pareto (II) distribution arises as a limiting distribution in certain record value contexts. An early reference in this development is Wilks (1959). Consider a sequence of independent identically distributed random variables, $X_1, X_2, \ldots$, whose common distribution is assumed to be continuous. Fix integers k and n, $(k < n)$, and consider a random variable N defined by

$$N = \min\{m : m \geq 1, X_{m+n} > X_{n-k+1:n}\}, \tag{3.8.16}$$

i.e., the waiting time after taking n observations until we get an observation bigger than the kth largest of the first n observations.

Clearly, N increases stochastically with n. One may verify that for n large, N/n is approximately distributed as a $P(\text{II})(0,1,k)$ random variable. Specifically, using square brackets to denote integer part,

$$\begin{aligned}\lim_{n\to\infty} P(N/n > x) &= \lim_{n\to\infty} P(N > [nx]) = \lim_{n\to\infty} \binom{n}{k} \Big/ \binom{n+[nx]}{k} \\ &= (1+x)^{-k} \end{aligned}\tag{3.8.17}$$

for every $x > 0$.

The above development focused on outstandingly large observations. One may also wait for observations which are outstanding in either direction. Thus, again considering a sequence, $X_1, X_2, \ldots$, of i.i.d. random variables with a common continuous distribution, fix integers k_1, k_2 and n, $(k_1 + k_2 < n)$, and define a random variable $\tilde{N}$ by

$$\tilde{N} = \min\{m : m \geq 1, X_{m+n} < X_{k_1:n} \text{ or } X_{m+n} > X_{n-k_2+1:n}\}. \tag{3.8.18}$$

In this case, for $x > 0$,

$$\begin{aligned}\lim_{n\to\infty} P(\tilde{N}/n > x) &= \lim_{n\to\infty} P(\tilde{N} > [nx]) = \lim_{n\to\infty} \binom{n}{k_1+k_2} \Big/ \binom{n+[nx]}{k_1+k_2} \\ &= (1+x)^{-(k_1+k_2)}, \end{aligned}\tag{3.8.19}$$

again a Pareto (II) limiting distribution. The case where $k_1 = k_2 = 1$ corresponds to waiting for a "two-sided" record value. Here we wait for the first observation to fall outside the interval $(X_{1:n}, X_{n:n})$, which is the convex hull (in $\mathbb{R}$) of $X_1, X_2, \ldots, X_n$. It is intriguing to consider whether, in this form, the result can be extended to higher dimensions. Thus, suppose $X_1, X_2, \ldots$ are i.i.d. k-dimensional random variables with common continuous distribution and define, for n fixed, N to be the smallest index m such that X_{n+m} is not in the convex hull of $X_1, X_2, \ldots, X_n$. Can we identify the limiting distribution of N (as $N \to \infty$)? Can we even assert that it will not depend on the particular common distribution of the X_i's?

3.9 Characterizations

The intimate relationship between the exponential distribution and the classical Pareto distribution would permit us to begin this section with a suitably translated catalog of the many existing exponential characterizations. An excellent reference for such characterizations is the monograph by Galambos and Kotz (1978). Attention in the present section will be centered on those characterizations of the Pareto distribution which had their genesis in the economic literature and seem to have arisen from the study of Pareto random variables and not as spin-offs or translations of characterizations which are more naturally considered in the setting of exponential distributions. The literature on exponential characterizations is better developed, and some of the Pareto characterizations to be discussed can be much improved by switching over and drawing on the exponential literature. This will sometimes result in consideration of characterizations whose economic implications are difficult to envision. Undoubtedly, the list of characterizations to be discussed in this section is restricted in a highly idiosyncratic manner. The reader's personal favorites may have been inexplicably omitted. Almost all the characterizations deal with the Pareto (I) and (II) families, i.e., the Pareto distributions closely related to the exponential. The representation of the Pareto (III) distribution as log-logistic suggests that characteristic properties of the logistic distribution might have interesting economic interpretations when restated in Pareto (III) terms. The literature on logistic characterizations is not extensive, when compared with the exponential literature. Only two Pareto (III) characterizations are presented here.

3.9.1 Mean Residual Life

It appears that the first published characterization of the Pareto distribution was that of Hagstroem (1925). Hagstroem considered the mean income of all individuals whose income exceeds t (dollars). Now suppose that, due to inflation, everyone's income is multiplied by a factor k. If the mean income of individuals whose income exceeds t is unchanged, then the original income distribution must have been standard Pareto. There are some obscurities in the paper. Are we assuming that the mean income of those whose income exceeds t will be unchanged for every t or just one value of t? Are we assuming that a k-fold inflation will not change the mean income of those whose income exceeds t for every k or just one value of k? It would appear from Hagstroem's proof that just one value of k is considered and that all values of

t are considered above some lower bound for incomes, say t_0. However, if t_0 were positive and k were bigger than 1 (as it would be expected to be), then the stated invariance could only hold for $t > kt_0$. The simplest resolution of this difficulty is to assume $t_0 = 0$. With this assumption, Hagstroem's claim is as follows. If for a positive random variable X and some positive constant k we have

$$E(X \mid X > t) = E(kX \mid kX > t), \quad \forall t > 0 \tag{3.9.1}$$

then $X \sim P(\text{II})(0, \sigma, \alpha)$ for some $\sigma > 0$ and $\alpha > 0$. Hagstroem argued that (3.9.1) implies

$$E(X \mid X > t) = ct \tag{3.9.2}$$

for some $c > 1$. This is not necessarily true. See for example Arnold (1971) or J. S. Huang (1974) where non-linear solutions to $g(x) = kg(x/k)$ are discussed. One way out is to assume that (3.9.1) holds for every $k > 0$ or (more pedantically) for two values k_1,k_2 such that $\{k_1^\alpha / k_2^\beta\}_{\alpha,\beta=1}^\infty$ is dense in $I\!R^+$.

The plot thickens somewhat when one realizes that (3.9.2) cannot hold for every $t > 0$. If it held for t arbitrarily close to zero, it would force X to be degenerate at 0. It seems that Hagstroem was heading in the direction of a characterization of the form: if

$$E(X \mid X > t) = ct, \quad \forall t > t_0, \tag{3.9.3}$$

then $X \sim P(\text{II})(t_0, t_0, \alpha) = P(\text{I})(t_0, \alpha)$ for some $\alpha > 0$. This is easily rephrased as a statement regarding the mean residual life function which is generally known to characterize the distribution (as discussed in Section 3.7). If we let $\mu(x) = \int_0^\infty \overline{F}^x(y)dy$ be the mean residual life function, then (3.9.3) may be rewritten as

$$\mu(x) = (c-1)x, \quad x > t_0, \tag{3.9.4}$$

and we have already seen this as the mean residual life function of a $P(\text{I})(t_0, c/(c-1))$ distribution in (3.7.6). A more general linear mean residual life function of the form

$$\mu(x) = a + bx, \quad x > t_0 \tag{3.9.5}$$

with $b > 0$ leads, via the results of Laurent (1974), to a $P(\text{II})(t_0, \frac{a}{b} + t_0, (b+1)/b))$ distribution. Characterizations of the Pareto (II) via linear mean residual life recur in the literature. Following Hagstroem's initial foray we find the result alluded to, for example, in d'Addario (1939), Bryson (1974), Vartak (1974), Sullo and Rutherford (1977), Patil and Ratnaparkhi (1979) and Xekalaki and Dimaki (2005). Morrison (1978) proved the result, subject to the initial prior restriction that the distribution in question be a mixture of exponential distributions.

There are several other closely related Pareto characterizations based on variations of the residual life theme. Huang and Su (2012) show that, provided adequate moments exist, any of the following three conditions is sufficient to characterize the Pareto (I) distribution.

$$E((X-x)^2|X>x) = c[E(X-x|X>x)]^2, \tag{3.9.6}$$

$$E((X-x)^k|X>x) = (a+bx)E((X-x)^{k-1}|X>x), \tag{3.9.7}$$

$$E((X-x)^k|X>x) = (a+bx)^2E((X-x)^{k-2}|X>x). \tag{3.9.8}$$

Sullo and Rutherford (1977) also considered higher moments of residual life. They showed that a constant coefficient of variation of residual life (with the constant greater than 1) can only arise with Pareto (II) distributions.

Hamedani (2002) identified conditions on the functions g,h and λ that will guarantee that the condition

$$E(g(X)|X > x) = \lambda(x)E(h(X)|X > x) \tag{3.9.9}$$

will characterize the Pareto distribution.

Instead of considering the conditional expectation of X given $X > x$, Ruiz and Navarro (1996) consider the following three functions

$$m(x,y) = E(X|x \leq X \leq y), \tag{3.9.10}$$

$$e_1(x,y) = E(X - x|x \leq X \leq y), \tag{3.9.11}$$

$$e_2(x,y) = E(y - X|x \leq X \leq y), \tag{3.9.12}$$

Knowledge of any one of these three functions for all (x,y) such that $F(x) < F(y)$ will completely determine the distribution function F of the random variable X. For example , if

$$m(x,y) = \frac{\alpha}{1-\alpha}\frac{x^\alpha y - y^\alpha x}{y^\alpha - x^\alpha}$$

for some $\alpha > 1$, then X has a Pareto (I)$(1,\alpha)$ distribution.

Schmittlein and Morrison (1981) discussed the possibility of characterizing the Pareto distribution via a linear median residual life function. Unfortunately percentile residual life functions do not characterize a distribution. See Arnold and Brockett (1983), Gupta and Langford (1984) and Joe (1985) for further discussion of this issue. Joe points out that the Pareto distribution is characterized by the property that it has one of its percentile residual life functions that coincides with its mean residual life function.

There is a remarkable relationship between the failure rate function and the mean residual life function of a Pareto (II) distribution (see equations (3.7.4) and (3.7.9)), namely one is a constant multiple of the reciprocal of the other, i.e.,

$$r(x)\mu(x) = \alpha/(\alpha - 1), \tag{3.9.13}$$

provided $\alpha > 1$. Sullo and Rutherford (1977) observed that this is a characteristic property of the Pareto (II) distribution. Thus, if $r(x)\mu(x) = c > 1$ then from (3.7.7) we conclude $\mu'(x) = (c-1)$, so that we have increasing linear mean residual life and, consequently, a Pareto (II) distribution. There is no apparent significance to the fact that the constant $\alpha/(\alpha-1)$ appearing in (3.9.13) is exactly Gini's δ index of concentration in the classical Pareto case.

In a variety of contexts, the concept of size (or length) biased distributions arises.

For a density f defined on $(0,\infty)$ with finite mean μ, the corresponding size biased density is given by

$$f_1(x) = \frac{xf(x)}{\mu} I(x > 0). \tag{3.9.14}$$

The corresponding distribution function, in the Economics literature, is usually called the first moment distribution; see Chapter 4 where such distributions are discussed with relation to the Lorenz curve of the corresponding distribution.

The motivating idea for size biasing is that frequently big items are more likely to be observed than small items (see Rao (1985) for relevant discussion in the context of general weighted distributions). Denote the survival function of the size biased version of f by $\overline{F}_1$ and the corresponding unbiased survival function by $\overline{F}$. It is then quite plausible that the ratio $\overline{F}_1/\overline{F}$ will be informative about the nature of the distribution function F and indeed might be characteristic of F. Such is the case. Gupta and Keating (1986) observe that

$$\frac{\overline{F}_1(x)}{\overline{F}(x)} = \frac{x+\mu(x)}{\mu}, \tag{3.9.15}$$

where $\mu(x)$ is the mean residual life function of F. Since the mean residual life function determines F, it follows that knowledge of the ratio $\overline{F}_1/\overline{F}$ will determine F.

In the Pareto (II) case, $\mu(x)$ is linear in x and from (3.9.15), $\overline{F}_1/\overline{F}$ is linear in x with positive slope. Linearity of the ratio $\overline{F}_1/\overline{F}$ with positive slope is thus a characteristic property of the Pareto (I) and (II) distributions (Gupta and Keating, 1986). In similar fashion, if the ratio of the corresponding failure rate functions $r(x)/r_1(x)$ is linear then the distribution F must be of the Pareto (II) form.

Another relative of a continuous distribution F with $F(0) = 0$ and mean μ is the corresponding forward recurrence time density which arises when F is associated with a sequence of recurrent events. The density in question is of the form

$$f^*(x) = \frac{\overline{F}(x)}{\mu} I(x > 0). \tag{3.9.16}$$

Denote the corresponding failure rate functions and mean residual life functions of F and f^* by $r(x), \mu(x), r^*(x)$ and $\mu^*(x)$, respectively. Gupta (1979) noted that $r^*(x) = 1/\mu(x)$. Nair and Hitha (1990) used this observation to conclude that $\mu(x) = p\,\mu^*(x)$ for some $p \in (0,1)$ if and only if F is a Pareto (II) distribution. Similarly, the property $p\,r(x) = r^*(x)$ for some $p \in (0,1)$ characterizes the Pareto (II) distribution.

Hamedani (2004) reports a Pareto characterization involving a differential equation for its failure rate function. Specifically he observes that, if $r(x)$ satisfies

$$\frac{d}{dx} \log r(x) = -(c+x)^{-1}, \tag{3.9.17}$$

then the corresponding distribution must be of the Pareto (II) form.

3.9.2 *Truncation Equivalent to Rescaling*

Rather than speak of residual life, we may speak of truncation (from below). Several early characterizations of the Pareto distribution were described with reference to the effects of truncation. In the case of a classical Pareto distribution, truncation is equivalent to rescaling (cf. equation (3.4.20)). Specifically, for $y > x_0$,

$$P(X > y \mid X > x_0) = P\left(\frac{x_0}{\sigma}X > y\right). \tag{3.9.18}$$

That this is a characteristic property of the Pareto (I) distribution was apparently first noted by Bhattacharya (1963). He argued the case using Lorenz curves, but a direct proof is not difficult. Consider a distribution for X with support (μ, ∞). It is assumed that for some $c > 0$ and every $0 < \mu < x < y$ we have

$$P(X > y \mid X > x) = P(cxX > y). \tag{3.9.19}$$

If we define $Z = X/\mu$, $z = x/\mu$ and write $y = \omega x$ where $\omega > 1$, it follows that

$$P(Z > \omega z) = P(Z > z)P(\mu c Z > \omega) \tag{3.9.20}$$

for every $z, \omega > 1$. By letting $z \to 1$ in (3.9.20), we conclude that $c = 1/\mu$. Then defining $T = \log Z$, we find (3.9.20) translates to the usual lack of memory property which characterizes the exponential distribution. Working back, one concludes that $X \sim P(\mathrm{I})(\mu, \alpha)$ for some $\alpha > 0$.

If one had been willing to assume the existence of the first moment of X, then (3.9.19) directly implies increasing linear mean residual life from which the Paretian form of the distribution of X could be deduced. Dallas (1976) assumed the existence of rth moments and characterized the Pareto by the condition that the rth moment of the truncated distribution is the same as the rth moment of a suitable scaling of the original distribution. Thus, assuming the support of the distribution is (μ, ∞), and that

$$E(X^r \mid X > x) = E[(cxX)^r] \tag{3.9.21}$$

for some $c > 0$, Dallas showed that X is a Pareto (I) variable. The easy way to get this result is to consider $Y = X^r$, which by (3.9.21) has an increasing linear mean residual life function, and so is Pareto (I). It follows that X is also Pareto (I), since powers of Pareto (I) variables are again Pareto (I) variables.

The property (3.9.18) with $\sigma = 1$ can be rewritten as

$$P(X > uv | X > v) = P(X > u), \tag{3.9.22}$$

where $u, v > 1$. In this form it was called the multiplicative lack of memory property and it is readily verified that it characterizes the Pareto (I)$(1, \alpha)$ distribution. Asadi, Rao and Shanbhag (2001) considered a strong multiplicative lack of memory property of the form

$$P(X > uV | X > V) = P(X > u), \tag{3.9.23}$$

where V is a random variable satisfying $V > 1$ almost surely. This too is a characteristic property of the Pareto (I)$(1,\alpha)$ distribution.

In Asadi (2004), an extension of the multiplicative lack of memory property was introduced. For it, we define

$$\theta(x) = \mu(x)/\mu,$$

where $\mu(x)$ is the mean residual life function and μ is the mean of a distribution function F. Assume that $\theta(x)$ is invertible with inverse function $\theta^{-1}(u)$. The distribution function F is said to have the extended multiplicative lack of memory property if

$$\overline{F}(\theta^{-1}(uv)) = \overline{F}(\theta^{-1}(u))\overline{F}(\theta^{-1}(v))$$

for u, v, uv in the domain of θ^{-1}. This condition is also only satisfied by Pareto distributions.

Oakes and Dasu (1990) presented a characterization of the Pickands generalized Pareto distribution, obtained by assuming that truncation is equivalent to rescaling, or in reliability terminology, that residual life follows an accelerated failure model. They assume that for some positive function $\theta(x)$ and some $\gamma > 0$

$$\overline{F}(x+y) = \overline{F}(x)\overline{F}(y/\theta(x)), \tag{3.9.24}$$

and conclude that for some $\alpha > 0$,

$$\overline{F}(x) = (1+\gamma x)^{-\alpha/\gamma}, \tag{3.9.25}$$

where $x > 0$ if $\gamma > 0$ and $0 < x < -1/\gamma$ if $\gamma < 0$. Their proof assumes that F is absolutely continuous. Asadi, Rao and Shanbhag (2001) assume only that (3.9.24) holds for a convergent sequence of values of x but, in addition, assume that F has a finite mean. Under these assumptions, it also follows that $\overline{F}$ satisfies (3.9.25).

3.9.3 *Inequality Measures*

Bhattacharya (1963) alluded to the property of having a Gini index that is invariant under truncation from below as being characteristic of the Pareto distribution. A proof was supplied by Ord, Patil and Taillie (1981b). They wrote $G(c)$ to be the Gini index of the distribution of F truncated at c (the Gini index will be defined in Chapter 4), and setting $G'(c) = 0$ [to reflect the truncation invariance], they arrived at a differential equation for F which indicates that F is a classical Pareto distribution. They showed similar results for a whole family of Mellin transform measures of inequality (discussion of which is deferred to Chapter 4). Basically, truncation invariance of inequality measures appears to be, essentially, uniquely a Pareto property.

A related observation is that the Lorenz curve of X truncated below at x_0 does not depend on x_0 only in the Pareto (I) case (Bhattacharya, 1963). Bhandari (1986) shows that, for a broad class of inequality measures, say $I(X)$, if X_c denotes the random variable X truncated below at c, then knowledge of $I(X_c)$ for all c will determine the distribution of X up to a change of scale. In particular, if for some $k > 0$, $I(X_c) = k$ for every c, then X has a Pareto (I) distribution.

Moothathu (1993) considered a different Lorenz curve characterization. It involves what he calls an α-mixture Pareto distribution. A random variable has such a distribution if it can be represented in the form

$$X = \sigma_1 U_1^{\alpha_1} I(U_0 < \alpha) + \sigma_2 U_2^{-\alpha_2} I(U_0 > \alpha)$$

where U_0, U_1, U_2 are i.i.d. $Uniform(0,1)$ random variables, $\alpha_1, \alpha_2 > 0$ and $\sigma_1 < \sigma_2$. This distribution is characterized by the property that the Lorenz curve of X truncated above at a point $x_1 < \sigma_1$ does not depend on x_1, and the Lorenz curve of X truncated below at $x_2 > \sigma_2$ does not depend on x_2.

While temporarily on the topic of measures of inequality, we may note a characterization based on two early inequality measures: Pareto's α and Gini's index δ. Both were defined as slopes of certain fitted lines. Pareto's α comes from plotting $\log x$ against $\log \overline{F}(x)$, while Gini's δ comes from plotting $\log \overline{F}(x)$ against $\log \overline{F}_1(x)$ where $F_1(x) = \int_0^x y\, dF(y) / \int_0^\infty y\, dF(y)$ (the first moment distribution). Actually, a linear fit will obtain only in the classical Pareto case and, in that case, $\delta = \alpha/(\alpha - 1)$. Pietra (1935b) was an early observer of this characterization of the Pareto distribution.

3.9.4 *Under-reported Income*

Krishnaji (1970) described a characterization of the Pareto (II) distribution based on under-reporting. He envisioned that reported income Y is related to actual income X in a multiplicative manner:

$$Y = RX \tag{3.9.26}$$

where X and $0 \le R \le 1$ are independent random variables. We assume that X has support (x_0, ∞) and that $P(RX > x_0) > 0$. If the distribution of $Y = RX$ truncated below at x_0 is the same as the distribution of X, what can we say about the distribution of X? In the special case where X is assumed to be absolutely continuous and R has as its density

$$f_R(r) = \delta r^{\delta - 1}, \quad 0 < r < 1 \tag{3.9.27}$$

where $\delta > 0$, Krishnaji was able to show that X must have a Pareto (II) distribution. He argued as follows. If $y > x_0$, then we have by hypothesis

$$\overline{F}_Y(y)/\overline{F}_Y(x_0) = \overline{F}_X(y), \tag{3.9.28}$$

while using (3.9.26) we have

$$\begin{aligned} \overline{F}_Y(y) &= \int_0^1 \overline{F}_X(y/r) \delta r^{\delta-1} dr \\ &= \delta y^\delta \int_y^\infty \overline{F}_X(u) u^{-(\delta+1)} du. \end{aligned} \tag{3.9.29}$$

Since $\overline{F}_X$ was assumed to be absolutely continuous, we may differentiate with respect to y in both (3.9.28) and (3.9.29) obtaining in this manner the relation

$$\frac{\overline{F}_Y(x_0)}{\delta} \frac{d}{dy} \left[\overline{F}_X(y) y^{-\delta} \right] = -\frac{1}{y} \left[\overline{F}_X(y) y^{-\delta} \right], \quad y > x_0. \tag{3.9.30}$$

Solving this differential equation and recalling that $\overline{F}_X(x_0) = 1$ we conclude that X has a Pareto (II) distribution. Krishnaji (1971) subsequently initiated work towards showing that the density (3.9.27) for R is not special and that almost any non-degenerate distribution could be put in its place without upsetting the characterization result.

J. S. Huang (1978) provided the latest word in this quest. His paper was written in the context of exponential random variables, but, as usual, a straightforward translation to the Pareto (I) case is possible. For convenience assume that the possible values of X are $(1,\infty)$. We will say that X has the truncation-scaling property at level $x_0 > 1$ if

$$P(X > y \mid X > x_0) = P(x_0 X > y), \quad \forall y > 1 \tag{3.9.31}$$

or equivalently if

$$\overline{F}_X(x_0 y) = \overline{F}_X(x_0)\overline{F}_X(y), \quad \forall y > 1. \tag{3.9.32}$$

As Bhattacharya (1963) observed, having the truncation-scale property at level x_0 for every $x_0 > 1$ is a sufficient condition for X to be a Pareto (I) variable. Actually (cf. Marsaglia and Tubilla, 1975, in the exponential context), it is enough to have (3.9.32) hold for two values x_0 and x_1 such that their ratio is irrational. See also remark (iii) in Section 3.12. Huang observed that Krishnaji's characterization was based on having (3.9.32) hold on the average, i.e., when the level x_0 is random. Thus, we say that X has the truncation-scale property at the random level $\tilde{X}$ if

$$\int_1^\infty \overline{F}_X(x_0 y)dG(x_0) = \int_1^\infty \overline{F}_X(x_0)\overline{F}_X(y)dG(x_0), \quad \forall y > 1 \tag{3.9.33}$$

where G is the d.f. of the random level $\tilde{X}$. Krishnaji's R can be identified with $\tilde{X}^{-1}$. Huang was able to show (not easily) that provided $\tilde{X}$ has a continuous distribution function, then we can conclude that if X is non-degenerate, it has a Pareto (I) distribution. He did not even have to assume a continuous distribution for $\tilde{X}$, but his minimal conditions do not translate comfortably into the Pareto framework. They are: (i) that $\log\tilde{X}$ be non-lattice and (ii) $P(X \geq \tilde{X}) > P(\tilde{X} = 1)$. Alternative treatments of this problem may be found in Ramachandran (1977), Shimizu (1979) and Rao and Shanbhag (1986). We remark, parenthetically, that there is an inherent weakness in such characterizations based on the identical form of the reported and actual income distributions. We never see actual income distributions, only reported ones. How are we to opine whether the two kinds of distributions are likely to be of the same type?

There are further interesting features of the Krishnaji model with $Y = RX$ in which R has a density of the form (3.9.27). Patil and Ratnaparkhi (1979) verify that in this model, the condition $E(X|Y = y) = cy$, for $y > \sigma$ is also sufficient to guarantee that X has a Pareto (I)(σ, α) distribution.

Korwar (1985) provides the following interesting characterization involving both of the distributions of X and R in the under-reported income model (3.9.26).

Assume that $Y = RX$ where R and X are independent and that there exists x_0 such that $0 < P(Y > x_0) < 1$. The following are equivalent.

- The distribution of Y truncated above at x_0 coincides with the distribution of x_0R, and the distribution of Y truncated below at x_0 coincides with the distribution of X.
- $X \sim$ Pareto (I)(σ, α) and R has a density of the form (3.9.27).

Revankar, Hartley and Pagano (1974) proposed a different characterization of the Pareto distribution related to under-reporting. The scenario is as follows. The variables, income (X), reported income (Y) and under-reporting error (U), are related by the equation

$$Y = X - U. \tag{3.9.34}$$

It is assumed that the average amount of under-reporting, given income level $X = x$, is a linear function of x and that the average amount of under-reporting given $X > x$ is a (different) linear function of x. Under such circumstances it is argued that X must have a Pareto (II) distribution. The two assumptions about the conditional expectations of U yield

$$\frac{1}{\overline{F}(y)} \int_y^\infty (a + bx) dF(x) = \alpha + \beta y, \quad y > \mu \tag{3.9.35}$$

where F is the distribution function of X, assumed to have support $[\mu, \infty)$. Note that (3.9.35) is, in reality, an assumption of linear mean residual life, and if we (as the authors did) assume $\beta > b > 0$, the distribution of X is necessarily of the Pareto (II) type.

3.9.5 *Functions of Order Statistics*

A large number of Pareto (I) characterizations are based on the behavior of functions of order statistics. These are frequently translatable into corresponding exponential characterizations, but it is probably worthwhile listing a selection of the better known results to give at least the flavor of the material. Let us assume that we have n observations from an absolutely continuous distribution F and let us denote the order statistics by $X_{1:n}, X_{2:n}, \ldots, X_{n:n}$. The regularity conditions can sometimes be relaxed, but the assumption of absolute continuity is often convenient. Random variables of the form $X_{k+1:n}/X_{k:n}$ may be called geometric spacings. It is convenient to consider $X_{1:n}$ to be the first geometric spacing. For an arbitrary absolutely continuous parent distribution F, the geometric spacings can be expected to be dependent and, even when scaled, to have distributions which are different from each other. In the case of samples from a Pareto (I) distribution, the geometric spacings are independent and, up to a scale parameter, identically distributed (see equation (3.4.7) and the accompanying discussion). That these remarkable properties might characterize the Pareto distribution was an idea which undoubtedly arose quite early. The first results in this direction, assuming independence of geometric spacings, are those of Fisz (1958) and Rogers (1959). The line of research culminates in the results of Rossberg (1972b). These authors all, in general, wrote in the context of the exponential distribution with passing references to the corresponding results for the Pareto distribution. The result:

$$X_{i:n}, X_{i+1:n}/X_{i:n} \text{ independent } \Rightarrow X's \sim \text{ Pareto (I)} \tag{3.9.36}$$

can be found in Rogers (1963), Malik (1970d) and Govindarajulu (1966). A more general statement:

For some $k_2 > k_1$, $X_{k_1:n}$ and $X_{k_2:n}/X_{k_1:n}$ *independent*

$$\Rightarrow X's \sim \text{ Pareto (I)} \tag{3.9.37}$$

was proposed by Ahsanullah and Kabir (1973) and is a special case of Rossberg's (1972b) results. Actually, it is not necessary to assume full independence. It suffices that the conditional expectation of $X_{i+1:n}/X_{i:n}$ given $X_{i:n}$ should be constant (Rogers, 1963). Ferguson (1967) considered related characterization problems. We may illustrate the type of analysis often used in these studies by proving (3.9.36) under the simplifying assumption that $E(X_{i+1:n})$ exists. Assume without loss of generality that the support of the X's is $[1,\infty)$. It follows from the assumed independence that

$$E\left(\frac{X_{i+1:n}}{X_{i:n}}\middle| X_{i:n} = x\right) = c > 0,$$

and, thus, $E(X_{i+1:n} \mid X_{i:n} = x) = cx$. However, applying Theorem 3.4.1, this yields

$$\frac{\int_x^\infty y d[1-(1-F(y))^{n-i}]}{[1-F(x)]^{n-i}} = cx.$$

Thus, $[1-(1-F(y))^{n-i}]$ has linearly increasing mean residual life. It, consequently, is a Pareto (I) distribution, and it follows that F, itself, is a Pareto (I) distribution. Once more we were able to fall back on the linear mean residual life property of the Pareto distribution. Result (3.9.37) is somewhat more tiresome to verify, while the following general result due to Rossberg (1972b) appears to require use of sophisticated analytic techniques,

$$X_{k:n} \text{ and } \prod_{i=k}^{n}(X_{i:n})^{c_i} \text{ independent } \left(\text{where } \sum_{i=k}^{n} c_i = 0\right)$$
$$\Rightarrow X's \text{ Pareto (I).}$$

Dallas (1976) made an observation which makes the moment assumption in the above proof a little more palatable. It is enough to assume a fractional moment exists. Specifically he demonstrated that:

$$\text{For some } r > 0, \quad E\left[\left(\frac{X_{i+1:n}}{X_{i:n}}\right)^r \middle| X_{i:n} = x\right] = c$$
$$\Rightarrow X's \sim \text{ Pareto (I).} \tag{3.9.38}$$

A related result is described in remark (iv) in Section 3.12.

A special role for the sample minimum in Pareto characterizations early became apparent. Srivastava (1965) announced the result

$$\begin{aligned} &X_{1:n} \text{ and } \sum_{i=1}^{n} X_{i:n}/X_{1:n} \text{ independent} \\ &\Rightarrow X's \sim \text{ Pareto (I).} \end{aligned} \tag{3.9.39}$$

A weaker related result was that of Samanta (1972):

$$\begin{aligned} &X_{1:n} \text{ and } (X_{2:n}/X_{1:n}, X_{3:n}/X_{1:n}, \ldots, X_{n:n}/X_{1:n}) \text{ independent} \\ &\Rightarrow X's \sim \text{ Pareto (I).} \end{aligned} \tag{3.9.40}$$

Samanta's result is actually also subsumed by either (3.9.36) or (3.9.37). Another result related to Srivastava's is one assuming finite rth moments due to Dallas:

$$E\left(\sum_{i=1}^{n} \frac{X_{i:n}^r}{X_{1:n}^r} \middle| X_{1:n} = x\right) = c \quad \Rightarrow X's \sim \text{ Pareto (I).} \tag{3.9.41}$$

These observations can be put in a better perspective, if we regard $X_{1:n}$ and $\underline{X}/X_{1:n}$ as size and shape variables of the sample, to use the nomenclature introduced by Mosimann (1970). In this setting, for a vector $\underline{X}$ of n independent positive random variables we say that $G(\underline{X})$ is a size variable, if the function G satisfies $G(c\underline{x}) = cG(\underline{x})$ for every $c \in \mathbb{R}^+$, $\underline{x} \in \mathbb{R}^{+n}$. To a given size variable G there corresponds a shape vector $\underline{X}/G(\underline{X})$. Mosimann (1970) proved, inter alia, that if some particular size variable $G(\underline{X})$ is independent of some particular shape variable (possibly constructed using a different size variable), then every shape variable is independent of the size variable $G(\underline{X})$. So we can speak of independence of size $G(\underline{X})$ and shape. It is important to observe that X_i's do not have to be identically distributed. James (1979) used Mosimann's results to show that (in the continuous case)

$$X_{1:n} \text{ independent of shape} \Rightarrow X's \sim \text{ Pareto (I).} \tag{3.9.42}$$

This is readily seen to be true in the identically distributed case since $\sum_{i=1}^{n} X_{i:n}/X_{1:n}$ is a function of a shape vector, and Srivastava's result can be applied. James (1979) proved an analogous result when the X_i's may have different distributions:

$$\begin{aligned} &\min(X_1/\sigma_1, \ldots, X_n/\sigma_n) \text{ independent of shape} \\ &\Rightarrow X_i \sim P(\text{I})(k\sigma_i, \alpha_i), \quad i = 1, 2, \ldots n. \end{aligned} \tag{3.9.43}$$

To do this he appealed to a result of Crawford (1966) who proved that for independent X, Y, independence of $X - Y$ and $\min(X, Y)$ implies that X and Y are exponential variates with possibly different scale parameters. This result allowed James to conclude that (3.9.43) is true for $n = 2$. The general result follows since we can write

$$\min_{1 \le j \le n} \{X_j/\sigma_j\} = \min\left\{X_i/\sigma_i, \min_{j \ne i}\{X_j/\sigma_j\}\right\}.$$

James discussed the Pareto distribution in its role as a limiting case of Ferguson's (1962) generalized gamma distributions, i.e. distributions of powers of gamma variables. The James result suggests that a fertile source of possible Pareto characterizations might be assumptions of independence of $X_{1:n}$ and some function of a shape vector, or constant regression of some function of a shape vector on $X_{1:n}$. Characterizations (3.9.39)–(3.9.41) were, indeed, of this form. Another of this genre is that due to Beg and Kirmani (1974):

$$E\left(\frac{X_1}{X_{1:n}}|X_{1:n}=x\right)=c>0 \Rightarrow X's \sim \text{ Pareto (I)}. \tag{3.9.44}$$

To prove this, we first observe that

$$E\left(\frac{X_1}{X_{1:n}}|X_{1:n}=x\right)=\frac{1}{n}+\frac{n-1}{n}\left[\int_x^\infty tdF(t)\right]\Big/x\overline{F}(x). \tag{3.9.45}$$

If this is to be a constant, then we must have

$$x\overline{F}(x)=a\int_x^\infty tdF(t)$$

which, upon differentiating, implies that the density $f(x)$ must satisfy

$$\overline{F}(x)=(1-a)xf(x),$$

from which it follows that the distribution must be Pareto (I).

What can we say about the parent distribution if certain geometric spacings have functionally related distributions? For example, we know that if the observations come from a Pareto (I) distribution, then

$$X_{k+1:n}/X_{k:n}\overset{d}{=}cX_{1:n-k}, \tag{3.9.46}$$

$$X_{k+1:n}/X_{k:n}\overset{d}{=}c(X_{1:n})^{n/(n-k)}, \tag{3.9.47}$$

$$X_{k+1:n}/X_{k:n}\overset{d}{=}(X_{j+1:n}/X_{j:n})^{(n-j)/(n-k)}. \tag{3.9.48}$$

Is it possible to derive characterizations of the Pareto (I) distribution based on any of these three properties? The answer is most likely to be affirmative under mild regularity conditions, but several open questions remain. A good discussion of this problem (in the exponential context) is provided by Galambos and Kotz (1978, pp. 41–46). Let us consider the implications of (3.9.46). Let F denote the common distribution function of the X_i's. The conditional distribution of $X_{k+1:n}/X_{k:n}$ given $X_{k:n}=x$ is like that of x^{-1} times the minimum of a sample of size $n-k$ taken from the distribution function F truncated from below at x (cf. Theorem 3.4.1). Now, if we denote by G the distribution of $X_{k:n}$ and by H the distribution of $X_{1:n-k}$, we may rewrite (3.9.46) when $c=1$ in the form

$$\int_1^\infty[\overline{H}(yx)-\overline{H}(y)\overline{H}(x)]\frac{dG(x)}{\overline{H}(x)}=0. \tag{3.9.49}$$

A comparison with (3.9.33), assuming F is continuous, permits the conclusion that if H is non-degenerate, it is necessarily a Pareto (I) distribution. It follows that F is Pareto (I). If the constant c in (3.9.46) is not equal to 1, we still get a Pareto (I) distribution, but with a scale parameter equal to $1/c$. This result was first obtained by Rossberg (1972a). Our proof relied on Huang's "truncation-scale property at a random level" characterization (3.9.33). The present development has in a sense put the cart before the horse. Huang, in the proof of his characterization, used a technique developed by Rossberg to prove the characterization based on (3.9.46).

The following related Pareto (I) characterization was derived by Dimaki and Xekalaki (1993). Suppose that for fixed integers r and n and two distinct integers k_1 and k_2 satisfying $1 \leq r < k_1 < k_2 \leq n$ we have

$$\frac{X_{k_i:n}}{X_{r:n}} \stackrel{d}{=} X_{k_i-r:n-r}, \quad i = 1,2. \tag{3.9.50}$$

It follows that the X_i's have a Pareto (I)$(1,\alpha)$ distribution for some $\alpha > 0$.

A further related problem was treated by Rossberg. Suppose that $X_{k+1:n}/X_{k:n}$ has a Pareto (I) distribution,; can we conclude that the parent distribution is Pareto (I)? In general, we cannot. Rossberg (1972a) provided a counterexample in the exponential context. In the Pareto setting a counterexample is readily derived as follows. Let Z_1,Z_2 be i.i.d. random variables with common density $f(z) = z^{1/2}e^{-z}/\Gamma(1/2)$, $z > 0$ (i.e. $\Gamma(1/2, 1)$ variables). Let $X = e^{(Z_1-Z_2)}$ and let X_1,X_2 be a sample of size two from the distribution F_X. It is easily verified that these X_i's do not have a Pareto (I) distribution, yet $X_{2:2}/X_{1:2}$ does indeed have a Pareto (I) distribution. Rossberg was able to verify that such anomalous cases can be ruled out, if we require that the Mellin transform of $F_X^k(x)$ has no zeros with non-negative real parts. Instead of requiring that the parent distribution have a well behaved Mellin transform, one might require that $X_{k+1:n}/X_{k:n}$ have a Pareto distribution for more than one value of k. Characterizations of this genre assuming absolute continuity were discussed by Thirugnanasambanthan (1980).

Pareto characterizations based on (3.9.47) and (3.9.48) are even more elusive. Ahsanullah (1976) has an exponential characterization related to (3.9.47), but he assumed a monotone hazard rate for the parent distribution. Not much else is available. Galambos and Kotz (1978, p. 45) alluded to the existence of several variants of Ahsanullah's result, but, presumably, these variants also involve somewhat unsatisfactory regularity conditions. The problem can be reformulated as follows. Characterize the class of parent distributions for which for some fixed $\beta > 0$ and some integers $1 \leq j < k < n$, we have

$$(X_{k+1:n}/X_{k:n}) \stackrel{d}{=} (X_{j+1:n}/X_{j:n})^{\beta} \tag{3.9.51}$$

(by convention $X_{0:n} = 1$). The case $\beta > 1$ might be expected to lead to Pareto (I) distributions. The case $\beta = 1$ can, by taking logarithms, be transformed to a problem of characterizing distributions based on identically distributed spacings. It would seem plausible that such a phenomenon can only occur when the parent distribution is uniform. Some, but not much, progress has been made in this direction (see Huang,

Arnold and Ghosh, 1979). The case $\beta < 1$ can by reciprocation be reduced to the case $\beta > 1$ and "should" characterize power function distributions.

Pareto characterizations involving random dilation of order statistics have recently attracted attention. They typically begin with an assumption that, for some integers i_1, n_1, i_2 and n_2, the order statistics corresponding to a parent distribution F satisfy

$$X_{i_1:n_1} \stackrel{d}{=} UX_{i_2:n_2}, \tag{3.9.52}$$

where $U \sim$ Pareto (I)$(1, \alpha)$ is assumed to be independent of the X_i's.

For example, Wesolowski and Ahsanullah (2004) show that, if for a fixed $k \leq n-1$, we have

$$X_{k:n} \stackrel{d}{=} UX_{k-1:n-1}, \tag{3.9.53}$$

or if we have

$$X_{k:n} \stackrel{d}{=} UX_{k-1:n}, \tag{3.9.54}$$

then the common distribution of the X_i's must be of the Pareto (I) form.

Oncel, Ahsanullah, Aliev and Aygun (2005) considered a similar characterization. They assume that

$$X_{k:n-1} \stackrel{d}{=} UX_{k:n}, \tag{3.9.55}$$

and conclude that the X_i's have a Pareto (III) distribution.

Arnold, Castillo and Sarabia (2008) assume that for some $n_1 < n_2$ we have

$$X_{1:n_1} \stackrel{d}{=} UX_{1:n_2}, \tag{3.9.56}$$

and conclude that $X_i \sim$ Pareto (IV)$(0, \sigma, n_1/(n_2 - n_1), 1/(n_2 - n_1))$.

Note that, since $U > 1$ with probability 1, the relationship (3.9.52) cannot hold for all sets of integers $i_1 \leq n_1, i_2 \leq n_2$. However, if (3.9.52) holds, then denoting the distribution and density of $X_{i:n}$ by $F_{i:n}$ and $f_{i:n}$, respectively, we have

$$F_{i_1:n_1}(x) = \int_1^\infty F_{i_2:n_2}(x/u)\alpha u^{-(\alpha+1)}\,du.$$

Setting $y = x/u$ in the integral this becomes

$$x^\alpha F_{i_1:n_1}(x) = \int_0^x F_{i_2:n_2}(y)\alpha y^{\alpha-1}\,dy,$$

which, upon differentiating with respect to x (assuming that F has a density), yields

$$\alpha x^{\alpha-1} + x^\alpha f_{i_1:n_1}(x) = \alpha x^{\alpha-1} F_{i_2:n_2}(x).$$

Rearranging, this becomes

$$x f_{i_1:n_1}(x) = \alpha[F_{i_2:n_2}(x) - F_{i_1:n_1}(x)]. \tag{3.9.57}$$

This equation has been successfully solved for F in the four cases described above. Delineation of the general class of sets of integers $i_1 \leq n_1, i_2 \leq n_2$ for which it can be solved awaits determination. In addition it will be interesting to determine which configurations of the integers $i_1 \leq n_1, i_2 \leq n_2$ will yield a solution of the generalized Pareto form. Note that in attempting to solve (3.9.57) one may, by raising each side of the equation to the α power, assume that $\alpha = 1$ and consequently that $U \sim$ Pareto (I)$(1,1)$ in (3.9.57) (since, if $U \sim$ P(I)$(1,\alpha)$ then $U^\alpha \sim$ P(I)$(1,1)$).

To conclude this Section, a selection of specialized Pareto characterizations involving order statistics is provided. In these examples we begin with $X_1, X_2, ..., X_n$ i.i.d. with common distribution function F, and corresponding order statistics $X_{1:n}, X_{2:n}, ..., X_{n:n}$.

Wang and Srivastava (1980), assuming that F is continuous, show that if, for some $k \in \{1,2,...,n-1\}$, we have

$$E(\frac{1}{n-k}\sum_{i=k+1}^{n}(X_{i:n} - X_{i-1:n})|X_{k:n} = x) = \alpha x + \beta \tag{3.9.58}$$

for some $\alpha > 0$, then F is a Pareto (II) distribution.

Asadi (2004) considers the condition

$$\frac{1}{c}[(cX_{1:n}+1)^n - 1]] \stackrel{d}{=} X_1. \tag{3.9.59}$$

If F is continuous and condition (3.9.59) holds for two distinct values of $n > 1$, say n_1 and n_2, for which $\log n_1 / \log n_2$ is irrational, or if (3.9.59) holds for one value of $n > 1$ and $\lim_{x \to 0} F(x)/x = \gamma \in (0, \infty)$, then F has a Pickands generalized Pareto distribution. Essentially, a special case of this result was obtained earlier by Dimaki and Xekalaki (1993). They assumed that $X_i > 1$ almost surely and that $X_{1:n}^n \stackrel{d}{=} X_1$ for every $n = 1, 2, ...$, and concluded that F must be a classical Pareto distribution.

In the same paper, Dimaki and Xekalaki argue that, if we define $U = 2X_{1:2}$ and $V = X_1 X_2$, then a necessary and sufficient condition for X_i to have a classical Pareto distribution is that

$$f_{U|V}(u|v) = \frac{2}{u \log v} I(1 < u < v). \tag{3.9.60}$$

Papathanasiou (1990) shows that, for $0 < q < 1$,

$$cov(X_{1:2}, X_{2:2}) \leq B(2,q) \int_0^1 u(1-u)^{3-q} g^2(u)\, du, \tag{3.9.61}$$

where $g(u) = [f(F^{-1}(u))]^{-1}$, and that equality occurs only in the Pareto (II) case.

Lin (1988), assuming the existence of appropriate moments, shows that the X_i's have a classical Pareto distribution if and only if for some $\ell < m$

$$E(X_{k:n+\ell}^m) = \sigma^\ell E(X_{k:n}^{m-\ell}), \forall n \geq k. \tag{3.9.62}$$

Kamps (1991) assumes the existence of appropriate moments and the existence of constants m and M such that

$$m \leq [F^{-1}(u)]^{\alpha_1 - \alpha_2} t^{\beta_1 - \beta_2} (1-t)^{\gamma_1 - \gamma_2} \leq M.$$

In this case it is possible to obtain a lower bound for $E(X_{k:n}^{\alpha_1+\alpha_2})$ that is a linear function of the $2\alpha_1$-moment of one order statistic and the $2\alpha_2$-moment of another order statistic. Equality occurs in certain cases, only when F is a classical Pareto distribution. For example, one finds

$$(M+m)E(X_{k:n}^{\alpha_1+\alpha_2}) \geq \frac{n-k+1}{n+1}E(X_{k:n+1}^{2\alpha_1}) + Mm\frac{n}{n-k}E(X_{k:n-1}^{2\alpha_2}), \tag{3.9.63}$$

with equality only in the Pareto (I) case.

3.9.6 Record Values

Recall from Section 3.5 that for a sequence of i.i.d. X_i's with common continuous distribution function F_X, the corresponding sequence of record values is denoted by $R_0, R_1, R_2, \ldots$. A convenient representation of the n'th record is

$$R_n \stackrel{d}{=} F_X^{-1}(1-e^{R_n^*}),$$

where $R_n^* = \sum_{i=0}^n X_i^* \sim \Gamma(n+1,1)$ in which the X_i^*'s are i.i.d. $exp(1)$ variables.

For classical Pareto records we then have

$$R_{n+k}/R_n \text{ and } R_n \text{ are independent,} \tag{3.9.64}$$

$$E(R_n|R_{n-1}=r) = \alpha r, \text{ for } \alpha > 1, \tag{3.9.65}$$

$$\frac{R_n}{R_{n-1}} \stackrel{d}{=} cR_0. \tag{3.9.66}$$

If $X \sim P(I)(\sigma,\alpha)$, then $R_n \stackrel{d}{=} \sigma e^{R_n^*/\alpha}$ and the three conditions above transform readily into properties associated with record values for exponential sequences. Arnold, Balakrishnan and Nagaraja (1998, pp. 102-103) provide a catalog of exponential characterizations via record values, using which we may confirm that each of the conditions (3.9.64)–(3.9.66) can be used to characterize the classical Pareto distribution. Needless to say, only minor modifications are required to obtain analogous characterizations of the Pareto (II) distribution.

It is well known that the sequence of expected record values, $\{E(R_n)\}_{n=0}^{\infty}$, determines the parent distribution $F_X(x)$. Consequently, provided that $E(X_1)$ exists, then the condition

$$E(R_n) = ab^n, \quad n = 0,1,2,\ldots$$

for some $a, b > 0$ will be sufficient to ensure that $X_1 \sim P(I)(\sigma,\alpha)$ for some $\sigma > 0$ and $\alpha > 1$. Refer to (3.5.15) for clarification of this observation. Some related moment based characterizations may be found in Lin (1988).

Oncel, Ahsanullah, Aliev and Aygun (2005) observed that if we have, for a fixed value of n,

$$R_n \stackrel{d}{=} VR_{n-1}, \tag{3.9.67}$$

where $V \sim P(I)(\sigma, \alpha)$ is independent of R_{n-1}, then necessarily the X_i's have a Pareto (I) distribution. The necessity of condition (3.9.67) is obvious from the representation (3.5.15). A related characterization is one that assumes that, for some n,

$$R_n \stackrel{d}{=} R_0 R_{n-1}, \tag{3.9.68}$$

where R_0 and R_{n-1}, R_n correspond to independent identically distributed record sequences.

The condition

$$R_0 \stackrel{d}{=} R_n / R_{n-1}, \tag{3.9.69}$$

for some n was used by Lee and Kim (2011) to characterize the classical Pareto distribution.

Condition (3.9.65), that $E(R_n|R_{n-1} = r) = \alpha r$, is essentially equivalent to an assumption regarding mean residual life functions, since the distribution of R_n given $R_{n-1} = r$ is merely the parent distribution, F_X, truncated below at r. Two additional variants of this characterization have appeared. Lee (2003) showed that if for some integer m

$$E(R_{n+k}|R_m = r) = c^k E(R_n|R_m = r), \tag{3.9.70}$$

for every $n \geq m+1$, then necessarily X_1 has a Pareto (I) distribution.

The second variant is a characterization of the Pickands generalized Pareto distribution based on $E(R_n|R_{n-1} = r)$. Wu and Lee (2001) show that if

$$E(R_n|R_{n-1} = r) = (1-\gamma)(1+\gamma r), \tag{3.9.71}$$

then, necessarily, $\overline{F}_{X_1} = (1+\gamma x)^{-1/\gamma}$. Here γ is permitted to be positive or negative, and if it is negative, then $0 < x < -1/\gamma$. Wu and Lee (2001) also discuss a characterization based on expressions for $E(R_n|R_m = r)$ and $E(R_{n-1}|R_m = r)$ where $n \geq m+2$.

3.9.7 *Generalized Order Statistics*

Recall from Section 3.6 that a set of n generalized order statistics $X_1, X_2, ...X_n$ with base distribution F can be represented as

$$X_i = \Psi_F\left(\sum_{j=1}^{i} X_j^* / \tau_j\right), \quad i = 1, 2, ..., n, \tag{3.9.72}$$

where $\Psi_F(u) = F^{-1}(1-e^{-u})$, the X_i's are i.i.d. standard exponential random variables and the τ_j's are positive parameters. Included within this generalized order statistics model are: order statistics, record values, k-record values, progressively censored data sets, m-generalized order statistics and Pfeifer records of the F^{α} form (see Arnold, Balakrishnan and Nagaraja (1998) for discussion of such Pfeifer records).

In the case in which the base distribution is of the classical Pareto form (i.e.,

$P(I)(\sigma,\alpha)$), one finds $\Psi_F(u) = \sigma e^{u/\alpha}$. Consequently the corresponding generalized order statistics admit the representation

$$X_i = \sigma \prod_{j=1}^{i} Z_j^{1/\tau_j}, \quad i = 1,2,...,n, \tag{3.9.73}$$

where the Z_j's are i.i.d. Pareto (I)$(1,\alpha)$ random variables.

If, instead, we assume a Pickands generalized Pareto base distribution, i.e., that

$$\overline{F}(x) = (1+\gamma x)^{-1/\gamma}, \quad x > 0 \text{ and } (1+\gamma x) > 0, \tag{3.9.74}$$

where $\gamma \in (-\infty,\infty) - \{0\}$, it is then a simple exercise to verify that if X has (3.9.74) as its survival function, then the random variable

$$Y = (1+\gamma X)^{1/\gamma} \tag{3.9.75}$$

has a Pareto (I)$(1,1)$ distribution. If Y is defined as in (3.9.75) then

$$X = (Y^\gamma - 1)/\gamma$$

and it follows that a set of n generalized order statistics, $X_1, X_2, ..., X_n$, with a Pickands generalized Pareto base distribution, can be represented in the form

$$X_i = \frac{\left[\prod_{j=1}^{i} Z_j^{\gamma/\tau_j}\right] - 1}{\gamma}, \quad i = 1,2,...,n, \tag{3.9.76}$$

where the Z_j's are i.i.d. Pareto (I)$(1,1)$ random variables.

The representations (3.9.73) and (3.9.76) naturally give rise to a spectrum of possible characterizations of the classical Pareto distribution and the Pickands distribution.

From (3.9.73) it is evident that a set of Pareto(I) generalized order statistics $X_1, X_2, ..., X_n$ will satisfy

$$X_n/X_{n-1} \text{ and } X_{n-1} \text{ are independent}, \tag{3.9.77}$$

$$E(X_n|X_{n-1} = x) = c_n x, \text{ for } \alpha > 1, \tag{3.9.78}$$

$$\frac{X_n}{X_{n-1}} \stackrel{d}{=} X_1^{\tau_1/\tau_n}, \tag{3.9.79}$$

$$X_1, \frac{X_2}{X_1}, \frac{X_3}{X_2}, ..., \frac{X_n}{X_{n-1}} \text{ are independent.} \tag{3.9.80}$$

For a set of Pickands generalized Pareto generalized order statistics, it is, using (3.9.76), the case that

$$(X_n+\gamma^{-1})/(X_{n-1}+\gamma^{-1}) \text{ and } X_{n-1} \text{ are independent,} \tag{3.9.81}$$

$$E(X_n|X_{n-1}=x)=a_n+b_nx, \text{ for } \alpha>1, \tag{3.9.82}$$

$$\frac{X_n+\gamma^{-1}}{X_{n-1}+\gamma^{-1}} \stackrel{d}{=} (X_1+\gamma^{-1})^{\tau_1/\tau_n}, \tag{3.9.83}$$

$$X_1+\gamma^{-1}, \frac{X_2+\gamma^{-1}}{X_1+\gamma^{-1}}, \ldots, \frac{X_n+\gamma^{-1}}{X_{n-1}+\gamma^{-1}} \text{ are independent.} \tag{3.9.84}$$

Proofs of characterizations using these properties are scattered in the literature. Sometimes regularity conditions such as continuity or absolute continuity of F are necessarily assumed. Some representative references follow.

Marohn (2002), in the context of progressive censoring, provides a proof that condition (3.9.84) characterizes the Pickands distribution. As a simple consequence (3.9.80) characterizes the classical Pareto distribution. See also Hashemi and Asadi (2007) and Hashemi, Tavangar and Asadi (2010). Cramer, Kamps and Keseling (2004) verify that the regression condition (3.9.82) characterizes the Pickands distribution; included in this is the fact that (3.9.78) characterizes the classical Pareto law. Related characterization may be found in Samuel (2003).

Tavangar and Asadi (2007, 2008) discuss several characterizations of the Pickands distribution involving the mean residual life function of the base distribution F. If X has distribution function F, they define

$$\theta_F(x)=E(X-x|X>x)/E(X). \tag{3.9.85}$$

Then in the context of generalized order statistics $X_1,X_2,\ldots,X_n$ with base distribution F, they verify that either of the following two conditions will characterize the Pickands distribution.

$$(i) \quad \text{For some } i<j, \quad \frac{X_j-x}{\theta_F(x)}|X_1>x \stackrel{d}{=} X_j \tag{3.9.86}$$

$$(ii) \quad \text{For some } i<j, \quad \frac{X_j-X_i}{\theta_F(x)} \text{ and } X_i \text{ are independent.} \tag{3.9.87}$$

Asadi, Rao and Shanbhag (2001) introduce a concept of extended neighboring order statistics which subsumes and extends the idea of two neighboring generalized order statistics. They present several characterizations of the Pickands distribution (and a forteriori) of the classical Pareto distribution using this concept.

Finally we mention characterizations based on the concepts of random contraction or dilation in the generalized order statistics context. A useful reference here is

Beutner and Kamps (2008). Suppose that, for a set of n generalized order statistics $X_1, X_2, ..., X_n$, it is the case that

$$X_n \stackrel{d}{=} VX_{n-1}, \tag{3.9.88}$$

where $V \sim P(I)(1,\alpha)$ is independent of the X_i's. Note that, using the representation (3.9.73), this condition will clearly be true for the case in which the base distribution is of Pareto (I) form. Beutner and Kamps (2008) verify that (3.9.88) holds only in the Pareto (I) case.

3.9.8 *Entropy Maximization*

The Pareto (I) and (II) distributions can be characterized as ones which maximize entropy. Ord, Patil and Taillie (1981b) suppose that we seek a density on $[c,\infty)$ which will maximize the Shannon entropy

$$\int_c^\infty f(x) \log f(x) dx \tag{3.9.89}$$

subject to the density having a fixed value for its geometric mean. The extremal distribution subject to this constraint is shown to be of the Pareto (I) form.

Rather than use the Shannon entropy to quantify uncertainty reduction, we may consider the Renyi entropy of order α, defined by

$$H_\alpha(f) = (1-\alpha)^{-1} \log \left[\int_0^\infty [f(x)]^\alpha \, dx \right], \tag{3.9.90}$$

for $\alpha > 0$, $\alpha \neq 1$, where f is a density on $(0,\infty)$. Awad and Marzuq (1986) show that, for $\alpha \in (1/2, 1)$, in the class of all densities with a given value μ for the mean of the density, $H_\alpha(f)$ is maximized when f is the density of a Pareto (II)$(0,(2\alpha-1)\mu/(1-\alpha), \alpha/(1-\alpha))$ distribution.

Raqab and Awad (2000) consider the relationship between the Shannon entropy of the first n k-records in a sequence of i.i.d. absolutely continuous random variables $X_1, X_2, ...$, denoted by $\underline{R}^{(k)} = (R_1^{(k)}, R_2^{(k)}, ... R_n^{(k)})$, and the Shannon entropy of $X_{1:k}$, the minimum of k independent X_i's. They verify that if for every $n \geq 2$ we have

$$H(f_{\underline{R}^{(k)}}) - nH(f_{X_{1:k}}) = \frac{n(n-1)}{k}\theta, \tag{3.9.91}$$

for some $\theta \in (-\infty, \infty)$, then the X_i's must have a Pickands generalized Pareto distribution. In a subsequent paper (Raqab and Awad, 2001), they focus on the relationship between the Shannon entropy of the first n ordinary records $\underline{R} = (R_1, R_2, ..., R_n)$ and that of $\underline{X} = (X_1, X_2, ..., X_n)$. In this setting they show that if, for every $n \geq 2$, we have

$$H(f_{\underline{R}}) - H(f_{\underline{X}}) = n(n-1)\theta, \tag{3.9.92}$$

for some $\theta \in (-\infty, \infty)$, then again the X_i's must have a Pickands generalized Pareto distribution.

3.9.9 Pareto (III) Characterizations

We next turn to characterizations of the Pareto (III) or log-logistic distribution. The first of these, based on geometric minimization, is a simple consequence of Theorem 3.8.1 which, in turn, was the key component of the Arnold-Laguna income model (Section 2.9). Essentially, the result is as follows. Let $X_1, X_2, \ldots$ be i.i.d. with common distribution function F. If a geometric minimum of the X_i's has, except for a scale factor, the same distribution as an individual X_i, then, provided $\lim_{x \to 0} x^{-\lambda} F(x) = \eta > 0$, the common distribution of the X_i's is $P(\text{III})(0, \sigma, \gamma)$ where $\gamma = 1/\lambda$ and $\sigma = \eta^{-1/\lambda}$. To prove this assertion, let $N \sim$ geometric(p) be independent of the X_i's and define $Y = \min(X_1, X_2, \ldots, X_N)$. By hypothesis $cY \stackrel{d}{=} X_1$ for some c (which clearly must be greater than 1). It is readily argued that the support of F must be $[0, \infty)$, and if we define

$$\varphi(x) = F(x)/[1 - F(x)], \quad x > 0, \tag{3.9.93}$$

we may conclude (since $cY \stackrel{d}{=} X_1$) that for $x > 0$

$$\varphi(x) = p^{-1} \varphi(x/c),$$

and by iteration

$$\varphi(x) = p^{-k} \varphi(x/c^k), \quad \forall k. \tag{3.9.94}$$

Since $\lim_{x \to 0} x^{-\lambda} F(x) = \eta > 0$, there is only one possible value for c such that (3.9.94) can hold, namely $c = p^{-1/\lambda}$. For this choice of c it follows that $\varphi(x) = \eta x^{\lambda}$ and, consequently, $X_1 \sim P(\text{III})(0, \eta^{-1/\lambda}, 1/\lambda)$.

Several variations of this characterization have been described in the literature. Baringhaus (1980) considered a sequence of i.i.d. non-negative random variables $X_1, X_2, \ldots$ and an independent positive integer valued random variable N. Assuming that $\min_{i \leq N} X_i \stackrel{d}{=} a + bX_1$ for some a and b and that $X_i \stackrel{d}{=} 1/X_i$, then he showed that N must be a geometric random variable and the X_i's must have a Pareto (III) distribution.

Arnold, Robertson and Yeh (1986) assumed i.i.d. non-negative X_i's and an independent geometric (p) random variable N. They noted that, for every $p \in (0,1)$,

$$p^{-\gamma} \min_{i \leq N} X_i \stackrel{d}{=} p^{\gamma} \max_{i \leq N} X_i \stackrel{d}{=} X_1, \tag{3.9.95}$$

if the X_i's have a Pareto (III)$(0, \sigma, \gamma)$ distribution. Under the assumption that $\lim_{x \to 0} x^{-1/\gamma} F(x) = \sigma^{-1/\gamma}$, they show that if for some $p \in (0,1)$ any one of the three equidistribution statements in (3.9.95) holds, then it must be the case that $X_i \sim P(III)(0, \gamma, \sigma)$. In fact it is enough to assume that one of the equidistribution statements in (3.9.95) holds for two values p_1 and p_2 with $\log p_1 / \log p_2$ irrational to conclude that the X_i's have a Pareto (III) distribution, without assuming anything about the behavior of F as $x \to 0$.

Arnold, Robertson and Yeh (1986) also discuss a Pareto (III) characterization based on geometric minima of geometric maxima of random variables.

Janjic (1986) considered geometric maxima of i.i.d. random variables and verified that if $\max_{i \le N} X_i \stackrel{d}{=} a(p) + b(p)X_1$ for every $p \in (0,1)$, then for non-negative X_i's it must be the case that $X_i \sim P(III)(\mu, \sigma, \gamma)$.

Voorn (1987) revisited the Baringhaus (1980) scenario and proved that it is not necessary to assume the self-reciprocal condition (i.e., $X_i \stackrel{d}{=} 1/X_i$) for the X_i's. Voorn also considers random maxima as well as minima.

Cifarelli, Gupta and Jayakumar (2010) also present a Pareto(III) characterization involving maxima and minima, but in their work the extremes correspond to samples of size 2. They assume that X_1, X_2, X_3, X_4 are i.i.d. positive random variables and that b_1 and b_2 are positive constants satisfying $b_1^\alpha + b_2^\alpha = 1$ for some $\alpha > 0$. If it is the case that

$$\min\left(\frac{X_1}{b_1}, \frac{X_2}{b_2}\right) \stackrel{d}{=} \min\left(X_1, \max\left(\frac{X_3}{b_1}, \frac{X_4}{b_2}\right)\right),$$

it then follows that $X_i \sim P(III)(\mu, \sigma, 1/\alpha)$.

The final Pareto (III) characterization is a translation of Galambos and Kotz's (1978, pp. 27–8) logistic characterization. Supposing that

$$\frac{\overline{F}(xy)}{\overline{F}(x)\overline{F}(y)} = \frac{F(xy)}{F(x)F(y)}, \quad \forall x, y > 0, \tag{3.9.96}$$

it follows that F corresponds to a $P(\text{III})(0,1,\gamma)$ distribution. To see that this is the case, consider $\varphi(x)$ as defined in (3.9.93). Clearly, (3.9.96) implies

$$\varphi(xy) = \varphi(x)\varphi(y), \quad \forall x, y > 0,$$

but this is a multiplicative version of the Cauchy functional equation. And, since φ is right continuous, it follows that $\varphi(x) = x^\lambda$ for some λ. The parameter λ evidently must be positive and the Pareto (III) nature of the distribution is established.

3.9.10 *Two More Characterizations*

Undoubtedly the list of characterizations thus far provided in this Section is incomplete. Each year new insights into properties of the Pareto distributions (I)–(IV) are documented and frequently generate characterizations. In this Section we present two closely related additional characterizations which, while they did not seem to fit naturally in the earlier discussion, do have some intrinsic interest.

Moothathu (1990) considered k i.i.d. random variables $Z_1, Z_2, ..., Z_k$ with $P(Z_i > 1) = 1$, together with a $k-1$ dimensional random vector $\underline{Y} = (Y_1, Y_2, ..., Y_{k-1})$ with support $(0, \infty)^{k-1}$ that is independent of $\underline{Z} = (Z_1, Z_2, ..., Z_k)$. He then defines

$$T = 1 + Y_1 + Y_2 + \cdots Y_{k-1} \tag{3.9.97}$$

and

$$V = \min\left\{Z_1^T, Z_2^{T/Y_1}, ..., Z_k^{T/Y_{k-1}}\right\}. \tag{3.9.98}$$

If the Z_i's have a Pareto(I) distribution then it may be verified that V and $\underline{Y}$ are independent and that $V \stackrel{d}{=} Z_1$. Moothathu (1990) shows that the converse is also true under mild regularity conditions. Thus, independence of V and $\underline{Y}$ is sufficient to guarantee that the Z_i's have a common Pareto (I) distribution.

Ravi (2010), in a paper discussing characterizations of generalized Pareto (Pickands sense) and generalized extreme value distributions, presents discussion of the $k = 2$ case of Moothathu's result. We will state this simpler result and sketch its proof.

Assume that X_1 and X_2 are i.i.d. and are independent of an absolutely continuous random variable Y with $f_Y(y) > 0$ on the interval $(0,1)$. Also assume that $P(X_1 > 1) = 1$. Consider

$$V = \min\left\{X_1^{1/Y}, X_2^{1/(1-Y)}\right\}$$

and assume that V and Y are independent. In this case we can conclude that $X_1 \sim P(I)(\sigma,\alpha)$ for some $\sigma > 0, \alpha > 0$.

To see this, we argue as follows. Since V and Y are independent, it follows that

$$[1-F_V(x)]F_Y(y) = P(V > x, Y \leq y) = \int_0^y [1-F_X(x^z)][1-F_X(x^{1-z})]dF_Y(z).$$

Differentiating with respect to y and dividing by $f_Y(y)$ yields

$$[1-F_V(x)] = [1-F_X(x^y)][1-F_X(x^{1-y})].$$

Taking the limit as $y \to 0$, we conclude that $1-F_V(x) = 1-F_X(x)$. Consequently

$$[1-F_X(x)] = [1-F_X(x^y)][1-F_X(x^{1-y})], \; y \in (0,1).$$

If we then define

$$K(u) = 1-F_X(e^u), \; u > 0,$$

it follows that

$$K(y) = K(xy)K(x(1-y)), \forall x > 0, y \in (0,1),$$

whence it follows that $K(y) = e^{cy}$ and, from this, we may conclude that X has a Pareto (I) distribution.

3.10 Related Distributions

In Section 2.11, a list was provided of several parametric families of densities that have been proposed as being more flexible alternatives to the Pareto distributions for modeling income. In many cases, only the form of the density was described and few, if any, distributional properties of the models were investigated. Fitting of the models frequently made use of maximum likelihood, the method of moments, or some variation on quantile matching. Several of the models are intimately related to either the classical or one of the more general Pareto models discussed in this chapter. Indeed many of them include Pareto distributions as special cases. In this Section we will present a more detailed discussion of some of these "extensions" of the Pareto and generalized Pareto models.

We begin with a remark that, extending the concept of income power introduced by Champernowne (1937), we may extend the basic Pareto models by assuming that, instead of income obeying the given law, some monotonic function of income does so. It is possible to add even more flexibility by considering a parametric family of possible monotonic transformations of the income data whose parameters must be estimated from the data. For example, we might begin with a parametric family of increasing functions $\psi(x;\underline{\tau})$ with corresponding inverse functions $\psi^{-1}(x;\underline{\tau})$ and assume that $\psi(X;\underline{\tau})$ has a Pareto(IV)$(\mu,\sigma,\gamma,\alpha)$ distribution. If we denote the corresponding $P(IV)(\mu,\sigma,\gamma,\alpha)$ distribution by $F_{\mu,\sigma,\gamma,\alpha}(x)$ then the distribution of X will be

$$F_X(x;\mu,\sigma,\gamma,\alpha,\underline{\tau}) = F_{\mu,\sigma,\gamma,\alpha}(\psi^{-1}(x;\underline{\tau})). \tag{3.10.1}$$

A variation on this theme, which has attracted considerable interest, is one in which the distribution functions in (3.10.1) are replaced by quantile functions. Thus the quantile function of X is assumed to be of the form

$$F_X^{-1}(u;\mu,\sigma,\gamma,\alpha,\underline{\tau}) = F_{\mu,\sigma,\gamma,\alpha}^{-1}(\widetilde{\psi}(u;\underline{\tau})), \tag{3.10.2}$$

where $\widetilde{\psi}(u;\underline{\tau})$ is a parametric family of monotone functions mapping $(0,1)$ onto $(0,1)$. A popular version of this type of construction was introduced by Jones (2004). It will be recalled that a random variable Z with distribution function $F_Z(z)$ can be viewed as being equivalent in distribution to $F_Z^{-1}(U)$ where $U \sim Uniform(0,1)$. Instead of the model $F_Z^{-1}(U)$, Jones proposed use of the model $F_Z^{-1}(B)$ where B has a beta distribution with parameters $\lambda_1 > 0$ and $\lambda_2 > 0$ (use of λ's as the parameters of the Beta distribution instead of the more customary α and β avoids subsequent confusion with the α parameter in the Pareto (IV) model).

If, in this scenario, $Z \sim P(IV)(\mu,\sigma,\gamma,\alpha)$ and if we define $X = F_Z^{-1}(B)$, we find that

$$\begin{aligned} F_X(x;\mu,\sigma,\gamma,\alpha,\lambda_1,\lambda_2) &= P(X \le x) = P(F_{\mu,\sigma,\gamma,\alpha}^{-1}(B) \le x) \\ &= P(B \le F_{\mu,\sigma,\gamma,\alpha}(x)) \\ &= F_{\lambda_1,\lambda_2}(F_{\mu,\sigma,\gamma,\alpha}(x)), \end{aligned} \tag{3.10.3}$$

where F_{λ_1,λ_2} denotes a Beta(λ_1,λ_2) distribution function. From (3.10.3), it follows that

$$F_X^{-1}(u;\mu,\sigma,\gamma,\alpha,\lambda_1,\lambda_2) = F_{\mu,\sigma,\gamma,\alpha}^{-1}(F_{\lambda_1,\lambda_2}^{-1}(u)),$$

i.e., a model of the form (3.10.2).

Clearly, there is no need to restrict attention to models in which B has a Beta distribution. The random variable B can be endowed with any distribution with support $[0,1]$, or rather we can assume that B has as its distribution some member of a large parametric family of distributions defined on $(0,1)$, to obtain even more flexibility. However, the Jones-Beta model has been the most popular and has been called the Beta-generalized Pareto distribution.

The density function of the Beta-generalized Pareto distribution (3.10.3), with $\mu = 0$ for simplicity, is given by

$$f_X(x;\sigma,\gamma,\alpha,\lambda_1,\lambda_2)$$
$$= \frac{\alpha\left[1-\left(1+\left(\frac{x}{\sigma}\right)^{1/\gamma}\right)^{-\alpha}\right]^{\lambda_1-1}\left(1+\left(\frac{x}{\sigma}\right)^{1/\gamma}\right)^{-\alpha(\lambda_2-1)}\left(\frac{x}{\sigma}\right)^{(1/\gamma)-1}}{\sigma\gamma B(\lambda_1,\lambda_2)},$$

where $x \in (0,\infty)$.

It is possible to obtain expressions for the low order moments of this distribution when $\gamma = 1$, since, in that case

$$E\left[\left(1+\frac{X}{\sigma}\right)^k\right] = \frac{B(\lambda_1,\lambda_2-k)}{B(\lambda_1,\lambda_2)}.$$

Further details on this distribution may be found in Akinsete, Famoye and Lee (2008) and Mahmoudi (2011).

Another popular model of the form (3.10.2) involves the simple choice $\widetilde{\psi}(u) = u^{1/\theta}$ where $\theta > 0$. In such a case we have

$$F_X(x;\mu,\sigma,\gamma,\alpha,\theta) = [F_{\mu,\sigma,\gamma,\alpha}(x)]^\theta, \quad x > \mu, \tag{3.10.4}$$

and the distribution is usually called the exponentiated generalized Pareto distribution. It can be recognized as a special case of the Beta-generalized Pareto model with the parameters chosen to be $\lambda_1 = \theta$ and $\lambda_2 = 1$. The simplicity of the exponentiated Pareto model accounts, to some degree, for its popularity. Recalling the definition of the Pareto (IV) distribution function, the exponentiated generalized Pareto distribution is of the form

$$F_X(x;\mu,\sigma,\gamma,\alpha,\theta) = \left[1-\left(1+\left(\frac{x-\mu}{\sigma}\right)^{1/\gamma}\right)^{-\alpha}\right]^\theta, \quad x > \mu. \tag{3.10.5}$$

Most of the discussion of this distribution in the literature is focused on the case in which $\sigma = \gamma = 1$ and $\mu = 0$. Shawky and Abu-Zinadah (2008) discuss record values from this distribution. Their results were extended by Khan and Kumar (2010) who obtained expressions for and recurrence relations between moments of generalized order statistics from the distribution (3.10.5), with $\sigma = \gamma = 1$ and $\mu = 1$. Afify (2010) developed maximum likelihood estimates and Bayes estimates using non-informative priors. Further properties of exponentiated Pareto distributions are discussed in Nadarajah (2005). Behboodian and Tahmasebi (2008) discuss the exponentiated Pareto entropy based on order statistics. Note that some simplification occurs in the model if either or both of α and θ are assumed to be integer valued.

Instead of the Beta distribution, one might use the Kumaraswamy distribution to obtain an alternative generalized Pareto distribution. We say that X has a Kumaraswamy(λ_1,λ_2) distribution if its density and distribution functions are:

$$f_K(x) = \lambda_1\lambda_2 x^{\lambda_1-1}(1-x^{\lambda_1})^{\lambda_2-1}, \quad 0 < x < 1, \tag{3.10.6}$$

and

$$F_K(x) = 1 - (1 - x^{\lambda_1})^{\lambda_2}, \; 0 < x < 1. \tag{3.10.7}$$

See Jones (2009) for a comprehensive introduction to the Kumaraswamy distribution. Let $F_{\mu,\sigma,\gamma,\alpha}(x)$ denote the $P(IV)(\mu,\sigma,\gamma,\alpha))$ distribution function and suppose that K has a Kumaraswamy(λ_1,λ_2) distribution. Define $Y = F^{-1}_{\mu,\sigma,\gamma,\alpha}(K)$, then Y has a Kumaraswamy-Pareto (IV) distribution with corresponding density

$$f_Y(y) = f_K(F_{\mu,\sigma,\gamma,\alpha}(y))f_{\mu,\sigma,\gamma,\alpha}(y).$$

Paranaiba, Ortega, Cordeiro and de Pascoa (2013) discuss this Kumaraswamy-generalized Pareto distribution. Submodels of this model are often of interest. For example, the exponentiated generalized Pareto distribution (3.10.4) is such a submodel.

It will be recalled from earlier discussion that, for a sequence $X_1, X_2, \ldots$ of i.i.d. Pareto(III)$(0,\sigma,\gamma)$ random variables, and an independent geometric(p) random variable N, we have, repeating part of (3.9.95), $p^{-\gamma}\min_{i\leq N} X_i \stackrel{d}{=} X_1$ and, under a regularity condition involving the behavior of $F_X(x)$ in a neighborhood of 0, that this property characterizes the Pareto(III)$(0,\sigma,\gamma)$ distribution. What happens if the regularity condition is not invoked? This question was addressed by Pillai (1991) who introduced what he called semi-Pareto distributions.

To motivate the introduction of such distributions, consider i.i.d. random variables $X_1, X_2, \ldots$ with common survival function $\overline{F}_X(x)$, and, for an independent geometric(p) random variable N, define

$$Y(p) = p^{-\gamma} \min_{i \leq N} X_i. \tag{3.10.8}$$

By conditioning on N, it may be verified that the survival functions of $Y(p)$ and X_1 are related by

$$\overline{F}_{Y(p)}(y) = \frac{p\overline{F}_{X_1}(p^\gamma y)}{1-(1-p)\overline{F}_{X_1}(p^\gamma y)}.$$

If we define

$$\varphi_{X_1}(x) = F_{X_1}(x)/[1 - F_{X_1}(x)]$$

as in (3.9.93), we can see that if $Y(p)$ and X_1 are to be identically distributed, it must be the case that

$$\varphi_{X_1}(x) = p^{-1}\varphi_{X_1}(p^\gamma y). \tag{3.10.9}$$

The general solution to the functional equation (3.10.9) is

$$\varphi_{X_1}(x) = x^{1/\gamma} h_{X_1}(x), \tag{3.10.10}$$

where $h_{X_1}(x)$ is a periodic function of $\log x$ with period $-2\pi/[\gamma \log p]$.

A random variable X is then said to have a semi-Pareto distribution if its survival function admits a representation of the form

$$\overline{F}_X(x) = [1 + x^{1/\gamma} h_X(x)]^{-1}, \tag{3.10.11}$$

where h_X is a periodic function of $\log x$, i.e., if, for some function $\varphi_X(x)$ which satisfies (3.10.10), it is of the form $\overline{F}(x) = [1+\varphi_X(x)]^{-1}$.

Observe that, while a Pareto(III) distribution will satisfy $Y(p) \stackrel{d}{=} X_1$ for every $p \in (0,1)$, a semi-Pareto variable, with a non-constant h_X, will only satisfy $Y(p) \stackrel{d}{=} X_1$ for values of p of the form $(p^*)^k$ for some fixed p^*. The class of semi-Pareto survival functions, after the introduction of location and scale parameters, is given by

$$\overline{F}(x;\mu,\sigma,\gamma,p) = \left[1+\left(\frac{x-\mu}{\sigma}\right)^{1/\gamma} h(\frac{x-\mu}{\sigma})\right]^{-1}, \quad x > \mu, \qquad (3.10.12)$$

where $\mu \in (-\infty,\infty)$, $\sigma,\gamma \in (0,\infty)$, $p \in (0,1)$ and $h(x)$ is a periodic function of $\log x$ with suitable period and with $h(0) = 1$. The case in which $h(x) \equiv 1$ for every x corresponds to the usual Pareto(III) model. Note that, in order for (3.10.12) to be a valid survival function, it must be the case that $x^{1/\gamma}h(x)$ is a non-decreasing function of x. An example, provided by Pillai (1991), of a function $h(x)$ which meets these requirements is

$$h(x) = \exp\{\beta \cos(\log x/\gamma)\},$$

where $0 \leq \beta < 1$. This choice of h produces a parametric extension of the Pareto (III) distribution which might possibly provide a better fit to certain data sets than would a simple Pareto (III) model.

As Pillai (1991) and several other researchers have observed (recently, for example, Cifarelli, Gupta and Jayakumar (2010)), it is possible to verify that many theoretical results involving Pareto (III) variables remain valid also for semi-Pareto variables. In addition, multivariate semi-Pareto distributions are available, together with a spectrum of semi-Pareto processes (see Chapter 6 for further discussion of these topics).

Cifarelli, Gupta and Jayakumar (2010) introduced a related more general class of distributions extending the semi-Pareto model as follows. For some positive integer k and a set of k constants $0 < p_i < 1, \; i = 1,2,...,k$, a random variable is said to have a generalized semi-Pareto distribution (Cifarelli, Gupta and Jayakumar just call it a generalized Pareto distribution, but that name is already overworked in this book) and we write $X \sim GSP(III)(0,\sigma,\gamma,p_1,p_2,...,p_k)$ if its survival function is of the form

$$\overline{F}_X(x) = [1+\psi(x/\sigma)]^{-1}, \qquad (3.10.13)$$

where $\psi(x)$ satisfies the following functional equation

$$\psi(x) = \sum_{i=1}^{k} \frac{1}{p_i} \psi((p_i/k)^{\gamma} x). \qquad (3.10.14)$$

Applications of such complex models are not discussed. The authors do provide a representation of a generalized semi-Pareto (III) distribution involving an arbitrary survival function $\overline{F}$. If, for positive $a_i, b_i, \; i = 1,2,...,k$ with $\sum_{i=1}^k a_i = 1$, a survival function admits the representation

$$\overline{G}(x) = \frac{\prod_{i=1}^k \overline{F}(b_i x)}{\sum_{i=1}^k a_i \left[\prod_{1 \leq j \leq k, j \neq i} \overline{F}(b_j x)\right]} \qquad (3.10.15)$$

then $\overline{G}$ has a $GSP(III)$ distribution, and moreover all $GSP(III)$ distributions admit a representation of the form (3.10.15)

In that same paper, in an autoregressive process context, they propose a semi-Pareto (IV) distribution with survival function of the form

$$\overline{F}(x) = [1+\psi(x)]^{-\alpha}, \ \ \alpha > 0,$$

where ψ satisfies the functional equation (3.10.10) or more generally the equation (3.10.14)

In a related paper, Pillai, Jose and Jayakumar (1995) introduced the class of distributions with universal geometric minima (u.g.m.). A distribution F is said to belong to the u.g.m. class if, for every $p \in (0,1)$, the function

$$\frac{\overline{F}(x)}{[p+(1-p)\overline{F}(x)]} \tag{3.10.16}$$

is a valid survival function. It is possible to prove a variety of closure properties of the u.g.m. class. The u.g.m. class is still occasionally referred to in the literature, often in contexts in which certain classes of distributions are argued to be subclasses of the u.g.m. class of distributions. Such results are undoubtedly correct, but they are of, at best, limited interest since, unfortunately, for any distribution function F and any $p \in (0,1)$ the function (3.10.16) is a valid survival function. It is non-decreasing, right continuous and of total variation 1. Thus the u.g.m. class coincides with the class of all distributions.

Sandhya and Satheesh (1996) introduced a generalized Pareto model of the form

$$F(x) = 1 - [g(x)]^{-1}, \tag{3.10.17}$$

where $g(x)$ has a completely monotone derivative and satisfies $g(0) = 1$ and $g(\infty) = \infty$. They denote this class of distributions by $GP(g)$. However, Sandhya and Satheesh (1996) argue that the $GP(g)$ class coincides with the u.g.m. class. This claim is surprising since (3.10.17) will be a valid distribution function for any monotone g; a completely monotone derivative is not required. It thus appears that the $GP(g)$ class is a proper subclass of the (overly extensive) u.g.m. class.

We next turn to discussion of hidden truncation or selection extensions of the Pareto (IV) distribution. Some motivation for consideration of such models is appropriate here. In many situations a variable, X, will only be observable if a covariable or concomitant variable, Y, satisfies certain constraints. For example, information about an individual's income in the year 2012, denoted by X, may only be available if that individual's 2011 income, Y, exceeded a given level, say y_0. In such a situation, X and Y are typically not independent so that the conditional distribution of X given that $Y > y_0$ will be different from the unconditional distribution of X. If we do not know the value of y_0 in such a setting, we may speak of X as having a distribution which has been modified by hidden truncation. It is quite evident that many data sets may be viewed as having been subject to such hidden truncation or some form of hidden selection, and it is appropriate to have models available that can accommodate such contingencies.

In a quite general setting involving a two-dimensional absolutely continuous random vector (X,Y), we focus on the conditional distribution of X, given that $Y \in M$ where M is a Borel subset of $(-\infty,\infty)$, often of the form (y_0,∞) or $(-\infty,y_0)$. In this situation we have

$$f_{X|\{Y\in M\}}(x) = \frac{f_X(x)P(Y\in M|X=x)}{P(Y\in M)}. \tag{3.10.18}$$

The form of such distributions is determined by

(i) the unconditional distribution of X, $f_X(x)$,

(ii) the conditional distribution of Y given $X = x$,

and

(iii) the selection mechanism, i.e., the choice of the restrictive subset M.

See Arnold and Beaver (2002) or Arellano-Valle, Branco and Genton (2006) for an extensive discussion of such models.

In the present context, we concentrate on such models in which the marginal distribution of X is of the Pareto (IV) form, and the conditional distribution of Y given $X = x$ is also of the Pareto form. A popular bivariate distribution with such structure (to be discussed in some detail in Chapter 6) has a joint survival function of the form

$$P(X>x,Y>y) = \left[1+\left(\frac{x-\mu}{\sigma}\right)^{1/\gamma}+\left(\frac{y-\nu}{\tau}\right)^{1/\delta}\right]^{-\alpha}, \quad x>\mu, y>\nu. \tag{3.10.19}$$

From (3.10.19) it may be verified that, for any $y > \nu$, we have

$$X|Y=y \sim P(IV)(\mu,\sigma_y,\gamma,\alpha+1), \tag{3.10.20}$$

where

$$\sigma_y = \sigma\left[1+\left(\frac{y-\nu}{\tau}\right)^{1/\delta}\right]^{\gamma}.$$

If we restrict (as we shall) attention to cases in which M is of the form (ν,y_0) or (y_0,∞) for some $y_0 > \nu$, we may identify the corresponding hidden truncation densities using (3.10.19) without reference to the conditional distribution (3.10.20).

First we consider truncation from below, i.e., $M = (y_0, \infty)$. In this case we have

$$P(X > x|Y > y_0) = \frac{P(X > x, Y > y_0)}{P(Y > y_0)}$$

$$= \frac{\left[1 + \left(\frac{x-\mu}{\sigma}\right)^{1/\gamma} + \left(\frac{y_0-\nu}{\tau}\right)^{1/\delta}\right]^{-\alpha}}{\left[1 + \left(\frac{y_0-\nu}{\tau}\right)^{1/\delta}\right]^{-\alpha}}$$

$$= \left[1 + \left(\frac{x-\mu}{\sigma_1}\right)^{1/\gamma}\right]^{-\alpha}, \quad x > \mu, \tag{3.10.21}$$

where

$$\sigma_1 = \sigma\left[1 + \left(\frac{y_0 - \nu}{\tau}\right)^{1/\delta}\right]^{\gamma}.$$

Thus, in this case, the hidden truncation distribution of X is still of the Pareto (IV) form, with merely a new scale parameter. Consequently, there would be no way that one could determine from a data set whether such hidden truncation had occurred. A more interesting result occurs if we assume that $M = (\nu, y_0)$, i.e., truncation from above. In this case we have

$$P(X > x|Y \leq y_0) = \frac{P(X > x, Y \leq y_0)}{P(Y \leq y_0)}$$

$$= \frac{P(X > x) - P(X > x, Y > y_0)}{1 - P(Y > y_0)}.$$

Using the expression (3.10.19) for the joint survival function of (X, Y), we obtain

$$P(X > x|Y \leq y_0)$$

$$\propto \left\{\left[1 + \left(\frac{x-\mu}{\sigma}\right)^{1/\gamma}\right]^{-\alpha} - \left[1 + \left(\frac{x-\mu}{\sigma}\right)^{1/\gamma} + \left(\frac{y_0-\nu}{\tau}\right)^{1/\delta}\right]^{-\alpha}\right\}.$$

Since it is assumed that y_0, ν, τ and δ are not known, we can introduce the following single unknown positive parameter

$$\theta = \left(\frac{y_0 - \nu}{\tau}\right)^{1/\delta}$$

and write

$$P(X > x | Y \leq y_0)$$

$$\propto \left\{ \left[1 + \left(\frac{x-\mu}{\sigma} \right)^{1/\gamma} \right]^{-\alpha} - \left[1 + \left(\frac{x-\mu}{\sigma} \right)^{1/\gamma} + \theta \right]^{-\alpha} \right\}. \tag{3.10.22}$$

Because, when $x = \mu$, this expression must equal 1, we can identify the required normalizing constant in (3.10.22). Upon differentiation, we obtain the following hidden truncation density (of X given $Y \leq y_0$)

$$f_{HT}(x; \mu, \sigma, \gamma, \alpha, \theta) = \frac{\alpha \left(\frac{x-\mu}{\sigma} \right)^{(1/\gamma)-1}}{\gamma \sigma [1 - (1+\theta)^{-\alpha}]} \tag{3.10.23}$$

$$\times \left[\left(1 + \left(\frac{x-\mu}{\sigma} \right)^{1/\gamma} \right)^{-(\alpha+1)} - \left(1 + \left(\frac{x-\mu}{\sigma} \right)^{1/\gamma} + \theta \right)^{-(\alpha+1)} \right] \quad x > \mu.$$

In this model, μ is a real valued parameter, often positive, while all of the other parameters, σ, γ, α and θ, are positive valued. An alternative representation of this density is possible. A little algebraic manipulation leads to the following expression.

$$f_{HT}(x; \mu, \sigma, \gamma, \alpha, \theta) \tag{3.10.24}$$

$$= \frac{1}{1 - (1+\theta)^{-\alpha}} \left[\frac{\alpha}{\gamma \sigma} \left(\frac{x-\mu}{\sigma} \right)^{(1/\gamma)-1} \left[1 + \left(\frac{x-\mu}{\sigma} \right)^{1/\gamma} \right]^{-(\alpha+1)} \right]$$

$$- \frac{(1+\theta)^{-\alpha}}{1 - (1+\theta)^{-\alpha}} \left[\frac{\alpha}{\gamma \sigma_1} \left(\frac{x-\mu}{\sigma_1} \right)^{(1/\gamma)-1} \left[1 + \left(\frac{x-\mu}{\sigma_1} \right)^{1/\gamma} \right]^{-(\alpha+1)} \right]$$

where $\sigma_1 = \sigma(1+\theta)^{\gamma}$. This is recognizable as a linear combination of two Pareto (IV) densities: the original P(IV)$(\mu, \sigma, \gamma, \alpha)$ density and the density of the Pareto (IV) survival function in (3.10.21) (which was associated with hidden truncation from above). Note that the density in (3.10.24) is a linear combination of two Pareto (IV) densities, but it is not a convex combination since, although the coefficients add up to 1, the second coefficient is negative.

Because we are dealing with a linear combination of Pareto (IV) densities, it is possible to obtain expressions for the moments of such a density, using known results for the Pareto (IV) moments (see equation (3.3.11)). Such expressions can be useful in the implementation of various estimation strategies for data suspected of having been subject to hidden truncation. More detailed discussion of these hidden truncation Pareto models, in the Pareto (II) case, may be found in Arnold and Ghosh (2011).

It is also possible to exhibit the information matrix corresponding to such models, although the expressions will be cumbersome. Some indication of this can be seen in the paper by Nadarajah and Kotz (2005) in which they discuss the information matrix for a mixture of two Pareto distributions.

Once it is recognized that a linear combination of two Pareto densities with one negative coefficient provides a plausible model for income, or other heavy-tailed data, it is a small step to consider more general k-component linear combinations as possible models. Thus we might consider models with survival functions of the form

$$\overline{F}(x;\mu,\underline{\sigma},\underline{\gamma},\underline{\alpha},\underline{\delta}) = \sum_{j=1}^{k} \delta_j \left[1+\left(\frac{x-\mu}{\sigma_j}\right)^{1/\gamma_j}\right]^{-\alpha_j}, \quad x>\mu, \tag{3.10.25}$$

where $\mu \in (-\infty,\infty)$ and $\underline{\sigma},\underline{\gamma},\underline{\alpha} \in (0,\infty)^k$, while $\underline{\delta}$ is a vector of k real valued coefficients (some of which may be negative) which sum to 1. It is possible to provide expressions for moments of distributions of the form (3.10.25), which would facilitate the implementation of some estimation strategies. If all of the δ_j's are positive, then we have a simple mixture model and there are no restrictions on the parameters $\mu,\underline{\sigma},\underline{\gamma},\underline{\alpha}$ and $\underline{\delta}$ needed to ensure that (3.10.25) represents a legitimate survival function. However, if negative values of the δ_j's are permitted, then there will be restrictions on the σ_j's,γ_j's,α_j's and δ_j's needed to ensure that (3.10.25) has a uniformly non-positive derivative which integrates to -1, in order to have a legitimate survival function. The precise nature of these restrictions has not yet been determined. Analogous questions regarding linear combinations of exponential densities are discussed in Bartholomew (1969) and Arnold and Gokhale (2014).

As a final entry in our list of Pareto-related distributions that include the Pareto distribution as a special case, we consider Dimitrov and von Collani's (1995) (aptly named?) contorted Pareto model. They define a random variable

$$X = Yb^Z, \tag{3.10.26}$$

where Y and Z are independent random variables and $b \in (1,\infty)$. Assume that Z has a geometric distribution with support $\{0,1,2,...\}$, i.e., assume that $P(Z=n) = p(1-p)^n$, $n=0,1,2,\ldots$. Also assume that the density of Y is of the form

$$f_Y(y) = \frac{\alpha y^{-\alpha-1}}{1-b^{-\alpha}} I(1<y<b), \tag{3.10.27}$$

i.e., Y has a truncated Pareto$(1,\alpha)$ distribution. In such a case, X is said to have a contorted Pareto distribution with density of the form

$$f_X(x) = p(1-p)^m \frac{\alpha x^{-\alpha-1} b^{(m+1)\alpha}}{b^{\alpha-1}}, \quad u \in [b^m, b^{m+1}), \quad m = 0,1,2,...$$

$$= 0, \text{ otherwise.} \tag{3.10.28}$$

It is only in the special case in which $1-p=b^{-\alpha}$ that the density (3.10.28) simplifies to yield a classical Pareto (I)(α, 1) distribution.

3.11 The Discrete Pareto (Zipf) Distribution

Heavy-tailed distributions with possible values $0,1,2,\ldots$ or sometimes $1,2,3\ldots$ have received considerable attention. As mentioned in Chapter 2, George Zipf cataloged an amazing variety of integer-valued random phenomena for which the probability of outcome i was approximately proportional to $i^{-(\alpha+1)}$ (for some $\alpha > 0$), at least for large values of i. Distributions in which this proportionality obtains exactly are sometimes called zeta distributions, since the required normalizing constant is the Riemann zeta function evaluated at $\alpha+1$. A more flexible family of distributions exhibiting Zipf tail behavior (i.e., $P(X=i) \sim i^{-(\alpha+1)}$) may be obtained by discretizing the various Pareto distributions introduced in Section 3.2.

There are several convenient ways of describing the distribution of an integer valued random variable X. Three commonly used descriptors are: (i) the discrete density, $P(X=j)$, $j=0,1,2,3,\ldots$; (ii) the survival function, $P(X \geq j)$, $j=0,1,2,3,\ldots$; and (iii) the ratio of successive terms in the density, i.e., $P(X=j)/P(X=j-1)$, $j=1,2,3,\ldots$. There are certainly others, but in the context of heavy-tailed discrete distributions, these three seem to be most frequently encountered. Thus, a simple algebraic form is assumed for (i), (ii) or (iii) depending on a few parameters, and fitting proceeds. The zeta distribution (exemplifying exact Zipf behavior) has a simple form of its density (i). The discretized Pareto distributions (exemplifying asymptotic Zipf behavior) have simple survival functions. The Yule distribution and other relatives arising from the work of Simon (cf. Section 2.6) have simple forms for the ratio of successive terms in the density function, (iii). We will reserve the title Zipf distribution for discretized Pareto distributions and utilize the names zeta and Simon for the other two families. There is minimal overlap between the families. One exception is the standard Zipf distribution which can be thought of as a member of both the discretized Pareto and Simon families. A rule of thumb is that if any one of (i), (ii) or (iii) is expressible in a simple form, then the other two will be complicated (but see the family of Waring distributions below for a violation of this rule). We now turn to our catalog of distributions exhibiting exact or asymptotic Zipf behavior.

3.11.1 Zeta Distribution

A random variable X will be said to have a zeta distribution with parameter α, if its discrete density is of the form

$$P(X=j) = [\zeta(\alpha+1)]^{-1} \quad j^{-(\alpha+1)}, \quad j=1,2,\ldots \tag{3.11.1}$$

where $\alpha > 0$ and ζ is the Riemann zeta function. In such a case we will write $X \sim \text{zeta}(\alpha)$. The moments of such a random variable are readily expressed in terms of the zeta function. Thus,

$$E(X^{\beta}) = \zeta(\alpha - \beta + 1)/\zeta(\alpha + 1), \quad \beta < \alpha. \tag{3.11.2}$$

3.11.2 Zipf Distributions

Next, consider the discretized Pareto distributions. In the discrete case the location parameter will be an integer, while the scale-like parameter and shape parameter will be positive real-valued. An analog of the Pareto (I) whose location and scale parameters are equal would involve an awkward restriction that the scale-like parameter be integer valued. We will omit discussion of such a distribution. Discretized versions of the Pareto II, III and IV and the Feller-Pareto distribution are readily defined as follows. We will say that a random variable X has a Zipf(II)(k_0, σ, α) distribution, if

$$P(X \geq k) = \left[1 + \left(\frac{k - k_0}{\sigma}\right)\right]^{-\alpha}, \quad k = k_0, k_0 + 1, k_0 + 2, \ldots . \tag{3.11.3}$$

Analogously $X \sim$ Zipf(III)(k_0, σ, γ), if

$$P(X \geq k) = \left[1 + \left(\frac{k - k_0}{\sigma}\right)^{1/\gamma}\right]^{-1}, \quad k = k_0, k_0 + 1, k_0 + 2, \ldots, \tag{3.11.4}$$

and $X \sim$ Zipf(IV)$(k_0, \sigma, \gamma, \alpha)$, if

$$P(X \geq k) = \left[1 + \left(\frac{k - k_0}{\sigma}\right)^{1/\gamma}\right]^{-\alpha}, \quad k = k_0, k_0 + 1, k_0 + 2, \ldots . \tag{3.11.5}$$

Finally a Feller-Zipf random variable would be of the form: k_0 plus the integer part of a Feller-Pareto random variable with $\mu = 0$ (cf. (3.2.11)). Moments of these Zipf distributions are not generally obtainable in closed form. If σ is an integer then the mean of a Zipf (II) distribution is expressible in terms of the Riemann zeta function. Order statistics of discrete random variables are typically intractable except for the extremes. Such is the case for these Zipf distributions. Minima are particularly well behaved. For example the minimum of n independent identically distributed Zipf (II) (respectively Zipf (IV)) variables is itself distributed as Zipf (II) (respectively Zipf (IV)). The Zipf (III) family is closed under geometric minimization as is the Pareto (III) family. The result for the Pareto (III) distribution was implicit in the development of the Arnold-Laguna model (Section 2.9), and a proof can be extracted from the details in the proof of Theorem 3.8.1. It was not explicitly stated, however. For completeness, we will give in detail the result in the Zipf (III) case.

Let $X_1, X_2, \ldots$ be independent identically distributed Zipf(III)(k_0, σ, γ) random variables and let N be a geometric random variable (i.e., $P(N = n) = pq^{n-1}$, $n = 1, 2, 3, \ldots$) independent of the X_i's. Define

$$U = \min(X_1, X_2, \ldots, X_N).$$

To obtain the distribution of U, we argue as follows: for $k \geq k_0$,

$$\begin{aligned} P(U \geq k) &= E(P(U \geq k \mid N)) = E([P(X_1 \geq k)]^N) \\ &= \sum_{n=1}^{\infty} \left[1 + \left(\frac{k-k_0}{\sigma}\right)^{1/\gamma}\right]^{-n} pq^{n-1} \\ &= \left[1 + \left(\frac{k-k_0}{\sigma p^{\gamma}}\right)^{1/\gamma}\right]^{-1}. \end{aligned} \tag{3.11.6}$$

Thus, $U \sim \text{Zipf(III)}(k_0, \sigma p^{\gamma}, \gamma)$. This result can be utilized to construct multivariate Zipf(III) distributions by considering multivariate geometric minima of i.i.d. Zipf (III) variables (see Section 6.6).

The name "standard Zipf distribution" will be reserved for the distribution of the form

$$P(U \geq k) = (1+k)^{-1}, \quad k = 0, 1, 2, \ldots, \tag{3.11.7}$$

i.e. the Zipf (IV)(0,1,1,1) distribution. The standard Zipf distribution has a discrete density which is simply expressible:

$$P(U = k) = [(k+1)(k+2)]^{-1}, \quad k = 0, 1, 2, \ldots . \tag{3.11.8}$$

A translated standard Zipf distribution arises in the study of record times in sequences of an i.i.d. random variables. Suppose $X_1, X_2, \ldots$ is a sequence of i.i.d. continuous random variables and that we define a random variable U to be the subscript of the first X_i which exceeds X_1 in value. Clearly $U > k+1$, if and only if, X_1 is the largest of the set $X_1, X_2, \ldots, X_{k+1}$. By symmetry, we conclude

$$P(U \geq k+2) = (k+1)^{-1}, \quad k = 0, 1, 2, \ldots, \tag{3.11.9}$$

i.e., $U \sim \text{Zipf(IV)}(2, 1, 1, 1)$. Feller (1971, p. 40) gives an example in which the standard Zipf distribution arises in connection with platoon formation in traffic on an infinite road.

The standard Zipf distribution arises as a uniform mixture of geometric distributions (i.e., if, given p, $P(X = k) = p(1-p)^k$ where p uniform(0,1) then, unconditionally, $X \sim \text{Zipf}(0, 1, 1, 1)$). This result can be considered to be a consequence of the possibility of representing a Pareto (IV) distribution as a gamma mixture of Weibull distributions.

3.11.3 Simon Distributions

Consideration of distributions which are mixtures of geometric distributions leads naturally to the study of the Simon distributions. Suppose that, given p, X has a geometric distribution and that p has a beta distribution with parameters α and β. It follows that, for $k = 0, 1, 2, \ldots$,

$$\begin{aligned} P(X = k) &= \left[\int_0^1 p(1-p)^k p^{\alpha-1}(1-p)^{\beta-1} dp\right] \Big/ B(\alpha, \beta) \\ &= B(\alpha+1, \beta+k)/B(\alpha, \beta). \end{aligned} \tag{3.11.10}$$

This distribution when translated to have support $1,2,3,\ldots$ is sometimes called the Waring distribution. If X has discrete density given by (3.11.10), the ratio of successive terms in its discrete density has the simple bilinear form for $k=1,2,\ldots$

$$P(X=k)/P(X=k-1) = (\beta-1+k)/(\alpha+\beta+k). \tag{3.11.11}$$

As described in Section 2.6, it is natural to extend (3.11.11) to a more general bilinear form. A random variable will be said to have a Simon distribution (with support $0,1,2,\ldots$), if for $k=1,2,\ldots$

$$P(X=k)/P(X=k-1) = (k+\alpha)/(\gamma k+\beta), \tag{3.11.12}$$

where $\alpha > -1$, $\beta > -\gamma$ and $\gamma \geq 1$ (if $\gamma = 1$, we must assume $\beta \geq \alpha+1$). Care should be taken when comparing the present section with Section 2.6. In that earlier section the distributions were assumed to have support $1,2,\ldots$. In the present section, to facilitate comparison with the Zipf distribution, the support has been taken to be $0,1,2,\ldots$. It is a simple matter to consider translated versions of the Simon distributions, as defined by (3.11.12). If in (3.11.12) we set $\alpha=0$, $\gamma=1$ and $\beta=\rho+1$, we get the Yule distribution with parameter ρ. Further specialization by setting $\rho=1$ leads to the standard Zipf distribution (3.11.8). Thus, (3.11.12) can qualify as a family of generalized Zipf distributions. It should be observed that Zipf tail behavior (i.e., $P(X=k)$ asymptotically behaving like a negative power of k) will be encountered in (3.11.12) only if $\gamma=1$ (and $\beta>\alpha$). The moments of the Simon distribution do not assume a simple form, except in the special subclass of Waring distributions (which includes the Yule distributions). In the Waring case we utilize the fact that the jth factorial moment of a geometric (p) random variable is given by $j!(1-p)^j p^{-j}$. Recalling that $p \sim B(\alpha,\beta)$, it follows that, for the Waring distribution (3.11.10), we have

$$\text{(Waring)} \quad E[X(X-1)\ldots(X-j+1)] = j!B(\alpha-j,\beta+j)/B(\alpha,\beta), \quad j<\alpha. \tag{3.11.13}$$

The special case of the Yule distribution is obtained by setting $\beta=1$ in (3.11.10). Its factorial moments are, thus, given by

$$\text{(Yule)} \quad E[X(X-1)\ldots(X-j+1)] = \left[\prod_{i=1}^{j}(\rho-i)\right]^{-1}, \quad j<\alpha. \tag{3.11.14}$$

Johnson, Kemp and Kotz (2005, p. 288) gave expressions for the ascending factorial moments of the Yule distribution with support $1,2,\ldots$. Utilizing the representation of the Waring distribution as a beta mixture of geometric distributions, we can readily obtain a simple form for its survival function: for $k=0,1,2,\ldots,$

$$\begin{aligned}\text{(Waring)} \quad P(X\geq k) &= \left[\int_0^1 (1-p)^k p^{\alpha-1}(1-p)^{\beta-1}dp\right] \Big/ B(\alpha,\beta) \\ &= B(\alpha,\beta+k)/B(\alpha,\beta).\end{aligned} \tag{3.11.15}$$

A particularly simple expression occurs when $\alpha = 1$, i.e., $P(X \geq k) = (1 + \frac{k}{\beta})^{-1}$, which corresponds to a Zipf(II)$(0, \beta, 1)$ distribution (cf. Johnson, Kemp and Kotz, 2005, p. 289–291). Further distributional results for the Simon distribution are to be found in Section 5.8.

Distributions for which the ratio of successive terms in the discrete density is biquadratic rather than bilinear have been considered in the literature. Johnson, Kemp and Kotz (2005, p. 256–7) provided some references in this context. The best known such distributions were introduced by Irwin (1968, 1975a,b,c) and called generalized Waring distributions. In such a distribution the ratio of successive terms in the discrete density is given by: for $k = 1, 2, \ldots$,

$$P(X = k)/P(X = k-1) = [(\beta + k - 1)(\gamma + k - 1)]/[k(\alpha + \beta + \gamma + k - 1)]. \quad (3.11.16)$$

The usual Waring distribution (3.8.11) corresponds to the choice $\gamma = 1$.

3.11.4 *Characterizations*

The apparent dearth of characterization results related to Zipf distributions is striking. The frequent appearance of Zipf-like distributions in modeling situations would seem to cry out for a battery of characterization results to provide the modeler with heuristic checks on the plausibility of his assumptions. We will report in this section some of the available characterizations of this type.

Theorem 3.11.1. *Let U be a non-negative integer-valued random variable, and suppose that U satisfies*

$$P(U = k)/P(U = k-1) = P(U \geq k+1)/P(U \geq k-1) \quad (3.11.17)$$

for every $k = 1, 2, 3, \ldots$, then $U \sim$ Zipf(II)$(0, \beta, 1)$ for some $\beta > 0$.

Proof. Let $q_k = P(U \geq k)$. From (3.11.17) we find $q_{k+1}^{-1} - q_k^{-1} = q_k^{-1} - q_{k-1}^{-1}$ from which it follows that $q_k = (1 + \frac{k}{\beta})^{-1}$. □

Theorem 3.11.2. *Let U be a non-negative integer-valued random variable, and suppose that U satisfies*

$$P(U = k \mid U \geq k) = \alpha P(U \geq k+1) \quad (3.11.18)$$

for every $k = 0, 1, 2, \ldots$ and some $\alpha > 0$, then $U \sim$ Zipf(II)$(0, \alpha^{-1}, 1)$.

Proof. Equation (3.11.18) implies $q_{k+1}^{-1} - q_k^{-1} = \alpha$ for $k = 0, 1, 2, \ldots$ (where $q_k = P(U \geq k)$). It is, then, clear that $q_k = (1 + \alpha k)^{-1}$. □

Neither of these characterization results is appealing. Both of the hypotheses (3.11.17) and (3.11.18) appear somewhat contrived. Equation (3.11.18) can be recognized as the condition under which the failure rate function and the survival function are proportional.

In Xekalaki and Dimaki (2005), the Waring distribution is characterized among positive integer valued distributions by the fact that it has a linear mean residual life function, i.e., that

$$E(X-k|X>k)=a+bk,\ \ k=0,1,2,...,$$

where $a>1$ and $b>0$. They also discuss parallel characterizations involving the hazard function, $h_X(k)=P(X=k|X\geq k),\ \ k=0,1,2,3,...$, and the vitality function $E(X|X>k)$. In addition they consider the case in which the hazard function and the mean remaining life function are proportional. Thus, if $h_X(k)=cE(X-k|X>k)$ for $k=0,1,2,...$, then X must have a Waring distribution.

3.12 Remarks

(i) Pederzoli and Rathie (1980) supply a general expression for the distribution function of a product of independent classical Pareto variables. The corresponding α_i's may or may not be distinct. Equations (3.3.29) and (3.3.30) are obtainable as special cases. They also consider ratios of independent Pareto (I) variables. The resulting density is one which arises in the under-reported income context (cf. (5.8.21)). It corresponds to a random variable whose logarithm has an asymmetrical Laplace distribution. Coelho and Mexia (2007) provide representations of the densities of products and ratios of independent Pareto (IV) variables with $\mu=0$. They refer to these Pareto (IV) random variables as generalized gamma-ratio variables. The representations are useful for accurate numerical computation.

(ii) Malik (1980) gives an exact (though complicated) expression for the distribution function of the rth quasi-range $(X_{n-r:n}-X_{r+1:n})$ of the logistic distribution. A straightforward transformation will then yield the distribution for $X_{n-r:n}/X_{r+1:n}$ based on $P(\text{III})(0,\sigma,\gamma)$ samples.

(iii) Talwalker (1980) refers to the truncation scaling property, (3.9.32), as a dullness property. In that paper it is shown that if an absolutely continuous random variable X has a concave distribution function which has the truncation-scaling property at one level $x_0>1$, then X is necessarily a classical Pareto random variable.

(iv) It is possible to consider the parameter space of a Pareto model as a statistical manifold (Amari, 1985), which can be analyzed using concepts from differential geometry. Two references in which this approach is discussed for the Pareto (I) model are Abdell-All, Mahmoud and Abd-Ellah (2003) and Peng, Sun and Jiu (2007).

(v) Xekalaki and Panaretos (1988) describe a distributional scenario which links the Yule distribution, the Poisson distribution and the Pareto (II) distribution. In it, it is assumed that for each $z\in(0,\infty)$ the conditional distribution of Y given $Z=z$ is a Poisson distribution with mean z. Then if, unconditionally, Y has a Yule distribution, then necessarily Z has a Pareto (II) distribution.

Chapter 4

Measures of inequality

4.1 Apologia for Prolixity

Early writers on inequality rarely distinguish clearly between sample and population statistics. A distribution for them might refer to some generic random variable or to the sample distribution of the finite number of observations actually available. We will try to be precise with regard to distinguishing between the "theoretical" distribution and the sample distribution. This will result in some repetition, since certainly the sample c.d.f. is a well defined distribution function. In many cases, one definition of a measure of inequality could serve in both settings. The sample measure of inequality arises when the sample c.d.f. is used, and the parent or theoretical measure of inequality arises when the evaluation is made with respect to the theoretical distribution. The increase in clarity is judged to be worth the price in redundancy. Actually, some early measures of inequality really only make sense when applied to sample data, and no amount of ingenuity seems to yield a plausible population version of the measure. In this class will be found measures which are slopes of certain fitted lines (sometimes fitted by eye!). When discussing population measures, we will speak of a single non-negative random variable X with distribution function $F(x)$ and survival function $\overline{F}(x)$. When we speak of sample measures of inequality, we will deal with n quantities $X_1, X_2, \ldots, X_n$ and their corresponding sample distribution function:

$$F_n(x) = \frac{1}{n}\sum_{i=1}^{n} I(x, X_i) \tag{4.1.1}$$

where

$$I(a,b) = \begin{matrix} 0 & \text{if} & a < b \\ 1 & \text{if} & a \geq b. \end{matrix}$$

The X_i's may be realizations of some random phenomena or may be just n numbers whose genesis is not specified but whose distribution is of interest with regard to its variability, etc. When we deal with the distribution of sample measures of inequality, it will be assumed that $X_1, X_2, \ldots, X_n$ are independent identically distributed, i.e., they constitute a random sample of size n from some (parent) distribution F. In such cases we are often interested in inferring properties of F based on properties of the sample. Thus forewarned about the possible excess of detail and apparent repetitiousness of the material, we will begin with a discussion of theoretical or population measures of inequality.

4.2 Common Measures of Inequality of Distributions

Complete equality would presumably be associated with a degenerate distribution, while departures in the direction of inequality would be evidenced by increasing spread or variability. Most of the measures of inequality to be discussed will be well defined only if the parent population has a finite mean. Some will require higher moments. We will adopt the convention that whenever a moment is written down it is assumed to be finite for the distribution under study. Generally, we will restrict attention to non-negative random variables. Occasionally, mention will be made when a particular concept makes sense and is of possible interest even when negative values are possible.

The first candidate measure of inequality is the mean deviation or absolute mean deviation:

$$\tau_1(X) = E(|X - E(X)|) = \int_0^\infty |x - m|\, dF(x) \tag{4.2.1}$$

where (and subsequently) $m = E(X)$. We will sometimes use the notation $\tau_1(F)$ for (4.2.1) or simply τ_1, if no confusion will result. Analogous notational liberties will be taken with the other measures of inequality to be introduced. The mean deviation is not scale invariant. In fact we have

$$\tau_1(cX) = c\tau_1(X). \tag{4.2.2}$$

It is translation invariant, i.e.,

$$\tau_1(X + c) = \tau_1(X). \tag{4.2.3}$$

The pros and cons regarding the desirability of scale invariance were sketched in Section 1.2. Similarly, arguments may be made for and against location invariance. Computational simplicity, as a consideration for inequality measures, has become less important with the advent of improved computational facilities, but it may sometimes still be justifiably invoked, since ease of interpretation may go hand in hand with ease of computation. The mean deviation is not as easy to evaluate as its simple form (4.2.1) might indicate. For example the expressions for τ_1 for the generalized Pareto distributions introduced in Chapter 3 involve incomplete Beta functions.

If one standardizes by dividing by the mean, one obtains the relative mean deviation

$$\tau_2(X) = \tau_1(X)/E(X) = \frac{1}{m}\int_0^\infty |x - m|\, dF(x). \tag{4.2.4}$$

It is scale but not translation invariant and is about equivalent to τ_1 in level of computational complexity. This measure except for a factor of $1/2$ is known as Pietra's index. We will discuss it in relation to the Lorenz curve later.

Computational considerations and a proclivity to argue via least squares have, in many contexts, suggested use of the (absolute) standard deviation as a measure of variability.

$$\tau_3(X) = \sqrt{E[(X-m)^2]} = \sqrt{\int_0^\infty (x-m)^2 dF(x)}. \tag{4.2.5}$$

Analogous to τ_1, τ_3 exhibits translation but not scale invariance. Its close relative, the relative standard deviation or coefficient of variation defined by

$$\tau_4(X) = \tau_3(X)/E(X), \tag{4.2.6}$$

behaves much like τ_2, being scale but not translation invariant.

As mentioned in Chapter 1, the above measures which relate inequality or variability to the average difference between one observation and the population mean are almost ubiquitous in the scientific, engineering and experimental literature. In the domain of economics and certain other social sciences, the Italian measures of inequality are more prevalent. These measures can be conveniently ascribed to Gini and his co-workers (though earlier references exist, see e.g., David, 1968, for a historical discussion). Basically they measure variability by considering the average difference between two independent observations from the distribution.

Gini's mean difference is defined by

$$\tau_5(X) = E|X_1 - X_2| = E(X_{2:2}) - E(X_{1:2}) \tag{4.2.7}$$

where X_1, X_2 are independent identically distributed random variables whose common distribution is that of X, and $X_{1:2}$, $X_{2:2}$ are the corresponding order statistics. The measure τ_5 has invariance properties in common with τ_1 and τ_3, while its standardized version shares invariance properties with τ_2 and τ_4. The standardization is customarily performed by dividing by twice the mean. The resulting measure is the Gini index or ratio of concentration:

$$G(X) = \tau_6(X) = \tau_5(X)/2E(X). \tag{4.2.8}$$

The Gini index is susceptible to a variety of possible interpretations. We will see later that it can be related to the area between the population Lorenz curve and the egalitarian line (as mentioned in Chapter 1). Some alternative expressions for G are possible and sometimes useful. Thus one may write

$$G(X) = 1 - [E(X_{1:2})/E(X_{1:1})] = [E(X_{2:2})/E(X_{1:1})] - 1, \tag{4.2.9}$$

which expresses G in terms of expectations of minimas (or maximas) of samples of sizes one and two. Equation (4.2.9) is to be found in Arnold and Laguna (1977). A more convenient reference is Dorfman (1979). Since it is obviously true that

$$E(X_{1:2}) \leq E(X_{1:1}) \leq E(X_{2:2}),$$

equation (4.2.9) provides a quick proof of the fact that the Gini index is always a number between 0 and 1.

Ord, Patil and Taillie (1978) related G to the covariance between X and $F(X)$. We have, using the second expression in (4.2.9):

$$\begin{aligned} G(X) &= \left[\int_0^\infty x dF^2(x)/m\right] - 1 = 2m^{-1}\int_0^\infty xF(x)dF(x) - 1 \\ &= 2m^{-1}\text{cov}(X, F(X)), \end{aligned} \tag{4.2.10}$$

since $E(F(X)) = 1/2$. A related result is obtainable in which the Gini index is related to the covariance between the value of a variable and its rank. To this end let $X_1, X_2, \ldots, X_n$ be a sample of size n from F, and let R_1 be the rank of X_1 in the sample. If one writes

$$R_1 = 1 + \sum_{i=2}^{n} J_i$$

where $J_1 = 1$ if $X_i < X_1$ and $J_i = 0$ otherwise, then we readily verify that for a continuous parent distribution F,

$$\text{cov}(X_1, R_1) = \left(\frac{n-1}{2}\right) mG(x) = \left(\frac{n-1}{4}\right) \tau_5(x). \tag{4.2.11}$$

See Kendall and Gibbons (1990, Chapter 9) where related results are credited to Stuart.

A convenient computational form for G exists when the distribution function is available in closed form:

$$G(X) = \int_0^\infty F(x)(1-F(x))\,dx \bigg/ \int_0^\infty (1-F(x))\,dx. \tag{4.2.12}$$

This is closely related to the form utilized by Wold (1935) in preparing one of the earliest lists of Gini indices for well-known distributions. Wold's formula was

$$G(X) = \int_0^\infty \int_0^t F(u)\,du\,dF(t)/m. \tag{4.2.13}$$

Variants of the standardized measures τ_2, τ_4 and τ_6 involve replacement of m, the mean of F, by $\tilde{m}$, the median of F. These modified measures have not received much attention in the literature. This is surprising, since for typically heavy-tailed distributions (such as income distributions) excessive dependence on the mean would seem risky.

Systematic statistics are sometimes proposed as measures of inequality. A popular instance is the inter-quartile range, sometimes standardized by dividing by the median. Alternatively, Bowley suggested the interquartile range divided by the sum of the quartiles as a suitable measure.

A general class of inequality measures are those of the form

$$\tilde{\tau}_g(X) = E[g(X/E(X))] \tag{4.2.14}$$

where g is a continuous convex function on $(0, \infty)$. Such measures were discussed in

Ord, Patil and Taillie (1978). They are scale invariant by construction. That they are monotone with respect to the Lorenz order (refer to Definition 4.4.5) is an argument in favor of their use. (A subfamily of (4.2.14) corresponds to the choice

$$g(x) = (x^{\gamma+1} - 1)/\gamma(\gamma+1) \tag{4.2.15}$$

where $\gamma \in (-\infty, \infty)$). Further specialization in (4.2.15) leads to certain frequently suggested inequality measures or indices. Thus $\gamma = -2, -1, 0$ and 1 can be associated with respectively: the ratio of the arithmetic to the harmonic mean, the ratio of the arithmetic to the geometric mean, the Theil index and the Herfindahl index (the squared coefficient of variation). Shorrocks (1980) argues in favor of inequality measures based on (4.2.15) on the grounds that they satisfy an additive decomposability criterion.

In addition to the large family of measures encompassed by (4.2.14), Ord, Patil and Taillie (1981b) reminded us that the whole available panoply of entropy measures can serve as well as measures of inequality. These are typically expressed in terms of the density of the distribution. A family of such measures which have been frequently used are those of the form

$$Entropy_\gamma(X) = \gamma^{-1} \int_0^\infty f_X(x)[1 - f_X^\gamma(x)]\, dx, \quad -1 < \gamma < \infty. \tag{4.2.16}$$

Included in this group is the Shannon measure of entropy corresponding to the choice $\gamma = 0$ (obtained from (4.2.16) by taking the limit as $\gamma \to 0$). Measures closely related to those included in (4.2.14) are obtainable by considering $E[g(X)]$, $E[g(X)]/g[E(X)]$ or $E[g(X)]/E(X)$ for some convex function g. Several economic measures based on additive utility functions are subsumed by this class. This issue will be addressed in Section 4.4 of the present chapter.

4.2.1 The Lorenz Curve

There is no doubt that, at least in the economics literature, the center of the stage in discussions of inequality is reserved for measures related directly or indirectly to the Lorenz curve. The generally accepted form of the Lorenz curve was first described for finite populations. It is a function $L(u)$ defined on $[0,1]$ such that, for fixed u, $L(u)$ represents the proportion of the total income in the population accounted for by the $100u\%$ poorest individuals in the population. Actually, as noted in Chapter 1, Lorenz's original curve was a reflection of the above described curve about the 45° line. The most convenient mathematical description of the Lorenz curve, appropriate for any parent distribution function F concentrated on $[0, \infty)$ with finite mean, can be attributed to Gastwirth (1971). Although with different notation, this version of the Lorenz curve may also be found in Pietra (1915). For any distribution function F we define the corresponding "inverse distribution function" or "quantile function" by

$$F^{-1}(y) = \sup\{x : F(x) \le y\}, \quad 0 < y < 1. \tag{4.2.17}$$

If the distribution function F has a finite mean, i.e., if

$$m = \int_0^1 F^{-1}(y)\, dy < \infty, \tag{4.2.18}$$

we may define the Lorenz curve by

$$L(u) = \left[\int_0^u F^{-1}(y)\,dy\right] \Big/ \left[\int_0^1 F^{-1}(y)\,dy\right], \quad 0 \le u \le 1. \tag{4.2.19}$$

Since the distribution is concentrated on $[0,\infty)$, it is clear that $L(u)$ is itself a continuous distribution function with $L(0) = 0$ and $L(1) = 1$. Typically a Lorenz curve is a "bow shaped" curve below the 45° line corresponding to $L(u) = u$. To use Lorenz's original colorful language, as the bow is bent, so concentration (i.e., inequality) increases. If one relaxes the assumption that the support of F is $[0,\infty)$, the corresponding Lorenz curve remains well defined. However, interpretative difficulties are encountered, since it will sag outside of the unit square. Wold (1935) provides an early example of this phenomenon (his Figure 2). As an example of a Lorenz curve, consider that of the exponential distribution $\overline{F}(x) = e^{-\lambda x}$, $x > 0$, where $\lambda > 0$. The corresponding mean is λ^{-1}, and we find

$$F^{-1}(y) = -\lambda^{-1}\log(1-y) \tag{4.2.20}$$

so that, using (4.2.19),

$$L(u) = u + (1-u)\log(1-u), \quad 0 \le u \le 1. \tag{4.2.21}$$

If F is a Pareto (IV) distribution (equation (3.2.8)) with parameters $(\mu,\sigma,\gamma,\alpha)$ then its inverse distribution is of the form

$$F^{-1}(y) = \mu + \sigma[(1-y)^{-1/\alpha} - 1]^{\gamma}.$$

It follows from (4.2.19), making a substitution of the form $z = (1-y)^{1/\alpha}$, that the corresponding Lorenz curve is given by

$$L(u) = \frac{\mu u + \sigma\alpha[B(\alpha-\gamma,\gamma+1) - I_{[(1-u)^{1/\alpha}]}(\alpha-\gamma,\gamma+1)]}{\mu + \sigma\alpha B(\alpha-\gamma,\gamma+1)} \tag{4.2.22}$$

where B indicates the Beta function and I_q the incomplete Beta function. In (4.2.22) it has been assumed that $\alpha > \gamma$ in order to have the requisite integrals converge (i.e., in order that the distributions have a finite mean). The Lorenz curves for other Pareto families introduced in Chapter 3 can be obtained by specialization in (4.2.22). In the classical Pareto case, i.e., $P(I)(\sigma,\alpha)$, we find the relatively simple expression

$$L(u) = 1 - (1-u)^{(\alpha-1)/\alpha}, \tag{4.2.23}$$

provided $\alpha > 1$. The expressions for the Pareto (II) and the Pareto (III) families simplify considerably when we assume $\mu = 0$ (i.e., support $[0,\infty)$), but still involve the incomplete Beta function.

Let us turn now to the general characteristics of Lorenz curves. Suppose that L is the Lorenz curve which corresponds to some fixed but arbitrary distribution function F. With what properties is L endowed? By construction we have seen that $L(0) = 0$ and $L(1) = 1$. Clearly, from (4.2.19), L is a non-decreasing function which is differentiable a.e. on $(0,1)$. More can be said. Since the inverse distribution function (which is a constant times the derivative of L, a.e.) is non-decreasing, it follows that the Lorenz curve is necessarily convex (remember Lorenz's "bow"). If the distribution function F is degenerate at some point, then it follows that the Lorenz curve is the 45° line $L(p) = p$. This is the case of complete equality. In general, we can only claim that $L(p) \leq p$, $0 \leq p \leq 1$. We can claim this by convexity since $L(0) = 0$ and $L(1) = 1$. As Thompson (1976) observed, if for some $p' \in (0,1)$ we have $L(p') = p'$, then the parent distribution is necessarily degenerate. This follows since in order to have $L(p') = p'$ we must have $L'(p) = 1$ for every $p \in (0, p')$. Additionally, in order to preserve convexity, the derivative must remain 1 on the interval $(p', 1)$. In the case where F corresponds to a discrete distribution with a finite number of possible values, its Lorenz curve assumes the form of a polygonal line.

Suppose L is an arbitrary convex non-decreasing function on $[0,1]$ with $L(0) = 0$ and $L(1) = 1$. Does there exist a distribution function on $[0,,\infty)$ with L as its Lorenz curve? An affirmative answer is appropriate since we may take F^{-1} to be determined a.e. by the derivative of L. If X has Lorenz curve L, what can we say about the Lorenz curve of cX? Since $F_{cX}^{-1}(y) = cF^{-1}(y)$, the Lorenz curves of X and cX coincide. Conversely, if two distributions have identical Lorenz curves, then they differ at most by a scale transformation. This is clear since, by differentiation, the Lorenz curve determines the inverse distribution function up to (but only up to) a scale factor. Further general characteristics of Lorenz curves are discussed in Pakes (1981).

4.2.2 *Inequality Measures Derived from the Lorenz Curve*

There are two well-known measures of inequality which are intimately related to the Lorenz curve. Both the Pietra and the Gini index have long histories. Both can be described as attempts to quantify the degree of "bowedness" exhibited by the Lorenz curve. That is, to what extent the Lorenz curve sags below the 45° line of complete equality. One candidate measure is the area between the egalitarian line and the Lorenz curve. This quantity multiplied by 2 (to give it a possible range of $[0,1]$) is the classical Gini index, i.e.,

$$G = 2\int_0^1 [u - L(u)]\,du. \tag{4.2.24}$$

This should be compared with the alternative expressions for G given in (4.2.8) through (4.2.13). To confirm that (4.2.24) is consistent with the earlier definitions, the following argument may be used.

$$
\begin{aligned}
2\int_0^1 [u-L(u)]\,du &= 1-2\int_0^1 L(u)\,du \\
&= 1-2m^{-1}\int_0^1\int_0^u F^{-1}(y)\,dy\,du \quad \text{(from 4.2.19)} \\
&= 1-2m^{-1}\int_0^1\left[\int_y^1 du\right]F^{-1}(y)\,dy \quad \text{(change order of integration)} \\
&= 1-m^{-1}\int_0^1 F^{-1}(y)2(1-y)\,dy \\
&= 1-m^{-1}E(X_{1:2})
\end{aligned}
$$

where $X_{1:2}$ is the smallest of a sample of size 2 from the distribution F. It is thus clear that (4.2.24) is equivalent to the earlier (4.2.9). The geometric interpretation of G, provided by the Lorenz curve, has undoubtedly accounted for its ubiquity. The earlier expressions (4.2.8)–(4.2.13) are usually considered to be computational devices, and not too much stock is put into their possible interpretations. While the Lorenz curve represents little compression of the information in the distribution function, the Gini index represents a considerable condensation. Markedly different distributions may share the same Gini index. Intermediate levels of compression are attainable, if one considers higher order Gini indices. These are most conveniently defined in a manner analogous to equation (4.2.9). Thus we may define G_n the nth order Gini index to be

$$G_n = 1-[E(X_{1:n+1})/E(X_{1:n})]. \tag{4.2.25}$$

The usual Gini index corresponds to the case $n=1$. The set of all such Gini indices $\{G_n : n=1,2,\ldots\}$ determines the parent distribution up to a scale factor.

Rather than use nth order Gini indices of the form (4.2.25), several authors have proposed alternative sequences of inequality indices.

Kakwani (1980a) suggested consideration of what he called generalized Gini indices defined by

$$\widetilde{G}_n = 1-[E(X_{1:n})/E(X)], \quad n=2,3,\ldots. \tag{4.2.26}$$

Clearly this sequence, like the sequence of nth order Gini indices (4.2.25), determines the parent distribution up to a scale factor. Muliere and Scarsini (1989) relate these indices to higher order stochastic dominance orderings of distributions.

Yitzhaki (1983) considered the following indexed family of inequality indices

$$G(\nu) = 1-\nu(\nu-1)\int_0^1 (1-u)^{\nu-2}L(u)du, \quad \nu\in(2,\infty). \tag{4.2.27}$$

The corresponding sequence $G(n)$, $n=2,3,\ldots$ will determine the Lorenz curve and thus the parent distribution up to a scale factor. These inequality indices, the $G(n)$'s, were called the extended family of Gini indices or S-Gini's when they were introduced by Donaldson and Weymark (1980,1983).

Klefsjö (1984) suggested the following sequence of indices

$$K_n = \int_0^1 (1-u)^n [u - L(u)] du, \quad n = 1.2, \ldots, \tag{4.2.28}$$

while Jain and Arora (2000) propose the use of

$$\widetilde{\widetilde{G}}_n = \int_0^1 u^n L'(u) du, \quad n = 1, 2, \ldots, \tag{4.2.29}$$

the sequence of moments of $L(u)$ viewed as a distribution function on the interval $(0, 1)$.

Aaberge (2001) considers the sequence

$$A_n = 1 - (n+1) \int_0^1 [L(u)]^n du, \quad n = 1, 2, \ldots, \tag{4.2.30}$$

which essentially corresponds to the moment sequence of the inverse Lorenz curve, viewed as a distribution function on $(0, 1)$.

All of these sequences of inequality indices respect the Lorenz order, to be introduced in Section 4.4.2 below, and in all cases, the full sequence suffices to essentially determine the parent distribution.

The majority of these indices, $G(n), K_n$, and $\widetilde{\widetilde{G}}_n$, are related to the moments of $L(u)$. The Aaberge indices A_n are linked to moments of $L^{-1}(v)$ viewed as a distribution function on $(0, 1)$. Two additional sequences of indices might be defined, We could consider the sequence, $E(X_{1:n})$, of expected minima corresponding to the distribution function $L(u)$, or the analogous sequence for $L^{-1}(v)$. Because of the simple form of the classical Pareto Lorenz curve, there is no difficulty in computing the values of all of these indices in the Pareto (I) case. For other distributions, the computation will be considerably more challenging.

To illustrate the computation of the Gini index (and higher order Gini indices, the G_n's), we turn to the exponential and Pareto families. Integration of the Lorenz curve is inadvisable in such cases as these, where expectations of sample minima are readily available. Thus (4.2.9) (and (4.2.25)) is the computational form which proves most convenient. Using it we get for the exponential case (in which $E(X_{1:n}) = (n\mu)^{-1}$)

$$\text{(exponential)} \quad G = 1/2 \tag{4.2.31}$$

and more generally

$$\text{(exponential)} \quad G_n = 1/(n+1). \tag{4.2.32}$$

While, referring to (3.4.19), we have for the P(IV)$(\mu, \sigma, \gamma, \alpha)$ distribution:

$$\text{P(IV)}(\mu, \sigma, \gamma, \alpha): \quad G = 1 - \frac{\mu + 2\sigma\alpha B(2\alpha - \gamma, \gamma + 1)}{\mu + \sigma\alpha B(\alpha - \gamma, \gamma + 1)}, \tag{4.2.33}$$

provided that $\alpha > \gamma$.

This expression when $\mu = 0$ and $\sigma = 1$ was also given in Fisk (1961a), Cronin (1979) and McDonald and Ransom (1979). Special subcases of (4.2.33) are

$$\mathrm{P(I)}(\sigma,\alpha): \quad G = (2\alpha - 1)^{-1}, \;\; \alpha > 1, \tag{4.2.34}$$

$$\mathrm{P(II)}(0,\sigma,\alpha): \quad G = \alpha/(2\alpha - 1), \;\; \alpha > 1, \tag{4.2.35}$$

and the remarkable result

$$\mathrm{P(III)}(0,\sigma,\gamma): \quad G = \gamma, \;\; 0 < \gamma < 1. \tag{4.2.36}$$

Equation (4.2.36) can be used to justify calling the parameter γ, in the Pareto (III) family, the Gini parameter or the inequality parameter. In (4.2.35) observe that $1/2 < G < 1$, so that all such distributions exhibit a rather high degree of inequality. The expression (4.2.34) for the Gini index of the classical Pareto distribution is, needless to say, of considerable antiquity. The corresponding expressions for the higher order Gini indices are

$$\mathrm{P(I)}(\sigma,\alpha): \quad G_n = [n(n+1)\alpha - n]^{-1}, \;\; \alpha > 1/n, \tag{4.2.37}$$

$$\mathrm{P(II)}(0,\sigma,\alpha): \quad G_n = \alpha/[(n+l)\alpha - 1], \;\; \alpha > 1/n, \tag{4.2.38}$$

$$\mathrm{P(III)}(0,\sigma,\gamma): \quad G_n = \gamma/n, \;\; \gamma < n. \tag{4.2.39}$$

It is often possible to utilize partial information regarding the distribution F to construct bounds on the corresponding Gini index. For example, from Gastwirth (1972) we have

Theorem 4.2.1. *Let F be a distribution function with support $[\underline{M},\overline{M}]$, $(0 < \underline{M} < \overline{M})$, and mean m, then the corresponding Gini index satisfies*

$$0 \le G \le (m - \underline{M})(\overline{M} - m)/m(\overline{M} - \underline{M}). \tag{4.2.40}$$

Proof. If we set $\mu_{i:n} = E(X_{i:n})$ in (4.2.9) we may write $G = (\mu_{2:2} - \mu_{1:2})/(\mu_{2:2} + \mu_{1:2})$ (observe that $\mu_{2:2} + \mu_{1:2} = 2\mu_{1:1}$). Since F has support $[\underline{M},\overline{M}]$, it follows that $\underline{M} \le \mu_{1:2} \le \mu_{2:2} \le \overline{M}$. It is obvious that $G \ge 0$. Let $a = \mu_{1:2} - \underline{M}$, $b = \mu_{2:2} - \mu_{1:2}$ and $c = \overline{M} - \mu_{2:2}$. To prove (4.2.40) we must show that

$$(\overline{M} - \underline{M})(\mu_{2:2} - \mu_{1:2}) \le (\mu_{1:2} + \mu_{2:2} - 2\underline{M})(2\overline{M} - \mu_{1:2} - \mu_{2:2})$$

which is equivalent to $2(a+b+c)b \le (2a+b)(b+2c)$, i.e., that

$$b^2 \le 4ac. \tag{4.2.41}$$

That this equation is always true is a consequence of the following result which may be found in Mallows (1973). □

Theorem 4.2.2. *For each $k = 1,2,\ldots$, let $X_{k:k}$ be the maximum of a sample of size k drawn from a distribution function F with support $[\underline{M},\overline{M}]$ and define $\mu_k = [\overline{M} - E(X_{k:k})]/[\overline{M} - \underline{M}]$. The sequence $\{\mu_k\}_{k=1}^{\infty}$ is the moment sequence of a distribution with support $[0,1]$.*

Proof. For each $k = 1,2,\dots$ we have, using integration by parts and letting $y = F(x)$,

$$\mu_k = \frac{\overline{M} - E(X_{k:k})}{\overline{M} - \underline{M}} = \frac{\overline{M} - \int_{\underline{M}}^{\overline{M}} x dF^k(x)}{\overline{M} - \underline{M}}$$

$$= \int_{\underline{M}}^{\overline{M}} F^k(x)\,dx/(\overline{M} - \underline{M})$$

$$= \int_0^1 y^k dF^{-1}(y)/(\overline{M} - \underline{M}).$$

It is evident that μ_k is the kth moment of the distribution function $F^{-1}(y)/\overline{M} - \underline{M})$ on $[0,1]$. □

To complete the proof of Theorem 4.2.1 we merely observe that with a, b and c as defined in that theorem and with μ_1, μ_2 as defined in Theorem 4.2.2, the statement $b^2 \leq 4ac$ (i.e., (4.2.41)) is equivalent to the evidently true statement $\mu_2 \geq (\mu_1)^2$. Gastwirth (1972) proved Theorem 4.2.1 by a different technique using an inequality in Hardy, Littlewood and Polya (1959). Ord (personal communication) has suggested a further alternative proof of Theorem 4.2.1. It hinges on the observation that the upper bound for G is associated with a two-point distribution with mean m, i.e., $P(X = \underline{M}) = 1 - P(X = \overline{M}) = (\overline{M} - m)/\overline{M} - \underline{M})$, the case in which everyone is either rich ($\overline{M}$) or poor ($\underline{M}$).

If the distribution F has support $[\underline{M}, \infty)$, a bound on the Gini index is still obtainable. One can conjecture the nature of the bound by letting $\overline{M}$ get arbitrarily large in (4.2.40). The conjecture is trivially verified in

Theorem 4.2.3. *Let F be a distribution function with support $[\underline{M}, \infty)$ for some $\underline{M} > 0$ and mean m, then the corresponding Gini index satisfies*

$$0 \leq G \leq (m - \underline{M})/m. \tag{4.2.42}$$

Proof. $0 \leq G = 1 - (\mu_{1:2}/\mu_{1:1}) \leq 1 - (\underline{M}/m)$. □

As another example in which partial information about the distribution F yields a bound on the Gini index, consider the following in which, in addition to assuming bounded support, a bounded density is assumed. The key tool is the following result first proved by Klamkin and Newman (1976). See Arnold (1976) for the necessary modifications to make the Klamkin and Newman result valid in a more abstract setting than they proposed. The forms of the theorem and of the proof included here are very close to those in the original version.

Theorem 4.2.4. *Let F be an absolutely continuous distribution function with support $[\underline{M}, \overline{M}]$. Suppose that the corresponding density function satisfies $f(x) \leq C$ a.e. on $[\underline{M}, \overline{M}]$ for some finite C. Then for any two integers $j < k$ we have*

$$\left[C(j+1)\int_{\underline{M}}^{\overline{M}} F^j(x)\,dx\right]^{1/(j+1)} \geq \left[C(k+1)\int_{\underline{M}}^{\overline{M}} F^k(x)\,dx\right]^{1/(k+1)}. \tag{4.2.43}$$

Note that $\underline{M}$ could be $-\infty$, and $\overline{M}$ could be $+\infty$.

Proof. Observe that $fF^j \leq CF^j$ (a.e.), so that

$$F^{j+1}(x) \leq C(j+1) \int_{\underline{M}}^{x} F^j(y)\,dy.$$

It follows that for $r > 0$,

$$F^{r(j+1)+j}(x) \leq C^r (j+1)^r F^j(x) \left[\int_{\underline{M}}^{x} F^j(y)\,dy \right]^r$$

for every x in $[\underline{M}, \overline{M}]$. Consequently,

$$(r+1) \int_{\underline{M}}^{\overline{M}} F^{r(j+1)+j}(y)dy \leq C^r (j+1)^r \left[\int_{\underline{M}}^{\overline{M}} F^j(y)\,dy \right]^{r+1}.$$

The desired result follows upon setting $k = r(j+l) + j$ and rearranging. □

From the Klamkin and Newman result we readily obtain:

Theorem 4.2.5. *Let F be an absolutely continuous distribution function with support $[0,1]$ and mean m, whose density f is bounded a.e. by C. The corresponding Gini index satisfies*

$$G \geq 1 - \frac{2}{3}\sqrt{2\mathrm{Cm}}. \tag{4.2.44}$$

Proof. Let $\mu_{1:n}$ be the expected value of $X_{1:n}$, the minimum of a sample of size n from the distribution F. We can write $X_{1:n} = 1 - Y_{n:n}$ where $Y_{n:n}$ is the maximum of a sample of size n from the distribution $1 - F(1-y)$. It follows that

$$\begin{aligned} \mu_{1:n} &= 1 - E(Y_{n:n}) = 1 - \int_0^1 y d[1 - F(1-y)]^n \\ &= \int_0^1 [1 - F(1-y)]^n dy \end{aligned}$$

where the last step involves integration by parts. Since F is absolutely continuous with a derivative a.e. less than or equal to C, the same is true of $1 - F(1-y)$. Theorem 4.2.4 thus applies. When in that theorem, we choose $j = 1$, $k = 2$, we conclude that $[2C\mu_{1:1}]^{1/2} \geq [3C\mu_{1:2}]^{1/3}$. Since $G = 1 - (\mu_{1:2}/\mu_{1:1})$, the desired result follows. □

The inequality provided by (4.2.44) may be trivial. For example, with $F(x) = x^2$, $0 \leq x \leq 1$, (4.2.44) asserts that $G \geq 1 - \sqrt{32/27}$ which is obviously true since $G \geq 0$. An example in which (4.2.44) yields a non-trivial inequality is provided by $F(x) = x^{3/2}$, $0 \leq x \leq 1$. The inequality (4.2.44) is sharp. Equality obtains when F corresponds to the uniform distribution on $[0,1]$.

Cerone (2009a) provides a survey of bounds for the Gini index obtained by utilizing many well-known functional inequalities. He treats both the bounded and unbounded support cases. In a subsequent parallel paper, Cerone (2009b) provides bounds for the Gini index and for the Lorenz curve, this time using Young's inequality.

Pietra's index of inequality is obtained by considering the maximum vertical deviation between $L(u)$, the Lorenz curve, and the 45° line of complete equality. Assuming that F is strictly increasing on its support, the convex function $u - L(u)$ will be differentiable everywhere on $(0,1)$ and its maximum value will occur when its derivative $1 - [F^{-1}(u)/m]$ is zero, i.e. when $u = F(m)$. The corresponding value of $u - L(u)$, i.e., the maximum vertical deviation between u and $L(u)$, is given by (refer to (4.2.19))

$$F(m) - m^{-1}\left[\int_0^{F(m)} [m - F^{-1}(y)]\,dy\right] = m^{-1}\left[\int_0^{F(m)} [m - F^{-1}(y)]\,dy\right]$$
$$= m^{-1}\left[\int_0^{m} (m-z)\,dF(z)\right] = m^{-1}\int_0^{\infty} |z-m|\,dF(z)/2.$$

Thus Pietra's index is in fact one-half of the relative mean deviation, $\tau_2(X)$, as defined in (4.2.4). If the parent distribution is not strictly increasing, the Lorenz curve may have linear and/or polygonal sections. It can happen that $L(u)$ is not differentiable at the point of maximum deviation between u and $L(u)$. Another contingency arises if there is a segment of $L(u)$ which is parallel to the line of equality. In such cases the maximum deviation is attained at all points along the corresponding interval of values of u, including $u = F(m)$. Even in these anomalous cases the deviation between u and $L(u)$ is a maximum at $u = F(m)$, and the value of that maximum is $\tau_2(X)/2$ or Pietra's index. The index is also associated with the names Yntema (1933) and Schutz (1951).

Elteto and Frigyes (1968) proposed new measures of inequality obtained by relating the average incomes of individuals above and below average to the overall average income. With slight notational changes their development is as follows. Define, for a given distribution F on $[0,\infty)$,

$$m = \int_0^\infty x\,dF(x) \tag{4.2.45}$$

$$m_1 = \left[\int_0^m x\,dF(x)\right] \Big/ F(m) \tag{4.2.46}$$

$$m_2 = \left[\int_m^\infty x\,dF(x)\right] \Big/ (1 - F(m)). \tag{4.2.47}$$

The interpretations being: m is average income, m_1 is the average income of the poorer than average individuals, and m_2 is the average income of the richer than average individuals. Elteto and Frigyes proposed the three inequality measures

$$U = m/m_1, \quad V = m_2/m_1, \quad W = m_2/m. \tag{4.2.48}$$

Kondor (1971) observed that the Pietra index can be written as a simple function of U, V and W namely:

$$\tau_2 = 2(U-1)(W-1)/(V-1). \tag{4.2.49}$$

He claimed without explicit justification that τ_2 contains all the essentials of

(U,V,W). It is not clear that this is the case, since τ_2 represents a definite condensation, i.e., τ_2 is a function of (U,V,U) but not conversely. Put another way, the Elteto and Frigyes vector (U,V,W) gives us more information about the parent distribution than does τ_2. The Elteto and Frigyes vector is scale invariant and does admit an interpretation in terms of the Lorenz curve. Since $V = UW$ we need only interpret U and W. The quantity U represents the reciprocal of the slope of the line joining $(0,0)$ to $(F(m), L(F(m)))$ on the Lorenz curve. The other index W gives the slope of the line joining $(F(m), L(F(m)))$ to $(1,1)$. Since $F(m)$ is the point at which $u - L(u)$ is maximized, we are actually dealing with the slopes of the sides of the maximal triangle which can be inscribed within the Lorenz curve. Pietra originally described his index $(\tau_2/2)$ as two times the area of such a maximal triangle. The relation between the Elteto-Frigyes index and the Pietra index now becomes clearer. One side of the maximal triangle is, of course, the line of equality, $(0,0) - (1,1)$. Elteto and Frigyes tell us essentially the two angles of this triangle, and thus determine it completely. Pietra only tells us its area. The possibility of constructing analogous indices based on the area of the largest $(k+2)$-sided polygon which can be inscribed in the area between the Lorenz curve and the line of equality does not seem to have been exploited. Such indices can be described as follows.[1] We wish to select $0 = u_0 < u_1 \ldots < u_k < u_{k+1} = 1$ such that the $(k+2)$-sided polygon with corners $\{u_i, L(u_i)\}_{i=0}^{k+1}$ inscribed within the Lorenz curve has maximal area. It follows that the u_i's must satisfy

$$\int_{u_{i-1}}^{u_{i+1}} [F^{-1}(y) - F^{-1}(u_i)]\, dy = 0, \quad i = 1,2,\ldots k. \tag{4.2.50}$$

Alternatively if we define $v_i = F^{-1}(u_i)$, we will seek v_i's which satisfy

$$E(X|X \in (v_{i-1}, v_{i+1}]) = v_i, \quad i = 1,2,\ldots,k. \tag{4.2.51}$$

The Pietra index of order k, P_k, can then be defined as two times the area of the resulting $(k+2)$-sided polygon. In the uniform case (where $F(x) = x$, $0 \le x \le 1$), which may indeed be the only case in which analytic expressions for P_k are obtainable, one finds

$$v_i = i/(k+1) \tag{4.2.52}$$

and consequently

$$P_k = k(k+2)/3(k+1)^2. \tag{4.2.53}$$

It is interesting to speculate to what degree the sequence $\{P_k\}_{k=1}^{\infty}$ determines the underlying distribution (via constraints on its Lorenz curve). It does not even determine F up to a scale transformation. For example, two Lorenz curves which are mutually symmetric in the sense described by Taguchi (1968) will have the same sequence of Pietra indices. In this context L_1 and L_2 are said to be mutually symmetric, if one

[1] The material on Pietra indices of higher order is based on discussions with C. A. Robertson.

can be obtained from the other by a reflection about the line $u+L(u)=1$. That is to say, if

$$L_1(u) = 1 - L_2^{-1}(1-u), \quad 0 \leq u \leq 1. \tag{4.2.54}$$

If the corresponding distribution functions have densities f_1, f_2 with means m_1, m_2, then (4.2.34) is equivalent to (Taguchi, 1968)

$$m_2 f_2(m_2/y) = y^3 m_1 f_1(m_1 y), \quad \forall y. \tag{4.2.55}$$

Self-symmetry occurs if a Lorenz curve is symmetric about the line $u+L(u)=1$. Necessary and sufficient conditions for self-symmetry are obtainable by erasing the subscripts in (4.2.54) and (4.2.55). Candel Ato and Bernadic (2011) provide a characterization of distributions with self-symmetric Lorenz curves based on the doubly truncated mean function

$$m_X(x,y) = E(X|x \leq X \leq y). \tag{4.2.56}$$

The Lorenz curve corresponding to a random variable X will be self-symmetric if $m_X(x, m^2/x) = m$ for every x such that x and m^2/x are in the support of X.

Taguchi also considers the following issue related to symmetry. Under what conditions will the point $(F(m), L(F(m))$ lie on the line $u+L(u)=1$? The point $(F(m), L(F(m))$ is, of course, the vertex of Pietra's maximal inscribed triangle. We know that $F(m) - L(F(m))$ is one-half the relative mean deviation, i.e., $\tau_2(X)/2$. In order for this point to fall on the line $u+L(u)=1$, we clearly require that

$$F(m) = \frac{1}{2} + \frac{\tau_2(X)}{4}. \tag{4.2.57}$$

We can illustrate these considerations regarding the Pietra index and symmetry of Lorenz curves by considering the family of classical Pareto Lorenz curves of the form given in (4.2.23)

$$L(u) = 1 - (1-u)^{(\alpha-1)/\alpha}, \quad 0 \leq u \leq 1.$$

Recall that for such a distribution $m = \alpha/(\alpha-1)$ (the scale parameter has without loss of generality been set equal to 1 and α is assumed greater than 1 so that the mean exists). The maximal deviation between u and $L(u)$ occurs when

$$u = 1 - \left(\frac{\alpha}{\alpha-1}\right)^{-\alpha} = F(m). \tag{4.2.58}$$

The corresponding value of the Pietra index is then

$$F(m) - L(F(m)) = (\alpha-1)^{(\alpha-1)}/\alpha^{\alpha}. \tag{4.2.59}$$

Recall that the Gini index for a standard Pareto distribution is $(2\alpha-1)^{-1}$. It is geometrically evident that the Pietra index is necessarily smaller than the Gini index (since the area of a triangle inscribed between the Lorenz curve and the egalitarian line clearly cannot exceed the area between the two curves). It is not algebraically transparent that the expression (4.2.59) is always less than $(2\alpha-1)^{-1}$. Turning to the Elteto and Frigyes indices, we must compute m, m_1 and m_2 (as defined in (4.2.45)–(4.2.47)). For the classical Pareto (with $\sigma=1$) these are of the form

$$\begin{aligned} m &= \alpha/(\alpha-1) \\ m_1 &= \left(\frac{\alpha}{\alpha-1}\right)\left[1-\left(\frac{\alpha}{\alpha-1}\right)^{-\alpha+1}\right] \Big/ \left[1-\left(\frac{\alpha}{\alpha-1}\right)^{-\alpha}\right] \\ m_2 &= [\alpha/(\alpha-1)]^2. \end{aligned} \tag{4.2.60}$$

The resulting expressions for U and W are

$$\begin{aligned} U &= \left[1-\left(\frac{\alpha}{\alpha-1}\right)^{-\alpha}\right] \Big/ \left[1-\left(\frac{\alpha}{\alpha-1}\right)^{-\alpha+1}\right] \\ W &= \alpha/(\alpha-1). \end{aligned} \tag{4.2.61}$$

The Lorenz curve which is mutually symmetric with the classical Pareto Lorenz curve is of the form

$$L^*(u) = u^{\alpha/(\alpha-1)}, \quad 0 \le u \le 1, \tag{4.2.62}$$

which corresponds to the distribution function

$$F^*(x) = x^{\alpha-1}, \quad 0 \le x \le 1. \tag{4.2.63}$$

An example of a distribution with a self-symmetric Lorenz curve is provided by the Pareto-like density (cf., Taguchi, 1968)

$$f(x) = 2^{-1/2}x^{-3/2}, \quad 1/2 < x < 2. \tag{4.2.64}$$

This is most easily verified by using (4.2.55) without the subscripts. Taguchi, in addition, observed that the log-normal distribution has a self-symmetric Lorenz curve. This can be verified using (4.2.55). Self-symmetry of the Lorenz curve is a sufficient (though not necessary) condition for the point $(F(m),L(F(m))$ to fall on the line $u+L(u)=1$. Consequently, the density (4.2.64) provides an example satisfying the "symmetry" condition (4.2.57).

Hagerbaumer (1977) proposed the minor concentration ratio as a useful supplement to the Gini index. It is intended to reflect the relative position of the poor. It is only described geometrically, and no acceptable analytic interpretation is presumably available. Consider the Lorenz curve $L(u)$ denoted by the broken line in Figure 4.2.1. The polygonal lines $(0,0),(G,0),(1,1)$ and $(0,0),(0,1-G),(1,1)$ represent extremal Lorenz curves having the same Gini index G as does $L(u)$. The minor concentration ratio is the ratio of the area of the shaded region in the diagram to the area of the small triangle which encloses it (the area of the small triangle is

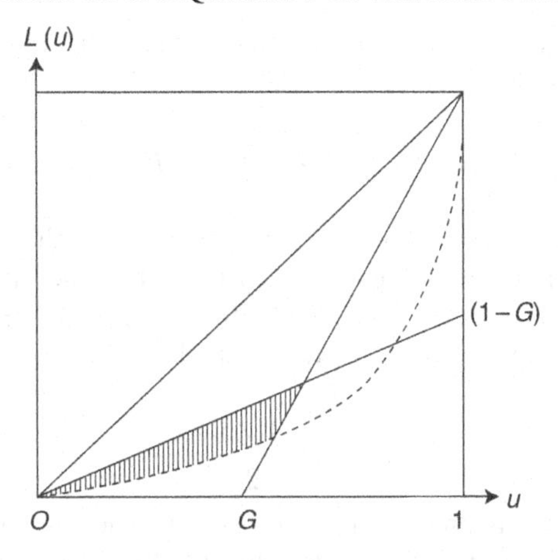

Figure 4.2.1 *Minor concentration ratio.*

$G(1-G)/[2(2-G)]$). The larger this ratio becomes, the smaller is the share of total income which devolves on the poor. An alternative approach to assessing poverty was proposed by Hamada and Takayama (1978). They assume a specific poverty line exists, and as an index of poverty, use the Gini index of the distribution censored at the poverty line.

There are other ways of assessing the deviation between the Lorenz curve and the equality line. Mehran (1976) proposed the general class of "linear measures" of the form

$$L_W = \int_0^1 [u - L(u)]\, dW(u), \tag{4.2.65}$$

where we can allow W to be an arbitrary distribution function. For example, the Klefsjö inequality indices, K_n in (4.2.28), are of this form.

The choice $W(u) = u$, $0 \leq u \leq 1$, clearly leads to the Gini index divided by 2. It is not possible to express the Pietra index in this form, unless we allow W to depend on F. The choice of W required is

$$W(u) = \begin{cases} 0 & \text{if } u > F(m) \\ 1 & \text{if } u \geq F(m). \end{cases}$$

There are computational advantages to expressing the Pietra index (illegitimately) in the form (4.2.65), even though it is not strictly speaking a linear measure. In the final analysis, it is remarkable that the use of linear indices (other than the Gini index) has been so much neglected. Suitable choices of the distribution W could emphasize income ranges of particular interest.

Amato (1968) and, independently, Kakwani (1980) suggested use of the length of the Lorenz curve as a measure of inequality. We will refer to this measure as the Amato index and, for a given non-negative random variable X, we will denote the index by $A(X)$. It is possible to represent the Amato index in the form (4.2.14) for a particular choice of the function $g(x)$. Thus

$$A(X) = E(\sqrt{1+(X/E(X))^2}) = E(g_A(X/E(X))), \tag{4.2.66}$$

where $g_A(x) = \sqrt{1+x^2}$. Perhaps the easiest way to verify this is to consider the case in which X is discrete with n equally likely possible values and then use a suitable limiting argument. For details see Arnold (2012). Since the length of a Lorenz curve will always be in the interval $(\sqrt{2}, 2)$, a standardized Amato index might be considered, defined by

$$\widetilde{A}(X) = \frac{A(X)-\sqrt{2}}{2-\sqrt{2}}.$$

It is often difficult to obtain analytic expressions for the Amato index corresponding to particular parametric families of densities for X. If X has finite moments of all orders then a series representation of the index is possible.

$$A(X) = \sum_{n=0}^{\infty} \binom{1/2}{n} \frac{E[X^{2n}]}{[E(X)]^{2n}}. \tag{4.2.67}$$

To justify that this series is convergent, observe that $0 \leq E(\sqrt{1+(X/E(X))^2}) \leq \sqrt{E(1+(X/E(X))^2)}$, and X^2 is assumed to be integrable.

As a specific example, suppose that $X \sim \Gamma(\alpha, \beta)$. In this case we have:

$$A(X) = \sum_{n=0}^{\infty} \binom{1/2}{n} \frac{\Gamma(\alpha+2n)}{\Gamma(\alpha)\alpha^{2n}}. \tag{4.2.68}$$

It is possible to use the value of the Lorenz curve at a particular point $u_0 \in (0,1)$ as a measure of inequality, thus putting all the emphasis at a specific income level. Such a procedure has been proposed by Alker and Russett (1966). Alternatively, one can focus on a particular value of the inverse Lorenz curve. The quantity $L^{-1}(1/2)$ has been discussed by Alker (1965) as a possible measure of inequality (called, in a political science setting, the minimal majority). Marshall, Olkin and Arnold (2011, pp. 561–566) provide an extensive list of inequality and diversity measures drawn from a variety of fields.

Certain income share ratios, intimately related to the Lorenz curve, are used as convenient summary measures of inequality. For $0 < \alpha, \beta < 0.5$, the $(1-\alpha, \beta)$ income share ratio is defined by

$$R(1-\alpha, \beta) = \frac{1-L(1-\alpha)}{L(\beta)}. \tag{4.2.69}$$

A popular choice for (α, β) is $(0.2, 0.2)$. Thus $R(0.2, 0.2)$ represents the ratio of the total income accruing to the richest 20% of the population to the total income accruing to the poorest 20%. Recently, Palma (2011) has advocated use of $R(0.9, 0.4)$ as a suitable inequality measure. An enthusiastic endorsement of this choice may be found in Cobham and Sumner (2014).

4.2.3 The Effect of Grouping

A problem of recurring interest is the effect of grouping on the Lorenz curve and Gini index. Suppose X has distribution function $F(x)$ (assumed continuous for convenience) and that a "grouped" version of X is defined by

$$\tilde{X} = b_i \quad \text{if } X \in [a_{i-1}, a_i) \tag{4.2.70}$$

where $0 = a_0 < a_1 < \ldots < a_n < \ldots$.. Frequently, only a finite number of grouping intervals are involved, i.e., for some n, $a_n = \infty$. We will use the tilde to indicate characteristics of the distribution of the grouped variable $\tilde{X}$. Thus $\tilde{F}$ will be its distribution function, $\tilde{L}$ its Lorenz curve, etc. The corresponding symbols without a tilde refer to X. The relationships between the distribution functions and the Lorenz curves are straightforward. Thus

$$\tilde{F}(x) = F(a_k), \quad x \in [b_{k-1}, b_k), \quad k = 1, 2, \ldots \tag{4.2.71}$$

while the Lorenz curve $\tilde{L}$ is the polygonal line joining the points $(0,0)$, $(p_1, L(p_1)), \ldots, (p_n, L(p_n)), \ldots, (1,1)$ where $p_i = F(a_i)$, provided we assume, as we do, that the b_i's are chosen to be equal to the corresponding stratum means. Since the original Lorenz curve was convex, it follows that $\tilde{L}(u) \geq L(u)$, $0 \leq u \leq 1$. Consequently $\tilde{G}$, the Gini index based on the grouped distribution, will underestimate the Gini index of the original distribution. The required integration to evaluate $\tilde{G}$ is trivial, yielding

$$\tilde{G} = 1 - \sum_{i=1}^{\infty} (p_i - p_{i-1})[L(p_i) + L(p_{i-1})]. \tag{4.2.72}$$

The difference between G and $\tilde{G}$ may be approximated to varying degrees of accuracy. A crude but sometimes adequate bound relies on the fact that, since L is convex, it cannot on the interval (p_{i-1}, p_i) ever fall below the extension of the line joining $(p_{i-2}, L(p_{i-2}))$ and $(p_{i-1}, L(p_{i-1}))$ nor below the extension of the line joining $(p_i, L(p_i))$ and $(p_{i+1}, L(p_{i+1}))$. The resulting polygonal line provides a lower bound (albeit not a convex one) for the Lorenz curve and consequently an upper bound for the Gini index. Mehran (1975) carried this approach somewhat further. Assuming that the grouping involves only a finite number of intervals (i.e., for some k, $p_{k+1} = 1$), Mehran defined β_i to be the slope of the line joining $(p_{i-1}, L(p_{i-1}))$ and $(p_i, L(p_i))$, $i = 1, 2, \ldots, k+1$. In addition, he let β_i^* represent the unknown slope of the tangent to $L(u)$ at the point p_i, $(i = 0, 1, 2, \ldots, k+1)$. Since L is convex, we have

$$0 \leq \beta_0^* \leq \beta_1 \leq \beta_1^* \leq \beta_2 \leq \cdots \leq \beta_{k+1} \leq \beta_{k+1}^*. \tag{4.2.73}$$

The tangent lines with slopes β_1^* yield an upper bound on the Gini index of the form

$$\tilde{G} + \sum_{i=1}^{k+1} (p_i - p_{i-1})^2 (\beta_1^* - \beta_i)(\beta_i - \beta_{i-1}^*)/(\beta_i^* - \beta_{i-1}^*) \tag{4.2.74}$$

where $\tilde{G}$ is as defined in (4.2.72). Our desired upper bound for the Gini index, which does not require the unavailable knowledge of the β_i^*'s, is obtained by maximizing (4.2.74) with respect to the β_i^*'s subject to the constraint (4.2.73). Mehran erroneously assumed that necessarily $\beta_0^* = 0$ and $\beta_{k+1}^* = \infty$, but only trivial modification of his argument is required to correct for this. Mehran illustrated the technique with some census data originally analyzed by Gastwirth (1972). The geometric bound fared surprisingly well.

An alternative approach to estimating G from grouped data was described by Gastwirth (1972). It relies on Theorems 4.2.1 and 4.2.3. We assume that X has a continuous distribution function F and that a grouped version $\tilde{X}$ of X is defined by (4.2.70). The true Gini index (i.e., corresponding to X) may be written in the form

$$\begin{aligned} G &= \frac{1}{2m}\int_0^\infty \int_0^\infty |x-y|\, dF(x)\, dF(y) \\ &= \frac{1}{2m}\sum_{i=1}^{\infty}\int_{a_{i-1}}^{a_i}\int_{a_{i-1}}^{a_i} |x-y|\, dF(x)\, dF(y) \\ &\quad + \frac{1}{m}\sum_{i>j}\int_{a_{i-1}}^{a_i}\int_{a_{j-1}}^{a_j} (x-y)\, dF(x)\, dF(y) \\ &= \frac{1}{2m}\sum_{i=1}^{\infty}\gamma_i^2\Delta_i + \frac{1}{2m}\sum_{i\neq j}\gamma_i\gamma_j|m_i - m_j| \end{aligned} \tag{4.2.75}$$

where

$$\gamma_i = p_i - p_{i-1}$$

$$\gamma_i m_i = \int_{a_{i-1}}^{a_i} x\, dF(x)$$

and

$$\gamma_i^2\Delta_i = \int_{a_{i-1}}^{a_i}\int_{a_{i-1}}^{a_i} |x-y|\, dF(x)\, dF(y).$$

Thus γ_i represents the probability of falling in the ith stratum, while m_i and Δ_i, respectively, represent the mean and the mean deviation of observations in the ith stratum. The effect of using grouped data to estimate G is equivalent to a procedure which ignores the mean deviations within strata and assumes all observations in the ith stratum are equal. We can expect that such a procedure will give a reasonable indication of the true value of the second term in (4.2.75), i.e., the term $\frac{1}{2m}\sum_{i\neq j}\gamma_i\gamma_j|m_i - m_j|$. Further precision could be obtained by using grouping corrections. See, for example, Kendall, Stuart and Ord (1994, pp. 95–102) where Sheppard's corrections are described. Some minor modifications are necessary, since in Kendall, Stuart and Ord's discussion the grouped observations were assumed to be concentrated at the mid-point rather than the mean of their corresponding stratum. Even after applying such corrective measures, we will still be underestimating by ignoring the first term in (4.2.75), called by Gastwirth and earlier authors the "grouping corrections," i.e.,

$$D = \frac{1}{2m}\sum_{i=1}^{\infty}\gamma_i^2\Delta_i. \tag{4.2.76}$$

We can use Theorem 4.2.1 (and Theorem 4.2.3, which would be needed if some $a_k = \infty$) to bound the grouping correction. Thus,

$$D \leq \overline{D} = \frac{1}{m} \sum_{i=1}^{\infty} \gamma_i^2 (m_i - a_{i-1})(a_i - m_i)/(a_i - a_{i-1}). \qquad (4.2.77)$$

If we consider the sample version of the Gini index based on n i.i.d. observations which have been grouped, Giorgi and Pallini (1987b) suggest a refined version of the bound (4.2.77) taking into account constraints due to the fact that, with n observations, the upper bound will in general not be attainable (unless fractional individuals are admitted into the dialog). Giorgi and Pallini (1987b) also draw attention to the fact that, in an earlier contribution, Pizzetti (1955) essentially obtained the same upper bound, (4.2.77), as did Gastwirth.

Although it is somewhat out of place in this section, this appears to be a convenient point at which to draw the reader's attention to the fact that Gastwirth, Nayak and Krieger (1986) discuss the asymptotic joint normality of the usual estimate of the Gini index (ignoring any grouping effect) and the grouping corrected bound, (4.2.77), as the sample size tends to infinity.

A cautionary note is sounded by McDonald and Ransom (1981). They provide simulation based evidence (using a gamma parent distribution) to indicate that the sample versions of the Gastwirth bounds will frequently not include the population value of the Gini index. Moreover, when the number of strata is increased, the sample bounds become closer to each other, but the relative frequency with which they include the population Gini index decreases. If the true stratum means (the m_i's) were known, then (4.2.77) could be used to provide a strict upper bound for the true Gini index. An alternative procedure would involve estimating the grouping correction by fitting a smooth curve in each stratum. Gastwirth (1972) discussed strengthened versions of Theorems 4.2.1 and 4.2.3, obtaining tighter bounds by imposing hopefully plausible additional constraints on the underlying distribution function. The results are

Theorem 4.2.6. *Let F be a distribution function which is concave on its support $[\underline{M}, \overline{M}]$, and let m denote the mean of the distribution. The corresponding Gini index satisfies*

$$(m - \underline{M})/3m \leq G \leq (m - \overline{M})(3\overline{M} + \underline{M} - 4m)/3m(\overline{M} - \underline{M}). \qquad (4.2.78)$$

Theorem 4.2.7. *Let F be a distribution function with support $(\underline{M}, \infty)$, mean m and decreasing failure rate on its support, then the corresponding Gini index satisfies*

$$(m - \underline{M})/2m \leq G \leq (m - \underline{M})/m. \qquad (4.2.79)$$

It may be observed that P(IV)$(\mu, \sigma, \gamma, \alpha)$ distributions have decreasing failure rate, provided that $\gamma \geq 1$. If $\gamma < 1$ (which may well be the common case for income data), we can conclude only that the failure rate is eventually decreasing. This restriction is not too disconcerting, however, since Theorem 4.2.7 would presumably only be used for the final stratum where decreasing failure rate can be reasonably expected. Theorem 4.2.6 would be applied to the lower strata. Focusing on the Pareto

(IV) family once more, we find that the hypothesis of concavity of F will only be satisfied throughout the support of the distribution if we have $\gamma \geq 1$. For $\gamma < 1$, we can only conclude that F is eventually concave (i.e., the density eventually is a decreasing function). This suggests that the improved bound given by Theorem 4.2.6 should not be used for the first few strata.

Gastwirth (1972) discussed the problem of bounding the entire Lorenz curve (rather than just the Gini index) based on grouped data. The resulting technique depends on knowledge of the stratum means and, as mentioned earlier, does not seem to be markedly better than Mehran's (1975) geometric technique. As in the case when we sought a bound for the Gini index, we proceed by considering bounds for the individual strata.

Theorem 4.2.8 (Gastwirth). *Let F be a distribution with mean m and support $[\underline{M}, \overline{M}]$. The corresponding Lorenz curve, $L(u)$, satisfies*

$$B(u) \leq L(u) \leq u \tag{4.2.80}$$

where

$$\begin{aligned} B(u) &= \underline{M}u/m, \quad 0 \leq u \leq (\overline{M}-m)/(\overline{M}-\underline{M}) \\ &= (m-\overline{M}+\overline{M}u)/m, \quad (\overline{M}-m)/(\overline{M}-\underline{M}) \leq u \leq 1. \end{aligned} \tag{4.2.81}$$

A tighter bound is obtainable if we are willing to assume concavity of F.

Krieger (1979) provided bounds for the Lorenz curve based on grouped data from a presumed unimodal distribution. Gastwirth and Glauberman (1976) suggested Hermite polynomial interpolation for sample Lorenz curves derived from grouped data.

An alternative to estimating the grouping correction (4.2.76) is to choose the stratum boundaries in a manner which will minimize it and then ignore it. Construction of such optimal stratum boundaries requires, unfortunately, knowledge of the underlying distribution F. For example, if the underlying distribution is uniform $[a, b]$, then the optimal division into k strata will have strata of equal widths whose boundaries are thus equally spaced fractiles of the distribution. This would not be expected to be optimal for other distributions. As Gastwirth remarked, this casts doubt on the common practice of dividing income data by means of equally spaced fractiles.

In connection with the problem of estimating the Gini index from grouped data, some observations of Peterson (1979) are worth recalling. The Gini index is by definition scale invariant (and thus, in an economic setting, unchanged by uniform inflation). Peterson pointed out that this will not be the case, if the estimate of G is based on grouped data, unless the stratum boundaries are also suitably inflated. A simple example would consider a large population whose incomes are (approximately) uniformly distributed over the interval $(0,5)$. Suppose they are grouped into cells given by $(0,1)$, $(1,2)$, $(2,4)$ and $(4,\infty)$. The Gini index of the grouped distribution with observations in the ith stratum set equal to the ith stratum mean will be .304. If inflation changes the underlying distribution first to a uniform $(0,6)$ and then to a uniform $(0,7)$ distribution, the revised values of the Gini index (based on the grouped data) become .306 and .297. Further inflation would further decrease the Gini index due to a "piling up" of observations in the cell $(4,\infty)$. Less well-behaved parent distributions will be affected by inflation in a less predictable fashion (see Peterson). Peterson's message was that because of the lack of scale invariance induced by the use of grouped data, care should be used in comparing Gini indices computed from data involving more than one time period. If we knew the level of inflation (and that it was uniform), we could correct our boundaries accordingly. Failing that, Peterson suggested using class boundaries which are equally spaced on a logarithmic scale with occasional introduction of new uppermost income classes following periods of rapid inflation.

It should be remarked that all the above discussion of grouped data involved an assumption that observations in a stratum be set equal to the true mean of the stratum. In practice the mean is not known, and the common value of the members of a stratum is usually taken to be the lower boundary or perhaps the midpoint of the stratum. When this is done, the results of Gastwirth, Mehran and Petersen are only approximately correct. The degree of approximation depends on the actual within stratum distributions which are typically not known.

Aghevli and Mehran (1981) discuss optimal grouping of income data. Specifically they ask how should income data be grouped into k groups in order to minimize the difference between the Gini indices of the original and the grouped data. This is equivalent to asking how does one determine the largest $(k+2)$-sided polygon which can be inscribed in the area between the Lorenz curve of the original data and the line of equality; i.e., finding the kth order Pietra index. They include equations equivalent to (4.2.50)–(4.2.52) and describe a tentative scheme for solving (4.2.51).

Aggarwal (1984) addressed this optimal grouping problem in the particular case of distributions whose Lorenz curves are of the form

$$L(u) = \frac{(1-\theta)^2 u}{(1+\theta)^2 - 4\theta u}, \quad 0 \leq u \leq 1, \tag{4.2.82}$$

where $\theta \in (0,1)$. Such Lorenz curves correspond to certain truncated Pareto (I) distributions. In this case, analytic expressions for the optimal group boundary points are obtained. These boundaries coincide with those which would be obtained by partitioning according to a cumulative $\sqrt{x}$ scale. Aggarwal (1984) suggests that such $\sqrt{x}$ partitioning will be good, though not optimal, for other parent distributions.

4.2.4 Multivariate Lorenz Curves

It was inevitable that multivariate versions of the Lorenz curve be developed. The surprise, perhaps, is the paucity of literature on the topic prior to the breakthrough paper, Koshevoy(1995), in which the Lorenz zonoid was introduced.

As will be discussed in the following subsection, there is a representation, (4.2.106), of the univariate Lorenz curve that does not involve order statistics nor does it involve the quantile function. As a consequence, it would appear to be a natural candidate for use in developing a concept of a multivariate version of the Lorenz curve. If X is assumed to be an absolutely continuous non-negative random variable with density $f_X(x)$ and with $0 < E(X) < \infty$, then equation (4.2.106) implicitly defines the Lorenz curve of X, parameterized by $x \in (0,\infty)$, as the set of points with coordinates

$$\left\{F(x), \frac{\int_0^x y f_X(y)dy}{E(X)}\right\}.$$

Lunetta (1972) and Taguchi (1972a) independently suggested the following natural extension of this concept known as a two-dimensional Lorenz curve or Lorenz surface. For it, consider an absolutely continuous non-negative bivariate random variable (X_1,X_2) with $0 < E(X_1), E(X_2) < \infty$ and joint density $f_{\underline{X}}(x_1,x_2)$. The surface is parameterized by $(x_1,x_2) \in (0,\infty)^2$ and is the set of points in $(0,\infty)^3$ with coordinates

$$\left\{F_{\underline{X}}(x_1,x_2), \frac{\int_0^{x_1}\int_0^{x_2} y_1 f_{\underline{X}}(y_1,y_2)dy_1dy_2}{E(X_1)}, \frac{\int_0^{x_1}\int_0^{x_2} y_2 f_{\underline{X}}(y_1,y_2)dy_1dy_2}{E(X_2)}\right\}. \quad (4.2.83)$$

It is not difficult to envision an analogous k-dimensional Lorenz set in $(0,\infty)^{k+1}$ corresponding to a k-dimensional random vector $(X_1,X_2,...,X_k)$ with finite positive marginal means. It is a set of points parameterized by $(x_1,x_2,...,x_k) \in (0,\infty)^k$ with first coordinate $F_{\underline{X}}(\underline{x})$ and $(i+1)$'th coordinate given by

$$\frac{\int \cdots \int_{\{\underline{y} \le \underline{x}\}} y_i f_{\underline{X}}(\underline{y})d\underline{y}}{E(X_i)}$$

for $i = 1,2,...,k$. Alternatively the points in this Lorenz set can be represented in the form

$$\left\{F_{\underline{X}}(\underline{x}), \frac{E(X_1 I(\underline{X} \le \underline{x}))}{E(X_1)}, \frac{E(X_2 I(\underline{X} \le \underline{x}))}{E(X_2)}, ..., \frac{E(X_k I(\underline{X} \le \underline{x}))}{E(X_k)}\right\}. \quad (4.2.84)$$

It turns out to be possible to provide an interpretation of this $(k+1)$-dimensional set (and the corresponding Lunetta-Taguchi surface (4.2.83)) in terms of the Lorenz zonoid, to be introduced later in this Section.

Taguchi (1972a,b, 1973) investigated some properties of this Lorenz surface, with $k = 2$, and evaluated its form in certain special cases. A good economic or inequality interpretation was, however, not provided.

An alternative definition in the case $k = 2$ can be set up as follows. It also generalizes readily to higher dimensions and, in the case of independence, is simply related to the marginal Lorenz curves.

For it, we define the two-dimensional Lorenz surface of the non-negative random variable (X_1,X_2) with density f_{X_1,X_2}, marginals f_{X_1} and f_{X_2} and $0 < E(X_1X_2), \infty$, to be that function $L(u,v)$ defined implicitly by

$$\begin{aligned} u &= \int_0^{x_1} f_{X_1}(y_1)dy_1 \\ v &= \int_0^{x_2} f_{X_2}(y_2)dy_2 \\ L(u,v) &= \int_0^{x_1}\int_0^{x_2} y_1y_2 f_{X_1,X_2}(y_1,y_2)dy_1\,dy_2/E(X_1X_2). \end{aligned} \quad (4.2.85)$$

Observe that if X_1 and X_2 are independent random variables with corresponding univariate Lorenz curves $L_1(u)$ and $L_2(v)$, respectively, the corresponding bivariate Lorenz surface function satisfies:

$$L(u,v) = L_1(u)L_2(v). \tag{4.2.86}$$

As an example, consider a bivariate density of the form

$$f_{X_1,X_2}(x_1,x_2) = \begin{cases} x_1 + x_2, & 0 \le x_1 \le 1, \quad 0 \le x_2 \le 1 \\ 0, & \text{otherwise.} \end{cases} \tag{4.2.87}$$

Straightforward computations yield a bivariate Lorenz surface of the form

$$L(u,v) = (1+4u-\sqrt{1+8u})(1+4v-\sqrt{1+8v})(\sqrt{1+8u}+\sqrt{1+8v}-2)/16. \tag{4.2.88}$$

What should be the bivariate analog of the Gini index in this setting? A plausible candidate is

$$G^{(2)} = 4\int_0^1\int_0^1 [uv - L(u,v)]\,du\,dv. \tag{4.2.89}$$

Does this index have any reasonable interpretation in terms of inequality? In the case of independence it does since then

$$1 - G^{(2)} = (1-G_1)(1-G_2) \tag{4.2.90}$$

where G_1 and G_2 are the marginal Gini indices. There are other possibilities. Recall that a convenient definition of the univariate Gini index involves independent identically distributed random variables X_1 and X_2 with common distribution function F. The Gini index of the distribution F can be expressed in the form

$$G = E[d(X_1,X_2)]/2E(d(X_1,0)) \tag{4.2.91}$$

(this is essentially (4.2.8)) where d indicates distance. Obviously (4.2.91) remains sensible and remains a plausible measure of inequality in the k-dimensional case provided d is a suitable metric on IR^k. Let $G_d^{(k)}$ represent the k-dimensional Gini index obtained from (4.2.91) using the distance function d. A particularly simple choice corresponds to the metric

$$d^*(\underline{x},\underline{y}) = \sum_{i=1}^k |x_i - y_i|. \tag{4.2.92}$$

In this case one finds, for $\underline{X} = (X_1, X_2, \ldots, X_k)$, a k-dimensional Gini index of the form

$$G_{d^*}^{(k)} = \sum_{i=1}^k \alpha_i G_i \tag{4.2.93}$$

where the G_i's are the marginal Gini indices and $\alpha_i = E|X_i|/\sum_{j=1}^k E|X_j|$. This choice of metric appears to be unsatisfactory, since it is unaffected by any dependence among the coordinates of $\underline{X}$. Such dependence will not be masked by other choices of d, although expressions as tractable as (4.2.93) cannot be expected.

Blitz and Brittain (1964) suggested the use of Lorenz curves to display correlation in a bivariate population. In this context let (X,Y) be a non-negative random variable with distribution function F_{12}, marginal distribution functions F_1 and F_2 and suppose $E(X) < \infty$. The Lorenz curve $L_{X|Y}(u)$ displaying the "correlation" of X on Y is defined implicitly by

$$
\begin{aligned}
u &= \int_0^y dF_2(\eta) \\
L_{X|Y}(u) &= \int_0^\infty \int_0^y \xi dF_{12}(\xi,\eta)/E(X)
\end{aligned}
\tag{4.2.94}
$$

or explicitly by

$$
L_{X|Y}(u) = \int_0^\infty \int_0^{F_2^{-1}(u)} \xi dF_{12}(\xi,\eta)/E(X). \tag{4.2.95}
$$

Such a curve will coincide with the Lorenz curve for X, if X and Y have perfect positive rank correlation. If X and Y are negatively correlated, then $L_{X|Y}(u)$ may be above the 45° line. Perfect negative rank correlation would yield a correlation curve which could be obtained by rotating the Lorenz curve of X through 180°. As a measure of association one might thus consider

$$
C_{X|Y} = 2\int_0^1 [u - L_{X|Y}(u)]\, du/G \tag{4.2.96}
$$

where G is the Gini index of X. It may be remarked that no moment assumption was made for Y in this development. A distressing aspect of this measure of association is its lack of symmetry, i.e., in general $C_{X|Y} \neq C_{Y|X}$.

Taguchi (1987) provided a detailed discussion of geometric features of the Lunetta-Taguchi surface (4.2.83). He identifies two boundary curves $L_1(x_1) = L(x_1,\infty)$ and $L_2(x_2) = L(\infty,x_2)$ and shows that suitable projections of these curves represent the marginal Lorenz curves of X_1 and X_2 (with the usual definitions). Other projections produce the correlation curves of Blitz and Brittain. In addition, Taguchi considers related surfaces obtained by changing the sense of one or both of the inequalities $X_1 \leq x_1$ and $X_2 \leq x_2$ in the $k = 2$ case of (4.2.84). In the latter sections of Taguchi (1987), a variety of Gini indices associated with two-dimensional random variables are defined and their values are computed for several well-known bivariate distributions. Such Gini indices were earlier discussed in Taguchi (1981).

An alternative multivariate extension of the Lorenz curve was proposed by Koshevoy (1995). For full details of this zonoid approach, see Mosler (2002). As mentioned earlier, to extend the Lorenz curve concept to higher dimensions, it is desirable to have definition of the Lorenz curve which is free of allusions to order statistics and/or quantile functions since such ideas do not extend readily to higher dimensions. In the one-dimensional setting, the Koshevoy approach can be described as follows.

Consider a set of non-negative numbers (individuals's incomes perhaps) $x_1, x_2, ..., x_n$ with a corresponding ordered set $x_{1:n} \leq x_{2:n} \leq ... \leq x_{n:n}$. The Lorenz curve for this set of numbers is a linear interpolation of the $(n+1)$ points in the unit square consisting of the points $(0,0)$, $(1,1)$ and the points

$$\left(\frac{i}{n}, \frac{\sum_{j=1}^{i} x_{j:n}}{\sum_{j=1}^{n} x_{j:n}}\right), \quad i = 1, 2, ..., n-1. \tag{4.2.97}$$

If we think of the x_i's as the incomes of n individuals in a population, then the second coordinate of (4.2.97) corresponds to the proportion of the total income accounted for by the poorest i individuals in the population. Rather than considering only groups of poor individuals, we can plot, for every subset consisting of i of the individuals in the population, a point in the unit square whose first coordinate is the proportion of the total population accounted for by the i individuals, i.e., i/n, and whose second coordinate is the proportion of the total income accruing to these i individuals. We do this for each $i = 0, 1, 2, ..., n$. There are 2^n subsets of the set of n individuals, so that the resulting plot will consist of 2^n points in the unit square. The convex hull of these 2^n points will be called the Lorenz zonoid corresponding to the set $\underline{x} = \{x_1, x_2, ..., x_n\}$ and will be denoted by $L(\underline{x})$. The lower boundary of this convex hull is the usual Lorenz curve. The upper boundary is called the reverse Lorenz curve; for it we consider the richest i individuals instead of the poorest. It is immediately evident that for two sets of non-negative numbers $\underline{x} = \{x_1, x_2, ..., x_n\}$ and $\underline{y} = \{y_1, y_2, ..., y_n\}$, the Lorenz curve for $\underline{x}$ will be below that of $\underline{y}$ (indicating that $\underline{x}$ exhibits more inequality than $\underline{y}$) if and only if the Lorenz zonoid of $\underline{y}$ is nested within the Lorenz zonoid of $\underline{x}$. We emphasize again that the Lorenz zonoid of $\underline{x}$ is defined without reference to any ordering of the x_i's.

To extend this zonoid definition to cover the class of all non-negative random variables with positive finite expectations, we first consider computing income shares for subsets of the population involving fractional individuals. For a vector $\underline{\alpha} \in [0,1]^n$, consider the share of the income of the total population accruing to a corresponding set of n fractional individuals. The share of the total income in this case will be $\alpha_1 \times x_1$ plus $\alpha_2 \times x_2$ plus.....plus $\alpha_n \times x_n$. This income share accrues to the fraction $\sum_{i=1}^{n} \alpha_i / n$ of the total population size. We could then plot the set of all points of the form $(\sum_{i=1}^{n} \alpha_i / n, \sum_{i=1}^{n} \alpha_i x_i / \sum_{i=1}^{n} x_i)$. However we can immediately recognize that this set of points corresponds to all of the points in the interior and the boundary of the Lorenz zonoid, the convex set determined by its two boundaries, the Lorenz curve and the reverse Lorenz curve.

Since any non-negative random variable, X, can be viewed as the limit of a suitable sequence of discrete random variables, $\{X_n\}_{n=1}^{\infty}$, with each X_n having n equally likely possible values, there is a unique extension of the definition of the n-point Lorenz zonoid to the more general case. It is of the form:

Definition 4.2.1. For any non-negative random variable X with $0 < E(X) < \infty$ and distribution function $F_X(x)$, the Lorenz zonoid of X, denoted by $L(X)$, is defined to be following the set of points in the unit square:

$$L(X) = \left\{ \int_0^\infty \psi(x) dF_X(x), \int_0^\infty x\psi(x) dF_X(x)/E(X) \ : \ \psi \in \Psi \right\}$$
$$= \left\{ \left(E(\psi(X)), \frac{E(X\psi(X))}{E(X)} \right) \ ; \ \psi \in \Psi \right\}, \tag{4.2.98}$$

where Ψ is the set of all measurable mappings from $[0,\infty)$ to $[0,1]$.

Note that the Lorenz curves of two random variables X and Y will be nested if and only if their Lorenz zonoids are similarly nested, i.e.,

$$L_X(u) \leq L_Y(u) \ \forall u \in (0,1) \iff L(X) \supseteq L(Y).$$

The motivation for considering the Lorenz zonoid rather than its less complicated cousin, the Lorenz curve, is that it has the potential to extend to k-dimensions without any concern for the potentially difficult task of ordering k-dimensional vectors. The extended definition takes the form:

Definition 4.2.2. For any k-dimensional random vector $\underline{X} = (X_1, X_2, ..., X_k)$ with $0 < E(X_i) < \infty$, $i = 1, 2, ..., k$, the Lorenz zonoid of $\underline{X}$, denoted by $L(\underline{X})$, is defined to be the following set of points in $[0,1]^{k+1}$:

$$L(\underline{X}) = \left\{ \left(E(\psi(X)), \frac{E(X_1\psi(\underline{X}))}{E(X_1)}, ..., \frac{E(X_k\psi(\underline{X}))}{E(X_k)} \right) \ ; \ \psi \in \Psi^{(k)} \right\} \tag{4.2.99}$$

where $\Psi^{(k)}$ is the set of all measurable mappings from $[0,\infty)^k$ to $[0,1]$.

When $k = 2$ the Lorenz zonoid will be a convex (American) football shaped subset of the unit cube which includes the points $(0,0,0)$ and $(1,1,1)$. An example is displayed in Figure 4.2.2. If $k > 2$, a rather more abstract football shape is encountered. In the k-dimensional case, as in one dimension, a natural partial order involving nested zonoids can be envisioned. For further discussion of this partial order and its relation to other inequality orderings of random vectors, see Section 4.4.8.

It was remarked earlier that there is a link between the Lunetta-Taguchi Lorenz surface, (4.2.83), and its higher dimensional extension, (4.2.84), and the Lorenz zonoid. To see this link, suppose that $\underline{X} = (X_1, X_2, ..., X_k)$ is a k-dimensional random vector with $0 < E(X_i) < \infty$ for $i = 1, 2, ..., k$ and with distribution function $F_{\underline{X}}(\underline{x})$. For each point $\underline{x} \in (0,\infty)^k$, consider the measurable function

$$\psi_{\underline{x}}(\underline{u}) = I(\underline{u} \leq \underline{x}),$$

and let $\Psi^{(k)}_{L-T}$ denote the class of all such functions. It is then evident that the definition of the Lunetta-Taguchi surface, (4.2.84), is the same as the definition of the Lorenz zonoid, (4.2.99), except that $\Psi^{(k)}$ has been replaced by $\Psi^{(k)}_{L-T}$. Thus the

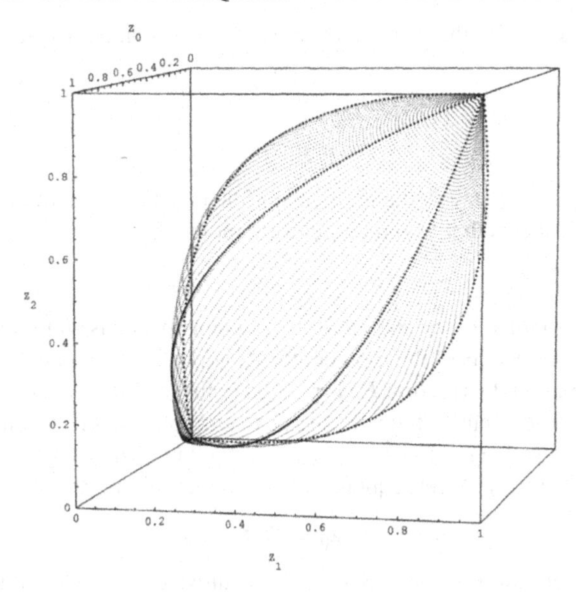

Figure 4.2.2 *The Lorenz zonoid for a bivariate Pareto (II) distribution with* $\alpha = 9$ *and parameters* $(\mu_1, \mu_2, \sigma_1, \sigma_2) = (0, 0, 1, 1)$.

Lunetta-Taguchi surface comprises a particular subset of the Lorenz zonoid. In one dimension the Lunetta-Taguchi surface coincided with the lower bound of the Lorenz zonoid, i.e., with the usual Lorenz curve. In k-dimensions, the Lunetta-Taguchi surface similarly plays the role of a lower bound of the Lorenz zonoid.

Note that Koshevoy and Mosler (1996) define a Lorenz surface associated with the Lorenz zonoid in a different way. However the Koshevoy-Mosler surface turns out to coincide with the Lunetta-Taguchi surface.

4.2.5 Moment Distributions

Hart (1975), in a useful survey paper, highlighted the central role of moment distributions in many discussions of inequality. He pointed out that many common measures have simple interpretations in terms of moments of moment distributions (see (4.2.100) below) and he argued against a proliferation of use of such ad hoc measures (of which he provided an extensive list). Inter alia Hart reiterated dissatisfaction with the Pareto distribution as an income model and seemed to favor the log-normal model in the sense that it seemed the best of a not very good lot. In the present section we are only concerned with abstracting some of his material on moment distributions.

To this end we define the rth moment distribution of a non-negative random variable X (with distribution function F) by

$$F_r(x) = \int_0^x t^r dF(t)/E(X^r) \tag{4.2.100}$$

when $E(X^r)$ exists. For the moments of the moment distributions we use the notation:

$$\begin{aligned}\mu_{(r)k} &= \int_0^\infty x^k dF_r(x)\\ \sigma^2_{(r)} &= \mu_{(r)2} - [\mu_{(r)1}]^2\end{aligned} \tag{4.2.101}$$

with $r = 0$ corresponding to the moments of the original distribution. From the definition of F_r it follows that, when the relevant moments exist,

$$\mu_{(r)k} = \mu_{(0)k+r}/\mu_{(0)r} \tag{4.2.102}$$

(this result is completely transparent if F has a density and is almost as transparent in general). Thus moments of the moment distributions are simple functions of the moments of the parent distribution F. Typically, both r and k in the above development are positive integers, but technically such a restriction is not necessary.

For a non-negative random variable X it is well known that $\log E(X^\gamma)$ is a convex function of γ, $(\gamma > 0)$. A consequence of this is that for any $\alpha < \beta$ and any $\delta > 0$

$$\mu_{(0)\alpha+\delta}/\mu_{(0)\alpha} \leq \mu_{(0)\beta+\delta}/\mu_{(0)\beta} \tag{4.2.103}$$

provided the relevant moments exist. This is equivalent to stating that for any δ the δth moment of the rth moment distribution is always a monotone function of r. The quantity $\mu_{(1)1}$ and a scalar multiple of $\mu_{(1)1}/\mu_{(0)1}$ have been proposed at various times as candidate measures of concentration (Hart referred to Niehans, 1958, Hirschman, 1945 and Herfindahl, 1950 in this context).

The quantity

$$\tau_H = 1 - [\mu_{(0)1}/\mu_{(1)1}], \tag{4.2.104}$$

which is a linear function of the reciprocal of the Herfindahl index, is thus a plausible measure of inequality. From (4.2.103) we may conclude that $0 \leq \tau_H \leq 1$ with $\tau_H = 0$ only if the underlying distribution is degenerate (the case of complete equality). Actually reference to (4.2.102) permits us to write τ_H directly in terms of the moments of the distribution F. One finds in fact that

$$\tau_H = C^2/(1+C^2) \tag{4.2.105}$$

where C is the coefficient of variation of the distribution F (cf. Hart, 1975). As an example, consider $X \sim P(I)(\sigma, \alpha)$ (a classical Pareto distribution). The corresponding value of τ_H is $(\alpha - 1)^{-2}$. Note that in order to have τ_H well defined, we must assume that F has a finite second moment (recall that only a finite first moment was required for the Gini index).

The Lorenz curve can also be related to moment distributions. Indeed, the curve can be described as the set of points in the unit square with coordinates $(F(x), F_1(x))$ where x ranges from 0 to ∞ (assuming the random variable in question is non-negative). The function $L(u)$ may be defined implicitly by the equations

$$\begin{aligned}u &= \int_0^x dF(\xi)\\ L(u) &= \int_0^x \xi dF(\xi)/E(X),\end{aligned} \tag{4.2.106}$$

or, equivalently,

$$L(u) = F_1(F^{-1}(u)), \ \ 0 \le u \le 1. \tag{4.2.107}$$

It will be recalled that it was such a definition which motivated the extension to a multivariate Lorenz surface described earlier (see (4.2.83) and (4.2.85)). It is a straightforward matter to relate the Gini index to the moment distributions. A convenient expression is

$$G = 1 - 2\int_0^\infty F_1(x)\,dF(x). \tag{4.2.108}$$

Hart claims that, as a consequence, G is "clearly governed by the moments of the moment distributions." It is, in the sense that if we know *all* of the moments of F, we will be able to compute G. It does represent a different condensation of the information in the full distribution than does any finite collection of moments of F. In fact as we saw in (4.2.9), it really is a simple function of the moments of order statistics from F.

Likewise, the Pietra index is related to the moment distributions. Letting m denote the mean of the distribution F, one may verify that the Pietra index is expressible in the form $F(m) - F_1(m)$ (in this form it is often associated with Schutz, 1951).

Hart chided "Italian statisticians" for continuing to develop measures of inequality based on differencing techniques. He would prefer to rely on measures based on the moments of the moment distributions or, equivalently, on the moments of the distribution F. It is not clear that his position can be justified beyond saying that the choice is a matter of taste.There is one telling piece of evidence which should be considered. If we are in a heavy-tailed distribution setting in which moment assumptions are not palatable, we can note that moment-based inequality measures typically require finite moments of higher order than do the corresponding "differencing" measures of inequality. An alternative view of the choice available is that Hart would have us measure inequality using the moments of F, while the Gini index and its relatives are based on $\{\mu_{1:n}\}$ (expected values of minima). These, following Kadane (1971), can be thought of as essentially moments of a distribution $\tilde{F}$ with support $[0,1]$ closely related to F. The distribution $\tilde{F}$ is defined by

$$\tilde{F}(x) = -\inf_{y \ge 0}\{y : F(y) \ge 1 - x\}.$$

The function $\tilde{F}$ determines F up to a scale parameter, and we must decide which of the two sets of moments (those of F or those of $\tilde{F}$) are best for describing inequality in the distribution F.

It is a remarkable fact that two of the most commonly used income distributions share a closure property with respect to the construction of moment distributions. For example if F is a $P(I)(\sigma,\alpha)$ distribution, then the kth moment distribution is also classical Pareto. In fact, it is $P(I)(\sigma,\alpha-k)$ [provided $k < \alpha$]. The corresponding result for the lognormal distribution is that if F is a log-normal distribution with parameters μ and σ^2 (i.e., if $\log X \sim N(\mu,\sigma^2)$), then F_k, the corresponding kth moment distribution is again log-normal but now with parameters $\mu + k\sigma^2$ and σ^2.

Although such a closure property is remarkable, it is not apparent that it can be interpreted. Such closure will be encountered whenever the density of $Y = \log X$ contains a factor of the form $e^{\theta y}$ for some parameter θ.

4.2.6 *Related Reliability Concepts*

Chandra and Singpurwalla (1978, 1981) pointed out that the Lorenz curve and the Gini index are intimately related to certain functions or transforms frequently studied in a reliability context. We will catalog these relationships here. The important implication of these relationships is that they may permit the borrowing of well-developed inferential techniques from the reliability literature for use in an income distribution or inequality setting.

As usual we consider a non-negative random variable X with distribution function F and mean m. The total time on test transform $H^{-1}(u)$ is defined by

$$H^{-1}(u) = \int_0^{F^{-1}(u)} \overline{F}(x)\,dx, \quad 0 \le u \le 1 \tag{4.2.109}$$

(this concept originated with Marshall and Proschan, 1965). Related to H^{-1} are the scaled total time on test transform

$$W(u) = H^{-1}(u)/H^{-1}(1), \quad 0 \le u \le 1 \tag{4.2.110}$$

and the cumulative total time on test transform

$$V = \int_0^1 W(u)\,du. \tag{4.2.111}$$

Note that $H^{-1}(1) = m$. As usual we denote the Lorenz curve of the distribution F by $L(u)$. The scaled total time on test transform (STTTT) can be bow shaped like the Lorenz curve. It can however be concave or convex. It will be concave if F has an increasing failure rate (IFR), and will be convex if F is DFR. See Klefsjö (1982) for details regarding the relationship between the STTTT curve and other aging properties of F.

If the STTTT had an economic interpretation, it would be a competitor to the Lorenz curve for displaying a distribution. As with the Lorenz curve, increasing inequality would be associated with a more pronounced bend in the bow. The STTTT and the Lorenz curve are related but not in an entirely trivial way. A change of variable in (4.2.109) (i.e., $v = F(u)$) followed by integration by parts yields the relation

$$W(u) = [(1-u)F^{-1}(u)/m] + L(u), \quad 0 \le u \le 1. \tag{4.2.112}$$

Integrating (4.2.112), recalling that $G = 1 - 2\int_0^1 L(u)du$ (see (4.2.24)), we conclude that V, the cumulative time on test transform, and G, the Gini index, are very simply related. In fact

$$G = 1 - V. \tag{4.2.113}$$

The integration of (4.2.112) can be done using integration by parts. Alternatively, the following observation may be used.

$$\begin{aligned}\frac{1}{m}\int_0^1 (1-u)F^{-1}(u)du &= \frac{1}{m}\int_0^1\int_u^1 F^{-1}(u)\,dv\,du \\ &= \frac{1}{m}\int_0^1\int_0^v F^{-1}(u)\,du\,dv \\ &= \int_0^1 L(v)\,dv.\end{aligned}$$

Chandra and Singpurwalla (1978) observed that both L and L^{-1} are distribution functions, one convex and one concave. They also related the Lorenz curve to the mean residual life function. The mean residual life function of the distribution F, $\mu(x)$, may be expressed in the form

$$\mu(x) = \int_x^\infty \overline{F}(u)\,du/\overline{F}(x). \tag{4.2.114}$$

From (4.2.112) it follows that $\mu(x)$ and the Lorenz curve $L(u)$ are related by

$$L(F(x)) = 1 - [\overline{F}(x)(\mu(x)+x)/m]. \tag{4.2.115}$$

Bhattacharjee (1993) provides a compendium of income interpretations of various aging concepts current in the reliability literature.

4.2.7 *Relations Between Inequality Measures*

Although it is difficult to interpret the significance of interrelationships between values of competing inequality measures, it is of interest to mention some of the well-known ones. It was earlier remarked that Pietra's index P and Gini's index G are clearly related in the sense that an inscribed triangle cannot have an area which exceeds that of the figure within which it is inscribed. Taguchi (1967) extended this result, again arguing geometrically (by drawing a line tangent to the Lorenz curve at the point $F(m)$), to obtain

$$P \le G \le P(2-P). \tag{4.2.116}$$

This result was independently rediscovered by Moothathu (1983).
Glasser (1961) observed that the Gini index G and the coefficient of variation τ_4 are related by

$$\sqrt{3}G \le \tau_4. \tag{4.2.117}$$

4.2.8 *Inequality Measures for Specific Distributions*

The following list provides a catalog of selected measures of inequality for several well-known distributions. Sources of the results are indicated where possible, but apology is made in advance for inevitable errors in assigning priority for results such as these. In the list, following the name of each parametric family of distributions, will be found (i) the density, distribution or survival function; (ii) the Lorenz curve $L(u)$; (iii) Gini index, G from (4.2.33); (iv) Pietra index, $\tau_2/2$ from (4.2.4); (v) coefficient of variation, τ_4 from (4.2.6) and (vi) Elteto and Frigyes indices, U, V, W from (4.2.48). Missing entries usually indicate that the given quantity is not available in simple closed form or that such a closed form expression is not known to the author.

(1) *Degenerate or equal distribution*

$$F(x) = \begin{cases} 0, & x < \mu \\ 1, & x \geq \mu \end{cases}$$

where $\mu > 0$.

Lorenz curve	$L(u) = u$
Gini index	0
Pietra index	0
Coefficient of variation	0
Elteto and Frigyes indices	$U = V = W = 1$

Reference: Discussed by Gastwirth (1972) but undoubtedly older.

(2) *Pareto* $(I)(\sigma,\alpha)$ *or classical Pareto*

$$\overline{F}(x) = (x/\sigma)^{-\alpha}, \quad x \geq \sigma$$

where $\sigma > 0$, $\alpha > 1$ to ensure finite mean.

Lorenz curve	$L(u) = 1-(1-u)^{(\alpha-1)/\alpha}$
Gini index	$(2\alpha-1)^{-1}$
Pietra index	$(\alpha-1)^{\alpha-1}/\alpha^{\alpha}$
Coefficient of variation	$(\alpha^2-2\alpha)^{-1/2}$ [if $\alpha > 2$]
Elteto and Frigyes indices	$U = [\alpha^{\alpha}-(\alpha-1)^{\alpha}]/[\alpha^{\alpha}-\alpha(\alpha-1)^{\alpha}]$
	$V = [\alpha^{\alpha}-(\alpha-1)^{\alpha}]/[(\alpha-1)\alpha^{\alpha}-(\alpha-1)^{\alpha}]$
	$W = \alpha/(\alpha-1)$

References: The indices U, V and W were computed by Elteto and Frigyes (1968); the other indices are of considerable antiquity.

(2a) *Truncated Pareto* $(I)(\sigma,\alpha)$

$$\overline{F}(x) = \frac{(x/\sigma)^{-\alpha} - b^{-\alpha}}{a^{-\alpha} - b^{-\alpha}}, \quad a\sigma < x < b\sigma,$$

where $\sigma, \alpha > 0$ and $1 < a < b$.

$$\text{Lorenz curve} \quad L(u) = \frac{1 - [1 - (1 - (a/b)^{\alpha})u]^{1-(1/\alpha)}}{1 - (a/b)^{\alpha-1}}$$

when $\alpha \neq 1$.

The curve for $\alpha = 1$ is obtained by taking the limit as $\alpha \to 1$.

This Lorenz curve simplifies in the case in which $\alpha = 1/2$. The corresponding Lorenz curve is then of the form

$$L(u) = \frac{u}{\sqrt{b/a} + (\sqrt{b/a} - 1)u}.$$

References: Arnold and Austin (1986), Aggarwal and Singh (1984).

(3) *Pareto(II)* $(0,\sigma,\alpha)$

$$\overline{F}(x) = [1 + (x/\sigma)]^{-\alpha}, \quad x > 0$$

where $\sigma > 0$ and $\alpha > 1$ to ensure finite mean. [Note: the parameter μ has been set equal to 0 to simplify the expressions; see the Pareto(IV) entry for the case $\mu \neq 0$.]

Lorenz curve	$L(u) = 1 - I_{[(1-u)^{1/\alpha}]}(\alpha - 1, 2)/B(\alpha - 1, 2)$
Gini index	$\alpha/(2\alpha - 1)$
Pietra index	$[(\alpha - 1)/\alpha]^{\alpha - 1}$
Coefficient of variation	$\alpha^{1/2}(\alpha - 2)^{-1/2}$ [if $\alpha > 2$]
Elteto and Frigyes indices	$W = (2\alpha - 1)/(\alpha - 1)$

(U, V omitted)

References: The Pareto (II) is simply a translated classical Pareto variate. Consequently, most of the above indices date back to the end of the nineteenth century.

(3a) *Truncated Pareto(II)* $(\mu, \sigma, 1/2)$

$$\text{Lorenz curve} \quad L(u) = \frac{u[1+(\eta-1)u]}{1+(\eta-1)u+\delta(1-u)}$$

where $\delta, \eta > 0$.

Reference: Arnold (1986).

(4) *Pareto(III)* $(0, \sigma, \gamma)$

$$\overline{F}(x) = [1 + (x/\sigma)^{1/\gamma}]^{-1}$$

where $\sigma > 0$, and $\gamma < 1$ to ensure finite mean. [Note: for $\mu \neq 0$, see the Pareto (IV) entry.] For $\mu = 0$:

$$\text{Lorenz curve} \quad L(u) = 1 - I_{(1-u)}(1-\gamma, \gamma+1)/B(1-\gamma, \gamma+1)$$

Gini index $\quad \gamma$

References: The fact that the parameter γ here can be identified with the Gini index was apparently first observed in Fisk (1961a) (His α is the present γ^{-1}).

(5) *Pareto(IV)*$(\mu, \sigma, \gamma, \alpha)$

$$\overline{F}(x) = \left(1 + \left(\frac{x-\mu}{\sigma}\right)^{1/\gamma}\right)^{-\alpha}, \quad x > \mu$$

where $\mu \geq 0$, $\sigma > 0$, $\gamma > 0$, $\alpha > 0$. Assume $\alpha > \gamma$ to ensure finite mean

$$\text{Lorenz curve } L(u) = \frac{\mu u + \sigma\alpha[B(\alpha-\gamma, \gamma+1) - I_{[(1-u)^{1/\alpha}]}(\alpha-\gamma, \gamma+1)]}{\mu + \sigma\alpha B(\alpha-\gamma, \gamma+1)}$$

$$\text{Gini index} \quad 1 - \frac{\mu + 2\sigma\alpha B(2\alpha-\gamma, \gamma+1)}{\mu + \sigma\alpha B(\alpha-\gamma, \gamma+1)}$$

References: Fisk (1961a), Arnold and Laguna (1977), Cronin (1979), McDonald and Ransom (1979), Singh and Maddala (1976).

(6) *Dagum (I)* (σ, γ, α)

$$F(x) = [1 + (x/\sigma)^{-1/\gamma}]^{-\alpha}, \quad x > 0$$

where $\sigma > 0, \alpha > 0$ and $0 < \gamma < 1$ to ensure finite mean. [Note that if $X \sim$

$P(IV)(0,1/\sigma,\gamma,\alpha)$ then $1/X$ has the Dagum$(I)(\sigma,\gamma,\alpha)$ distribution.

$$\text{Lorenz curve} \quad L(u) = \frac{I_{[u^{1/\alpha}]}(\gamma+\alpha,1-\gamma)}{B(\gamma+\alpha,1-\gamma)}$$

$$\text{Gini index} \quad \frac{\Gamma(\alpha)\Gamma(2\alpha+\gamma)}{\Gamma(2\alpha)\Gamma(\alpha+\gamma)} - 1$$

References: Dagum (1977), Sarabia (2008a).

(7) *Lognormal*

$$F(x) = \Phi([\log(x-\mu)-\nu]/\tau), \quad x > \mu$$

where Φ is the standard normal distribution function, $\mu \geq 0$, $\tau > 0$ and $-\infty < \nu < \infty$.

Gini index $\quad e^{[\nu+(\tau^2/2)]}(1-2\Phi(-\tau/\sqrt{2}))/(\mu+e^{[\nu+(\tau^2/2)]})$

Pietra index $\quad [2\Phi(\tau/\sqrt{2})-1]/[1+\mu e^{[\nu+(\tau^2/2)]}]$

Coefficient of variation $\quad [e^{\nu}\sqrt{e^{\tau^2}(e^{\tau^2}-1)}]/[\mu+e^{[\nu+(\tau^2/2)]}]$

Elteto and Frigyes indices

$$U = \frac{\left(1+\frac{\mu}{\alpha}\right)\Phi(\tau/\sqrt{2})}{1-\left(1-\frac{\mu}{\alpha}\right)\Phi(\tau/\sqrt{2})}$$

(here let $\alpha = e^{[\nu+(\tau^2/2)]}$)

$$V = \left[\frac{\Phi(\tau/\sqrt{2})}{1-\Phi(\tau/\sqrt{2})}\right]\left[\frac{\left(1-\frac{\mu}{\alpha}\right)\Phi(\tau/\sqrt{2})+\frac{\mu}{\alpha}}{1-\left(1-\frac{\mu}{\alpha}\right)\Phi(\tau/\sqrt{2})}\right]$$

$$W = \frac{1}{1+\frac{\mu}{\alpha}}\left[\frac{\left(1-\frac{\mu}{\alpha}\right)\Phi(\tau/\sqrt{2})+\frac{\mu}{\alpha}}{1-\Phi(\tau/\sqrt{2})}\right]$$

(When $\mu = 0$, we get the simple relationship: $U = W$ and $V = U^2$.)
References: Aitchison and Brown (1957) and Elteto and Frigyes (1968).

(8) *Exponential*

$$\overline{F}(x) = e^{-(x-\mu)/\sigma}, \quad x > \mu$$

where $\mu > 0$, $\sigma > 0$.

Lorenz curve $L(u) = u + \left(1 + \frac{\mu}{\sigma}\right)^{-1} (1-u)\log(1-u)$

Gini index $\sigma/[2(\mu+\sigma)]$

Pietra index $\sigma e^{-1}/(\mu+\sigma)$

Coefficient of variation $\sigma/(\mu+\sigma)$

Elteto and Frigyes indices $U = \dfrac{(\sigma+\mu)(1-e^{-1})}{[\sigma(1-2e^{-1})+\mu(1-e^{-1})]}$

$$V = \frac{(2\sigma+\mu)(1-e^{-1})}{[\sigma(1-2e^{-1})+\mu(1-e^{-1})]}$$

$$W = \frac{(2\sigma+\mu)}{(\sigma+\mu)}$$

References: Castellano (1933) and Gastwirth (1972).

(9) *Geometric*

$$P(X=j) = pq^{j-1}, \quad j = 1,2,\ldots$$

where $0 < p < 1$ and $p+q = 1$.

Lorenz curve $L(u) = 1 - iq^{i-1} + (i-1)q^i + pi[u - (1-q^{i-1})],$
$1 - q^{i-1} \leq u < 1 - q^i, \quad i = 1,2,\ldots$

Gini index $q/(1+q)$

Coefficient of variation $\sqrt{q}$

References: Gastwirth (1971) and Dorfman (1979).

(10) *Gamma*

$$f(x) = \frac{\left(\frac{x-\mu}{\sigma}\right)^{\alpha-1} e^{-\left(\frac{x-\mu}{\sigma}\right)}}{\Gamma(\alpha)\sigma}, \quad x > \mu$$

where $\mu \geq 0$, $\sigma > 0$, $\alpha > 0$.

$$\text{Gini index} \quad \sigma\Gamma(\alpha+1/2)/[\sqrt{\pi}\Gamma(\alpha)(\mu+\sigma\alpha)]$$

$$\text{Pietra index} \quad \frac{\sigma\alpha^{\alpha+1}e^{-\alpha}}{\Gamma(\alpha+2)(\mu+\sigma\alpha)}\,{}_1F_1\left[\begin{matrix} 2; \\ \alpha+2; \end{matrix} \quad \alpha\right]$$

where ${}_1F_1$ represents a confluent hypergeometric series.

$$\text{Coefficient of variation} \quad \sigma\alpha^{1/2}/(\mu+\sigma\alpha)$$

Since the first moment distribution corresponding to a gamma distribution is again a gamma distribution, a parametric form for the Lorenz curve is expressible in terms of incomplete gamma functions. Thus

$$(u, L(u)) = \left(\frac{\Gamma(\alpha,x)}{\Gamma(\alpha)}, \frac{\Gamma(\alpha+1,x)}{\Gamma(\alpha+1)}\right), \quad x > 0,$$

where

$$\Gamma(\alpha,x) = \int_0^x t^{\alpha-1}e^{-t}\,dt.$$

References: Wold (1935) discussed the case $\alpha = 2$. Elteto and Frigyes (1968) gave the Gini index. The expression for the Pietra index is due to MacDonald and Jensen (1979a). Sarabia (2008a) provided the parametric form of the Lorenz curve.

(11) *Weibull*

$$\overline{F}(x) = e^{-(\frac{x-\mu}{\sigma})^{\gamma}}, \quad x > \mu$$

where $\mu \geq 0$, $\sigma > 0$, $\gamma > 0$.

$$\text{Gini index} \quad \sigma(1-2^{-1/\gamma})\Gamma(\gamma^{-1}-1)/[\mu+\sigma\Gamma(\gamma^{-1}-1)]$$

(When $\mu = 0$, the Gini index simplifies to $1-2^{-1/\gamma}$.)

(12) *Beta*

$$f(x) = \left(\frac{x}{\sigma}\right)^{\alpha-1}\left[1-\left(\frac{x}{\sigma}\right)^{\beta-1}\right]/\sigma B(\alpha,\beta), \quad 0 < x < \sigma$$

where $\alpha > 0$, $\beta > 0$, $\sigma > 0$.

$$\text{Gini index} \quad \frac{\Gamma(\alpha+\beta)\Gamma(\alpha+1/2)\Gamma(\beta+1/2)}{\Gamma(\alpha+\beta+1/2)\Gamma(\alpha+1)\Gamma(\beta)\sqrt{\pi}}$$

$$\text{Coefficient of variation} \quad \sqrt{\frac{\beta}{\alpha(\alpha+\beta+1)}}$$

Reference: The Gini index was computed by MacDonald and Ransom (1979).

(12a) *Uniform*

$$f(x) = \sigma^{-1}, \quad \mu < x < \mu + \sigma$$

where $\mu \geq 0$, $\sigma > 0$.

Lorenz curve	$L(u) = [2\mu u + \sigma u^2]/(2\mu + \sigma)$
Gini index	$\sigma/(6\mu + 3\sigma)$
Pietra index	$\sigma/(8\mu + 4\sigma)$
Coefficient of variation	$\sigma/[\sqrt{3}(2\mu + \sigma)]$
Elteto and Frigyes indices	$U = (4\mu + 2\sigma)/(4\mu + \sigma)$
	$V = (4\mu + 3\sigma)/(4\mu + \sigma)$
	$W = (4\mu + 3\sigma)/(4\mu + 2\sigma)$

References: Castellano (1933), Gastwirth (1972), Elteto and Frigyes (1968).

(12b) *Elliptical*

$$f(x) = \frac{4}{\pi\sigma}\sqrt{\frac{x}{\sigma} - \left(\frac{x}{\sigma}\right)^2}, \quad 0 < x < \sigma$$

where $\sigma > 0$.

Lorenz curve	A parametric version is given by $u = (2\alpha - \sin 2\alpha)/(2\pi)$ $L(u) = [(2\alpha - \sin 2\alpha)/(2\pi)] - (2sin^3\alpha)/(3\pi)$ where $0 < \alpha < \pi$
Gini index	$128/(45\pi^2) = 0.2882024$
Coefficient of variation	$1/2$

Reference: Castellano (1933).

(12c) *Arc-sine*

$$f(x) = [\pi x(\sigma - x)]^{-1/2}, \quad 0 < x < \sigma$$

where $\sigma > 0$.

Lorenz curve	$L(u) = u - \frac{1}{\pi}\cos[\pi(u-1/2)]$
Gini index	$4/\pi^2$
Pietra index	$1/\pi$
Coefficient of variation	$1/\sqrt{2}$
Elteto and Frigyes indices	$U = \pi/(\pi-2)$
	$V = (\pi+2)/(\pi-2)$
	$W = (\pi+2)/\pi$

(13) *Linear*

$$f(x) = (\sigma + bx) \Big/ \left[\sigma^2\left(1+\frac{b}{2}\right)\right], \quad 0 < x < \sigma$$

where $\sigma > 0$, $b > 0$.

Gini index	$[1+b+(b^2/5)] \Big/ \left(3+\frac{7}{2}b+b^2\right)$
Pietra index	$\left(\frac{3}{2}+2b+\frac{5}{6}b^2+\frac{1}{9}b^3\right) \Big/ \left[6\left(1+\frac{b}{2}\right)^3\right]$

Reference: Castellano (1933).

The reader is referred to Castellano (1933) for discussion of several other specialized distributions. Taguchi (1968) supplied differential equations for the Lorenz curves corresponding to a great variety of distributions.

4.2.9 Families of Lorenz Curves

Modeling efforts for income distribution data sets, which often are only available in the form of a partial description of the corresponding Lorenz curve, frequently involve consideration of parametric families of Lorenz curves from which one selects a curve which is, in some sense, minimally discordant from the available data (i.e., the available points on the sample Lorenz curve). Each Lorenz curve does, of course, have associated with it a scale parameter family of distributions, but relatively few distributions have analytic closed form expressions available for both the distribution and the corresponding Lorenz curve. Simple distributions don't always have simple

Lorenz curves, and though simple Lorenz curves typically yield simple quantile functions, the required inversion to obtain the corresponding distribution function and/or density is likely to be difficult. As a consequence there has been considerable interest in proposing and investigating families of Lorenz curves which may or may not have analytic expressions available for their corresponding distribution functions. Moreover, methods for constructing new families of Lorenz curves from available families have received attention. In this direction, a good source of ideas for such construction is Sarabia (2008).

From the viewpoint of this book, with its emphasis on Pareto distributions, these families of Lorenz curves will frequently be popular competitors of Pareto models for fitting income data sets. Thus, some knowledge of their structure and properties will be desirable. Note also that, when building new Lorenz curves from old ones, the "old" ones in question might well be generalized Pareto Lorenz curves, and the resulting more complex models would still retain some degree of theoretical support inherited from the support usually advanced in favor of Paretian models. For example, considering mixtures of generalized Pareto Lorenz curves would seem to be a very natural avenue to follow in the search for more flexible models. With that said, we begin our survey of families of Lorenz curves and methods for "building" them.

First, it is well to review the characteristic properties of Lorenz curves. A function $L(u)$ mapping $[0,1]$ onto $[0,1]$ will be a Lorenz curve, in the sense that there exists a distribution F for which it is the associated Lorenz curve, if it is non-decreasing, continuous and convex on $[0,1]$ with $L(0)=0$ and $L(1)=1$. It will always be almost every where differentiable (recall the Gastwirth definition (4.2.19)). All of the examples of Lorenz curves to be discussed in this subsection will be twice differentiable on the interval $(0,1)$, many in fact will be infinitely differentiable. If we restrict attention to twice differentiable curves, then necessary and sufficient conditions for the curve to be a Lorenz curve are those attributed to Gaffney and Anstis by Pakes (1981), i.e.,

$$L(0)=0,\ \ L(1)=1,\ \ L^{'}(0+)\geq 0,\ \ L^{''}(u)\geq 0 \text{ for } u\in(0,1). \tag{4.2.118}$$

These conditions are readily checked for the examples cataloged in this subsection, and these are the conditions that must be checked when we are building new curves from old ones. Note that if $L(u)$ is a Lorenz curve corresponding to a distribution function F with mean μ, then

$$F^{-1}(u)=\mu L^{'}(u), \tag{4.2.119}$$

and information about the corresponding density function $f(x)$ is supplied by the following theorem.

Theorem 4.2.9. *(Arnold, 1987) If the second derivative $L^{''}(u)$ exists and is positive everywhere in an interval (u_1,u_2), then the corresponding distribution function $F(x)$ with mean μ has a finite positive density function in the interval $(\mu L^{'}(u_1),\mu L^{'}(u_2))$ which is given by*

$$f(x)=[\mu L^{''}(F(x))]^{-1}. \tag{4.2.120}$$

The following Lorenz-curve-building lemmas are typically readily proved, usually by using the Gaffney and Anstis conditions (4.2.118). In all cases $L(u)$ is defined on $[0,1]$ and when we say that a curve is a valid Lorenz curve, we mean that there is a scale family of distribution functions for which it is the corresponding Lorenz curve.

Lemma 4.2.1. *$L(u) = u^{\alpha}$ is a valid Lorenz curve provided that $\alpha \geq 1$.*

Lemma 4.2.2. *If $L_0(u)$ is a valid Lorenz curve, then $L(u) = [L_0(u)]^{\alpha}$ is a valid Lorenz curve if $\alpha \geq 1$.*

Lemma 4.2.3. *If $L_1(u)$ and $L_2(u)$ are valid Lorenz curves, then $L(u) = L_1(u)L_2(u)$ is a valid Lorenz curve.*

Note 4.2.1. *If we consider a curve of the form $L_{\alpha}(u) = u^{\alpha}L(u)$ where $L(u)$ is a valid Lorenz curve, then Lemma 4.2.1 and Lemma 4.2.3 together assure us that $L_{\alpha}(u)$ is a valid Lorenz curve when $\alpha \geq 1$. Sarabia, Castillo and Slottje (1999) point out that $L_{\alpha}(u)$ will also be a valid Lorenz curve for $0 < \alpha < 1$ provided that $L(u)$ has a non-negative third derivative.*

Lemma 4.2.4. *If $L_1(u)$ and $L_2(u)$ are valid Lorenz curves, then $\alpha L_1(u) + (1-\alpha)L_2(u)$ is a valid Lorenz curve if $\alpha \in [0,1]$.*

Lemma 4.2.5. *If $L_1(u)$ and $L_2(u)$ are valid Lorenz curves, then $L(u) = \max\{L_1(u), L_2(u)\}$ is a valid Lorenz curve.*

Lemma 4.2.6. *(Sarabia, Jordá and Trueba, 2013) If $L_1(u)$ and $L_2(u)$ are valid Lorenz curves, then $L(u) = L_2(L_1(u))$ is a valid Lorenz curve.*

Note 4.2.2. *Lemma 4.2.2 could be viewed as a corollary to Lemma 4.2.6.*

Lemma 4.2.7. *If $L_0(u)$ is a valid Lorenz curve, then $L(u) = 1 - L_0^{-1}(1-u))$ is a valid Lorenz curve.*

Extensions of these lemmas to involve more than one or two Lorenz curves are, of course, possible. Thus we have, from Lemma 4.2.1 and Lemma 4.2.3:

Lemma 4.2.8. *If $L_1(u), L_2(u), ..., L_k(u)$ are valid Lorenz curves, then $L(u) = \prod_{i=1}^{k}[L_i(u)]^{\alpha_i}$ is a valid Lorenz curve if $\alpha_i \geq 1,\ i = 1, 2, ..., k$.*

Using Lemma 4.2.4, we have:

Lemma 4.2.9. *If $L_1(u), L_2(u), ..., L_k(u)$ are valid Lorenz curves, then $L(u) = \sum_{i=1}^{k} \alpha_i L_i(u)$ is a valid Lorenz curve if $\alpha_i \geq 0, \;\; i = 1, 2, ..., k$ with $\sum_{i=1}^{k} \alpha_i = 1$.*

Note 4.2.3. *It is possible for a linear combination of Lorenz curves of the form $\sum_{i=1}^{k} c_i L_i(u)$ with some of the c_i's negative to be a valid Lorenz curve. However it is not easy to enunciate general sufficient conditions for this. It depends on the nature of the individual $L_i(u)$'s that are involved.*

Lemma 4.2.5 can be extended trivially:

Lemma 4.2.10. *If $L_1(u), L_2(u), ..., L_k(u)$ are valid Lorenz curves, then $L(u) = \max\{L_1(u), L_2(u), ..., L_k(u)\}$ is a valid Lorenz curve.*

It is inevitable that one would consider more general mixtures. We have (Sarabia, Castillo, Pascual and Sarabia, M., 2005) the following extended version of Lemma 4.2.9:

Lemma 4.2.11. *If $L(u;\delta)$ is an indexed collection of valid Lorenz curves where $\delta \in \Delta$, and if $\pi(\delta;\theta), \;\; \theta \in \Theta$ denotes a parametric family of densities on the set Δ, it follows that*

$$L_\theta(u) = \int_\Delta L(u;\delta)\pi(\delta;\theta)d\delta, \;\; \theta \in \Theta$$

is a parametric family of valid Lorenz curves.

Sarabia (2013) proposed a parallel extension of Lemma 4.2.8:

Lemma 4.2.12. *If $L(u;\delta)$ is an indexed collection of valid Lorenz curves where $\delta \in \Delta$, and if $\pi(\delta;\theta), \;\; \theta \in \Theta$ is a parametric family of densities on the set Δ, then*

$$L_\theta(u) = \exp\{\int_\Delta [\log L(u;\delta)]\pi(\delta;\theta)d\delta\}, \;\; \theta \in \Theta$$

is a parametric family of valid Lorenz curves.

Using the above array of lemmas, we can construct an extensive array of valid Lorenz curves and parametric families of valid Lorenz curves by beginning with perhaps the most trivial basic Lorenz curve, $L(u) = u$, the egalitarian curve. Thus, for example, using Lemma 4.2.2, we find that, for $\alpha \geq 1$, $L(u) = u^\alpha$ is a Lorenz curve. Then from Lemma 4.2.7, we conclude that $L(u) = 1 - (1-u)^\beta$, with $\beta \in (0,1)$, is a Lorenz curve (in fact it is a classical Pareto Lorenz curve). But then we also have valid Lorenz curves of the forms:

$$L(u;\alpha,\beta) = 1-(1-u^{\alpha})^{\beta}, \tag{4.2.121}$$

$$L(u;\beta,\gamma) = [1-(1-u)^{\gamma}]^{\beta}, \tag{4.2.122}$$

$$L(u;\alpha,\beta,\gamma) = [1-(1-u^{\alpha}0^{\gamma}]^{\beta}, \tag{4.2.123}$$

for suitable choices of α,β and γ. See Sarabia (2008a) for more details on this hierarchy of families of Lorenz curves. Note that if $\beta = 1/\alpha$ in (4.2.121) and if $\beta = 1/\gamma$ in (4.2.122) then the resulting Lorenz curves are said to be of the Lamé form (Sarabia, Jordá and Trueba, 2013).

Sarabia, Castillo and Slottje (2001) consider a parallel system of Lorenz curve families beginning with a basic Lorenz curve (which they call an exponential Lorenz curve, not to be confused with the Lorenz curve of the exponential distribution) of the form

$$L_0(u;c) = \frac{e^{cu}-1}{e^{c}-1}, \quad 0 \leq u \leq 1. \tag{4.2.124}$$

Sarabia and Pascual (2002) instead begin with a more flexible two parameter base curve,

$$L_0(u;a,b) = \frac{e^{cu}-bu-1}{e^{c}-b-1}, \quad 0 \leq u \leq 1. \tag{4.2.125}$$

An alternative was provided by Basmann, Hayes, Slottje and Johnson (1990). They built a hierarchy of models beginning with a basic curve of the form

$$L_0(u;b) = ue^{b(u-1)}, \quad 0 \leq u \leq 1, \tag{4.2.126}$$

a Lorenz curve earlier suggested by Kakwani and Podder (1973). The full Basmann, Hayes, Slottje and Johnson model is of the form

$$L_0(u;\alpha,\beta,\gamma,\delta) = u^{\alpha u+\beta}e^{\gamma(u-1)}e^{\delta(u^2-1)}, \quad 0 \leq u \leq 1. \tag{4.2.127}$$

The reader will have no difficulty in constructing ever more complicated models using any one of the above non-linear Lorenz curves as a basic curve (or any other Lorenz curve as desired) and building up using Lemmas 4.2.1– 4.2.12.

Yet more methods for building new Lorenz curves from old ones can be derived by considering simple operations on the underlying distribution functions. Two such results (presented in Arnold and Austin (1987), but undoubtedly well known much earlier) are:

Lemma 4.2.13. *Let X and Y be non-negative integrable random variables related by $Y = X + c$, then the corresponding Lorenz curves satisfy*

$$L_Y(u) = \frac{cu+E(X)L_X(u)}{c+E(X)}, \quad 0 \leq u \leq 1. \tag{4.2.128}$$

Lemma 4.2.14. *Let X be a non-negative integrable random variable and define Z to be X truncated above at c. The corresponding Lorenz curves are related by*

$$L_Z(u) = \frac{L_X(uF_X(c))}{L_X(F_X(c))}, \quad 0 \leq u \leq 1. \tag{4.2.129}$$

Note that the result in Lemma 4.2.13 can be viewed as representing $L_Y(u)$ as a mixture of the Lorenz curves $L_0(u) = u$, the egalitarian line, and $L_X(u)$. Lemma 4.2.14 can be recast in the following form: For any Lorenz curve $L_0(u)$ and any $\delta \in (0,1)$, the function $L(u) = L_0(\delta u)/L_0(\delta)$ is a valid Lorenz curve.

We turn now to a brief catalog of other families of Lorenz curves that have been introduced in the literature. The ordering will appear to be somewhat random since unifying themes and relationships among the models are often not available.

The generalized Tukey Lambda distribution (GTΛ) is one which is defined in terms of its quantile function rather than its distribution function. This is somewhat inconvenient if we are interested in the study of features of the corresponding distribution or density, but, provided that the random variable is non-negative and integrable, computation of the corresponding Lorenz curve is straightforward. The quantile function of the GTΛ distribution is of the form

$$F^{-1}(u) = \lambda_2[\lambda_1 + u^{\lambda_3} - (1-u)^{\lambda_4}], \quad 0 < u < 1, \tag{4.2.130}$$

where $\lambda_2 > 0$ is a scale parameter. Conditions must be imposed on the parameters λ_1, λ_3 and λ_4 in (4.2.130) to ensure that $F^{-1}(u)$ is non-negative and integrable on the interval $(0,1)$. If λ_3 and λ_4 are positive, then the only constraint required to ensure non-negativity and integrability is that $\lambda_1 \geq 1$. However, in the original formulation of the GTΛ distribution λ_1, λ_3 and λ_4 can take on any values in $(-\infty, \infty)$. Ramberg and Schmeiser (1974) provide information on the range of values assumed by F^{-1} for various values of the parameters. Sarabia (1997) in a quest for flexible families of Lorenz curves considered the integral of F^{-1} and was led to a family of Lorenz curves of the form

$$L(u) = \alpha u + \beta u^{\delta_1} + \gamma[1 - (1-u)^{\delta_2}], \quad 0 \leq u \leq 1, \tag{4.2.131}$$

where $\alpha + \beta + \gamma = 1$. If α, β and γ are all non-negative and if $\delta_1 > 1$ and $\delta_2 \in (0,1)$, this is just a convex combination of three well-known Lorenz curves. However, members of the family of curves (4.2.131) will be valid Lorenz curves for a more extensive set of choices for $\alpha, \beta, \gamma, \delta_1$ and δ_2. For example, δ_2 can be greater than 1 for certain negative values of γ. For details, see Sarabia (1997). It was pointed out following Lemma 4.2.9 that a linear combination of Lorenz curves with some negative coefficients can still be a Lorenz curve. In the present case we have an instance in which a Lorenz curve is a linear combination of two Lorenz curves and a concave function.

We turn next to consider an extensive family of Lorenz curves introduced by Villaseñor and Arnold (1984, 1989) and known as general quadratic Lorenz curves.

They began by considering the general quadratic form

$$ax^2 + bxy + cy^2 + dx + ey + f = 0. \quad (4.2.132)$$

Many of the curves of the form (4.2.132) pass through the points $(0,0)$ and $(1,1)$ and satisfy the conditions listed in (4.2.118). The subfamily of the curves (4.2.132) which satisfy these conditions is called the class of general quadratic Lorenz curves. These curves, joining the points $(0,0)$ and $(1,1)$, are segments of ellipses, parabolas or hyperbolas according as $b^2 - 4ac$ is < 0, $= 0$ or > 0. Note that, if the curve is to pass through the points $(0,0)$ and $(1,1)$, it must be the case that $f = 0$ and $e = -(a+b+c+d)$. So we may rewrite (4.2.132) in the form

$$(a+b+c+d)y = ax^2 + bxy + cy^2 + dx. \quad (4.2.133)$$

If $c = 0$ then the Lorenz curve is necessarily a hyperbola. If in addition $b = 0$ then the resulting family of parabolic Lorenz curves can be re-parameterized in the form

$$L(u) = \alpha u + (1-\alpha)u^2, \ 0 \leq u \leq 1, \quad (4.2.134)$$

where $\alpha \in [0,1]$. These are sometimes called quadratic Lorenz curves. It is readily verified that such curves correspond to either a degenerate distribution (if $\alpha = 1$) or a uniform distribution (if $\alpha \in [0,1)$). They are thus of limited interest for income modeling, even though simple expressions are available for the corresponding distribution and density functions.

If $c = 0$ and $b \neq 0$, then the corresponding Lorenz curve will be a hyperbola which, after re-parameterization, is of the form

$$L(u) = \frac{u(1+(\eta-1)u)}{1+(\eta-1)u+\delta(1-u)}, \ 0 \leq u \leq 1, \quad (4.2.135)$$

where $\delta, \eta > 0$. A particular case of this curve was discussed by Rohde (2009) and then extended by Sarabia, Prieto and Sarabia, M. (2010). The Lorenz curves introduced by Aggarwal and Singh (1984) correspond to the case in which $\eta = 1$. From (4.2.135) it is not difficult to identify the corresponding density function. It is of the form

$$f(x) \propto [\delta x + (\eta-1)(\mu - x)]^{-3/2}, \quad (4.2.136)$$

on the interval $(\mu(1+\delta)^{-1}, \mu(1+\delta\eta^{-1})$. This density will be decreasing on its support set if $\delta > \eta - 1$ and, in that case, a further re-parameterization of the density takes the form

$$f(x) \propto \left[1 + \frac{x-\lambda_2}{\lambda_1}\right]^{-3/2}, \ \lambda_2 < x < \lambda_3, \quad (4.2.137)$$

where $\lambda_1 > 0$ and $0 < \lambda_2 < \lambda_3 < \infty$.

If $\delta = \eta - 1$, the resulting density is uniform, while if $\delta < \eta - 1$, the density is increasing on its support set and thus would likely be judged to be inappropriate for modeling income.

If in (4.2.133) we have $c \neq 0$, then we can still have a hyperbolic Lorenz curve but the formula for the curve will be more complicated. See Villaseñor and Arnold (1984a,b) for further discussion of this case.

The class of general quadratic Lorenz curves also contains elliptical curves. These correspond to the case in which, in equation (4.2.133), $c \neq 0$ and without loss of generality equals 1, and $b^2 - 4ac < 0$. In addition it is required that $a+b+d+1 > 0$, $d \geq 0$ and $a+d \geq 1$. See Villaseñor and Arnold (1989) for details. The inequality in the last condition was inadvertently reversed in the Villaseñor and Arnold paper. The corrected equation, $a+d \geq 1$, was presented by Krause (2013). These elliptical Lorenz curves are of the form

$$L(u) = \frac{1}{2}\,[-bu + a + b + 1 + d - \{(b^2 - 4ac)u^2 - (2bc - 4d)u + (a+b+1+d)^2\}^{1/2}], \ 0 \leq u \leq 1, \quad (4.2.138)$$

which is somewhat more complicated than the form of the hyperbolic Lorenz curves (4.2.135). However, after re-parameterization, the corresponding densities exhibit an attractive simplicity (similar in complexity to the expression (4.2.136) for the hyperbolic case), i.e.,

$$f(x) \propto \left[1 + \left(\frac{x - \nu}{\tau}\right)^2\right]^{-3/2}, \quad \tau\eta_1 + \nu < x < \tau\eta_2 + \nu, \quad (4.2.139)$$

where $0 < \eta_1 < \eta_2 < \infty$, $\tau > 0$ and ν is such that $\tau\eta_1 + \nu \geq 0$.

A special subclass of elliptical Lorenz curves was discussed by Ogwang and Rao (1996). These are called circular Lorenz curves and are defined to be arcs of circles passing through $(0,0)$ and $(1,1)$. It is readily verified that the center of such a circle is located at the point $(-a, 1+a)$ where $a \geq 0$. The corresponding Lorenz curves are given by

$$L(u) = 1 + a - \sqrt{(1+a)^2 - 2au - u^2}, \ 0 \leq u \leq 1, \quad (4.2.140)$$

These Lorenz curves are self-symmetric, i.e., they satisfy $L(u) = 1 - L^{-1}(1-u)$.

Holm (1993) considers Lorenz curves corresponding to densities which have maximal entropy subject to side conditions on the value of the Gini index and on the distance between the mean income and the minimum income. The resulting densities have quantile functions which satisfy

$$\frac{d}{du}F^{-1}(u) \propto [1 + \delta_1(1-u) + \delta_2 u(1-u)]^{-1}, \ 0 \leq u \leq 1, \quad (4.2.141)$$

for various values of δ_1 and δ_2.

If $\delta_1 \neq 0$ and $\delta_2 = 0$, the corresponding Lorenz curves form a one parameter family of the form

$$L(u;r) = u + G\frac{(r-u)\log(1-u/r) - (r-1)\log(u-1/r)u}{r(r-1)\log(1-1/r) + r - 0.5}, \ 0 \leq u \leq 1, \quad (4.2.142)$$

where G is the Gini index and $r < 0$ or $r > 1$. If one takes the limit as $r \downarrow 1$ in (4.2.142), one obtains

$$L(u) = u + 2G(1-u)\log(1-u)$$

which can be recognized as the Lorenz curve corresponding to a shifted exponential distribution.

If $\delta_1 = 0$ and $\delta_2 \neq 0$, a two parameter family of curves is obtained, which includes, as a limiting case, the curve

$$L(u) = u + G[u\log(u) + (1-u)\log(1-u)], \quad 0 \leq u \leq 1. \tag{4.2.143}$$

If $\delta_1 \neq 0$ and $\delta_2 \neq 0$, then a two parameter family of curves is obtained which includes, as a limiting case, the Lorenz curve corresponding to a truncated Pareto(I)$(1,1)$ distribution.

Sarabia, Gómez-Déniz, Sarabia, M. and Prieto (2010) proposed a technique for generating Lorenz curves utilizing probability generating functions of positive integer valued random variables. Specifically they begin with a base Lorenz curve $L_0(u)$ and a positive integer valued random variable X with probability generating function

$$P_X(s) = E(s^X) = \sum_{j=1}^{\infty} P(X=j)s^j.$$

They then observe that

$$L(u) = P_X(L_0(u)) \tag{4.2.144}$$

is a valid Lorenz curve. This could be justified in more than one way. First, observe that the generating function P_X is increasing and convex and indeed is itself a Lorenz curve. The composition $P_X(L_0(u))$ is then a Lorenz curve, using Lemma 4.2.6. Second, note that for each positive integer j, $[L_0(u)]^j$ is a Lorenz curve, from Lemma 4.2.2. Then note that

$$L(u) = P_X(L_0(u)) = \sum_{j=1}^{\infty} P(X=j)[L_0(u)]^j,$$

which shows that $L(u)$ is a mixture of Lorenz curves and thus is a Lorenz curve, using 4.2.11.

For example, if X has a geometric distribution with $P(X=j) = p(1-p)^{j-1}$, $j = 1,2,\ldots$, then we have that

$$L(u) = P_X(L_0(u)) = \frac{pL_0(u)}{1-(1-p)L_0(u)} \tag{4.2.145}$$

is a valid Lorenz curve for every $p \in (0,1]$.

A second example is based on a random variable X with the property that $X-1$ has a Poisson(λ) distribution. In this case we obtain

$$L(u) = P_X(L_0(u)) = L_0(u)e^{\lambda(L_0(u)-1)}, \tag{4.2.146}$$

where $\lambda \in (0,\infty)$. In the special case in which $L_0(u) = u$, this becomes a family of Lorenz curves discussed by Gupta (1984).

If, instead, we assume that X is a non-negative integer valued random variable (with possible values $0,1,2,\ldots$) and generating function $P_X(s)$, then we may construct Lorenz curves of the forms

$$L_1(u) = [L_0(u)]^\alpha P_X(L_0(u))$$

and

$$L_2(u) = \frac{P_X(L_0(u)) - p_0}{1-p_0},$$

where $L_0(u)$ is the base Lorenz curve and $p_0 = P(X=0) \in (0,1)$. For details see Sordo, Navarro and Sarabia (2013).

As the final entry in our catalog of Lorenz curve generating techniques, we return to consider a remarkable property of the log-normal distribution. If X has a log-normal distribution, i.e., $\log X \sim N(\mu,\sigma^2)$, then the corresponding first moment distribution is again of the log-normal form. The corresponding Lorenz curve from (4.2.107) will be of the form $L(u) = F_1(F^{-1}(u))$ which simplifies to yield

$$L(u) = \Phi(\Phi^{-1}(u) - \sigma^2), \tag{4.2.147}$$

where Φ is the standard normal distribution function and σ^2 is the variance of $\log X$. Arnold, Robertson, Brockett and Shu (1987) considered what happens if Φ in (4.2.147) is replaced by another distribution function, say G. They show that a sufficient condition for the function

$$L(u) = G(G^{-1}(u) - \sigma^2) \tag{4.2.148}$$

to be a valid Lorenz curve is that G has a strongly unimodal density and has support that is unbounded above. For example, the classical Pareto Lorenz curve admits a representation of the form (4.2.148) in which G is an extreme value distribution. Several other examples are described in Arnold, Robertson, Brockett and Shu (1987).

The property of having a first moment distribution in the same family as the original distribution allows one to write down a corresponding parametric representation of the Lorenz curve (as $(F(x), F_1(x))$ where x ranges from 0 to ∞). Bladt and Nielsen (2011) show that the family of matrix-variate exponential distributions (a family that includes the important family of Phase-type distributions) has this closure property. As a consequence an albeit complicated parametric representation of the corresponding Lorenz curves is thus possible. See Bladt and Nielsen (2011) for more details on this flexible family of Lorenz curves.

4.2.10 *Some Alternative Inequality Curves*

Although the Lorenz curve undoubtedly continues to be the graphical summary of choice in inequality discussions, several alternative curves have been proposed as suitable inequality summarization devices. Many, but not all, are intimately related to the Lorenz curve. A selection of such curves will be described in this section.

In all cases, the curves are described for a particular distribution F corresponding to a non-negative random variable X with finite positive mean m. The corresponding quantile function and first moment distribution function of F are, as usual, denoted by F^{-1} and F_1, respectively.

For comparison purposes, recall the definition of the Lorenz curve.

$$L(u) = m^{-1}\int_0^u F^{-1}(v)dv, \quad 0 \le u \le 1.$$

Shorrocks (1983) proposed consideration of the generalized Lorenz curve defined by

$$GL(u) = mL(u) = \int_0^u F^{-1}(v)dv, \; 0 \le u \le 1. \tag{4.2.149}$$

This curve is clearly not scale invariant. The partial order determined by nested generalized Lorenz curves is known in the literature as second order stochastic dominance. This partial order is distinct from the Lorenz order. See Shorrocks and Foster (1987) for related economic implications of this generalized Lorenz order.

In the field of informetrics, the Leimkuhler curve is often used instead of the Lorenz curve. Sarabia (2008b) provides a useful representation of this curve which facilitates understanding of its relationship with the Lorenz curve. The Leimkuhler curve, $K(u)$, corresponding to the random variable X with distribution function F is defined by

$$K(u) = m^{-1}\int_{1-u}^1 F^{-1}(v)dv, \; 0 \le u \le 1. \tag{4.2.150}$$

It is a continuous non-decreasing concave function with $K(0) = 0$ and $K(1) = 1$. A Leimkuhler partial order, $\le_K$, can be defined as follows.

$$X \le_K Y \Leftrightarrow K_X(u) \le K_Y(u), \; \forall u \in [0,1]. \tag{4.2.151}$$

It will be recalled that the Lorenz order, based on nested Lorenz curves, i.e.,

$$X \le_L Y \Leftrightarrow L_X(u) \ge L_Y(u), \; \forall u \in [0,1],$$

is equivalent to the partial order based on nested Lorenz zonoids, defined in (4.2.98). However, it is not difficult to verify that, while the Lorenz curve corresponds to the lower boundary of the zonoid, the Leimkuhler curve coincides with the upper boundary of the zonoid. As a consequence, the Leimkuhler partial order is the same as the Lorenz order. Results obtained for the Lorenz order can thus be routinely translated into corresponding statements for Leimkuhler curves. A convenient representation of the close relationship between the two curves is provided by the observation that

$$K(u) = 1 - L(1-u), \; 0 \le u \le 1. \tag{4.2.152}$$

Bonferroni (1930) introduced an alternative inequality curve, which can be conveniently defined by

$$B(u) = L(u)/u, \; 0 < u \le 1. \tag{4.2.153}$$

Note that it is always the case that $L(u) \leq B(u)$. A corresponding Bonferroni partial order is defined by

$$X \leq_B Y \Leftrightarrow B_X(u) \geq B_Y(u), \ \forall u \in (0,1]. \tag{4.2.154}$$

The representation (4.2.153) makes it transparent that the Bonferroni partial order coincides with the Lorenz order (and the Leimkuhler order). An alternative expression for the Bonferroni curve is available:

$$B(u) = \int_0^u \left[\frac{F_1^{-1}(v)}{F^{-1}(v)}\right] dv, \ 0 < u \leq 1, \tag{4.2.155}$$

which can be used for an alternative argument for the scale invariance of the Bonferroni curve. The integrand in (4.2.155), i.e., $F_1^{-1}(v)/F^{-1}(v)$, is sometimes described as a local measure of inequality. It will arise again in the discussion of the Zenga-I curve below. There is a Bonferroni inequality index defined in terms of the integrated Bonferroni curve (parallel to the way that the Lorenz index is defined in terms of the integrated Lorenz curve). It is given by

$$B = 1 - \int_0^1 B(u)du, \ 0 < u \leq 1. \tag{4.2.156}$$

More detailed discussion of the Bonferroni curve may be found in Tarsitano (1990).

We will next discuss two Zenga curves. In 1984, Zenga proposed a new inequality curve which we will call the Zenga-I curve. At a later date Zenga (2007) proposed an alternative curve, which we will label the Zenga-II curve. Both curves endeavor to reflect inequality as evidenced by the relationship between the distribution F and the corresponding first moment distribution F_1. The Zenga-II curve is actually defined in terms of the Lorenz curve, as follows

$$Z^{II}(u) = 1 - \left[\frac{L(u)}{u}\right]\left[\frac{1-u}{1-L(u)}\right], \ 0 \leq u < 1. \tag{4.2.157}$$

Scale invariance of the Zenga-II curve follows from the scale invariance of the Lorenz curve. From the definition it is possible to represent the Zenga-II curve as a function of the Lorenz curve and the Bonferroni curve, thus

$$Z^{II}(u) = \frac{1-B(u)}{1-L(u)}, \ 0 \leq u < 1. \tag{4.2.158}$$

More importantly, it is possible to recover the Lorenz curve from the Zenga-II curve. Specifically, we may write

$$L(u) = \frac{u(1-Z^{II}(u)}{1-uZ^{II}(u)}, \ 0 \leq u < 1. \tag{4.2.159}$$

Consequently

$$L_X(u) \leq L_Y(u) \Leftrightarrow Z_X^{II}(u) \geq Z_Y^{ii}(u), \ 0 \leq u < 1, \tag{4.2.160}$$

and the Zenga-II and the Lorenz orderings coincide.

A Zenga-II inequality index is defined in terms of the integrated curve, as follows:

$$Z^{II} = 1 - \int_0^1 Z^{II}(u)du. \tag{4.2.161}$$

Zenga's 1984 curve reflects a somewhat different aspect of inequality. It is defined in terms of the quantile functions of F and F_1 as follows

$$Z^I(u) = 1 - \frac{F^{-1}(u)}{F_1^{-1}(u)}, \quad 0 < u < 1. \tag{4.2.162}$$

This curve is evidently scale invariant. A corresponding partial order can be defined.

$$X \leq_{Z^I} Y \Leftrightarrow Z_X^I(u) \geq Z_Y^I(u), \quad 0 < u < 1. \tag{4.2.163}$$

This Zenga-I partial order is not equivalent to the Lorenz order. Zenga-I curves have been computed for distributions in several well-known parametric families, and certain monotonicity properties with respect to the parameters in the distributions have been verified. A clear understanding of the nature of the Zenga-I order and its relationship with the Lorenz order would be desirable. For example, it is possible to have two essentially different distributions (i.e., not related by a scale change) with the same Zenga-I curve. See Arnold (2014) for discussion of distributions with constant Zenga-I curves.

The scaled total time on test transform (STTTT), introduced in the reliability literature, might also be considered as a candidate for quantifying inequality. It is defined by

$$\widetilde{T}(u) = m^{-1} \int_0^{F^{-1}(u)} [1 - F(v)]dv, \quad 0 \leq u \leq 1. \tag{4.2.164}$$

Chandra and Singpurwalla (1981) derived the following relationship between $\widetilde{T}(u)$ and the Lorenz curve in the case in which F is strictly increasing on its support.

$$L(u) = \widetilde{T}(u) - m^{-1}F^{-1}(u)[1-u], \quad 0 < u < 1. \tag{4.2.165}$$

The STTTT is related to the excess wealth transform, $W(u)$, introduced by Shaked and Shanthikumar (1998). It is defined by

$$W(u) = \int_{F^{-1}(u)}^{\infty} [1 - F(v)]dv, \quad 0 < u < 1. \tag{4.2.166}$$

Since $\int_0^\infty [1 - F(v)]dv = m$, it is evident that

$$\begin{aligned} W(u) &= m[1 - \widetilde{T}(u)] \\ &= m[1 - L(u)] - F^{-1}(u)[1-u], \quad 0 < u < 1, \end{aligned} \tag{4.2.167}$$

where the last equality assumes that F is strictly increasing. The excess wealth ordering is distinct from the Lorenz order. For example, it is location invariant rather than scale invariant. For more detailed discussion see Shaked and Shanthikumar (1998).

Earlier, Kakwani (1984) proposed use of what he called the relative deprivation curve to quantify inequality. It is defined as

$$\phi(u) = 1 - L(u) - L'(u)[1-u], \quad 0 \le u \le 1. \tag{4.2.168}$$

Since $L'(u) = F^{-1}(u)/m$, it is evident from (4.2.167) that

$$W(u) = m\phi(u), \quad 0 \le u \le 1, \tag{4.2.169}$$

so that the two concepts, excess wealth and relative deprivation, are essentially equivalent. However it should be noted that the relative deprivation curve, defined as a function of the Lorenz curve, is scale invariant, in contrast to the location invariance property of the excess wealth transform. The relative deprivation curve is a decreasing function on $[0,1]$, with $\phi(0) = 1$ and $\phi(1) = 0$. It can be convex, depending on the third derivative of $L(u)$.

4.3 Inequality Statistics

We now turn to the study and measurement of inequality in data sets. We will consider n data points $X_1, X_2, \ldots, X_n$ (which when written in increasing order are denoted by $X_{1:n}, X_{2:n}, \ldots, X_{n:n}$). Often the data points are associated with some concrete phenomenon. Thus, they might represent incomes of individuals in a population. As such, it cannot reasonably be assumed that they represent realizations of independent and identically distributed (i.i.d.) random variables from some parent distribution F. Surely individual incomes are not independent. Nevertheless, with a consequent loss of strict objectivity, such data sets are analyzed as if the observations were i.i.d. This will be especially true when we come to treat estimation problems in Chapter 5. It will also be true when we discuss the distribution of measures of inequality in the present section. For the moment we need only assume that we have n numbers available and that we can associate with these numbers a sample distribution function, a sample Lorenz curve, etc. We can plot them any way we wish. Without much loss of generality, in view of the areas in which we will apply the techniques, we may assume that the sample distribution is likely to be heavy-tailed (a property probably inherited from the parent distribution if such is assumed to exist). How should we measure inequality in such an array of numbers?

Many different approaches to this problem have been advocated. It will be convenient to divide the techniques into two classes, which conveniently may be labeled graphical and analytic. Generally speaking, it is only the analytic methods which are susceptible to an easy interpretation in the context of i.i.d. observations from a parent distribution. This is true because the graphical techniques typically involve fitting curves to the data points sometimes by eye, sometimes by least squares or by even more complicated methods. In some instances it would be possible to develop at least an asymptotic distribution for the fitted constants. Such an analysis is not usually performed except in cases where we are merely fitting the sample distribution function, and standard goodness-of-fit tests are available. Since such procedures will be, perhaps arbitrarily, dubbed analytic rather than graphical, it is possible that a satisfactory definition of the dichotomy is provided by saying that an inequality measuring procedure is analytic if some distribution theory can or is usually applied to it, otherwise it is graphical. The boundary is best left a little fuzzy.

We will denote the sample distribution function corresponding to the data points $X_1, \ldots, X_n$ by $F^{(n)}$. Thus

$$F^{(n)}(x) = \frac{1}{n}\sum_{i=1}^{n} I(x, X_i), \quad 0 \leq x < \infty, \tag{4.3.1}$$

where $I(a,b) = \begin{cases} 0 & \text{if } a < b, \\ 1 & \text{if } a \geq b. \end{cases}$

We also require a sample first moment distribution

$$F_1^{(n)}(x) = \left[\sum_{i=1}^{n} X_i I(x, X_i)\right] \Bigg/ \left[\sum_{i=1}^{n} X_i\right], \quad 0 \leq x < \infty. \tag{4.3.2}$$

The corresponding survivor functions are, as usual, denoted by bars:

$$\overline{F}^{(n)}(x) = 1 - F^{(n)}(x), \tag{4.3.3}$$

$$\overline{F}_1^{(n)}(x) = 1 - F_1^{(n)}(x). \tag{4.3.4}$$

Thus $\overline{F}_1^{(n)}(x)$, in an income setting, represents the fraction of the total income of the population that accrues to individuals whose incomes exceed the level x.

4.3.1 Graphical Techniques

For the application of graphical techniques, the available data sometimes are complete, i.e., includes all the points X_i, $(i = 1, 2, \ldots, n)$, and, thus, implicitly give the sample c.d.f. $F^{(n)}(x)$ and the sample first moment distribution $F_1^{(n)}(x)$. If the data are grouped, then we only have values available for $F^{(n)}(x)$ and $F_1^{(n)}(x)$ for selected values of x (i.e., for selected $X_{i:n}$'s). We will describe graphical techniques, assuming all the points X_i, $(i = 1, 2, \ldots, n)$, are available. When we speak of plotting and joining points or plotting and fitting points, we envision having as many points as observations (i.e., n). In any application in which for any reason (grouping or missing data, etc.) points are missing, the graphical techniques are merely adjusted to plot and fit or plot and join the available points. Bowman's (1945) survey of graphical techniques remains a valuable source. Much of the following discussion has been influenced by her observations.

The earliest graphical measure of inequality is probably the Pareto chart. It is described as a plot of $\log x$ (on the horizontal axis) against $\log \overline{F}^{(n)}(x)$ (on the vertical axis). Since $\overline{F}^{(n)}(x)$ is discontinuous and $\log x$ is continuous such a plot would, strictly speaking, consist of a series of horizontal lines. The continuous curve drawn by Pareto is obtained by plotting the points $(\log X_{i:n}, \log \overline{F}^{(n)}(X_{i:n}))$ and joining them by linear interpolation. We will adopt this convention for all the graphical techniques. If we speak of a plot of, say, $g(x)$ against $h(x)$, we mean a linear interpolation of the points $(g(X_{i:n}), h(X_{i:n}))$, $i = 1, 2, \ldots, n$.

Pareto in 1897 published the observation that such Pareto charts (as we shall call them) typically were well fitted by a straight line. We have already discussed in Chapter 1 the extent to which this observation needs to be modified in practice. Ironically, Pareto in his more dogmatic moods argued that both the form (essentially linear) and the critical parameter (the slope) were immutable, or more or less so. As such, there would be little interest in computing the slope of the fitted line, since it would fluctuate but little from sample population to sample population. That it does vary from population to population was quickly observed and even admitted by Pareto. The slope of a straight line fitted (usually by least squares) to the Pareto chart is called the Pareto index of inequality. Actually, the Pareto chart is more often used to compare populations. It is lamentable that confusion early entered the field with regard to the question of whether the population whose Pareto chart exhibited the larger negative slope was the population with the greater or lesser degree of inequality. Pareto felt that large negative slope was associated with a large degree of inequality. Since in an actual classical Pareto population (i.e., one with survival function (3.2.1)), the

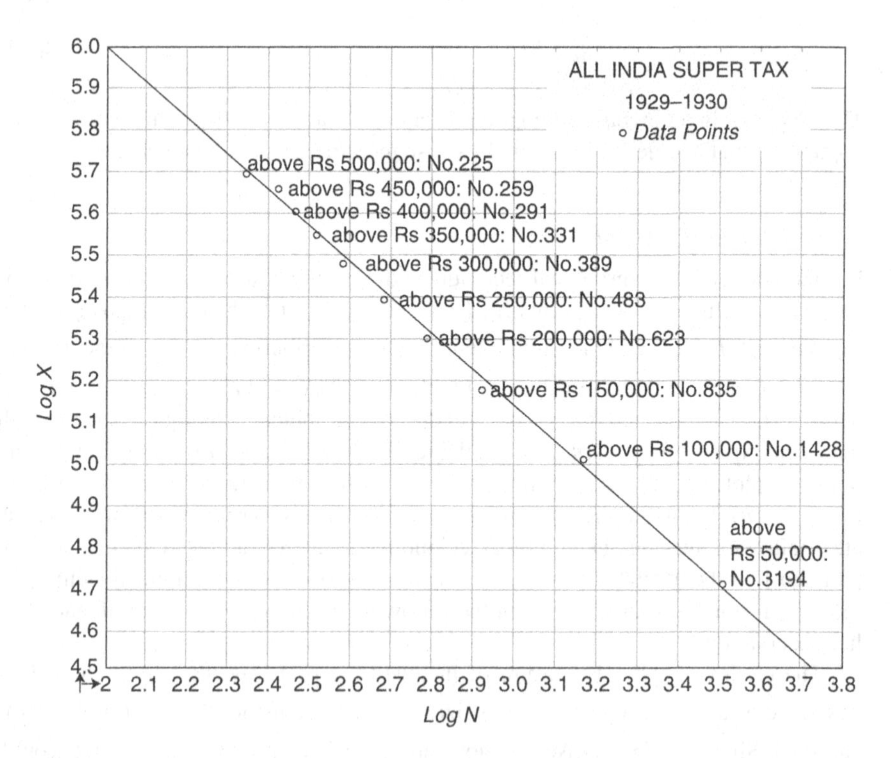

Figure 4.3.1 *A Pareto chart for Indian income data (based on super tax records) from Shirras (1935).*

plot of $\log x$ against $\log \overline{F}(x)$ will be linear with slope given by $(-\alpha)$, and since, as we saw at the end of Section 4.2, almost all commonly used measures of inequality for a classical Pareto distribution are monotonically decreasing functions of α, it is evident why the majority of subsequent statisticians have taken a different view from Pareto's. The magnitude of the slope of the line fitted to the Pareto chart will be taken to be negatively associated with inequality. Thus, an excessive degree of inequality will be associated, in the present work, with a nearly horizontal Pareto chart. Bowman (1945) provided a discussion of the genesis of the confusion. As a parting remark, we observe a proclivity to interchange the axes in a Pareto chart. The interpretation of inequality based on the magnitude of the slope then becomes reversed, i.e., extreme inequality now becomes associated with a nearly vertical chart. Care must be taken to observe how the axes have been labeled before one interprets the slope of the graph (a trite observation). As an illustration of a Pareto chart we reproduce in Figure 4.3.1 one published by Shirras (1935) based on Indian income tax data. Note that he had the axes reversed with respect to the original Pareto scheme. It is to be recalled that Figure 4.3.1 was part of the evidence marshaled by Shirras in his efforts to discredit(!) the Pareto model. Before one becomes too enchanted by the observable linearity in the figure, one should remark that it is based on the upper tail of the distribution. Bowman (1945) provided examples which indicate that linearity can be hoped for only if we restrict attention to incomes above the mean.

A popular alternative graphical technique was originated by Gini (1909). For income data it was apparent that not only did $\log \overline{F}^{(n)}(x)$ appear to be a linear function of $\log x$, but also a similar relationship appeared to hold between $\log \overline{F}_1^{(n)}(x)$ and $\log x$ (the corresponding slopes being different). Consequently, a plot of the points $(\log \overline{F}_1^{(n)}(X_{i:n}), \log \overline{F}^{(n)}(X_{i:n}))$ can also be expected to be approximately linear. Such a plot (with the points joined by linear interpolation) is now known as a Gini chart. The slope of the corresponding fitted straight line is Gini's δ (not to be confused with the Gini index G). Large values of δ are associated with a high level of inequality. It is only in the case of a true classical Pareto distribution that such plots will be exactly linear. In such a case the relationship $\delta = \alpha/(\alpha - 1)$ (where α is Pareto's index) is readily verified. The discussion regarding which of the two, Pareto's α or Gini's δ, was a better measure of inequality was important, because in hammering away at the observed inconsistencies, important insights into the nature of inequality were developed. The relationship $\delta = \alpha/(\alpha - 1)$ only holds for a true Pareto distribution. In practice it does not obtain. That did not mean that one of the indices was inherently better than another, but it was used to fuel the controversy. Harter (1977) abstracted numerous articles dealing with this issue. As remarked in Chapter 1, both α and δ are questionable as inequality measures in the absence of an underlying Pareto distribution. In the presence of such a distribution, there would appear to be little to choose between them. In practice, empirical Gini charts seem to be linear throughout a greater part of their range than do empirical Pareto charts. They, thus, make prettier graphs, and linear fitting seems more plausible. Effectively, the Gini chart focuses on higher incomes, and the lack of linear fit at the lower end is masked. Consequently, the Gini index probably emerged victorious in the lists. The strength of Pareto's name (and undoubtedly his personal influence) kept the battle waging longer than one might have predicted. Bowman (1945) pointed out that one could, if one wished, focus on lower incomes rather than higher ones. The resulting plot of $(\log F_1^n(X_{i:n}), \log F(X_{i:n}))$ is called a reversed Gini chart.

Where does the Lorenz curve fit in here? Chronologically, Lorenz's (1905) description of his graphical device for analyzing inequality seems to fit between the introduction of Pareto's α (in 1897) and Gini's δ (in 1909). It is not really clear who first thought of the Lorenz curve or even when. Footnotes in Dalton (1920) and Bowman (1945) advanced the claims of several statisticians and did not pin the date down with any precision. Gini's δ probably predated the 1909 paper in which it was discussed, and probably both it and the Lorenz curve can be ascribed to approximately the year 1905. As was remarked in Chapter 1, Lorenz's original curve is "upside down" by modern standards. But except for that, it has survived unchanged and virtually unchallenged as a device for comparing distributions with regard to level of inequality.

In its modern form, the Lorenz curve is a plot of $F^{(n)}(x)$ (on the x axis) against $F_1^n(x)$ (on the y axis). In practice we plot the points $(i/n, F_1^{(n)}(X_{i:n}))$ and interpolate linearly. The resulting bow shaped convex curve always falls below the 45° line of equality. The more bowed it is, the more unequal is the distribution it describes. Often a direct visual assessment of the Lorenz curves suffices to determine an ordering on the basis of increasing inequality. If two such curves are plotted on the same graph and if they are nested, i.e., one is uniformly closer to the 45° line than the other, then it is obvious which exhibits the greater degree of inequality. Only ambiguous comparisons are possible if the Lorenz curves cross. (The ordering induced by nested Lorenz curves will be discussed in more detail in Section 4.4.)

How should we associate a numerical measure with the Lorenz curve? The two most popular measures are the sample Gini index (twice the area between the sample Lorenz curve and the egalitarian line) and the sample Pietra index (the maximum vertical deviation between the sample Lorenz curve and the egalitarian line). Another appealing measure is the sample Amato index (the length of the sample Lorenz curve). We will later discuss these in more detail as analytic measures of inequality. Hagerbaumer's (1977) minor concentration ratio introduced in Section 4.2 (see Figure 4.2.1) is a less well known and understood means of associating a numerical value to a sample Lorenz curve. It intends to reflect, in a sense, the position of the poorer individuals in the population.

Before turning to analytic measures of inequality, we mention two other graphical techniques which have been proposed. A semi-log graph, i.e., a plot of $\log x$ against $\overline{F}^{(n)}(x)$, was ascribed by Bowman (1945) to Durand (1943). It presumably focuses on the lower income levels. Bowman (1945) is enthusiastic about its use in conjunction with the Lorenz curve. Bryson (1974) proposed a plot of x against $\log \overline{F}^{(n)}(x)$ as a visual check for heavy-tailed distributions. It could be used as an indicator of inequality. The Durand and the Bryson charts are minor variants of the Pareto chart. They suffer in comparison with the Pareto chart in that they do not retain its simple appealing form (a straight line) when applied to classical Pareto data. Analogous ad hoc graphical techniques can be suggested with reference to the Gini chart. Thus one could plot $\overline{F}^{(n)}(x)$ against $\log \overline{F}_1^{(n)}(x)$ or, if preferred, $\log \overline{F}^{(n)}(x)$ against $\overline{F}_1^{(n)}(x)$. Again we lose simplicity in the case of an underlying Pareto distribution; in return we get a closer look at intermediate income ranges.

4.3.2 *Analytic Measures of Inequality*

In dealing with analytic measures of inequality we will adopt a convention of using a subscripted T to denote the sample version of the corresponding population measure of inequality denoted by a subscripted τ and introduced in Section 4.2. Exceptions are the sample Gini, Amato, Bonferroni and Zenga-II indices, for which we use the same notation, G, A and Z^{II} for the sample and the population measures. The sample size will, when needed, be indicated by a superscript n. Naturally, all the analytic measures can be thought of as functions of the sample distribution function $F^{(n)}(x)$ defined in (4.3.1) or as functions of the order statistics. Considerable analytic machinery exists for dealing with the distribution (especially for n large) of such functions when the X_i's are i.i.d.. Specifically, in such a case we know that $F^{(n)}(x)$ converges uniformly almost surely to the population distribution F. The normalized deviation process $\sqrt{n}(F_n(x) - F(x))$ converges in distribution to a normal process with mean zero and covariance function $F(x_1)(1 - F(x_2))$, $x_1 \leq x_2$. It is not surprising that analogous results should hold for sample Lorenz curves, that nice functions (like inequality measures) of the sample distribution function (or sample Lorenz curve) should converge almost surely to the corresponding function of the population distribution function (or population Lorenz curve) and that for large n the deviations will be approximately normal. There is often a considerable amount of hard work required checking regularity conditions and working out the sometimes complicated limiting variances, but generally there are no surprises. To some extent, the utility of the asymptotic results is limited by the complicated way in which the asymptotic variances depend on the usually unknown parent distribution F. The detailed theoretical underpinnings of the material in this section involve careful consideration of the appropriate space in which the empirical process is to converge to the limiting normal process. The basic reference for such concepts remains Billingsley (1968). An excellent detailed discussion of this material, specifically in the context of Lorenz curves, is provided by Goldie (1977). Contributions in this context by Gastwirth, Gail and Gastwirth, and Chandra and Singpurwalla will be cited as they appear. As appropriate, we will on occasion recognize the possibility of developing asymptotic theory, using representations involving linear functions of order statistics whose limiting behavior is well understood.

Turning to the list of analytic inequality statistics, we may begin with the (absolute) mean deviation

$$T_1(\underline{X}) = \frac{1}{n}\sum_{i=1}^{n}|X_i - \overline{X}|. \tag{4.3.5}$$

This measure is location but not scale invariant. Two other representations of (4.3.5) merit attention. In terms of the sample d.f. we may write

$$T_1^{(n)}(\underline{X}) = \int_0^\infty |x-m| dF^{(n)}(x) \tag{4.3.6}$$

where $m = \int_0^\infty x dF^{(n)}(x) = \overline{X}$. The mean deviation is not expressible as a linear combination of order statistics. A useful representation involving order statistics is provided by (Gastwirth, 1974)

$$T_1^{(n)}(\underline{X}) = 2\left[N\overline{X} - \sum_{X_i<\overline{X}} X_i\right]/n \tag{4.3.7}$$

where N is the (random) number of observations which are less than $\overline{X}$. If the X_i's are i.i.d. from a parent distribution with finite mean, then $T_1^{(n)}(\underline{X})$ will provide an asymptotically unbiased estimate of the population mean deviation $\tau_1(F)$. In the case of a normal parent distribution (which is of only peripheral interest given our emphasis on non-negative random variables) the exact distribution of $T_1^{(n)}$ has been obtained (Godwin, 1945), and closed form expressions for the exact mean and variance are available (dating back to Helmert, 1876!). A convenient reference is Kendall, Stuart and Ord (1994, p. 362). For non-normal parent distributions only asymptotic results are available. Rather than try to approach the problem using the sample distribution function representation (4.3.6), we follow Gastwirth and utilize (4.3.7). Provided that the parent distribution F has a continuous density in a neighborhood of its mean m and a finite variance σ^2, we conclude that $T_1^{(n)}(\underline{X})$ is asymptotically normally distributed. Specifically we have (Gastwirth, 1974)

$$\sqrt{n}[T_1^{(n)}(\underline{X}) - \tau_1(F)] \xrightarrow{d} N(0, \sigma_1^2) \tag{4.3.8}$$

where

$$\begin{aligned}\sigma_1^2 = 4\left\{[F(m)]^2 \int_m^\infty (x-m)^2 dF(x) + [\overline{F}(m)]^2 \int_{-\infty}^m (x-m)^2 dF(x)\right\} \\ -[\tau_1(F)]^2.\end{aligned} \tag{4.3.9}$$

The proof of (4.3.8) can be broken into two steps. First show that the effect of replacing each X_i by $X_i - m, N$ by $nF(m)$ and $\overline{X}$ by m where it appears under the summation

Table 4.3.1 *Asymptotic variance* (σ_1^2) *of the mean deviation in the case of a* $P(I)(\sigma,\alpha)$ *distribution (from Gastwirth, 1974).*

α	σ_1^2
2.5	2.5072
3.0	0.7972
3.5	0.3820
4.0	0.2212
5.0	0.1000

sign in (4.3.7) is negligible (in fact it is $0_p(n^{-1})$). Then apply the central limit theorem to

$$\left(F(m)\sum_{i=1}^{n}(X_i - m), \sum_{i=1}^{n} I_i(X_i - m)\right) \tag{4.3.10}$$

where I_i is the indicator random variable corresponding to the event $\{X_i < m\}$. Gastwirth (1974) provided the details. He also remarked on a discrepancy between the asymptotic variance given by (4.3.9) and an earlier published approximate variance for $T_i^{(n)}$ (Kendall, Stuart and Ord, 1994, p. 361). The discrepancy can be non-negligible in the case of an asymmetric parent distribution. For some parent distributions the asymptotic variance (4.3.9) is computable in closed form. In general, it is not. For example, if the X_i's $\sim \Gamma(1,\sigma)$ (i.e., an exponential parent distribution), then one may verify that the asymptotic variance is given by $4\sigma^2[2e^{-1} - 4e^{-2}]$. In the classical Pareto case (i.e., $X_i \sim P(I)(\sigma,\alpha)$), Gastwirth provided a brief table for the asymptotic variance of $T_1^{(n)}(\underline{X})$. [Note that he treated $T_1^{(n)}(\underline{X})/2$, and so his V^2 must be multiplied by 4 to get σ_1^2 in Table 4.3.1.]

To obtain scale invariance, standardization by the sample mean leads us to the relative mean deviation statistic

$$T_2(\underline{X}) = T_1(\underline{X})/\overline{X}. \tag{4.3.11}$$

Location invariance is sacrificed. The sample Pietra index is, perhaps, more often encountered in discussions of income distributions. It is given by

$$P(\underline{X}) = T_2(\underline{X})/2, \tag{4.3.12}$$

i.e., one-half the relative mean deviation. We can identify the Pietra index as the ratio of the area of the maximal inscribed triangle within the sample Lorenz curve (to be defined later exactly in the way one would predict) to the area under the egalitarian line. Thus, the Pietra index is two times and the relative mean deviation is four times the area of the maximal inscribed triangle. The advantage of the Pietra index is that its range is in the interval $[0,1]$. The value 1 is not achievable. One may verify that

$$0 \le P(\underline{X}) \le (n-1)/n. \tag{4.3.13}$$

As is often the case, the inequality measure assumes its largest value when all but one of the X_i's are zero (i.e., when all the wealth is in one individual's hands).

The small sample distributional behavior of the relative mean deviation is predictably intractable, but large sample theory can be developed. Again, a convenient reference is Gastwirth (1974). Provided that the parent distribution has a finite variance (σ^2) and a continuous density in a neighborhood of its mean (m), we may conclude that $T_2^{(n)}(\underline{X})$ is asymptotically normal. Specifically,

$$\sqrt{n}[T_2^{(n)}(\underline{X}) - \tau_2(F)] \xrightarrow{d} N(0,\sigma_2^2) \tag{4.3.14}$$

where

$$\sigma_2^2 = \frac{\sigma_1^2}{m^2} + \frac{\sigma^2[\tau_2(F)]^2}{m^2} - \frac{4\tau_2(F)}{m^2}\left[F(m)\sigma^2 - \int_{-\infty}^{m}(x-m)^2 dF(x)\right] \quad (4.3.15)$$

where σ_1^2 is given by (4.3.9). This result is also a consequence of the central limit theorem applied to (4.3.10) (and the discussion which precedes that expression).

The sample Amato index, the length of the sample Lorenz curve, admits the representation

$$A(\underline{X}) = \frac{1}{n}\sum_{i=1}^{n}\sqrt{1+\left(\frac{X_i}{\overline{X}}\right)^2}. \quad (4.3.16)$$

The extremal configurations are $(1,1,\ldots,1)$ and $(0,0,\ldots,0,1)$ and the range of possible values of the sample Amato index for population of size n satisfies

$$\sqrt{2} \leq A_n(X) \leq 1 - \frac{1}{n} + \sqrt{1+\left(\frac{1}{n}\right)^2}. \quad (4.3.17)$$

Asymptotic normality of this index is readily verified.

The sample Bonferroni index can be represented in the form

$$B_n(\underline{X}) = 1-(n-1)^{-1}\sum_{i=1}^{n-1}\frac{m_i}{m}, \quad (4.3.18)$$

where $m = n^{-1}\sum_{i=1}^{n} X_{i:n}$ and $m_i = i^{-1}\sum_{j=1}^{i} X_{j:n}$. For this index also, asymptotic normality can be proved.

The sample Zenga-II index is given by

$$Z_n^{II}(\underline{X}) = 1 - n^{-1}\sum_{i=1}^{n}\frac{i^{-1}\sum_{k=1}^{i} X_{k:n}}{(n-i)^{-1}\sum_{k=i+1}^{n} X_{k:n}}. \quad (4.3.19)$$

Greselin and Pasquazzi (2009) confirmed the asymptotic normality of this index and discussed related asymptotic confidence intervals.

The sample standard deviation

$$T_3(\underline{X}) = \sqrt{\frac{1}{n}\sum_{i=1}^{n}(X_i - \overline{X})^2} \quad (4.3.20)$$

has well-known sampling properties. It is location but not scale invariant. Though quite well-behaved for normal samples ($[T_3(\underline{X})]^2$ having a gamma distribution), closed form distributional results are generally not available. Since it is a straightforward function of the first two sample moments, asymptotic normality is inevitable, provided that the parent population has finite fourth moments. Specifically, one may verify that

$$\sqrt{n}[T_3(\underline{X}) - \tau_3(F)] \xrightarrow{d} N(0, \sigma_3^2) \quad (4.3.21)$$

where

$$\sigma_3^2 = \{E(X^4) - [E(X^2)]^2\}/4E(X^2). \tag{4.3.22}$$

The corresponding standardized measure is the coefficient of variation

$$T_4(\underline{X}) = T_3(\underline{X})/\overline{X}. \tag{4.3.23}$$

One may verify that

$$0 \leq T_4(\underline{X}) \leq \sqrt{n-1}. \tag{4.3.24}$$

Again asymptotic normality occurs provided fourth moments exist. One finds

$$\sqrt{n}[T_4(\underline{X}) - \tau_4(F)] \xrightarrow{d} N(0, \sigma_4^2) \tag{4.3.25}$$

where

$$\sigma_4^2 = \frac{[E(X)]^2\{E(X^4) - [E(X)]^2\} - 4\{[E(X^2)]^3 - E(X)E(X^2)E(X^3)\}}{4[E(X)]^4\{E(X^2) - [E(X)]^2\}} \tag{4.3.26}$$

If we consider differences instead of deviations, we can develop a parallel set of inequality measures. Gini's mean difference has the form

$$T_5(\underline{X}) = \frac{1}{n(n-1)} \sum_{i=1}^{n} \sum_{j \neq i} |X_i - X_j|. \tag{4.3.27}$$

This is the average of all the pairwise differences between the observations. The form (4.3.27) is most popular. There is good reason to consider a variant form of the definition

$$T_5'(\underline{X}) = \frac{1}{n^2} \sum_{i=1}^{n} \sum_{j=1}^{n} |X_i - X_j|. \tag{4.3.28}$$

The advantage to the use of (4.3.28) is that when it is used to subsequently define a sample Gini index, that sample index will be precisely two times the area of the concentration polygon (defined by the sample Lorenz curve). That is to say, the Gini index thus defined would be related to the sample Lorenz curve (and the sample distribution function) in exactly the way that the population Gini index was related to the population Lorenz curve (and the distribution function). Wold (1935) advocated use of (4.3.28) for this reason. The difference between (4.3.27) and (4.3.28) is slight for reasonable sample sizes. Custom dictates use of (4.3.27). The advantage, in terms of consistency between sample and population statistics, of (4.3.28) has been ignored. Obviously the two are asymptotically equivalent. Somewhat reluctantly, we will bow to custom and use the formula involving distinct values, i.e., (4.3.27). There are several equivalent variants of the Gini statistic in terms of order statistics. David (1968) is a useful reference here. One may write

$$\frac{n(n-1)}{2}T_5(\underline{X}) = \sum_{i=1}^{[n/2]}(n-2i+1)\{X_{n-i+1:n} - X_{i:n}\} \tag{4.3.29}$$

$$= \sum_{i=1}^{n}\sum_{j=i+1}^{n}\{X_{j:n} - X_{i:n}\} \tag{4.3.30}$$

$$= \sum_{i=1}^{n}(2i-n-1)X_{i:n}. \tag{4.3.31}$$

A useful representation in terms of spacings is

$$\frac{n(n-1)}{2}T_5(\underline{X}) = \sum_{i=1}^{n-1} i(n-i)\{X_{i+1:n} - X_{i:n}\}. \tag{4.3.32}$$

One property that $T_5(\underline{X})$ does enjoy is that of unbiasedness. It is a U statistic (in the sense of Hoeffding, 1948) and, in fact, is the unique unbiased estimate of the population Gini mean difference, $\tau_5(F) = E(X_{2:2}) - E(X_{1:2})$. The variance of the mean difference was first obtained by Nair (1936) and subsequently by Lomnicki (1952). One first computes the second moment, thus

$$E([T_5(\underline{X})]^2) = \frac{1}{n^2(n-1)^2}E\left[\sum_{i\neq j}|X_i - X_j|^2 + \sum_{i\neq j\neq k}|X_i - X_j||X_i - X_k| + \sum_{i\neq j\neq k\neq \ell}|X_i - X_j||X_k - X_\ell|\right]. \tag{4.3.33}$$

If we then use the notation

$$\xi^2 = \frac{1}{2}E(X_1 - X_2)^2$$
$$\lambda = E(|X_1 - X_2||X_1 - X_3|) \tag{4.3.34}$$

and

$$\tau_5 = E(|X_1 - X_2|),$$

we readily obtain an expression for the variance subtracting τ_5^2 from (4.3.33):

$$\mathrm{var}(T_5(\underline{X}) = \frac{1}{n(n-1)}[4\xi^2 + 4(n-2)\lambda - 2(2n-3)\tau_5^2]. \tag{4.3.35}$$

In particular, from Nair (1936), we have the expressions

$$[X_i\text{'s} \sim \Gamma(1,\sigma)] \quad E(T_5(\underline{X})) = \sigma$$
$$\text{var}(T_5(\underline{X})) = \sigma^2(4n-2)/(3n^2-3n) \tag{4.3.36}$$

$$[X_i\text{'s} \sim \text{Uniform }(0,\sigma)] \quad E(T_5(\underline{X})) = \sigma/3$$
$$\text{var}(T_5(\underline{X})) = \sigma^2(n+3)/[45(n^2-n)]. \tag{4.3.37}$$

In the case of a $P(\text{I})(\sigma,\alpha)$ parent distribution the relevant quantities in (4.3.34) assume the values

$$\begin{aligned} & \xi^2 = \sigma^2\alpha/[(\alpha-1)^2(\alpha-2)] \\ [X_i\text{'s} \sim P(\text{I})(\sigma,\alpha)] \quad & \tau_5 = 2\alpha\sigma/[(\alpha-1)(2\alpha-1)] \\ & \lambda = \frac{\sigma^2[8\alpha^3-11\alpha^2+2\alpha]}{[(\alpha-1)^2(\alpha-2)(2\alpha-1)(3\alpha-2)]}. \end{aligned} \tag{4.3.38}$$

These may be substituted in (4.3.35) to obtain the variance of $T_5(\underline{X})$, the Gini mean difference. The quantity λ defined in (4.3.34) may alternatively be expressed in the form

$$\lambda = \int_0^\infty \left[\int_0^\infty (x_1-x_2)dF(x_2)\right]^2 dF(x_1) \tag{4.3.39}$$

(Fraser, 1957, p. 230, for example, uses this representation).

The analysis analogous to that which led to the variance of $T_5(\underline{X})$ can be developed for higher moments. An example is provided by Kamat's (1953) derivation of the third moment. Kamat also considered the approximate distribution of Gini's mean difference for the case where the underlying distribution is normal.

In fact, in unusual circumstances the actual distribution of Gini's statistic can be directly computed. Since it has a representation as a linear function of spacings, it is not surprising that its distribution should prove to be (reasonably) tractable in the case of an underlying exponential distribution. Kabe (1975) discussed the distribution of the Gini statistic in both the exponential and uniform cases.

Recall (4.3.32) which can be written as

$$T_5(\underline{X}) = \frac{2}{n(n-1)}\sum_{i=1}^{n-1} i(n-i)Y_{i+1:n}$$

where $Y_{i+1:n} = X_{i+1:n} - X_{i:n}$ is the $(i+1)$st spacing. However, in an exponential population, the spacings are themselves exponentially distributed and are independent (refer to the Sukhatme Theorem 3.4.2). In fact $Y_{i:n} \sim \Gamma(1,\sigma/(n-i+1))$ where σ is the scale parameter of the parent exponential distribution. It follows that

$$T_5(\underline{X}) = \sum_{i=1}^{n-1} U_i \tag{4.3.40}$$

where the U_i's are independent and

$$U_i \sim \Gamma\left(1, \frac{2i\sigma}{n(n-1)}\right), \quad i = 1, 2, \ldots, n-1. \tag{4.3.41}$$

Since all the scale parameters in (4.3.41) are distinct, we may use the result from Feller (1971, p. 40) to write the exact form of the density of the Gini statistic in the exponential case. Thus, if the X_i's $\sim \Gamma(1, \sigma)$,

$$f_{T_5(\underline{X})}(x) = \frac{n}{2\sigma(n-2)!} \sum_{i=1}^{n-1} \left[\prod_{\substack{j=1 \\ j \neq i}}^{n-1} \left(\frac{1}{j} - \frac{1}{i}\right) \right] e^{-xn(n-1)/2i\sigma}. \tag{4.3.42}$$

Kabe's equation (8) is equivalent to (4.3.42). Kabe uses the density of $T_5(\underline{X})$ to evaluate its first and second moments. An alternative, perhaps easier, approach is to utilize the representation (4.3.40). The results obtained in this way agree with those obtained earlier by Nair (1936) (see (4.3.36)). The expression given in Kabe (1975) for the second moment is incorrect by a factor of 48. Higher momenta of the Gini statistic in the exponential case are, of course, obtainable using the representation (4.3.40). For example, one finds

$$E([T_5(\underline{X})]^3) = \sigma^3(5n^2 - 3n + 4)/[n(n-1)]. \tag{4.3.43}$$

In the case of a rectangular parent distribution the representation (4.3.31) may be used. Wani and Kabe (1972) gave an expression for the density of a linear combination $\sum_{i=1}^{n} a_1 X_{i:n}$ of the order statistics from a uniform sample provided the a_i's are distinct. Using this, Kabe (1975) was able to write an expression of the density of $T_5(\underline{X})$ from which an expression for the moments is obtainable. The first and second moments had earlier been derived by Nair (1936) (see (4.3.37)).

Asymptotic normality of the Gini mean difference is typically justified by recognizing it to be a U statistic and using Hoeffding's (1948) result on the limiting distribution of such statistics. In fact, Hoeffding used $T_5(\underline{X})$ as an example in his original paper, and it remains the standard textbook example of the use of Hoeffding's theory (see e.g., Fraser, 1957, p. 230). Specifically, we have, provided second moments exist,

$$\sqrt{n}[T_5^{(n)}(\underline{X}) - \tau_5(F)] \xrightarrow{d} N(0, \sigma_5^2) \tag{4.3.44}$$

where

$$\sigma_5^2 = 4(\lambda - [\tau_5(F)]^2) \tag{4.3.45}$$

(for λ, see (4.3.34) or (4.3.39)).

Aaberge (2007) provides discussion of the asymptotic distribution of a spectrum of empirical rank dependent measures of inequality making use of their representation as functionals of the empirical Lorenz process to be defined in Section 4.3.4.

Giorgi and Pallini (1987a, 1990) present some preliminary simulation based results comparing the rates of convergence to normality of several of the sample inequality indices discussed in this subsection. More extensive and more detailed investigation of this issue would be desirable.

4.3.3 The Sample Gini Index

To obtain scale invariance, it would be plausible to divide T_5 by $\overline{X}$. Custom dictates division by $2\overline{X}$ to obtain the sample Gini index. As with the Gini mean difference there are two variants:

$$G_n = T_6(\underline{X}) = \left[\sum_{i=1}^{n}\sum_{j\neq i}|X_i - X_j|]/[2n(n-1)\overline{X}\right] \tag{4.3.46}$$

and

$$G_n' = T_6'(\underline{X}) = \left[\sum_{i=1}^{n}\sum_{j=1}^{n}|X_i - X_j|\right] \Big/ 2n^2\overline{X}. \tag{4.3.47}$$

In general, (4.3.46) is used (see the discussion following (4.3.28)). The Gini index can be written as a ratio of linear functions of order statistics. Thus,

$$G_n = \left[\sum_{i=1}^{n}(2i-n-1)X_{i:n}\right] \Big/ \left[\sum_{i=1}^{n}(n-1)X_{i:n}\right]. \tag{4.3.48}$$

Cicchitelli (1976) uses such a representation to obtain an expression for the density of G_n.

Although the maximal area between the sample Lorenz curve and the egalitarian line cannot exceed $(n-1)/2n$, the sample Gini index defined by (4.3.46) satisfies

$$0 \leq G_n \leq 1. \tag{4.3.49}$$

The extremal configurations are $(1,1,\ldots,1)$ and $(0,0,\ldots,0,1)$.

In the case of an underlying exponential distribution a relatively simple exact expression is available for the distribution of G_n. Cicchitelli (1976) treated this case. Gail and Gastwirth (1978a) gave two alternative expressions citing work of Dempster and Kleyle (1968). The key to the derivation is the Sukhatme representation. If we denote the spacings by $Y_{i:n} = X_{i:n} - X_{i-1:n}$, then we know that the $(n-i+1)Y_{i:n}$'s, $i = 1,2,\ldots,n$, are i.i.d. exponential random variables. Now, if we define

$$T = \sum_{i=1}^{n} X_i = \sum_{i=1}^{n}(n-i+1)Y_{i:n}, \tag{4.3.50}$$

it follows that G_n can be written as a linear combination of the coordinates of a symmetric Dirichlet $(1,1,1,1,\ldots,1)$ random vector. Thus,

$$G_n = \sum_{j=1}^{n} \frac{n-j}{n-1} D_j \tag{4.3.51}$$

where $(D_1,\ldots,D_n)$ has a Dirichlet distribution. From this Dempster and Kleyle (1968) obtained the following expression for distribution of G_n when the X_i's $\sim \Gamma(1,\sigma)$,

$$P(G_n \leq z) = \frac{z^{n-1}}{(\prod_{i=1}^{n} c_i)} - \sum_{j=1}^{n-1} \frac{[(z-c_j)_+]^{n-1}}{\left\{c_j \prod_{k\neq j}^{n-1}(c_k - c_j)\right\}}, \quad 0 \leq z \leq 1 \tag{4.3.52}$$

where $c_j = (n-j)/(n-1)$ and $(u)_+ = u$ if $u > 0$ and is zero otherwise. The variant form of the distribution is

$$P(G_n \leq z) = 1 - \sum_{j=1}^{n-1} \frac{[(c_j - z)_+]^{n-1}}{c_j \prod_{k \neq j}^{n-1} (c_j - c_k)}, \quad 0 \leq z \leq 1. \tag{4.3.53}$$

Gail and Gastwirth (1978a) tabulated the distribution of G_n and showed that the approach to normality is rapid. A sample of size 10 is already sufficiently large to permit use of the normal approximation. It is not obvious from (4.3.52) or (4.3.53) that G_n is a symmetric random variable (symmetric about $1/2$), but such is the case (see Gail and Gastwirth for a justification in which they wrote $G_n - 1/2$ as a linear combination of the coordinates of a symmetric Dirichlet vector). The representation of G_n (in the exponential case) as a linear combination of coordinates of a Dirichlet variable (4.3.51) permits easy computation of the mean and variance:

$$[X_i\text{'s} \sim \Gamma(1, \sigma)] \quad E(G_n) = \frac{1}{2} \tag{4.3.54}$$

$$\text{var}(G_n) = [12(n-1)]^{-1}. \tag{4.3.55}$$

Generally speaking, the moments of G_n are not expressible in simple form. Exceptional cases exist, e.g., the exponential distribution discussed above. It is also not usually feasible to obtain an unbiased estimate of the population Gini index. One contribution in this direction is that of Taguchi (1978). He described a sampling scheme for finite populations which would yield an unbiased estimate of G.

For a general parent distribution with finite second moments the asymptotic normality of the sample Gini index ($G_n = T_6(\underline{X})$) is readily verified. The technique of proof is to first show asymptotic joint normality of Gini's mean difference and the sample mean. With knowledge of the asymptotic covariance between the two statistics it is straightforward (though a bit tedious) to derive the asymptotic distribution of the ratio which, except for a factor of $1/2$, is the Gini index. One finds

$$\sqrt{n}[T_6^{(n)}(\underline{X}) - \tau_6(F)] \xrightarrow{d} N(0, \sigma_6^2) \tag{4.3.56}$$

where

$$\sigma_6^2 = \frac{\lambda}{m^2} - \frac{\rho \tau_5}{m^3} + \frac{\xi^2}{2m^4}. \tag{4.3.57}$$

Here λ, τ_5 and ξ^2 are defined in (4.3.34) while $m = E(X)$ and

$$\rho = \int_0^\infty \int_0^\infty x_1 |x_1 - x_2| \, dF(x_1) \, dF(x_2). \tag{4.3.58}$$

Chandra and Singpurwalla (1978) pointed out that the above result can alternatively be obtained by referring to Barlow, Bartholomew, Bremner and Brunk's (1972) result for the cumulative total time on test transform. Goldie (1977) obtained an equivalent result by regarding G_n as a function of the sample Lorenz curve. He used the variant definition (4.3.47) for the Gini index. Goldie's approach, though, in some ways the right way of looking at the problem, requires additional regularity conditions beyond the existence of second moments. We will mention these conditions below. They are needed to guarantee weak convergence of the Lorenz process.

4.3.4 Sample Lorenz Curve

In discussing the asymptotic behavior of sample Lorenz curves we must take into account two competing definitions. One, of course, is the Lorenz curve (as defined in Section 4.3) of the sample distribution function. Thus, for $0 \leq u \leq 1$, we define

$$L_n(u) = \int_0^u F_n^{-1}(y)\,dy \Big/ \int_0^1 F_n^{-1}(y)\,dy. \tag{4.3.59}$$

It is not difficult to verify that this definition describes a polygonal line joining the $(n+1)$ points $(j/n, (\sum_{i \leq j} X_{i:n} / \sum_{i=1}^n X_{i:n}))$. The variant definition used by Taguchi (1968), Gail and Gastwirth (1978b) and Chandra and Singpurwalla (1978) is defined by the equation

$$L'_n(u) = \sum_{i \leq [nu]} X_{i:n}/n\overline{X}, \quad 0 \leq u \leq 1 \tag{4.3.60}$$

where $[\cdot]$ denotes "integer part." The variant definition is discontinuous at the points $X_{i:n}$ and flat between these points. Goldie (1977) pointed out that asymptotically the two variant definitions are equivalent. This is because

$$\sup_{0 \leq u \leq 1} |L_n(u) - L'_n(u)| = X_{n:n}/n\overline{X}, \tag{4.3.61}$$

and the quantity $X_{n:n}/n\overline{X}$ converges almost surely to 0 (provided the first moment exists). Since they are asymptotically indistinguishable and since L_n, defined by (4.3.59), is consistent with the usual definition of a Lorenz curve, Goldie urged use of (4.3.59) as a definition. The variant appears to be more popular. Provided that $E(X)$ exists, the almost sure uniform convergence of F_n to F can be used in a straightforward manner to verify the almost sure uniform convergence of L_n to the population Lorenz curve, L (using definition (4.3.59)). The same is true even with the modified definition (4.3.60), but the proof is technically more complicated (see Goldie, 1977, where the inverses L_n^{-1} and $L_n'^{-1}$ are also shown to converge uniformly almost surely). Gail and Gastwirth (1978b) provided a relatively straightforward proof that, for a fixed $u \in [0,1], L'_n(u)$ converges almost surely to $L(u)$.

If we are willing to assume finite second moments, then we can reasonably hope to have asymptotic normality of suitably scaled deviations $(L_n(u) - L(u))$ (since such is the case for deviations between the sample and population distributions). To this end we use Goldie's notation and define the Lorenz process λ_n by

$$\lambda_n = \sqrt{n}(L_n - L). \tag{4.3.62}$$

In order to prove weak convergence of these processes to a limiting normal process, certain regularity conditions must be imposed. Several choices are available. We will quote one of Goldie's sufficient conditions to give the flavor of the results.

Theorem 4.3.1. *Let F be a continuous distribution function with finite second moment and connected support S with* $\sup S = \infty$. *In addition, assume there exists $a < 1$, $t_0 < \infty$, $A < \infty$ such that*

$$[F^{-1}(1-(vt)^{-1})/F^{-1}(1-t^{-1})] \leq Av^a, \quad v \geq 1, t > t_0. \tag{4.3.63}$$

Then, if we let λ_n denote the sample Lorenz process (4.3.62) based on a sample of size n, we have λ_n converging weakly in the space of continuous functions on $[0,1]$ to a normal process λ. The limiting normal process may be expressed in the form

$$\lambda(u) = m^{-1}\int_0^1 [L(u) - I(t \le u)]\beta(t)\,dF^{-1}(t), \quad 0 \le u \le 1 \tag{4.3.64}$$

where $I(t \le u) = 1$ if $t \le u$ and 0 otherwise and β is a Brownian bridge (a normal process with $E[\beta(u)] = 0 \forall u$ and $\operatorname{cov}(\beta(u), \beta(v)) = u(1-v)$ if $u \le v$).

Rao and Zhao (1995) and subsequently Csorgo and Zitikis (1996, 1997a) prove a law of the iterated logarithm for the sample Lorenz curve. Csorgo, Gastwirth and Zitikis (1998) discuss the development of asymptotic confidence bands for the Lorenz curve and for the Bonferroni curve. See also Csorgo and Zitikis (1997b). Earlier, Csorgo, Csorgo and Horvath (1987) discussed estimation of Lorenz curves (and total time on test transforms) based on data subject to random censoring from the right. Csorgo and Yu (1999) consider the consequences of relaxing the i.i.d. assumption for the sample $X_1, X_2, ... X_n$, replacing it by an assumption of mixing dependence. If only length biased data are available, it is still possible to estimate the Lorenz curve via a suitable normal process under appropriate regularity conditions (Fakoor, Ghalibaf and Azarnoosh, 2011).

If we are willing to forego proving weak convergence of the processes and merely consider the asymptotic distribution of $\lambda_n(u)$ for one fixed u (or a finite number of fixed values $u_1, \dots u_k$), then some regularity conditions can be dispensed with. For example from Gail and Gastwirth (1978b) we have, for the variant definition of the Lorenz process:

Theorem 4.3.2. *Let F correspond to a positive random variable with finite second moments. Assume that F has a unique uth quantile x_u (i.e., $F(x_u) = u$) and that F is continuous at x_u. It follows that $[L'_n(u) - L(u)]/\sqrt{\operatorname{var} L'_n(u)}$ converges in distribution to a standard normal variable. Equivalently, we may write*

$$\sqrt{n}[L'_n(u) - L(u)] \xrightarrow{d} N(0, \sigma^2_{\lambda(u)}) \tag{4.3.65}$$

where

$$\sigma^2_{\lambda(u)} = \frac{\tau_1^2}{m^2} + \frac{\tau_2^2\eta^2}{m^4} - \frac{2\tau_{12}\eta}{m^3} \tag{4.3.66}$$

in which

$$\begin{aligned}
m &= \int_0^\infty x\,dF(x) \\
\eta &= \int_0^{x_u} x\,dF(x) \\
\tau_1^2 &= 2\int_0^{F^{-1}(u)} \left[\int_0^{F^{-1}(y)} F(x)\,dx\right][1 - F(y)]\,dy \\
\tau_2^2 &= \int_0^\infty (x-m)^2\,dF(x) \\
\tau_{12} &= \tau^2 + \int_0^{F^{-1}(u)} F(x)\,dx \int_{F^{-1}(u)}^\infty [1 - F(y)]\,dy.
\end{aligned} \tag{4.3.67}$$

The proof relies on asymptotic theory of linear functions of order statistics. Gail (1977) gave the parallel result describing the asymptotic joint normality of $(L'_n(u_1), L'_n(u_2), \ldots, L'_n(u_k))$ for k fixed values $u_1, u_2, \ldots, u_k$. The expression (4.3.66) for the asymptotic variance can be troublesome to evaluate. One tractable case involves an exponential parent distribution. In this case Gail and Gastwirth (1978b) obtained

$$[X_i\text{'s} \sim \Gamma(1,\sigma)] \quad \sigma^2_{\lambda(u)} = 2(1-u)\log(1-u) + u(2-u) - [u + (1-u)\log(1-u)]^2. \tag{4.3.68}$$

Chandra and Singpurwalla (1978) also addressed the problem of determining the asymptotic distribution of $L'_n(u)$ for a fixed value of u. Their results were obtained assuming that F^{-1} has a non-zero continuous derivative on $(0,1)$. Sendler (1979) gave an alternative derivation, assuming a finite $(2+\delta)$'th moment for F.

Note 4.3.1. *Sample Lorenz curves are piecewise linear functions. One might prefer a smoothed version of L_n since most income models involve a differentiable population Lorenz curve. A kernel based non-parametric alternative to L_n is discussed in Anderson and Arnold (1993).*

As remarked earlier, it is possible to view the Gini index as a continuous function of the Lorenz curve and use this result to obtain its asymptotic distribution. Using this approach, Goldie showed that G'_n (see (4.3.47)), which is the Gini index of the sample Lorenz curve $L_n(u)$ (see (4.3.59)), will converge almost surely to G the Gini index of the population Lorenz curve $L(u)$. With additional assumptions he obtained the asymptotic normality of G_n, obtaining results which agree with (4.3.56). Analogous arguments can be used to study the asymptotic behavior of the sample Pietra index (or sample coefficient of variation).

If we take the variant definition of the sample Lorenz curve $L'_n(u)$ (see (4.3.60)) and compute twice the area between it and the egalitarian line, we obtain

$$G''_n = \left[X_{1:n} + \sum_{i=2}^{n-1}(2i-n)X_{i:n}\right] \Big/ \left[n\sum_{i=1}^{n} X_{i:n}\right]. \tag{4.3.69}$$

This third (!) variant of the sample Gini index differs from both the earlier ones [(4.3.46) and (4.3.47)]. However, all three variants are asymptotically equivalent (compare (4.3.69) and (4.3.48)). Since we know $L'_n(u)$ converges to $L(u)$, the functional G''_n will converge to G, so that we have almost sure convergence and asymptotic normality for G''_n. Since G_n and G''_n are asymptotically equivalent, we get the corresponding limit theorems for G_n. This is essentially the approach of Chandra and Singpurwalla (1978), who related G''_n to the cumulative total time on test statistic.

What can we say about the small sample behavior of the sample Lorenz curve? The relevant work here was done by Gail (1977) and Gail and Gastwirth (1978b) using the variant definition $L'_n(u)$. Results are only available in the case of an exponential parent distribution. The key observation is that when the X_i's are $\Gamma(1,\sigma)$ variables, then the vector $(U_1, \ldots, U_n)$ where $U_i = X_i/(\sum_{i=1}^n X_i)$ has a Dirichlet $(1,1,\ldots,1)$ distribution. Thus $L'_n(u)$, from (4.3.60), has the same distribution as the sum of the first [nu] ordered coordinates of a Dirichlet random vector. Since uniform spacings have such a Dirichlet distribution, we conclude that $L'_n(u)$ has the same distribution as the sum of the [nu] smallest spacings based on a sample of size $n-1$ from a uniform $(0,1)$ distribution. Using Mauldon (1951) this distribution may be expressed in the form

$$P(L'_n(u) > t)$$
$$= \sum_{j<(m-tn)/(1-t)} \frac{n!(m-j-t(n-j))^{n-1}(-1)^j}{(n-j)(n-m)^{m-1}(m-j)^{n-m-1}j!(m-i)!(n-m)!},$$
$$0 < t < m/n \tag{4.3.70}$$

where $m = [\text{nu}]$. Gail and Gastwirth (1978b) provided tables of percentiles of the distribution of $L'_n(0.5)$ for $n \leq 40$. For larger values of n, the normal approximation can be used.

Giorgi and Mondani (1995) derive analogous results for the small sample distribution of the sample Bonferroni index (4.3.18) in the case of an exponential parent distribution. They verify that, in this setting, $B_n(\underline{X})$ can be expressed as a ratio of linear combinations of exponential order statistics and consequently as a ratio of linear combinations of independent exponentially distributed spacings.

There remains one more group of measures related to the sample Lorenz curve. These are the sample versions of the Elteto and Frigyes indices (defined in (4.2.48)). We will use a subscript n to distinguish the sample versions from the population versions. Thus, we define

$$\begin{aligned} U_n &= \left[N\sum_{i=1}^{N} X_i\right] \Big/ \left[n\sum_{X_i<\overline{X}} X_i\right] \\ V_n &= \left[N\sum_{X_i>\overline{X}} X_i\right] \Big/ \left[(n-N)\sum_{X_i<\overline{X}} X_i\right] \\ W_n &= \left[n\sum_{X_i>\overline{X}} X_i\right] \Big/ \left[(n-N)\sum_{i=1}^{n} X_i\right] \end{aligned} \tag{4.3.71}$$

where N is the random number of X_i's less than $\overline{X}$. Gastwirth (1974) discussed the asymptotic distribution of these indices in the same paper in which he dealt with the sample mean deviation. The analysis is a little more delicate (which accounts for the fact that the initial computation of the asymptotic variance by Elteto and Frigyes was incorrect). The asymptotic result given by Gastwirth (1974) for the deviation between U_n and the population Elteto and Frigyes index U is

$$\sqrt{n}(U_n - U) \xrightarrow{d} N(0, \sigma_U^2) \tag{4.3.72}$$

where

$$\sigma_U^2 = [F(m)]^{-2}\left\{\frac{v_1}{m_1^2} + \frac{v_2 m^2}{m_1^4} - \frac{2cm}{m_1^3}\right\} \tag{4.3.73}$$

in which $m = E(X)$ and

$$m_1 = [F(m)]^{-1} \int_0^m x dF(x)$$

$$v_1 = m^2 F(m)\overline{F}(m) + \xi^2 (F(m) + mf(m))^2 \\ - 2m(mf(m) + F(m))F(m)(m - m_1)$$

$$v_2 = m^2 f^2(m)\xi^2 + \int_0^m x^2 dF(x) - [F(m)]^2 m_1^2 \\ + 2mf(m) \int_0^m x(x-m) dF(x)$$

$$c = mm_1 F(m)\overline{F}(m) - m^2 f(m)F(m)[m - m_1] \\ + mf(m)[mf(m) + F(m)]\xi^2 + [mf(m) + F(m)] \int_0^m x(x-m) dF(x).$$

In the above, $\xi^2 = \text{var}X$. Analogous results could be worked out for V_n and W_n. No finite sample size distribution theory appears to have been worked out for the Elteto and Frigyes indices.

4.3.5 *Further Sample Measures of Inequality*

Having discussed the major sample indices of inequality in some detail, we now turn to the task of listing several others which have been suggested. Some have been quite popular, others appear to be idiosyncratic choices of individual authors. In almost no case is tractable finite sample size distribution theory available, but in almost all cases, under suitable regularity conditions, asymptotic normality obtains.

Yntema (1933) proposed use of the mean deviation of the logarithms of the data points. Thus

$$T_7(\underline{X}) = \frac{1}{n} \sum_{i=1}^{n} |Y_i - \overline{Y}| \tag{4.3.74}$$

where $Y_i = \log X_i$. Alternatively, the standard deviation of the logarithms of the data points can be considered:

$$T_8(\underline{X}) = \sqrt{\frac{1}{n-1} \sum_{i=1}^{n} (Y_i - \overline{Y})^2} \tag{4.3.75}$$

where, again, $Y_i = \log X_i$. The asymptotic distribution theory is quite straightforward. We merely have the distribution of log X_i playing the role earlier played by the distribution of X_i. The use of the logarithmic transformation in preference to some other monotone transformation appears arbitrary, but is hallowed by years of practice. In addition, if the parent distribution is log-normal or almost so (as it well might be for income data), the logarithms have a particularly attractive distribution.

Bowley's interquartile measure has attractive robustness properties. It ignores outlying observations. This might be inappropriate in income studies where emphasis is often on the upper tail of the distribution. The inequality measure in question is of the form (see Dalton, 1920)

$$T_9(\underline{X}) = (X_{3n/4:n} - X_{n/4:n})/(X_{3n/4:n} + X_{n/4:n}). \tag{4.3.76}$$

If we denote the population quartiles by $\eta_{1/4}$ and $\eta_{3/4}$ and assume the population density, f, is continuous in neighborhoods of $\eta_{1/4}$ and $\eta_{3/4}$, then we know that

$$\sqrt{n}(X_{n/4:n}, X_{3n/4:n}) \xrightarrow{d} N^{(2)}((\eta_{1/4}, \eta_{3/4}), \Sigma) \tag{4.3.77}$$

where

$$\begin{aligned} \sigma_{11} &= 3[16f^2(\eta_{1/4})]^{-1}, \\ \sigma_{22} &= 3[16f^2(\eta_{3/4})]^{-1}, \\ \sigma_{12} &= [16f(\eta_{1/4})f(\eta_{3/4})]^{-1}. \end{aligned}$$

If we let $\tau_9 = (\eta_{3/4} - \eta_{1/4})/(\eta_{3/4} + \eta_{1/4})$, then it is a straightforward exercise to verify that as a consequence of (4.3.77),

$$\sqrt{n}(T_9^{(n)}(\underline{X}) - \tau_9) \xrightarrow{d} N(0, \sigma_9^2) \tag{4.3.78}$$

where

$$\sigma_9^2 = 3\xi_{1/4}^2 - 2\xi_{1/4}\xi_{3/4} + 3\xi_{3/4}^2 \tag{4.3.79}$$

in which

$$\begin{aligned} \xi_{1/4} &= \eta_{3/4}/[2f(\eta_{1/4})] \\ \xi_{3/4} &= \eta_{1/4}/[2f(\eta_{3/4})]. \end{aligned} \tag{4.3.80}$$

Among many other indices, Dalton (1920) discussed the ratio of the logarithms of the arithmetic and geometric means. Thus he considered

$$T_{10}(\underline{X}) = \log \overline{X} / \log \overline{X}_g \tag{4.3.81}$$

where $\overline{X}_g = (\prod_{i=1}^n X_i)^{1/n}$. To avoid problems with division by zero and possible negative values of T_{10}, we assume $X_i > 1$ for every i. By the arithmetic-geometric mean inequality T_{10} satisfies

$$T_{10}(\underline{X}) \geq 1, \tag{4.3.82}$$

with equality if all X_i's are equal. If one wished to have a related inequality measure whose range would be $(0, 1)$, the natural choice would be $1 - (T_{10})^{-1}$. To describe the asymptotic behavior of T_{10}, it is convenient to define

$$m = \int_1^\infty x\,dF(x) \tag{4.3.83}$$

and

$$\tilde{m} = \int_1^\infty \log x \, dF(x). \tag{4.3.84}$$

It then may be verified that

$$\sqrt{n}[T_{10} - (\log m)/\tilde{m}] \xrightarrow{d} N(0, \sigma_{10}^2) \tag{4.3.85}$$

where

$$\sigma_{10}^2 = \frac{\xi_1^2}{m^2 \tilde{m}^2} - \frac{2\xi_{12} \log m}{m \tilde{m}^3} + \frac{\xi_2^2 (\log m)^2}{\tilde{m}^4} \tag{4.3.86}$$

in which

$$\xi_1^2 = \text{var}(X), \quad \xi_2^2 = \text{var}(\log X), \quad \text{and} \quad \xi_{12} = \text{cov}(X, \log X).$$

For various convex functions g, indices of the forms

$$T_g(\underline{X}) = c \sum_{i=1}^n g(X_i), \tag{4.3.87}$$

$$\tilde{T}_g(\underline{X}) = \frac{1}{n} \sum_{i=1}^n g(X_i/\overline{X}) \tag{4.3.88}$$

and

$$\tilde{\tilde{T}}_g(\underline{X}) = \left[\frac{1}{n} \sum_{i=1}^n g(X_i) \right] / g(\overline{X}) \tag{4.3.89}$$

have been proposed. See, e.g., Ord, Patil and Taillie (1978). Such indices are asymptotically normal, provided adequate moments exist. Special cases of T_g, $\tilde{T}_g$ or $\tilde{\tilde{T}}_g$, which have received attention, correspond to the choices: (a) $g(x) = x^2$, Simpson (1949); (b) $g(x) = x \log x$, the negative of the entropy; (c) $g(x) = -U(x)$ where $U(x)$ is a concave utility function (Dalton, 1920, and Atkinson, 1970); (d) $g(x) = \sqrt{1+x^2}$, the Amato index.

Mehran (1976) proposed linear measures of inequality which take the form

$$T_W(\underline{X}) = \left[\sum_{i=1}^n X_{i:n} W\left(\frac{i}{n+1} \right) \right] \Big/ (n\overline{X}) \tag{4.3.90}$$

where W is a smooth weight function. Asymptotic normality of such measures is a consequence of Stigler (1974).

The sample Gini index is asymptotically equivalent to the linear measure corresponding to the weight function $W(u) = 2u - 1$ (see (4.3.44)). Few other choices of W seem to have been explored. Weymark (1979) has discussed generalized Gini indices which, when standardized, take the form

$$G^* = \sum_{i=1}^n a_{i,n} X_{i:n} / n\overline{X}. \tag{4.3.91}$$

To ensure Schur convexity, we need to assume $a_{i,n} \uparrow$ as $i \uparrow$ for each n. Subsequently, Donaldson and Weymark (1980) focussed on the particular case

$$a_{i,n} = [i^{\delta} - (i-1)^{\delta}]/n^{\delta}. \tag{4.3.92}$$

The case $\delta = 2$ is closely related to the usual Gini index.

Marshall, Olkin and Arnold (2011) provide a list of indices of inequality. Most of their entries have already been described. Among the remainder, the following may be of interest.

(i) Emlen's (1973) measure,

$$T_{11}(\underline{X}) = \sum_{i=1}^{n} \frac{X_i}{n\overline{X}} e^{-X_i//(n\overline{X})}. \tag{4.3.93}$$

(ii) Minimal majority (Alker, 1965),

$$T_{12}(\underline{X}) = L_n^{-1}(1/2) \tag{4.3.94}$$

where L_n is the sample Lorenz curve.

(iii) Top 100α percent (Alker and Russet, 1966),

$$T_{13}(X) = L_n(\alpha). \tag{4.3.95}$$

(iv) Fishlow's (1973) poverty measure,

$$T_{14}(\underline{X}) = \sum_{i=1}^{n} (c - X_i)^{+} \tag{4.3.96}$$

where $c > 0$ represents the poverty level.

(v) Quantile ratios of the form

$$T_{15}(1-\alpha,\beta) = \frac{X_{[(1-\alpha)n]:n}}{X_{[\beta n]:n}}, \tag{4.3.97}$$

where $0 < \alpha, \beta < 0.5$.

(vi) Sample income share ratios of the form

$$T_{16}(1-\alpha,\beta) = \frac{1 - L_n(1-\alpha)}{L_n(\beta)}, \tag{4.3.98}$$

where $0 < \alpha, \beta < 0.5$. The sample Palma index is included as the case in which $\alpha = 0.1$ and $\beta = 0.4$.

Hart (1980) includes some additional candidate inequality indices in his excellent survey of income inequality measurement.

4.3.6 Relations Between Sample Inequality Measures

There is not much theoretical justification for seeking interrelationships between indices. However, many of them do seem to be "almost measuring the same thing." The consensus of opinion is that no single measure really suffices to summarize inequality. It is usually suggested that several different indices be evaluated and reported. When pressed for a single number, the Gini index seems to be most often accepted (cf. Morgan, 1962). If one does propose to report several indices, it is desirable to select measures which focus on different aspects of inequality. Bowman (1945) in her useful survey of graphical measures discussed the sensitivity of such measures to various facets of inequality. Yntema (1933) computed values of eight different analytic measures of inequality for a variety of populations. Among the measures he considered, he identified four classes and determined that there were three classes within which inequality appeared to be measured and defined essentially equivalently. The four classes were:

(i) Measures based on first powers of deviations or differences.

(ii) Measures based on second powers of deviations.

(iii) Deviations of logarithms.

(iv) The Gini and Pareto graphical indices.

Between groups there is not much evidence of consistency in how populations are ranked in terms of inequality. Some of Yntema's conclusions can be put on a more solid theoretical basis in terms of whether the measures in question preserve or do not preserve the Lorenz ordering (based on nested Lorenz curves and to be discussed in detail in Section 4.4). Generally speaking, measures in Yntema's classes (i) and (ii) preserve the Lorenz ordering while class (iii) measures do not. As usual, class (iv) measures are hard to evaluate, because of their heuristic definitions (which tacitly assume almost Paretian tail behavior of the distribution in question).

We have already seen that some measures of inequality are necessarily ordered. Using the variant form of the sample Gini index (G'_n defined in (4.3.47)), Taguchi's (1967) inequality (4.2.116) remains valid for sample Gini and Pietra (4.3.12) indices:

$$P_n \leq G'_n \leq P_n(2 - P_n). \tag{4.3.99}$$

Glasser's inequality (4.2.117) also is valid for sample indices G'_n and T_4. It follows from (4.3.99) that

$$1 \leq G'_n/P_n \leq (2 - P_n) \tag{4.3.100}$$

and, consequently, (since $G_n = nG'_n/(n-1)$) we conclude that

$$n/(n-1) \leq G_n/P_n \leq n(2 - P_n)/(n-1), \tag{4.3.101}$$

a result obtained by Taguchi (1967). Glasser (1961) obtained an alternative upper bound for the ratio:

$$G_n/P_n \leq 2. \tag{4.3.102}$$

Benedetti (1961) provided an alternative lower bound

$$\begin{aligned} n/(n-1) &\le G_n/P_n \qquad (n \text{ even}) \\ (n+1)/(n-1) &\le G_n/P_n \quad (n \text{ odd}). \end{aligned} \tag{4.3.103}$$

Bhandari and Mukerjee (1986) derived the following bounds for the sample Gini index in terms of the sample coefficient of variation (4.3.23) based on a sample of size n:

$$\frac{T_4(\underline{X})}{\sqrt{n-1}} \le G_n' \le \sqrt{1-\frac{2(n+1)}{(n-1)^2}}\ \frac{T_4(\underline{X})}{\sqrt{3}} \tag{4.3.104}$$

(compare with equation (4.2.117) where the population Gini index and coefficient of variation are related). Bhandari and Mukerjee (1986) also discuss relations between other commonly used inequality indices. Note that throughout their paper they use the variant form of the sample Gini index (G_n' defined in (4.3.47)).

4.4 Inequality Principles and Utility

One convenient starting point in the quest to measure income inequality is consideration of the effects of simple changes in income patterns in fixed populations. In this section we will always speak of income (or wealth), rather than data from other sources, although the discussion carries over with only slight interpretational problems to other settings. Suppose that we consider a fixed population of n individuals with incomes $X_1, X_2, \ldots, X_n$. Changes in the income pattern can be effected by taxation, confiscation, philanthropy, etc. A revised income distribution will result. Any measure of inequality will reflect changes in the income pattern. Can we agree that certain kinds of changes necessarily increase or decrease inequality? For example, we might agree that adding 1 to each X_i (giving everyone a dollar, if you wish) would decrease inequality. Mathematically, it cuts down the relative variability. In this framework, Dalton (1920) initiated discussion of what can be called inequality principles. It is a reasonable conjecture that none of the inequality principles which have been enunciated command universal acceptance. One, known as the Pigou–Dalton transfer principle (P1 below), has attained wide acceptance. It is the principle referred to in Chapter 1 as the Robin Hood axiom. Simply stated, it avers that taking from the rich and giving to the poor decreases inequality. Clearly, this must be subject to a constraint that one doesn't take too much from the rich to transform them into new poor. This principle will turn out to be intimately related to the Lorenz ordering (based on nested Lorenz curves). Acceptance of a particular inequality principle logically forces one to restrict attention to inequality measures which satisfy certain invariance principles. Thus, if you really think that giving everyone a dollar will decrease inequality, then you will be unhappy with any inequality measure which is translation invariant. In our discussion of various inequality measures we have often noted which ones are scale invariant and which ones are location invariant. Such information will be useful in determining whether a particular measure is consistent with a given inequality principle.

4.4.1 Inequality Principles

Dalton (1920) proposed the following four inequality principles.

(P1) The principle of transfers.

(P2) The principle of proportionate additions to income.

(P3) The principle of equal additions to income.

(P4) The principle of proportional additions to persons.

The principle of transfers (P1, our old friend Robin Hood) is associated with the rich mathematical theory of majorization. Discussion of this principle will be deferred until last. The other principles, for some of which, in general, there is not unanimity regarding their acceptability, can be discussed quite quickly.

The second principle (P2) states that multiplication of incomes by a constant greater than 1 should decrease inequality. Acceptance of this principle leads to consideration of inequality measures which are *not* scale invariant. The remarkable aspect of Dalton's second principle is that the vast majority of the inequality measures described in Section 4.3 do not satisfy it! One measure which does is T_{10}, (4.3.81), a measure introduced by Dalton. Dalton also proposed a measure involving the arithmetic and harmonic mean of the data set which also satisfies the second principle. It is paradoxical that a plausible inequality principle such as (P2) should be at variance with so many intuitively acceptable inequality measures. The two instances, supplied by Dalton, of measures which do not violate the principle have been honored by neglect in the literature. The paradox is connected with the fact that the formulae for computing inequality indices will be unable to distinguish whether the "new incomes" came from the "old incomes" by multiplying by 100 (which makes everyone richer and, assuming a concave welfare function, plausibly reduces inequality) or by merely re-expressing the incomes in pennies rather than dollars (which hasn't changed anything of consequence). In fact arguments based on revaluation of currency would suggest that inequality measures *should* be scale invariant. Ironically, the inequality measures discussed in Section 4.3 can be, with very few exceptions, classified as either scale invariant or as having the property that they increase when incomes are multiplied by a factor greater than 1. That is to say they are either insensitive to proportionate additions to incomes or they behave in the opposite way to that predicted by the second principle. Dalton observed that relative measures "came closer" to satisfying his second principle than did absolute measures. He thus confined attention to relative measures, retaining a special niche in his affection for the measures he proposed which actually satisfy the second principle. "Coming close" to satisfying the principle actually means being scale invariant. The final verdict of the discussion seems to be that while one would dearly love to have Principle 2, scale invariance is the best one can do. If we rephrase the principle to the following weaker form, we can continue to use it in our discussion.

(P2′) Weak principle of proportional additions to income. Multiplication of incomes by a constant greater than 1 will not increase inequality.

Almost always, as we have seen, if $(P2')$ is satisfied it is because of scale invariance. Perhaps Dalton's two measures

$$\log \overline{X}_A / \log \overline{X}_G \tag{4.4.1}$$

$$[c - (\overline{X}_A)^{-1}]/[c(\overline{X}_H)^{-1}], \tag{4.4.2}$$

which do satisfy the strong form of the principle, should not have been set aside in a cavalier fashion. In (4.4.1) and (4.4.2), c is a constant while the subscripts on the $\overline{X}$'s refer to the arithmetic, geometric and harmonic means. Dalton identifies c as the reciprocal of the minimum income.

The third principle (P3) states that addition of the same constant to all of the incomes will decrease inequality. This may appear to be a compelling principle, but arguments involving existence of poverty lines can suggest that equal additions to incomes might lead to greater inequality (i.e., if before the addition all were below the poverty line and after the additions some were above and some below, then inequality has increased). Generally speaking, the absolute measures of inequality fail to satisfy the third principle, while those measures which are standardized (typically by division by a multiple of $\overline{X}$) do satisfy the criterion. Many of the absolute measures are in fact location invariant, and so satisfy a weak form of Principle 3. Relative measures of inequality, such as the Pietra index and the Gini index, would seem to be appropriate, if the principle is adopted in its strong form. The weak form of the principle (addition of a constant will not increase inequality) does not provide much discrimination among the indices. Dalton made the interesting observation that Principle 3 could be thought of as a corollary of Principles 1 and 2. Thus, equal additions could be obtained by proportionate additions followed by transfers from rich to poor. Since Principle 2 appears a bit shaky, it appears best to enunciate Principle 3 as being reasonably compelling on its own and not dependent on acceptance of Principle 2. It also can be verified that Principle 3 is a consequence of scale invariance and Principle 1.

Principle 4, the principle of proportional additions to persons, can be rephrased in the form: a measure of inequality should be invariant under "cloning." Specifically, consider a population of n individuals and a related population of kn individuals which consists of k identical copies (clones), with respect to income, of each of the n individuals in the original population. The population with kn individuals can be called the cloned population. Principle 4 requires that the measure of inequality should yield the same value for the cloned population as it does for the original population. This can be rephrased as a requirement that the measure of inequality for a population should be a function of the sample distribution function of the population. Most commonly used measures of inequality satisfy this principle.

4.4.2 Transfers, Majorization and the Lorenz Order

Now we return to Principle 1. It will be noted that it provides a criterion for comparing inequality in populations whose total income is the same. Robin Hood only redistributes; he neither creates or destroys. Principle P1 states that if an amount of money is transferred from one richer individual to a poorer individual in such a way as to not disturb their relative rank, then the resulting income distribution exhibits less inequality. Rothschild and Stiglitz (1973) called the reverse of such operations regressive transfers. The rich to poor transfers can be called progressive transfers. We will modify slightly the inequality ordering that they introduced. An income distribution $\underline{Y}$ can be obtained from a distribution $\underline{X}$ by transfers from rich to poor, if it can be obtained by a series of progressive transfers. In such a case we write

$$\underline{X} \geq_T \underline{Y}, \tag{4.4.3}$$

i.e., $\underline{X}$ exhibits more inequality (in the transfer sense) than $\underline{Y}$. If we accept the Pigou–Dalton transfer principle, then we should restrict attention to measures of inequality which preserve the $\geq_T$ ordering. Since the transfer principle relates income distributions for populations with the same number of members and the same total income, we can, without less of generality, assume that the n individuals share a total income of 1 unit, i.e., $\sum_{i=1}^{n} X_i = 1$. It is (perhaps) serendipitous that an ordering on vectors in $\mathbb{R}^n$ equivalent to $\geq_T$ has received much attention in the mathematical literature. The concept in question is that of majorization. Marshall, Olkin and Arnold's (2011) book is an indispensable reference for discussion of majorization, although many of the key ideas were already detailed in Hardy, Littlewood and Polya (1959). The usual definition of majorization involves decreasing order statistics (i.e., the X_i's are ranked in decreasing order). An equivalent definition is possible in terms of the usual (increasing) order statistics.

Definition 4.4.1. Let $\underline{X}$ and $\underline{Y}$ be two n-dimensional vectors. We will say that $\underline{X}$ majorizes $\underline{Y}$ and write

$$\underline{X} \geq_M \underline{Y} \tag{4.4.4}$$

if

$$\sum_{i=1}^{k} X_{i:n} \leq \sum_{i=1}^{k} Y_{i:n} \quad k = 1, 2, \ldots, n-1 \tag{4.4.5}$$

and

$$\sum_{i=1}^{n} X_{i:n} = \sum_{i=1}^{n} Y_{i:n}.$$

The last condition, in the income setting, states that the two populations should have equal total incomes (to be assumed equal to 1 where convenient). But condition (4.4.5) has a familiar look. If we plot the points $((k/n), \sum_{i=1}^{n} X_{i:n})$, we see that $X \geq_M Y$, if and only if the Lorenz curve of $\underline{Y}$ is nested within that of $\underline{X}$. Majorization is thus a Lorenz ordering. We will say that $\underline{X}$ is more unequal than $\underline{Y}$ (in the Lorenz ordering) and write $\underline{X} \geq_L \underline{Y}$, if the Lorenz curve of $\underline{Y}$ is wholly nested within that of $\underline{X}$. The orderings $\geq_M$ and $\geq_L$ are identical. We will speak of the Lorenz ordering or of majorization interchangeably. That the Lorenz ordering or majorization is equivalent to the progressive transfer ordering $\geq_T$ is well known but not transparent. However, the string of equivalent definitions of the progressive transfer ordering is just beginning. As we shall see, it will involve doubly stochastic matrices, Schur convex functions and other mathematical paraphernalia. Fortunately, many of the concepts which can appear quite arbitrary in a mathematical treatment of majorization have nice economic interpretations, if we operate in the context of income distributions. In the income context the connections with utility concepts date back to Dalton, while the relationship to majorization can be traced in a series of papers which appeared in the early 1970's in the *Journal of Economic Theory* beginning with Atkinson (1970) (see Newbery, 1970; Sheshinski, 1972; Dasgupta, Sen and Starrett, 1973; and Rothschild and Stiglitz, 1973). With one last reminder that Marshall, Olkin and Arnold (2011) and Hardy, Littlewood and Polya (1959) are the key references for any needed clarification, we will begin to untangle the web of equivalent orderings.

It is possible to identify a progressive transfer with a linear operator. Specifically the pre-transfer income vector $\underline{X}$ and the post-transfer vector $\underline{Y}$ are related by

$$\underline{Y} = P\underline{X} \tag{4.4.6}$$

where P is an $n \times n$ matrix, which, in fact, is a rather simple doubly stochastic matrix. In (4.4.6) we use column vectors. Specifically, a transfer from a richer individual, "k_1," with income X_{k_1} to the poorer individual, "k_2," with income X_{k_2} results in new incomes of the form

$$Y_{k_1} = \theta X_{k_1} + (1-\theta)X_{k_2} \tag{4.4.7}$$

$$Y_{k_2} = (1-\theta)X_{k_1} + \theta X_{k_2} \tag{4.4.8}$$

where, to preserve the income ordering within the population, $\theta \in (^1\!/_2, 1)$. Since it does not matter, as far as inequality is concerned, whether we relabel individuals, we can conveniently permit θ to be any number in the interval (0,1). The doubly stochastic matrix in (4.4.6) which corresponds to such a progressive transfer will have

$$p_{ij} = \begin{cases} \theta & \text{if } (i,j) = (k_1,k_1) \text{ or } (k_2,k_2) \\ 1-\theta & \text{if } (i,j) = (k_1,k_2) \text{ or } (k_2,k_1) \\ 1 & \text{if } i = j \notin \{k_1,k_2\} \\ 0 & \text{otherwise.} \end{cases} \tag{4.4.9}$$

Thus, progressive transfer represents an averaging process which can be expected to reduce inequality. It is plausible and, in fact, true that any doubly stochastic matrix P can be represented as a finite product of simple doubly stochastic matrices of the form (4.4.9). Thus, the effect of a finite string of progressive transfers is the same as multiplication by a doubly stochastic matrix. Therefore, we have:

Theorem 4.4.1. *$\underline{X} \geq_T \underline{Y}$, if and only if $\underline{Y} = P\underline{X}$ for some doubly stochastic matrix P.*

The important result that identifies the transfer ordering $\geq_T$ with the Lorenz ordering (or majorization) is due to Hardy, Littlewood and Polya (1929).

Theorem 4.4.2. *$\underline{X} \geq_T Y$ if and only if $\underline{X} \geq_L \underline{Y}$ (i.e., if and only if $\underline{X}$ majorizes $\underline{Y}$).*

In one direction the proof is not difficult. One may verify that any progressive transfer is equivalent to a possibly infinite string of progressive transfers involving money passing from the $(i+1)$th richest person to the ith richest person (i.e., to the individual immediately below him/her in the income ordering). That such a transfer will raise the Lorenz curve at just one point is geometrically transparent. It follows that $\underline{X} \geq_T \underline{Y}$ implies $\underline{X} \geq_L \underline{Y}$. The converse may be proved by verifying that for any doubly stochastic matrix P we have $P\underline{X} \leq_M \underline{X}$. (See Marshall, Olkin and Arnold's (2011) Theorem 2.B.2 for a proof, or going further back see the discussion preceding Hardy, Littlewood and Polya's (1959) Theorem 46.)

Thus, we have nested Lorenz curves if one income distribution is obtainable by a series of progressive transfers from the other. If we accept Dalton's first principle, we will seek inequality measures which preserve the Lorenz ordering (or majorization). Before trying to identify the class of functions which preserve the ordering, we will mention a few equivalent conditions that may be encountered in the literature. Muirhead (1903) discussed symmetric means (which include as special cases the arithmetic and geometric means). We may, for any $\underline{\alpha} \in \mathbb{R}^n$ with non-negative coordinates, define the $[\underline{\alpha}]$-symmetric mean of a vector $\underline{a} \in \mathbb{R}^n$ with positive coordinates by

$$[\underline{\alpha}](\underline{a}) = \frac{1}{n!} \sum_{\pi} \prod_{i=1}^{n} a_{\pi(i)}^{\alpha_i}, \tag{4.4.10}$$

where the summation is over all permutations of $(1,2,\ldots n)$. Muirhead's theorem stated that $[\underline{\alpha}](\underline{a}) \geq [\underline{\beta}](\underline{a})$ for every $\underline{a}$ (with positive coordinates), if and only if a $\underline{\alpha} \geq_M \underline{\beta}$. An interpretation of Muirhead's characterization in terms of income inequality does not appear to have been achieved. A geometric characterization of majorization is possible, since the class of doubly stochastic matrices is the convex hull of the class of permutations matrices. From this observation we may conclude that $\underline{X} \leq_L \underline{Y}$, if and only if $\underline{X}$ is in the convex hull of the points obtained by permuting the coordinates of $\underline{Y}$. Thus, $\underline{X}$ is an average of essentially equivalent income distributions and may be expected to exhibit less inequality. Rothschild and Stiglitz (1973) described an equality preferring ordering of income distributions. They proposed that an ordering is equality preferring, if whenever it is indifferent between two income distributions, it prefers a distribution which is a weighted average of both to either one alone. In the light of the above discussion it is not surprising that ordering by equality preference is another surrogate of majorization.

What kinds of functions preserve the Lorenz ordering (i.e., majorization)? The answer will be particularly important in our inequality measuring context, since we will, if we accept Dalton's transfer principle, wish to restrict attention to inequality measures which preserve the Lorenz ordering. The class of functions with continuous first partial derivatives which preserve majorization was identified by Schur (1923). They are now known as Schur convex functions.

Definition 4.4.2. A function $g : \mathbb{R}^{n+} \to \mathbb{R}$ with continuous first partial derivatives is said to be Schur convex if it is symmetric under permutations of the coordinates in $\mathbb{R}^{n+}$ and if

$$(X_1 - X_2)\left[\frac{\partial}{\partial X_1} g(\underline{X}) - \frac{\partial}{\partial X_2} g(\underline{X})\right] \geq 0, \quad \forall \underline{X} \in \mathbb{R}^{n+}. \tag{4.4.11}$$

Schur proved that the functions described in Definition 4.4.2 are the only ones (with continuous partial derivatives) which preserve majorization. Consequently, to check whether a given inequality measure is consistent with Dalton's transfer principle, we need only verify its Schur convexity (using condition (4.4.11)). An example is the Gini index. This could be checked by looking at partial derivatives; however, the definition in terms of the area between the Lorenz curve and the egalitarian line guarantees it is consistent with the Lorenz ordering, i.e., it is Schur convex. Other examples are not as transparent.

One important class of Schur convex functions is the class of separable convex functions. These are of the form $g(\underline{X}) = \sum_{i=1}^{n} h(X_i)$ where h is a convex function on $\mathbb{R}^+$. It is a simple matter to check condition (4.4.11) in this case. Such separable convex functions have a dual significance in our discussion of inequality measures. From a mathematical viewpoint they are important in their role of characterizing majorization.

Theorem 4.4.3. *$\underline{X} \leq_L \underline{Y}$, if and only if $\sum_{i=1}^{n} h(X_i) \leq \sum_{i=1}^{n} h(Y_i)$ for every continuous convex function h.*

From the economic viewpoint such separable convex functions are encountered in developments of inequality measures via utility. Specifically, it is frequently assumed that the value of an income X to an individual is governed by a utility function, U, which is assumed increasing and concave. Then social welfare is assumed to be a function, say W_0, of the individual utilities. Thus,

$$W(\underline{X}) = W_0(U(X_1), U(X_2), \ldots, U(X_n)). \tag{4.4.12}$$

Inequality can be thought of as the negative of welfare. The simplest possible welfare function is an additive one, i.e., $W(\underline{X}) = \sum_{i=1}^{n} U(X_i)$. Proponents of additive welfare functions would be enthusiastic about inequality measures that are separable convex functions. Such functions are Schur convex but, of course, not every Schur convex function is separable convex. Atkinson (1970) and Newbery (1970) were critical of the Gini index, basically because it could not be associated with an additive utility function, i.e., because it was not separable convex, only Schur convex. This is in contrast to the Pietra and Amato indices, which are separable convex. Sheshinski (1972) pointed out that the Gini index may well be identified with a reasonable welfare function which is not additive. He provided an example. Additive utility is such an appealing simplification that it becomes difficult to disassociate from it. Dalton's principle leads to Schur convexity, not to separable convexity. One might aver that the plausibility of the Gini index is actually evidence against an assumption of additive utility, rather than the converse.

Theorem 4.4.3 has an attractive interpretation in terms of additive utility. If we are comparing populations whose Lorenz curves are nested and if we assume an additive utility model, $W(\underline{X}) = \sum_{i=1}^{n} U(X)$, then it does not matter what choice we make for U (provided it is increasing and concave). For any choice of U the same ranking will be obtained. On the other hand, if two populations have Lorenz curves which intersect, it is always possible to find two choices of U which will rank the populations differently (Atkinson, 1970). Theorem 4.4.3 can be stated in the following form. A necessary and sufficient condition that two populations will be ranked independently of the choice of the utility function U is that they have nested Lorenz curves.

Rothschild and Stiglitz (1973) introduced the concept of a locally equality preferring welfare function, a slight weakening of quasi-concavity. They remarked that no quasi-convex inequality measure can rank populations as does the Gini index. Symmetric quasi-convex functions are Schur convex, but, except for mathematical considerations, restriction of attention to symmetric quasi-convex inequality measures (which include the separable convex measures) does not seem justifiable (especially since such a restriction would rule out the Gini index). The relevant definitions and theorems follow.

Definition 4.4.3. A function $g : \mathbb{R}^{n+} \to \mathbb{R}$ is locally inequality preferring, if for any vector $\underline{X}$ and any $\alpha \in [0,1]$

$$g(\underline{X}) \geq g(\alpha \underline{X} + (1-\alpha)\underline{X}') \tag{4.4.13}$$

where $\underline{X}'$ is any vector related to $\underline{X}$ by

$$X_j' = X_k' = (X_j + X_k)/2 \text{ for some } j,k$$

and $X_i' = X_i$ otherwise.

Definition 4.4.4. A function $g : \mathbb{R}^{n+} \to \mathbb{R}$ is quasi-convex, if for any vectors $\underline{X}$ and $\underline{Y}$ with $g(\underline{X}) \geq g(\underline{Y})$ and any $\alpha \in [0,1]$

$$g(X) \geq g(\alpha \underline{X} + (1-\alpha)\underline{Y}). \tag{4.4.14}$$

Theorem 4.4.4. *$\underline{X} \leq_L \underline{Y}$, if and only if $g(\underline{X}) \leq g(\underline{Y})$ for every symmetric monotone locally inequality preferring function g.*

Theorem 4.4.5. *$\underline{X} \leq_L \underline{Y}$ if and only if $g(\underline{X}) \leq g(\underline{Y})$ for every symmetric quasi-convex function g.*

The final verdict of Dalton's inequality principles can be summarized in the statement that, generally speaking, we should seek inequality measures which are relative rather than absolute, are scale invariant and are Schur convex. Referring to Section 4.3 a list of inequality measures which satisfy these criteria would include (4.3.11), (4.3.12), (4.3.23), (4.3.46), (4.3.71), (4.3.87), (4.3.88), (4.3.89), (4.3.90), (4.3.91) which is Schur concave, (4.3.92), (4.3.93) and (4.3.94). Dalton's principles do not really help us to select from this list. Additive utility proponents might replace the Schur convex requirement by a separable convex requirement. The most notable casualty from the list, if this more stringent requirement is invoked, is the Gini index.

Dalton's Principles 1, 2′, 3 and 4 can be enunciated for arbitrary populations or, if we wish, arbitrary income distributions. We no longer can speak of majorization in terms of progressive transfers, although (following Atkinson) one can argue in terms of infinitesimal transfers. It is simpler, however, to define the analog of majorization to be the Lorenz ordering via nested Lorenz curves.

Definition 4.4.5. We will say that a distribution F_1 is at least as unequal in the Lorenz sense as a distribution F_2, if the corresponding Lorenz curves (defined by (4.2.19)) satisfy

$$L_1(u) \leq L_2(u), \forall u \in [0,1],$$

and we write $F_1 \geq_L F_2$.

Moothathu (1991) provides some insight into the Lorenz ordering relationship in terms of the corresponding quantile functions. Consider two distributions F_1 and F_2 with corresponding Lorenz curves L_1 and L_2. Define $R(u) = F_1^{-1}(u)/F_2^{-1}(u)$. He verifies that, if $R(u)$ is increasing on $(0,1)$, then $L_1(u) < L_2(u)$ for all $u \in (0,1)$.

It is easy to check that in the case of finite populations $\underline{X}$ and $\underline{Y}$ with equal means and equal cardinality, we have $\underline{X} \geq_L \underline{Y}$, if and only if for the corresponding sample distribution functions we have $F_X^{(n)} \geq_L F_Y^{(n)}$ in the sense of Definition 4.4.5. An analog to Theorem 4.4.3 is available.

Theorem 4.4.6. *Let F_1 and F_2 be distributions with equal means. $F_1 \leq_L F_2$, if and only if*

$$\int_0^\infty h(u)dF_1(u) \leq \int_0^\infty h(u)dF_2(u) \tag{4.4.15}$$

for every continuous convex function h.

There does not appear to be a straightforward analog to Schur convexity in this more general setting. Nevertheless, we are led via Dalton's principles to seek inequality measures which, as functionals on the space of all distributions, preserve the Lorenz order. The Gini, Pietra and Amato indices are obvious examples. Mehran's (1976) linear measures are also of this class (see (4.2.65) and (4.3.90)). Of course, by Theorem 4.4.6, any inequality measure of the form $\int_0^\infty h(u)dF(u)$ (where h is continuous convex) will preserve the Lorenz order.

We will say that an inequality measure satisfies Dalton's first principle, if it preserves the Lorenz ordering (on the space of distributions). Principle $2'$ is again associated with scale invariance, while Principle 3 states that addition of a constant should reduce inequality, i.e., the inequality associated with the distribution $F(x)$ should be greater than the inequality associated with the distribution $F(x-c)$ for any $c > 0$. The list of population inequality measures which satisfy Principles 1, $2'$ and 3 includes (from Section 4.2) those measures defined by (4.2.4), (4.2.6), (4.2.8) and (4.2.65).

Remarks.

(i) Taillie (1979) illustrates potential uses of the Lorenz ordering in the study of species equitability.

(ii) Mosler and Muliere (1996) discuss a modification of the Robin Hood axiom involving a partition of the population into a rich class and a poor class. In their discussion, inequality will be reduced only if Robin Hood takes money from a rich individual (a member of the rich class) and gives it to a member of the poor class. Predictably this will broaden somewhat the class of inequality measures which are acceptable in the sense that they are decreased by any one of these more restrictive Robin Hood operations (though they are not necessarily Schur convex since they may fail to be monotone with respect to all Robin Hood operations). The class of such inequality measures is characterized by a requirement that a suitably restricted version of (4.4.11) should hold.

(iii) Shorrocks and Foster (1987) consider a different restricted class of redistributive operations involving composite income transfers. The resulting inequality ordering, in some cases, will provide a ranking of certain Lorenz curves which cross each other (i.e., are not nested). Aaberge (2009) discusses alternative approaches to the problem of inequality ranking when Lorenz curves cross.

(iv) What happens if Robin Hood siphons off a little money for himself when transferring money from a rich individual to a poorer one? Or, what happens if he adds a little of his own cash to the amount being transferred? We are no longer guaranteed that such modified Robin Hood operations will reduce inequality as measured by the Lorenz curve. Lambert and Lanza (2006) provide a detailed discussion of this issue. They even consider the case in which Robin takes money from both the rich individual and the poor individual, pocketing the entire amount for himself. Predictably, the effect of such operations, which involve changing the wealth of one or two individuals, on the Lorenz curve will depend on the sizes of the changes and on the relative positions of the individuals involved in the ranking of the entire population. For example, Lambert and Lanza (2006) address the question of what percentage of the amount to be transferred can be kept by Robin Hood without destroying the progressive nature of the operation.

4.4.3 How Transformations Affect Inequality

Suppose X is a random variable with distribution function $F(x)$ and Lorenz curve $L(u)$. Let us define a new random variable $Y = \psi(X)$. For example, Y could be obtained from X, as a consequence of taxation or as a consequence of replacement of actual money by its utility, etc. How are the Lorenz curves of X and Y related or, if you wish, what is the effect of the transformation ψ on the inequality in the population? Fellman (1976) has discussed such issues. They also arise in discussions of poverty indices. An example of Fellman's results is the following.

Theorem 4.4.7. *Let X be a positive random variable with finite mean and Lorenz curve $L_X(u)$. Suppose that $g:\mathbb{R}^+ \to \mathbb{R}^+$ is a monotone increasing function for which $g(x)/x$ is monotone increasing. Define $Y = g(X)$ and assume $E(Y)$ exists. The Lorenz curve corresponding to Y satisfies*

$$L_Y(u) \le L_X(u), \quad 0 \le u \le 1, \tag{4.4.16}$$

[i.e., $F_X \le_L F_Y$ or, in a slight abuse of notation, $X \le_L Y$].

Proof. Since g is increasing, it follows that $F_Y^{-1}(z) = g(F_X^{-1}(z))$, so we may write

$$\begin{aligned} L_Y(u) - L_X(u) &= \int_0^u \left[\frac{g(F_X^{-1}(z))}{E(Y)} - \frac{F_X^{-1}(z)}{E(X)} \right] dz \\ &= \int_0^{F_X^{-1}(u)} \left[\frac{g(x)}{E(Y)} - \frac{x}{E(X)} \right] dF_X(x) \\ &= \int_0^{F_X^{-1}(u)} \left[\frac{g(x)}{x} - \frac{E(Y)}{E(X)} \right] \frac{x}{E(Y)} dF_X(x). \end{aligned}$$

Since $g(x)/x$ is monotone increasing, the integrand is first negative and changes sign once becoming eventually positive. The maximum value of $L_Y(u) - L_X(u)$ will thus occur either at $u = 0$ or $u = 1$. However, $L_Y(0) = L_X(0)(= 0)$ and $L_Y(1) = L_X(1)(= 1)$, so we conclude that $L_Y(u) \leq L_X(u), \forall u \in [0, 1]$. □

Arnold and Villaseñor (1985) discuss the converse to Theorem 4.4.7. They show that if the function g fails to satisfy any one of the conditions in Theorem 4.4.7, then there exists a non-negative random variable X with two point support with $X \nleq_L g(X)$. The justification of the converse result makes use of the following two Lemmas concerning the Lorenz ordering of distributions with common two point support.

Lemma 4.4.1. *Suppose $0 < x_1 < x_2$. If X and Y are random variables defined by*

$$P(X = x_1) = p, \qquad P(X = x_2) = 1 - p,$$

$$P(Y = x_1) = q, \qquad P(Y = x_2) = 1 - q,$$

then X and Y are not comparable in the Lorenz ordering, except in the trivial cases in which $p = q$, $pq = 0$ or $(1 - p)(1 - q) = 0$.

Lemma 4.4.2. *Suppose $x > 0$. If X and Y are random variables defined by*

$$P(X = 0) = p, \qquad P(X = x) = 1 - p,$$

$$P(Y = 0) = q, \qquad P(Y = x) = 1 - q,$$

then $p \leq q \Rightarrow X \leq_L Y$.

If g is increasing and $g(x)/x$ is monotone decreasing, then the inequality in (4.4.16) is reversed. Naturally if $g(x)/x$ is a constant, then $L_Y(u) = L_Y(u)$ (since the Lorenz curve is scale invariant). A result analogous to that described in Theorem 4.4.7 is contained in Marshall, Olkin and Proschan (1967).

Note 4.4.1. *Several authors, when discussing inequality attenuation, assume that the function g in Theorem 4.4.7 is differentiable. Eichhorn, Funke and Richter (1984) provide a careful discussion of the result without assuming differentiability. They credit Jakobsson (1976) with having first noted that Fellman's conditions were necessary and sufficient, though they noticed some gaps in Jakobsson's arguments. The Eichhorn, Funke and Richter paper deals only with finite populations but limiting arguments can be used to extend their results to more general settings. Fellman (2009) provides a careful discussion of the need for an assumption of continuity for g in Theorem 4.4.7 and shows that it can be dispensed with. It will be observed that continuity of g was not used in the present proof of the theorem.*

Ord, Patil and Taillie (1978) discussed the effects of taxation on the Lorenz curve. They provided a result very similar, but not equivalent to, Theorem 4.4.7. Using the notation of Theorem 4.4.7, their result stated that a sufficient condition for $L_Y(u) \geq$

$L_X(u)$, $0 \leq u \leq 1$ is that $Y = g(X)$ where $g(x) = x + \tau - h(x)$ in which $h(x) \leq x$, h increases with x and $h'(x) < 1$ for all x. Under these hypotheses they also provided an expression relating the Gini indices of X and Y:

$$G_Y = G_X - 2[\text{cov}(h(X), F_X(X))/E(X)]. \tag{4.4.17}$$

As a simple example of the application of Theorem 4.4.7, consider the transformation $g(x) = x + c$ (i.e., an equal addition to incomes). Here $g(x)/x\downarrow$ as $x\uparrow$, and so, if we let $Y = X + c$, we find $L_Y(u) \geq L_X(u)$, i.e., $Y \leq_L X$. Equal additions to incomes thus reduce inequality in the Lorenz ordering sense. It will be recalled that Dalton's Principle 3 stated that equal additions to incomes should decrease inequality. The effect of a progressive tax can be discussed in the framework of Theorem 4.4.7. If the tax is progressive, then the post-tax income Y and the pre-tax income X are related by $Y = g(X)$ where $g(x)/x\downarrow$ as $x\uparrow$. From the theorem it follows that a progressive tax will reduce inequality (in the Lorenz ordering sense). The special case of a linear tax was considered by Levine and Singer (1970).

One important class of inequality measures are the so called poverty measures. Effectively, they depend on the identification of a poverty line and concentrate on the proportion of individuals below that line and, sometimes, how far below the line the poor individuals are. Takayama (1979) building on earlier work by Sen (1976) finally settled upon the Gini index of the population censored from above at the poverty line (i.e., any income above the poverty level X_0 is replaced by the value X_0).

It is a simple consequence of Theorem 4.4.7 that this index of poverty increases as the poverty line increases (as it clearly should). If one truncates rather than censors, the effect on the Lorenz curve is not as predictable. Ord, Patil and Taillie (1978) have identified necessary and sufficient conditions for the Lorenz curve to move towards the diagonal when the truncation point (from below) is increased. Their result is

Theorem 4.4.8. *Let X be a non-negative random variable. For each $c > 0$, define $F_c(x) = P(X \leq x | X \geq c)$ and let L_c be the corresponding Lorenz curve. $L_c(u)$ increases (decreases) for all u as c increases, if and only if $E(X|X \geq c)/c$ is a decreasing (increasing) function of c.*

Remarkably, a large number of distributions do satisfy the condition that $E(X|X \geq c)/c$ decreases as c increases. As examples, Ord, Patil and Taillie (1978) cited the log-normal, gamma, Weibull and others. The classical Pareto is, of course, a boundary case, since its Lorenz curve is truncation invariant. Generally speaking, we can expect that estimation of inequality using truncated data may well lead to underestimates.

In the above development attention has been directed to the effects on inequality of deterministic transformations. What happens when possibly random transformations are permitted? Some of the first results obtained in this direction were obtained in the process of studying underreported income models (cf. Sections 3.7 and 5.8.4).

McDonald (1981), reporting on work done in collaboration with Israelson and Newley, discussed several aspects of these models. He returned to Krishnaji's multiplicative formulation of the situation. Thus, we have true income X, reported income Y and underreporting fraction U related by the equation

$$Y = UX. \tag{4.4.18}$$

As usual, assume that U and X are independent. The presence of misreporting increases inequality, as measured by the coefficient of variation (τ_4, defined in (4.2.6)). In fact, one finds

$$\tau_4(Y) = \sqrt{[\tau_4(U)]^2([\tau_4(X)]^2 + 1) + [\tau_4(X)]^2} > \tau_4(X). \tag{4.4.19}$$

The derivation of (4.4.19) does not involve any assumption that U be less than 1, i.e., it is not really a result dealing with underreporting, rather it is a result for arbitrary scale mixtures of non-negative random variables. McDonald did not obtain analogous results for measures of inequality other than the coefficient of variation. It is possible to derive an analog to (4.4.19) for the Gini index, if one uses the representation (4.2.9) and applies the Schwarz inequality. Arnold (1980) provided unification to these concepts by proving that scale mixing increases inequality quite generally in the sense that it "bends Lorenz's bow." Less picturesquely stated we have the following theorem.

Theorem 4.4.9. *Suppose that $Y = UX$ where U and X are independent non-negative random variables with finite first moments. If $P(U = 0) \neq 1$, then the Lorenz curve of X is nested within that of Y, i.e., $L_Y(u) \leq L_X(u)$, $\forall u \in [0,1]$.*

Proof. Since the Lorenz curve is scale invariant, we may assume without loss of generality that $E(U) = 1$ so that $E(Y) = E(X)$. In such a case we may apply Theorem 4.4.6. Thus, it will suffice to prove that $E(h(Y)) \geq E(h(X))$ for every continuous convex function h. However, for such a function h we have

$$\begin{aligned} E(h(Y)) &= E(h(UX)) = E[E(h(UX)|X)] \\ &\geq E[h(E(UX|X)] \quad \text{(by Jensen's inequality)} \\ &= E[h(XE(U|X))] \\ &= E[h(XE(U))] \quad \text{(since } U,X \text{ are independent)} \\ &= E(h(X)) \quad \text{(since } E(U) = 1). \end{aligned}$$

As a consequence of Theorem 4.4.9, any misreporting (under or over) that is independent of the income level will cause an apparent increase in inequality as measured by any inequality index which preserves the Lorenz order. A close inspection of the proof of Theorem 4.4.9 reveals that the full strength of the independence assumption was not used. In fact, all that is needed is that $E(U|X = x)$ does not depend on x. Thus, any misreporting will increase inequality in the Lorenz order sense, provided only that the expected value of the ratio of reported income to true income is independent of the level of true income.

Theorem 4.4.9 is plausible in the light of the observation of Ord, Patil and Taillie (1978) that, intuitively, adding noise should increase inequality. Phrased another way, averaging should decrease inequality (cf. Arnold, 1981). A theorem related to this observation follows.

Theorem 4.4.10. *Let X be a non-negative random variable with finite first moment. Suppose that, for some random variable Z, $Y = E(X|Z)$. It follows that $Y \leq_L X$.*

Proof. Since clearly $E(Y) = E(X)$, it is sufficient, using Theorem 4.4.6, to prove that $E(h(Y)) \leq E(h(X))$ for every continuous convex function h. This follows by applying Jensen's inequality after conditioning on Z. □

Basically the same argument was used in Theorem 4.4.9. The two theorems are, on close inspection, actually equivalent. The present theorem tells us that if Y can be identified as the conditional expectation of X given some random variable Z, then $Y \leq_L X$. Thus, such averaging attenuates inequality as measured by any index of inequality that respects the Lorenz order.

Ord, Patil and Taillie (1981b) present another related result which can be clearly understood in the light of Theorem 4.4.10. They introduce a taxation scenario in which

$$Y = \psi(X,Z) \tag{4.4.20}$$

where X is original income, Z is random and ψ is a deterministic function. Suppose that the goal of the taxation policy is to transform from X to $g(X)$ but that the goal is only achieved on the average, i.e., $E(Y|X=x) = g(x)$ or equivalently $E(Y|X) = g(X)$. Ord, Patil and Taillie prove that $Y \geq_L g(X)$, i.e., the obtained distribution is more unequal than the goal. The deduced ordering is, using Theorem 4.4.10, a reflection of the fact that $g(X)$ is an averaging of Y.

Arnold and Villaseñor (1985) also consider the "random" taxation model. The post-tax income Y is assumed to be of the form (4.4.20). Inequality accentuation seems to be likely since the transformation involves addition of "noise." Sufficient conditions are provided in the following theorem.

Theorem 4.4.11. *Suppose the function ψ in (4.4.20) is such that $\psi(x,z))$ and $\psi(x,z)/x$ are non-decreasing in x for every z. Assume that X and Z are independent non-negative random variables such that $0 < E(X) < \infty$ and $0 < E(\psi(X,Z)) < \infty$. It follows that $X \leq_L \psi(X,Z)$.*

The result is obtained by conditioning on Z and then considering expectations of continuous convex functions.

Theorem 4.4.9 which states that misreporting of income accentuates inequality is a simple consequence of Theorem 4.4.11. The model used relates reported income Y to true income X in the following manner

$$Y = UX \tag{4.4.21}$$

where U and X are independent and U is the misreporting factor. Theorem 4.4.11 applies and we may conclude that $X \leq_L Y$.

Misreporting can play havoc with inequality attenuating efforts. Suppose that we apply an inequality attenuating function to reported income. The true post-tax income will be

$$Y = X - UX + g(UX).$$

There is no guarantee that $Y \leq_L X$; indeed it may even be the case that $Y >_L X$.

Theorem 4.4.10, in essence, gives necessary and sufficient conditions. The relevant result (due to Strassen, 1965), restated in terms of the Lorenz order, is as follows.

Theorem 4.4.12. *For non-negative random variables with $E(X) = E(Y)$ we have $Y \leq_L X$ if and only if there exist jointly distributed random variables X', Z' such that $X \stackrel{d}{=} X'$ and $Y \stackrel{d}{=} E(X'|Z')$. (The notation $X \stackrel{d}{=} X'$ is to be read "X and X' are identically distributed.")*

Although Strassen's theorem identifies completely the class of all inequality attenuating transformations, theorems like 4.4.7 and 4.4.8 remain of interest since the random variables X', Z' appearing in Strassen's theorem are not always easy to identify. For example, from Theorem 4.4.6 we know that $\sqrt{X} \leq_L X$. How can this be obtained using Theorem 4.4.12?

Rather than focus on inequality attenuating transformations (described in Theorem 4.4.7), it is of interest to identify what kinds of transformations will preserve the Lorenz order. Denote the class of all such transformations by $\mathscr{G}$, thus

$$\mathscr{G} = \{g : X \leq_L Y \Rightarrow g(X) \leq_L g(Y)\}.$$

Note that if $g \in \mathscr{G}$, it must be the case that g maps $[0,\infty)$ into $[0,\infty)$ and also it must be the case that if $E(X) \in (0,\infty)$ then $E(g(X)) \in (0,\infty)$. The class $\mathscr{G}$ is not extensive.

Theorem 4.4.13. *If $g \in \mathscr{G}$ (i.e., if g preserves the Lorenz order), then it must be of one of the following three forms.*

$$\begin{aligned} g_{1,a}(x) &= ax, \; x \geq 0, \; \textit{where } a \in (0,\infty), \\ g_{2,b}(x) &= b, \; x \geq 0, \; \textit{where } b \in (0,\infty), \\ g_{3,c}(x) &= 0, \; x = 0, \\ &= c, \; x > 0, \; \textit{where } c \in (0,\infty). \end{aligned}$$

The proof of this theorem involves use of Lemmas 4.2.1 and 4.4.2 and an argument that any member of $\mathscr{G}$ must be non-decreasing (Arnold and Villaseñor, 1985).

In contrast, functions which preserve majorization (rather than the Lorenz order) must, if measurable, be linear. Consequently, there are non-measurable functions which preserve majorization but do not preserve the Lorenz order, while functions of the form $g_{3,c}$ provide examples which preserve the Lorenz order but not majorization.

A strong Lorenz order may be defined as follows:

$$X <_L Y \Leftrightarrow X \leq_L Y \text{ and } Y \nleq_L X.$$

The class of functions that preserve the strong Lorenz order is even more restricted. Such functions must be of the form $g_{1,a}(x) = ax, \; x \geq 0$, where $a \in (0,\infty)$.

4.4.4 *Weighting and Mixing*

Parallel to the discussion of inequality attenuating and preserving transformations, one can consider what happens when distributions are weighted rather than transformed. It is not difficult to envision situations in which an individual income will be observed with a probability that depends on the size of the income. Such size-biased distributions and more general weighted distributions have received much attention in the literature, following early work summarized by Rao (1985). Instead of observing a random variable with density $f(x)$, we in fact observe a random variable with density $g(x)f(x)$ for some non-negative weight function $g(x)$. Though, of course, we do not have to assume the existence of densities in the discussion.

Suppose that X is non-negative with $E(X) \in (0,\infty)$. For a suitable non-negative function g, the g-weighted version of X, denoted by X_g, is defined to be a random variable with a distribution function of the form

$$P(X_g \le x) = \left[\int_0^x g(y)\, dF(x)\right] / E(g(X)),$$

provided that $E(g(X)) < \infty$. Note that, if X_g is to have a positive finite expectation, it must be the case that $E(Xg(X)) \in (0,\infty)$.

Let $\mathscr{W}_1$ be the class of all inequality preserving weightings and let $\mathscr{W}_2$ denote the class of all inequality attenuating weightings. Thus

$$\mathscr{W}_1 = \{g : X \le_L Y \Rightarrow X_g \le_L Y_g\}$$

and

$$\mathscr{W}_2 = \{g : X \ge 0 \text{ and } E(X) \in (0,\infty) \Rightarrow X_g \le_L X\}.$$

Repeated use of Lemmas 4.4.1 and 4.4.2 rule out all but very simple possibilities for g in $\mathscr{W}_1$ and $\mathscr{W}_2$. Eventually one finds that $g \in \mathscr{W}_1$ if and only if

$$\begin{aligned} g(0) &= \alpha, \\ g(x) &= \beta, \ x > 0, \end{aligned}$$

where $\alpha \ge \beta > 0$.

Similarly, $g \in \mathscr{W}_2$ if and only if

$$\begin{aligned} g(0) &= \alpha, \\ g(x) &= \beta, \ x > 0, \end{aligned}$$

where $\beta > 0$ and $0 \le \alpha \le \beta$.

We turn next to consideration of mixtures. In studies of income distributions in finite populations, it is of interest to consider the effects of pooling populations on inequality. Consider a situation in which we have n_1 individuals in population 1 with corresponding empirical income distribution $F_1(x)$ together with n_2 individuals in population 2 with corresponding empirical income distribution $F_2(x)$. The pooled population has empirical income distribution

$$\widetilde{F}(x) = \frac{n_1}{n_1+n_2} F_1(x) + \frac{n_2}{n_1+n_2} F_2(x).$$

Lam (1986) discussed the question of when will the pooled population have less inequality than one of the two component populations.

We can extend Lam's results to consider mixtures of non-negative random variables with finite positive expectations, i.e., with well defined Lorenz curves. Let X and Y be two such random variables. For $\alpha \in (0.1)$, an $(\alpha, 1-\alpha)$ mixture of X and Y is a random variable X_α defined by

$$X_\alpha = I_\alpha X + (1 - I_\alpha)Y \tag{4.4.22}$$

where X and Y are independent and I_α is a Bernoulli random variable independent of X and Y with $P(I_\alpha = 1) = \alpha$.

When can we conclude that $X_\alpha \leq_L X$?

Theorem 4.4.14 (Lam, 1986)**.** *Suppose that X and Y have well defined Lorenz curves, and that $E(X) = E(Y)$ and $Y \leq_L X$, then X_α, defined in (4.4.22), exhibits less inequality than X, i.e., $X_\alpha \leq_L X$.*

Proof. Without loss of generality assume that $E(X) = E(Y) = 1$, so that $E(X_\alpha) = 1$. It will suffice to show that, for any continuous convex function g, $E(g(X_\alpha)) \leq E(g(X))$. This is true since

$$\begin{aligned} E(g(X_\alpha)) &= \alpha E(g(X)) + (1-\alpha)E(g(Y)) \\ &\leq \alpha E(g(X)) + (1-\alpha)E(g(X)) \quad (\text{since } Y \leq_L X) \\ &= E(g(X)). \end{aligned}$$

□

A partial converse to Theorem 4.4.14 is available.

Theorem 4.4.15. *Suppose that X and Y have well defined Lorenz curves and that X_α is as defined in (4.4.22). Assume that $F_X^{-1}(0) > 0$. If $X_\alpha \leq_L X$ then $E(X) = E(Y)$ and $Y \leq_L X$.*

Inequality attenuating transformations applied within subpopulations do not necessarily reduce inequality in the pooled population. Inequality will of course be reduced in the pooled population if the same inequality attenuating transformation is applied within each subpopulation. It is not unusual for different taxation schedules to be used in different subpopulations and even if they are all inequality attenuating, it is possible that the overall effect on the pooled population is an increase in inequality! In general this phenomenon is not encountered in the real world. Lambert (1993) provides some sufficient conditions for attenuation in the pooled population. Just one of these sufficient conditions will be described here.

Each of the k subpopulations can be associated with a non-negative random variable with a well defined Lorenz curve. For $i = 1, 2, ..., k$, X_i will be the random variable associated with the ith subpopulation. Assume that within population i an inequality attenuating transformation g_i is applied (corresponding to a progressive tax schedule). The pooled untransformed incomes (pre-tax) can be associated with a mixture random variable

$$X_{\underline{\alpha}} = \sum_{i=1}^{k} I_{\alpha_i} X_i, \tag{4.4.23}$$

a natural extension of (4.4.22) to the case of k subpopulations. The pooled post-tax incomes will be associated with the random variable

$$\widetilde{X}_{\underline{\alpha}} = \sum_{i=1}^{k} I_{\alpha_i} g_i(X_i). \tag{4.4.24}$$

When can we conclude that $\widetilde{X}_{\underline{\alpha}} \leq_L X_{\underline{\alpha}}$? To identify a sufficient condition, Lambert defines the average tax-rate associated with the application of g to the random variable X to be

$$T(g,X) = \frac{E(X - g(X))}{E(X)}. \tag{4.4.25}$$

Using this concept, we have

Theorem 4.4.16 (Lambert, 1993). *If each g_i, $i = 1,2,...,k$ is inequality attenuating and if $T(g_1,X_1) = T(g_2,X_2) = \cdots = T(g_k,X_k)$ (equal average tax-rates within subpopulations) then $\widetilde{X}_{\underline{\alpha}} \leq_L X_{\underline{\alpha}}$ (overall inequality attenuation).*

The proof of this result makes use of Strassen's Theorem 4.4.12. Lambert (1993) has other results on this topic, but there is scope for further development.

4.4.5 Lorenz Order within Parametric Families

The family of log-normal distributions with Lorenz curves of the form (4.2.147), i.e., $L_{\sigma^2}(u) = \Phi(\Phi^{-1}(u) - \sigma^2)$, provides us with a classic example of a family of distributions with nested Lorenz curves. As σ^2 increase, inequality increases, i.e., if $\sigma_1^2 \leq \sigma_2^2$ then $L_{\sigma_1^2}(u) \geq L_{\sigma_2^2}(u)$. Analogous Lorenz ordering is encountered in the more general family $L_\tau(u) = G(G^{-1}(u) - \tau)$.

There are other parametric families which are Lorenz ordered by their parameters. We will focus only on families that have been used as income models, including the generalized Pareto models introduced in Chapter 3.

The family of gamma distributions was shown by Taillie (1981) to be Lorenz ordered by its shape parameter. Thus if $X_1 \sim \Gamma(\alpha_1, \beta_1)$ and $X_2 \sim \Gamma(\alpha_2, \beta_2)$ with $\alpha_1 < \alpha_2$, then $X_2 \leq_L X_1$. Taillie also discussed Lorenz ordering of generalized gamma variables. A random variable X is said to have a generalized gamma distribution if it can be represented in the form $X = V^{1/\delta}$ where $V \sim \Gamma(\alpha, \beta)$. Wilfling (1996a) provided a simple necessary and sufficient condition for Lorenz ordering in the generalized gamma family with positive δ's. If $X_i = V_i^{1/\delta_i}$, $i = 1,2$, where $V_i \sim \Gamma(\alpha_i, \beta_i)$, $i = 1,2$, then $X_1 \leq_L X_2$ if and only if $\delta_1 \geq \delta_2$ and $\delta_1 \alpha_1 \geq \delta_2 \alpha_2$. Taillie's discussion also allows the δ's to be negative.

Kleiber (1999) considered Lorenz ordering of Feller-Pareto variables with $\mu = 0$. See (3.2.19) for the corresponding density function. Kleiber, following McDonald (1984), uses the name generalized beta of the second kind (GB2) for the distribution rather than the name Feller-Pareto. If $X \sim FP(0, \sigma, \gamma, \gamma_1, \gamma_2)$ and $X' \sim FP(0, \sigma', \gamma', \gamma_1', \gamma_2')$, then a sufficient condition for $X \leq_L X'$ is that $\gamma \leq \gamma'$, $(\gamma_1/\gamma) \geq (\gamma_1'/\gamma')$ and $(\gamma_2/\gamma) \geq (\gamma_2'/\gamma')$. Kleiber verifies this result by utilizing results of Taillie (1981) on Lorenz ordering of generalized gamma variables, and the fact that products of independent Lorenz ordered variables are Lorenz ordered. The result falls a little short of providing a necessary and sufficient condition for Lorenz ordering in the Feller-Pareto family, but it covers many cases of interest and Kleiber conjectures that his sufficient condition is also necessary.

Wilfling (1996b) provides some sufficient conditions for Lorenz ordering within the generalized beta of the first kind distribution(GB1).

Finally, reference to the examples of Section 4.2.8 will identify several additional one parameter families of distributions that are easily seen to be Lorenz ordered by their respective parameters.

4.4.6 The Lorenz Order and Order Statistics

In the study of Lorenz ordering of order statistics, the following density crossing sufficient condition due to Shaked is often of use.

Theorem 4.4.17 (Shaked, 1980). *Suppose that X and Y are absolutely continuous non-negative random variables with $E(X) = E(Y) \in (0,\infty)$ and densities f_X and f_Y. A sufficient condition for $X \leq_L Y$ is that $f_X(x) - f_Y(x)$ changes sign twice on $(0,\infty)$ and the sequence of signs for $f_X - f_Y$ is $-$, $+$, $-$.*

Suppose that $X_1, X_2, ..., X_n$ is a sample of size n from a distribution F. Denote the corresponding order statistics by $X_{1:n}, X_{2:n}, ..., X_{n:n}$. In some cases these order statistics are ordered with respect to their variances. It is natural to ask whether there are also any Lorenz ordering relationships among them. The question at issue is then: for a given parent distribution F, for which pairs (i_1, n_1) and (i_2, n_2) do we have $X_{i_1:n_1} \leq_L X_{i_2:n_2}$?

Arnold and Villaseñor (1991) addressed this question in the case in which F is a Uniform$(0,1)$ distribution. Denoting these uniform order statistics by $U_{i:n}$, they proved the following results.

$$U_{i+1:n} \leq_L U_{i:n}, \ \forall i,n, \tag{4.4.26}$$

$$U_{i:n} \leq_L U_{i:n+1}, \ \forall i,n, \tag{4.4.27}$$

$$U_{n-j+1:n+1} \leq_L U_{n-j:n} \ \forall j,n, \tag{4.4.28}$$

$$U_{n+2:2n+3} \leq_L U_{n+1:2n+1}, \ \forall n. \tag{4.4.29}$$

All four of these results can be verified using Shaked's density crossing criterion (Theorem 4.4.17) and the fact that $U_{i:n} \sim Beta(i, n+1-i)$. Alternatively, using the Markov property of order statistics, it can be verified that $U_{i:n}$ has the same distribution as a product of two suitably chosen independent Beta variables and this will allow us to derive (4.4.26) and (4.4.27) and, with a little more work, (4.4.28) as consequences of Strassen's Theorem 4.4.12.

The four relations (4.4.26)–(4.4.29) can be summarized as saying that, in the uniform case:

- Small order statistics exhibit more inequality than larger order statistics, within the same sample.
- Small order statistics exhibit more inequality as sample size increases.
- Large order statistics exhibit less inequality as sample size increases.
- Sample medians exhibit less inequality as sample size increases.

These results are not always in accord with intuition, although the fourth one is, since the sample median will be more concentrated around $1/2$ as the sample size increases.

Results analogous to (4.4.26) and (4.4.27) can be obtained for X's having either a power function distribution or a classical Pareto distribution. This is because such random variables can be represented as $X = \sigma U^{\delta}$ for some $\sigma > 0$ and $\delta > -1$ (so that $E(X) < \infty$). If δ is positive, then $X_{i:n} = \sigma U_{i:n}^{\delta}$, while if δ is negative $X_{i:n} = \sigma U_{n-i+1:n}^{\delta}$. Thus we have

Theorem 4.4.18 (Arnold and Villaseñor, 1991b). *(a) If X has a power function distribution [i.e., $F_X(x) = (x/\sigma)^{\gamma}$, $0 \leq x, \sigma$, $\gamma > 0$], then*

$$X_{i+1:n} \leq_L X_{i:n}, \quad \forall i, n$$

and

$$X_{i:n} \leq_L X_{i:n+1}, \quad \forall i, n.$$

(b) If $X \sim P(I)(\sigma, \alpha)$ [i.e., $\overline{F}_X(x) = (x/\sigma)^{-\alpha}$, $x > \sigma > 0$, $\alpha > 0$], then

$$X_{i:n} \leq_L X_{i+1:n}, \quad \forall i, n$$

and

$$X_{n-j:n} \leq_L X_{n-j+1:n+1}, \quad \forall j, n.$$

Sample medians can be verified to be Lorenz ordered, with inequality decreasing as sample size increases, for any symmetric parent distribution using a density crossing argument (Arnold and Villaseñor, 1986). A density crossing argument can also be used to verify that (4.4.28) holds in the case of power function distribution, but is reversed for a classical Pareto distribution.

If X has a power function distribution then $X =^d \sigma U_{i:n}^{\delta}$, which is a generalized Beta variable of the first kind (GB1). Wilfling's sufficient conditions for Lorenz ordering within the GB1 family can then be used to identify certain Lorenz order relations between power function order statistics, even when the samples come from power function distributions with different δ's (Wilfling, 1996b). Kleiber (2002) discusses necessary conditions for Lorenz ordering of order statistics from distributions with regularly varying tails. These results allow one to identify certain pairs (i_1, n_1) and (i_2, n_2) for which $X_{i_1:n_1} \not\leq_L X_{i_2:n_2}$.

The special feature of the power function distribution and the classical Pareto distribution that was useful in the above developments is the fact that the corresponding quantile functions are so simple. Few results are available for other choices of F. One case that can be completely resolved is that of an exponential parent distribution.

Theorem 4.4.19 (Arnold and Nagaraja, 1991). *If X has an exponential distribution and $i \leq j$, then the following are equivalent.*

(i) $X_{j:m} \leq_L X_{i:n}$.

(ii) $(n-i+1)E(X_{i:n}) \leq (n-j+1)E(X_{j:m})$.

This result is obtained by making use of the representation of $X_{i:n}$ as a linear combination of i.i.d. exponential random variables (the normalized spacings). Karlin and Rinott (1988) used a similar argument to show that, with an odd sample size and an exponential parent distribution, we have

$$\overline{X}_{2n+1} \leq_L X_{n+1:2n+1}, \quad n = 1,2,\dots . \tag{4.4.30}$$

Thus, in this setting, the sample mean exhibits less inequality than does the sample median. See also Arnold and Villaseñor (1986) for an alternative derivation of this result.

Undoubtedly (4.4.30) holds for other parent distributions, but sufficient conditions on F for (4.4.30) to hold are not known. Viewed from another perspective, no cases in which (4.4.30) fails to hold have been described.

Remark. The Lorenz order for sample medians, i.e.,

$$X_{n+2:2n+3} \leq_L X_{n+1:2n+1},$$

which is true for a symmetric parent distribution, has an appealing parallel result involving sample means. Thus

$$\overline{X}_n \leq_L \overline{X}_{n-1}, \quad n > 1.$$

However this result will hold for any parent distribution for which $F(0) = 0$ and $E(X) \in (0, \infty)$, as may be verified using Strassen's Theorem 4.4.12 (Arnold and Villaseñor, 1986).

4.4.7 Related Orderings

A bewilderingly extensive array of partial orders on the class of non-negative integrable random variables is available in the literature. The best current guide to this maze of interrelated orderings is provided by Shaked and Shanthikumar (2007). See also Chapter 17 of Marshall, Olkin and Arnold (2011) for a discussion more focused on the Lorenz order. In this subsection we will discuss a small selection of these orderings. The motivation is that the conditions for several of the partial orders are easier to check than the corresponding conditions for the Lorenz order. In such a case, if the given ordering between X and Y can be verified, and if that ordering implies Lorenz ordering, then we will have a relatively simple argument for Lorenz ordering of X and Y.

It will be convenient in this subsection to denote the class of all non-negative random variables with finite positive expectations by $\mathcal{L}$. The letter $\mathcal{L}$ is appropriate since these are the random variables for which the Lorenz curve is well defined. We begin by recalling the definition of the Lorenz order on $\mathcal{L}$.

Definition 4.4.6. For $X,Y \in \mathscr{L}, X \leq_L Y \Leftrightarrow L_X(u) \geq L_Y(u), \forall u \in [0,1]$.

If the quantile functions of X and Y are available in analytic form, it is possible to make use of star-ordering to confirm Lorenz ordering.

Definition 4.4.7. For $X,Y \in \mathscr{L}$, we say that X is star-shaped or star-ordered with respect to Y and write $X \leq_* Y$ if $F_X^{-1}(u)/F_Y^{-1}(u)$ is a non-increasing function of u.

Clearly star-ordering is scale invariant. Also a sufficient condition for star-ordering is that $F_Y^{-1}(F_X(x))/x$ is increasing. This is sometimes used as a definition of star-ordering, and it reflects the genesis of the name of the ordering, since such monotonicity is equivalent to the function $F_Y^{-1}(F_X(x))$ being star-shaped. One advantage of considering $F_Y^{-1}(F_X(x))/x$ rather than $F_X^{-1}(u)/F_Y^{-1}(u)$ is that, using it, an analytic expression is required for only one of the quantile functions. In some cases the required monotonicity can be confirmed by differentiation.

Theorem 4.4.20. *Suppose $X,Y \in \mathscr{L}$. If $X \leq_* Y$, then $X \leq_L Y$.*

Proof. Without loss of generality $E(X) = E(Y) = 1$, so that

$$L_X(u) - L_Y(u) = \int_0^u [F_X^{-1}(v) - F_Y^{-1}(v)]/dv. \tag{4.4.31}$$

Since $F_X^{-1}(v)/F_Y^{-1}(v)$ is non-increasing, the integrand is first positive and then negative as v ranges from 0 to 1. Consequently, the integral assumes its smallest value when $u = 1$. It follows that

$$L_X(u) - L_Y(u) \geq L_X(1) - L_Y(1) = 0, \ \forall u \in [0,1].$$

Thus $X \leq_L Y$. □

The key to the proof of Theorem 4.4.20 was that, with $E(X) = E(Y) = 1$, the function $F_X^{-1}(v) - F_Y^{-1}(v)$ had only one sign change $(+,-)$ on $(0,1)$. This sign change property is not scale invariant, but it does motivate introduction of the sign-change ordering.

Definition 4.4.8. For $X,Y \in \mathscr{L}$, we say that X is sign change ordered with respect to Y and write $X \leq_{sc} Y$ if the function

$$\frac{F_X^{-1}(u)}{E(X)} - \frac{F_Y^{-1}(u)}{E(Y)} \tag{4.4.32}$$

has at most one sign change, from $+$ to $-$, as u varies from 0 to 1.

Clearly, $X \leq_* Y$ implies $X \leq_{sc} Y$ which, in turn, implies $X \leq_L Y$, since the argument in the proof of Theorem 4.4.20 only made use of the sign change property.

Earlier it was noted that, if densities for X and Y exist, one can infer Lorenz ordering using a density crossing argument (recall Theorem 4.4.17). This motivates the introduction of a density crossing ordering.

Definition 4.4.9. For $X, Y \in \mathscr{L}$ with corresponding densities f_X and f_Y, we say that X is density-crossing ordered with respect to Y and write $X \leq_{dc} Y$ if the function

$$E(X)f_X(E(X)x) - E(Y)f_Y(E(Y)x) \tag{4.4.33}$$

has either two sign changes $(-,+,-)$ or one sign change $(+,-)$ as x varies from 0 to ∞.

Theorem 4.4.21. *Suppose $X, Y \in \mathscr{L}$ with densities f_X and f_Y. If $X \leq_{dc} Y$, then $X \leq_{sc} Y$.*

Proof. Since $X \leq_{dc} Y$, the density of $X/E(X)$ minus the density of $Y/E(Y)$ [given by (4.4.33)] has at most two sign changes. Consequently $F_X(E(X)x) - F_Y(E(Y)x)$ has a single sign change, i.e., the two distributions cross only once. But this implies that the corresponding quantile functions cross only once, i.e., that $X \leq_{sc} Y$. □

Theorem 4.4.21 has the attractive feature that it will permit confirmation of sign-change ordering, and hence Lorenz ordering, using only densities and not requiring analytic expressions for either the corresponding distribution functions or quantile functions. Note that, if analytic expressions for the distribution functions are available (rather than the quantile functions), the sign-change ordering criterion can be rewritten in the form: $X \leq_{sc} Y$ if $F_X(E(X)x) - F_Y(E(Y)x)$ has at most one sign change $(-,+)$ as x varies from 0 to ∞.

Dispersion ordering, introduced by Doksum, also has potential for identifying cases in which Lorenz ordering obtains.

Definition 4.4.10 (Doksum, 1969)**.** For real valued random variables X and Y, X is said to be dispersion ordered with respect to Y, written $X \leq_{disp} Y$ if for any $0 < \alpha < \beta < 1$

$$F_X^{-1}(\beta) - F_X^{-1}(\alpha) \leq F_Y^{-1}(\beta) - F_Y^{-1}(\alpha).$$

There is an intimate relationship between star-ordering and dispersion ordering. The key insight involves rewriting both definitions in terms of distribution functions. It is the case that $X \leq_* Y$ if, for any $c > 0$, the distribution functions of X and cY are such that $F_X(x) - F_Y(x/c)$ changes sign at most once, and if it changes sign, it is from $-$ to $+$. In contrast, $X \leq_{disp} Y$ if, for every $c \in (-\infty, \infty)$, the distribution functions of X and $Y + c$ are such that $F_X(x) - F_Y(x - c)$ changes sign at most once, and if it changes sign, it is from $-$ to $+$ (Shaked, 1982). These observations permit the conclusion that $X \leq_* Y$ if and only if $\log X \leq_{disp} \log Y$. Thus $\log X \leq_{disp} \log Y$ is sufficient for $X \leq_L Y$.

4.4.8 Multivariate Extensions of the Lorenz Order

In Subsection 4.2.4, two possible extensions of the Lorenz order to higher dimensions were discussed. One was based on a Lorenz surface which consisted of the lower boundary of the Lorenz zonoid. An alternative surface, defined in the two-dimensional case in (4.2.85), although it has some advantage in terms of ease of computation, does not appear to have been much used or advocated (but see Sarabia and Jorda (2013) for some discussion of the potential utility of this surface).

The Lorenz zonoid defined in (4.2.99) appears to be an attractive tool for extending the Lorenz order to higher dimensions. It was verified in Subsection 4.2.2 that, in the one-dimensional case, nested Lorenz curves occur whenever the corresponding zonoids are nested. For convenience we repeat equation (4.2.99) here. It provides the definition of the Lorenz zonoid, $L(\underline{X})$, associated with a k-dimensional random variable $\underline{X}$ with distribution function $F_{\underline{X}}$ supported in $(0,\infty)^k$.

$$L(\underline{X}) = \left\{ \left(E(\psi(X)), \frac{E(X_1\psi(\underline{X}))}{E(X_1)}, \ldots, \frac{E(X_k\psi(\underline{X}))}{E(X_k)} \right) ; \psi \in \Psi^{(k)} \right\}$$

where $\Psi^{(k)}$ is the set of all measurable mappings from $[0,\infty)^k$ to $[0,1]$.

We then define the k-dimensional Lorenz order in terms of nested Lorenz zonoids. It is defined on the class $\mathscr{L}^k$ of random variables $\underline{X} = (X_1, X_2, \ldots, X_k)$ where each coordinate random variable is non-negative with a positive finite expectation. It will be denoted by $\leq_L$ without ambiguity, since it coincides with the usual Lorenz order in the one-dimensional case.

Definition 4.4.11. For $\underline{X}, \underline{Y} \in \mathscr{L}^k$,

$$\underline{X} \leq_L \underline{Y} \Leftrightarrow L(\underline{X}) \subseteq L(\underline{Y}). \tag{4.4.34}$$

There are several competing definitions of multivariate majorization viewed as a partial oder on the set of $m \times n$ matrices with real elements (see Chapter 15 in Marshall, Olkin and Arnold (2011)). No one of these partial orders commands acceptance as the "right" extension of the concept of majorization for vectors. In the Lorenz order case, the nested zonoid order defined above appears to be a very natural extension. As we will see below there are however here also some viable competitors.

It will be recalled that there was more than one way to define the Lorenz order when X is one-dimensional. An attractive, and often convenient, definition was one involving expectations of convex functions of $X/E(X)$. Thus, essentially repeating Theorem 4.4.6, $X \leq_L Y$ if and only if $E(g(X/E(X))) \leq E(g(Y/E(Y)))$ for all continuous convex functions g for which the indicated expectations are finite. The class of continuous convex functions on $(0,\infty)^k$ is well defined and this motivates consideration of an alternative Lorenz order on $\mathscr{L}^k$, to be denoted by $\leq_{L_1}$ below. Three other plausible k-dimensional extensions of the univariate Lorenz order are also included in the following set of definitions.

Definition 4.4.12. Four partial orders on $\mathscr{L}^k$ are identified by subscripts below. In each case, $\underline{X}, \underline{Y} \in \mathscr{L}^k$.

(i) $\underline{X} \leq_{L_1} \underline{Y}$ if

$$E(g(X_1/E(X_1), X_2/E(X_2), \ldots, X_k/E(X_k)) \\ \leq E(g(Y_1/E(Y_1), Y_2/E(Y_2), \ldots, Y_k/E(Y_k))$$

for all continuous convex functions g for which the indicated expectations are finite.

(ii) $\underline{X} \leq_{L_2} \underline{Y}$ if $\sum_{i=1}^k a_i X_i \leq_{GL} \sum_{i=1}^k a_i Y_i, \ \forall \, \underline{a} \in (-\infty, \infty)^k$.

(iii) $\underline{X} \leq_{L_3} \underline{Y}$ if $\sum_{i=1}^k c_i X_i \leq_L \sum_{i=1}^k c_i Y_i, \ \forall \, \underline{c} \in [0, \infty)^k$.

(iii) $\underline{X} \leq_{L_4} \underline{Y}$ if $X_i \leq_L Y_i, \ i = 1, 2, ..., k$.

In the definition of $\leq_{L_2}$ above it should be noted that use has been made of a generalized one-dimensional Lorenz order. This is required since, for some choices of the vector $\underline{a}$, the random variable $\sum_{i=1}^k a_i X_i$ can take on negative values and the usual Lorenz order will not be well defined. For arbitrary one-dimensional random variables X and Y, not necessarily non-negative, we write $X \leq_{GL} Y$ if $\int_0^u F_X^{-1}(u)\, du \geq \int_0^u F_Y^{-1}(u), \ \forall \, u \in (0,1)$.

Definitions 4.4.11 and 4.4.12 describe what appear to be a total of five partial orders on $\mathscr{L}^k$. However, there are actually only four, not five. It may be verified that the partial orders $\leq_L$ and $\leq_{L_2}$ are identical. Thus the nested zonoid order on $\mathscr{L}^k$ can be reinterpreted as an ordering which corresponds to generalized Lorenz ordering of all linear combinations of the coordinate random variables.

All four partial orders in Definition 4.4.12 are distinct and they are actually listed in the order of decreasing strength. Thus we have

$$\underline{X} \leq_{L_1} \underline{Y} \Rightarrow \underline{X} \leq_{L_2} \underline{Y} \Rightarrow \underline{X} \leq_{L_3} \underline{Y} \Rightarrow \underline{X} \leq_{L_4} \underline{Y}. \tag{4.4.35}$$

Names can be associated with these k-dimensional Lorenz orders. The last order on the list, $\leq_{L_4}$, the weakest of the group, is naturally called marginal Lorenz ordering. It ignores any dependence relations between the coordinate random variables. The first in the list, $\leq_{L_1}$, can be called convex ordering. The partial order, $\leq_{L_2} \equiv \leq_L$, will be called the zonoid ordering or the linear combinations ordering. The partial order $\leq_{L_3}$ is sometimes called the price Lorenz order, or the positive combinations order, or the exchange rate Lorenz order. The genesis of the last of these names is as follows. Suppose that the coordinates of $\underline{X}$ and $\underline{Y}$ represent holdings (or earnings) in k different currencies. Suppose that we exchange all of the holdings into one currency, perhaps Euros, according to k exchange rates $c_1, c_2, ..., c_k$, then it is natural to compare $\sum_{i=1}^k c_i X_1$ and $\sum_{i=1}^k c_i Y_1$ with regard to inequality by the usual univariate Lorenz order. The ordering $\leq_{L_3}$ requires that $\sum_{i=1}^k c_i X_1 \leq_L \sum_{i=1}^k c_i Y_1$ for every vector of exchange rates $\underline{c}$. In this context, it is natural to require that the c_i's be positive. How would you interpret a negative exchange rate?

This last observation highlights a possible lacuna in the arguments in support of the zonoid ordering. Since $\leq_L$ and $\leq_{L_2}$ are equivalent, the zonoid ordering requires generalized Lorenz ordering of the univariate random variables $\sum_{i=1}^k a_i X_1$ and $\sum_{i=1}^k a_i Y_1$ even when some of the a_i's are negative (corresponding to negative exchange rates!). See Koshevoy and Mosler (1996) for an insightful introduction to the Lorenz zonoid order, including discussion of the role of the univariate generalized Lorenz order in the zonoid order. A more extensive discussion may be found in Mosler (2002).

A positive feature of the convex order, $\leq_{L_1}$, is that the averaging theorem (i.e., Theorem 4.4.10) survives intact. Moreover, Strassen's Theorem 4.4.12 is still true in the k-dimensional setting. Thus

Theorem 4.4.22. *For $\underline{X},\underline{Y} \in \mathscr{L}^k$, $\underline{Y} \leq_{L_1} \underline{X}$ if and only if there exist jointly distributed random variables $\underline{X}'$ and Z' such that $\underline{X} \stackrel{d}{=} \underline{X}'$ and $\underline{Y} \stackrel{d}{=} E(\underline{X}'|Z')$.*

In fact Strassen proved the result in a much more abstract setting than we require (see Meyer, 1966, for further related discussion). Whitt (1980) discussed application of these ideas in a reliability context.

Despite the fact that the zonoid ordering will fail to have Strassen's balayage equivalence theorem (Theorem 4.4.12) and will involve comparisons of generalized Lorenz curves with curious exchange rate interpretations, it appears to this author that the zonoid order ($\leq_L \equiv \leq_{L_2}$) is the most defensible of the competing partial orders. Nevertheless it cannot be ruled out that, in some applications, the other three partial orders might be more appropriate.

Note 4.4.2. If one considers multivariate majorization instead of multivariate Lorenz ordering, it is not necessary to introduce the concept of a "generalized Lorenz curve" into the discussion. Majorization is well defined for vectors in $(-\infty,\infty)^n$, rather than just $[0,\infty)^n$ for Lorenz ordering. Consequently there is no concern about linear combinations taking on negative values. Refer to Chapter 15 of Marshall, Olkin and Arnold (2011) for more details.

There are several available choices for summary measures of inequality in the k-dimensional case. One natural choice which respects the convex order, $\leq_{L_1}$, would be associated with a particular choice of continuous convex function, say h, defining the inequality measure to be

$$E(h(X_1/E(X_1), X_2/E(X_2), ..., X_k/E(X_k))). \tag{4.4.36}$$

It is however difficult to argue for any one particular choice of h, and it must be remembered that each h will completely order $\mathscr{L}^k$, but different choices of h will generally lead to different complete orders.

Another possibility is motivated by a suggested two-dimensional Gini index defined in (4.2.91). However, let us rewrite that formula in the form

$$G(X) = E[d(\widetilde{X}^{(1)}, \widetilde{X}^{(2)})] \tag{4.4.37}$$

where $\widetilde{X}^{(1)}$ and $\widetilde{X}^{(2)}$ are i.i.d. copies of a normalized version of X, defined by $\widetilde{X} = X/E(X)$. Here d is a suitable metric on $(0,\infty)$. A natural extension of this measure to k-dimensions takes the form, for $\underline{X} \in \mathscr{L}^k$,

$$G(\underline{X}) = E[d(\widetilde{\underline{X}}^{(1)}, \widetilde{\underline{X}}^{(2)})] \tag{4.4.38}$$

where $\widetilde{\underline{X}}^{(1)}$ and $\widetilde{\underline{X}}^{(2)}$ are i.i.d copies of $\widetilde{\underline{X}}$ which is the vector $\underline{X}$ rescaled so that all marginal means are equal to 1. The particular choice of Euclidean distance in (4.4.38) has the advantage that it is not difficult to identify the required standardization of the index to assure that it will range in value from 0 to 1. With the k-dimensional Euclidean norm, $\|\cdot\|$, the standardized version of (4.4.38) is

$$G(\underline{X}) = \frac{1}{2^k} E(\| \widetilde{\underline{X}}^{(1)} - \widetilde{\underline{X}}^{(2)} \|). \tag{4.4.39}$$

Mosler (2002) provides a proof that if $\underline{X} \leq_L \underline{Y}$ then $G(\underline{X}) \leq G(\underline{Y})$ where $G(\underline{X})$ is as defined in (4.4.39). Of course, other norms could be used if so desired.

If we decide to use the zonoid ordering, $\leq_L$, which corresponds to nested zonoids, several theoretically attractive geometric summary measures of inequality in $\mathscr{L}^k$ can be considered. These are natural analogs of popular inequality measures that are used in the one-dimensional case. Three such measures are:

(i) The $(k+1)$-dimensional volume of the Lorenz zonoid.

(ii) The k-dimensional volume of the boundary of the Lorenz zonoid.

(iii) The maximal distance between a point on the boundary of the Lorenz zonoid and the egalitarian line joining the points $(0,0,...,0)$ and $(1,1,...,1)$.

These are k-dimensional analogs of, respectively, the one-dimensional Gini, Amato and Pietra indices. Although quite simple expressions for these indices are available when $k=1$, this is in general not so in higher dimensions. Koshevoy and Mosler (1997) do provide an attractive expression for the volume of the Lorenz zonoid, though in practice it will be generally difficult to evaluate. They derive the expression as follows. For $\underline{X} \in \mathscr{L}^k$, first define the corresponding standardized variable $\widetilde{\underline{X}}$ with $\widetilde{X}_i = X_i/E(X_i)$, $i = 1,2,...,k$. Next consider $k+1$ i.i.d. copies of $\widetilde{\underline{X}}$, denoted by $\widetilde{\underline{X}}^{(1)}, \widetilde{\underline{X}}^{(2)}, ..., \widetilde{\underline{X}}^{(k+1)}$. It will be recalled that in the case $k=1$, inequality of X was often characterized by measures depending on two independent copies of X, so the choice of $k+1$ copies in the k dimensional case is, in a sense, quite natural. Now define a $(k+1)\times(k+1)$ random matrix Q whose i'th row is $(0, \widetilde{\underline{X}}^{(i)})$, $i = 1,2,...,k+1$. Koshevoy and Mosler verify that, with this construction of the random matrix Q, we have

$$\begin{aligned} \widetilde{G}(\underline{X}) &= (k+1)\text{-dimensional volume of } L(\underline{X}) \\ &= \frac{1}{(k+1)!} E(|\det(Q)|). \end{aligned} \tag{4.4.40}$$

This measure can assume the value 0 for certain non-degenerate distributions. Mosler (2002) proposes some variant definitions which do not have this negative feature. The expression in (4.4.40), when $k=1$, simplifies to yield $(1/2)E(|\widetilde{X}^{(1)} - \widetilde{X}^{(2)}|)$ which is recognizable as one of the several available expressions for the classic one-dimensional Gini index. However, as mentioned previously, the task of evaluating $\widetilde{G}(\underline{X})$ defined by (4.4.40) for any particular distribution for $\underline{X}$ is daunting indeed. But it can be approximated as long as it is possible to simulate realizations from the distribution of $\underline{X}$. This same approach can be used to generate an approximate version of the corresponding Lorenz zonoid, and to generate approximations of essentially arbitrary accuracy of other multivariate measures of inequality of interest.

4.5 Optimal Income Distributions

How should income be distributed? There is ample evidence that governmental authorities and international agencies feel capable of recognizing undesirable income distributions. Vide their efforts to promote changes in income structures by economic and legislative intervention. In fact, there is a tendency to correlate the ethereal quality of "development" with another ethereal quality of "improved income distribution." Perhaps we are fortunate that intervention programs appear to be able to only make minor changes in prevailing income distributions. If we were capable of massive changes, it would become more crucial to be able to answer the question: "Towards what optimal distribution should we shift the present distribution?" As it stands, concern is focused on prevailing high levels of inequality in lesser developed countries and with efforts to diminish the inequality, if ever so slightly. Pareto would, of course, argue that, generally speaking, we might not expect to be able to alter the form of the distribution; the Paretian tail behavior will ever reappear. He might agree that the inequality level (or the slope on the Pareto graph) could be manipulated. Proponents of other income models might propose that the class of income distributions within which we seek the optimum should be restricted in other ways (perhaps log-normal, etc.).

If we seek to reduce inequality, then the answer is straightforward. Using almost any reasonable measure of inequality, the level of inequality is minimized by degenerate distributions (i.e., equal incomes), for which the Lorenz curve corresponds to the 45° line $L(u) = u$ (called the egalitarian line for this reason). Such degenerate distributions are attainable only as limiting cases, if we restrict attention to, for example, Pareto distributions or log-normal distributions. If we assume an additive utility model for welfare, then either using Jensen's inequality or appealing to Theorem 4.4.6, we conclude that for a given mean income m, the optimal distribution is degenerate at m. Unless we are willing to accept such a distribution as optimal (and it is doubtful that we are), this result casts doubt either on the concavity of individual utilities or on the additivity assumption.

There is an enormously complex literature on aspects of optimal income distributions and optimal taxation policies. Some of it involves poverty line considerations. Other approaches suggest modification and restrictions on the acceptable activities of Robin Hood. Entropy maximization approaches have been considered. The interested reader is encouraged to enter the phrase "optimal income distribution" into his/her favorite search engine to begin an exciting quest.

We will end this subsection with a reference to one of the few papers which specifically identify an optimal distribution that is not egalitarian. Adams (1979) suggested the intriguing possibility that an optimal value of α for the classical Pareto distribution might be 1.618 (the golden section). The corresponding optimal Gini index would then be 0.447. The appearance of the golden ratio in this prescription gives it a somewhat Pythagorean aspect.

Chapter 5

Inference for Pareto distributions

5.1 Introduction

The development of inference procedures for Pareto distributions and their close relatives has been predictably uneven. Initially, the classical Pareto distribution was heavily emphasized. A two-fold explanation exists. First, the classical Pareto distribution has been around the longest and second (and probably more importantly), by a simple transformation, inference for the Pareto (I) (and the Pareto (II)) is reducible to inference for exponential variables (a very well-plowed field!). The techniques however can be (and generally were) derived without explicit use of the exponential transformation. As soon as one moves to study more complex Pareto distributions, "nice" results vanish. Some authors have assumed one or several parameters to be known in order to be able to make progress. If all parameters are assumed unknown, then even such tried and true techniques as maximum-likelihood and the method of moments will involve iterative solution of systems of distinctly non-linear equations. Often the quantity of interest is the value of some inequality measure for the population (i.e., one of the measures described in Section 4.2). One could, of course, use as an estimate the corresponding sample measure of inequality (cf. Section 4.3), but alternative estimates which utilize the assumed parametric model for the population can be, and sometimes have been, developed.

During the last 20–30 years considerable attention has been devoted to inference for generalized Pareto distributions in the Pickands sense. Much of this material is relevant for discussion of inference for the P(II), P(III), P(IV) and FP generalized Pareto models introduced in Chapter 3. Discussion of parameter estimation for these models will be presented in Sections 5.2.6–5.2.8.

Interval estimation for Pareto populations has not been extensively investigated. Results are obtainable for the Pareto (I). A lack of convenient pivotal quantities hampers efforts in the case of more general Pareto distributions. Basically, the same situation occurs in the setting of testing parametric hypotheses, i.e., some results for the Pareto (I) and less for the more general distributions. Of course this parallel is a reflection of the natural interrelationship between tests of hypotheses and confidence regions (i.e., thc possibility of testing hypotheses by "inverting" good confidence intervals).

A basic question, which by rights should precede discussion of inference techniques which assume a Pareto model, is how is one to select between competing models? Put another way, how can we decide whether a Pareto (or other) model adequately fits the data at hand? Several more or less qualitative approaches are suggested in the literature as supplements to the always available (but perhaps not too useful) Kolmogorov-Smirnov and χ^2 tests with parameters estimated from the data. The work of Quandt (1966) is useful in this context. More recently Gastwirth and his coworkers have contributed techniques utilizing the sample Lorenz curve and related statistics. Other early contributions to the problem of testing Pareto versus "something else" are Bryson (1974), DuMouchel and Olshen (1975) and Arnold and Laguna (1976). Several more recent contributions on goodness of fit tests are discussed in Section 5.5.

Much income data are only available in grouped form. The resulting multinomial likelihood is susceptible to several kinds of analysis. Early contributors to this field were Fisk (1961b) and Aigner and Goldberger (1970). Zellner (1971) presented the corresponding Bayesian analysis as an example in his text. The relatively recent dates on these references certainly do not indicate that grouped data are a recent phenomenon. What happened was that earlier researchers apparently were not disturbed by the "lumpy" character of their data. For large samples the corresponding continuous theory can probably be safely used, but with the computing facilities currently available, analyses which take into account the grouped character of the data are more appealing.

The final section of the present chapter will be devoted to inference for distributions related to Pareto distributions. Here will be found discussion of the discrete Pareto, under-reporting income models, etc. The emphasis is on point estimation with some attention paid to model selection tests.

5.2 Parameter Estimation

The simplest and the most thoroughly investigated situation is that in which we have a sample of size n from a classical Pareto distribution (i.e., the Pareto (I) distribution whose survival distribution is given in (3.2.1)). Let $\underline{X} = (X_1, \ldots, X_n)$ denote the sample and $X_{i:n}$, $(i = 1, 2, \ldots, n)$, the corresponding order statistics. The corresponding likelihood function is of the form

$$L(\sigma, \alpha) = \alpha^n \sigma^{n\alpha} \left(\prod_{i=1}^{n} X_i \right)^{-(\alpha+1)}, \quad \alpha > 0, \quad \sigma \le X_{1:n}. \tag{5.2.1}$$

The minimal sufficient statistics for (σ, α), namely $(X_{1:n}, \prod_{i=1}^{n} X_i)$, can be read off from (5.2.1). An alternative, sometimes more convenient, form is $(X_{1:n}, \sum_{i=1}^{n} \log X_i)$.

5.2.1 *Maximum Likelihood*

To obtain the maximum likelihood estimate $(\hat{\sigma}, \hat{\alpha})$ of (σ, α), we consider the log-likelihood function

$$\ell(\sigma, \alpha) = n \log \alpha + n\alpha \log \sigma - (\alpha + 1) \sum_{i=1}^{n} \log X_i. \tag{5.2.2}$$

For a fixed α this is maximized when σ is set equal to $X_{1:n}$. Differentiation with respect to α can then be used to obtain the maximizing value of α. Thus we find

$$\text{(Pareto I)} \qquad \hat{\sigma} = X_{1:n}$$

$$\hat{\alpha} = \left[\frac{1}{n} \sum_{i=1}^{n} \log(X_i/X_{1:n}) \right]^{-1} \tag{5.2.3}$$

(cf. Muniruzzaman, 1957, and Quandt, 1966). The maximum likelihood estimates are consistent (in fact strongly consistent).

This follows since evidently $X_{1:n} \xrightarrow{\text{a.s.}} \sigma$ and, by the strong law of large numbers, $\frac{1}{n}\sum_{i=1}^{n} \log X_i \xrightarrow{\text{a.s.}} E(\log X_1) = \log\sigma + \alpha^{-1}$.

It turns out that we can derive the exact form of the joint density of the maximum likelihood estimates $(\hat{\sigma}, \hat{\alpha})$ (Muniruzzaman, 1957). First we recognize that $\hat{\sigma} = X_{1:n} \sim P(I)(\sigma, n\alpha)$ (from (3.4.5)). Thus we require the conditional distribution of $\hat{\alpha}$ given $\hat{\sigma}$. Malik (1970b) showed that in fact $\hat{\alpha}$ and $\hat{\sigma}$ are independent (a result implicit in Muniruzzaman, 1957; see also Baxter, 1980). If we define $Y_i = \log X_i$, then it follows that $Y_i - \log\sigma \sim \Gamma(1, \alpha^{-1})$ (i.e., an exponential distribution). The independence of $\hat{\alpha}$ and $\hat{\sigma}$ follows readily from the independence of exponential spacings (recall Theorem 3.4.2). For completeness we write the actual densities

$$f_{\hat{\sigma}}(u) = n\alpha\sigma^{n\alpha} u^{-(n\alpha+1)}, \quad u > \sigma \tag{5.2.4}$$

and

$$f_{\hat{\alpha}}(v) = \frac{(\alpha n)^{n-1}}{\Gamma(n-1)v^n} e^{-(n\alpha/v)}, \quad v > 0. \tag{5.2.5}$$

The reciprocal of $\hat{\alpha}$ has a gamma distribution, i.e., $\hat{\alpha}^{-1} \sim \Gamma(n-1, (\alpha n)^{-1})$ or equivalently

$$2\alpha n/\hat{\alpha} \sim \chi^2_{2n-2}. \tag{5.2.6}$$

Note also that

$$2\alpha n \log(\hat{\sigma}/\sigma) \sim \chi^2_2. \tag{5.2.7}$$

The above observations permit one to easily determine the means and variances of the maximum likelihood estimates:

$$E(\hat{\sigma}) = \sigma\left(1 - \frac{1}{n\alpha}\right)^{-1}, \tag{5.2.8}$$

$$\text{var}(\hat{\sigma}) = \sigma^2 n\alpha(n\alpha - 1)^{-2}(n\alpha - 2)^{-1}, \tag{5.2.9}$$

$$E(\hat{\alpha}) = \alpha n(n-2)^{-1}, \tag{5.2.10}$$

$$\text{var}(\hat{\alpha}) = (\alpha n)^2(n-2)^{-2}(n-3)^{-1}. \tag{5.2.11}$$

The biased character of the maximum likelihood estimates is evident from the above. The estimate $\hat{\sigma}$ always overestimates σ. Considerations such as these have prompted investigation of modified (but asymptotically equivalent) versions of the maximum likelihood estimates. Before discussing these modifications we may quickly dispose of the question of the asymptotic distribution of $\hat{\sigma}$ and $\hat{\alpha}$. From (3.6.4) we find

$$\lim_{n\to\infty} P((\alpha n)[\hat{\sigma}/\sigma) - 1] > z) = e^{-z}, \tag{5.2.12}$$

i.e., an asymptotically exponential distribution. The asymptotic normality of $\hat{\alpha}$ is a simple consequence of (5.2.6) (since the reciprocal of a χ^2 variable with a large number of degrees of freedom is approximately normal).

It will be observed at the end of Section 5.6 below that a wide variety of Pareto (I) data configurations have log-likelihood functions essentially of the form (5.2.2). These include generalized order statistics, certain truncated data scenarios, record values, progressive censoring, etc. All these data configurations will yield maximum likelihood estimates parallel in structure to the estimates in (5.2.3).

Remark. *Rather than estimating the parameters (σ,α), it may be of interest to estimate certain functions of (σ,α) which are meaningful descriptors of the Pareto(I) distribution. For example we might wish to estimate:*

(i) $f_X(x_0) = \frac{\alpha}{\sigma}\left(\frac{x_0}{\sigma}\right)^{-(\alpha+1)}$, *the density at x_0,*

(ii) $\overline{F}_X(x_0) = \left(\frac{x_0}{\sigma}\right)^{-(\alpha)}$, *the survival function at x_0,*

(iii) $F_X^{-1}(u_0) = \sigma(1-u_0)^{-1/\alpha}$, $u_0 \in (0,1)$, *the quantile function at u_0,*

(iv) $E(X)) = \frac{\sigma\alpha}{\alpha-1}$, *the mean of X,*

(v) $G(X) = (2\alpha-1)^{-1}$, $\alpha > 1$, *the Gini index of X,*

(vi) $L_X(u_0) = 1-(1-u_0)^{(\alpha-1)/\alpha}$, $u_0 \in (0,1)$, *the Lorenz curve at u_0,*

(vii) $c.v.(X) = (\alpha^2-2\alpha)^{-1/2}$, $\alpha > 2$, *the coefficient of variation of X,*

(viii) $r_X(x_0) = \alpha/x_0$, *the failure rate function at x_0.*

In all cases, of course, the corresponding maximum likelihood estimates are obtainable by substituting $(\tilde{\sigma},\tilde{\alpha})$ for (σ,α) in the definition of the parametric function of interest. For a fixed value of the sample size n, the distributions of the maximum likelihood estimates will be troublesome for those parametric functions in the list *which involve both σ and α. For functions (v)–(viii), which are functions of α only, a simple change of variable in (5.2.5) will produce the required density. Asymptotic normality of all these maximum likelihood estimates can be confirmed using the facts that $\hat{\sigma}_n \xrightarrow{a.s.} \sigma$ and $\hat{\alpha}_n$ is asymptotically normal. See Moothathu (1985) for some related discussion. Of course, estimation strategies distinct from maximum likelihood might be employed with reference to the parametric function (i)–(viii). We will consider some examples of this in subsequent subsections.*

5.2.2 *Best Unbiased and Related Estimates*

How might we modify $\hat{\alpha}$ and $\hat{\alpha}$ to improve their behavior for a fixed finite value of n? An unbiased modification of $\hat{\alpha}$ is readily available namely

$$\hat{\alpha}_U = (n-2)n^{-1}\hat{\alpha} \tag{5.2.13}$$

(cf. Saksena, 1978; Baxter, 1980; and Cook and Mumme, 1981).

Alternatively, we might seek that constant multiple of $\hat{\alpha}$ which minimizes the mean squared error (a la Goodman, 1953). The resulting estimate, apparently first studied by A. M. Johnson (see Saksena, 1978), is of the form

$$\hat{\alpha}_J = (n-3)n^{-1}\hat{\alpha}. \tag{5.2.14}$$

The corresponding mean squared error functions are

$$\text{m.s.e.}(\hat{\alpha}) = \alpha^2(n+6)(n-2)^{-1}(n-3)^{-1}, \tag{5.2.15}$$

$$\text{m.s.e.}(\hat{\alpha}_U) = \alpha^2(n-3)^{-1}, \tag{5.2.16}$$

$$\text{m.s.e.}(\hat{\alpha}_J) = \alpha^2(n-2)^{-1}. \tag{5.2.17}$$

Note that

$$\text{m.s.e.}(\hat{\alpha}) > \text{m.s.e.}(\hat{\alpha}_U) > \text{m.s.e.}(\hat{\alpha}_J). \tag{5.2.18}$$

Remark. *However, the Johnson estimate (5.2.14) is itself inadmissible with squared error loss (Nagata, 1983). For related inadmissibility discussion see Arnold (1970) and Zidek (1973).*

Modification of $\hat{\sigma}$ requires only a little more ingenuity. Using the independence of $\hat{\alpha}$ and $\hat{\sigma}$ it is not difficult to obtain an unbiased estimate of σ of the form

$$\hat{\sigma}_U = [1-(n-1)^{-1}\hat{\alpha}^{-1}]\hat{\sigma} \tag{5.2.19}$$

with mean squared error (i.e., variance) given by

$$\text{m.s.e.}(\hat{\sigma}_U) = \sigma^2\alpha^{-1}(n-1)^{-1}(\alpha n-2)^{-1}, \tag{5.2.20}$$

(see Saksena, 1978, and Baxter, 1980). It is not difficult to verify that $\hat{\sigma}_U$ has a uniformly smaller mean squared error than $\hat{\sigma}$, provided that $n \geq 3$ and $n\alpha > 2$. A. M. Johnson had earlier considered estimates of the form

$$\hat{\sigma}(c) = \hat{\sigma}(1-c\hat{\alpha}^{-1}) \tag{5.2.21}$$

and had tried to determine the optimal value of c (to minimize the mean squared error). The corresponding mean squared error function is

$$\text{m.s.e.}(\hat{\sigma}(c)) = \frac{\sigma^2[c^2 n(n-1)(n\alpha-1) - 2c(n-1)n\alpha + 2n\alpha]}{n\alpha(n\alpha-1)(n\alpha-2)} \tag{5.2.22}$$

so that the optimal value of c, namely

$$c = \alpha(n\alpha - 1)^{-1}, \tag{5.2.23}$$

depends on the unknown parameter α. Two possible resolutions of this dilemma were discussed in Saksena (1978). One approach would replace α in (5.2.23) by the corresponding maximum likelihood estimate $\hat{\alpha}$, which on substituting back into (5.2.21) yields the estimate

$$\hat{\alpha}_s = [1 - (n\hat{\alpha} - 1)^{-1}]\hat{\sigma}. \tag{5.2.24}$$

Saksena did not investigate the properties of $\hat{\sigma}_s$ since it does not have a finite mean. The second avenue open to us relies on the observation that, for reasonably large values of n and for values of α which are not too small, the quantity in (5.2.23) is well approximated by n^{-1}. This choice of c leads to A. M. Johnson's estimate

$$\hat{\sigma}_J = (1 - n^{-1}\hat{\alpha}^{-1})\hat{\sigma}. \tag{5.2.25}$$

Saksena observed that

$$\text{m.s.e.}(\hat{\sigma}_U) < \text{m.s.e.}(\hat{\sigma}_J) \Leftrightarrow \alpha < 2 - n^{-1}. \tag{5.2.26}$$

For reasonably large values of n, the choice between the unbiased estimate and the Johnson estimate hinges on whether α is believed to be less than or greater than 2. Since (see e.g., Champernowne, 1952) "typical" values of α lie in the interval 1.5 to 2.5, the choice between $\hat{\sigma}_U$ and $\hat{\sigma}_J$ remains difficult. Comparison of (5.2.19) and (5.2.25) reminds us that $\hat{\sigma}_U$ and $\hat{\sigma}_J$ are barely distinguishable, and so the problem of deciding which to use is somewhat of a tempest in a teapot. Saksena (1978) provided some Monte Carlo evidence which suggests that the estimates $(\hat{\alpha}_U, \hat{\sigma}_U)$ and $(\hat{\alpha}_J, \hat{\sigma}_J)$ are essentially equivalent and are better (in terms of generalized mean squared error) than the maximum likelihood estimates $(\hat{\alpha}, \hat{\sigma})$ and than the method of moments and least squares estimates which we will describe later in this section.

Cook and Mumme (1981) considered some estimates which are closely related to the unbiased estimates $(\hat{\alpha}_U, \hat{\sigma}_U)$. They propose the "unbiased equivalents" to the maximum likelihood estimates to be

$$\tilde{\alpha} = n^{-1}(n-2)\hat{\alpha} \tag{5.2.27}$$

$$\tilde{\sigma} = (n\tilde{\alpha} - 1)(n\tilde{\alpha})^{-1}\hat{\sigma}. \tag{5.2.28}$$

The estimate $\tilde{\alpha}$ is indeed unbiased (and is our old friend $\hat{\alpha}_U$ defined in (5.2.13)). The estimate $\tilde{\sigma}$, though worthy of consideration, fails to be unbiased. It is actually a close relative of A. M. Johnson's estimate. It can be obtained by replacing $\hat{\alpha}$ by $\tilde{\alpha}$ in (5.2.25). We can therefore expect $\tilde{\sigma}$, $\hat{\sigma}_J$ and $\hat{\sigma}_U$ to be basically equivalent estimates.

A closely related innovative approach was also discussed by Cook and Mumme (1981). They observed that if σ were known, then the best unbiased estimate of α would be of the form

$$\alpha^* = \left[\frac{1}{n-1}\sum_{i=1}^{n}\log(X_i/\sigma)\right]^{-1}, \tag{5.2.29}$$

while if α were known, the best unbiased estimate of σ would be

$$\sigma^* = (n\alpha - 1)(n\alpha)^{-1}X_{1:n}. \tag{5.2.30}$$

They proposed using an initial estimate of σ, say $\sigma_0 = X_{1:n}$, and then iterating using (5.2.29) and (5.2.30), i.e., the kth stage estimates will satisfy

$$\alpha_k = \frac{1}{n-1}\sum_{i=1}^{n}\log(X_i/\sigma_k), \quad k = 1,2,\ldots \tag{5.2.31}$$
$$\sigma_k = (n\alpha_{k-1}{}^{-1})(n\alpha_{k-1})^{-1}X_{1:n}.$$

Convergence is guaranteed, since the sequences $\{\alpha_k\}$ and $\{\sigma_k\}$ defined in (5.2.31) are monotone decreasing. The iterated estimate obtained from (5.2.31) apparently is competitive with the maximum likelihood estimate (this observation is based on the small Monte Carlo study described in Cook and Mumme).

Iwinska (1980) suggested an alternative estimate of σ, assuming that α is known. She considered estimates of the form $b\overline{X}$ and then chose b to minimize the mean squared error of the estimate. The resulting estimate is of the form

$$\hat{\sigma}_I = \frac{n(\alpha-2)(\alpha-1)}{n\alpha^2 - 2n\alpha + 1}\overline{X}. \tag{5.2.32}$$

It is not likely that such an estimate will be recommended, though a modification involving computation of its conditional expectation given $X_{1:n}$ might be of interest.

In the discussion of the Cook and Mumme iterative scheme, reference was made to "best" unbiased estimates. We mean best in the sense of having minimum variance. The family of distributions of the maximum likelihood estimates for samples from the classical Pareto distribution can be shown to be complete. This is true whether we assume that both parameters are unknown or just that (either) one of the parameters is unknown and the other known. See Saksena and Johnson (1984) for details of this. It follows that an unbiased estimate which is a function of the maximum likelihood estimate(s) will be the unique minimum variance unbiased estimate (UMVUE). Thus (5.2.13), (5.2.19), (5.2.29), (5.2.30) are UMVUE's (cf. Baxter, 1980, and Likes, 1969). Completeness makes it possible, at least formally, to find UMVUE's for many parametric functions. We merely find an available unbiased estimate and compute its conditional expectation given the maximum likelihood estimate. Lwin (1972) illustrated this technique in finding the UMVUE for the survival probability $(t/\sigma)^{-\alpha}$ (where t is a fixed positive number). He considered all three cases, i.e., σ known, α

known and both α and σ unknown. In cases where we wish to estimate a function which has a representation as a power series in σ and α, it is sometimes possible to find an unbiased estimate by using our knowledge of the moments of the estimates $\hat{\sigma}$ and $\hat{\alpha}$. By completeness, that unbiased estimate is the UMVUE. For example we might wish to estimate the Gini index, i.e., $(2\alpha-1)^{-1}$. If we assume both parameters are unknown, then the maximum likelihood estimate of α, given by (5.2.3), has a known density (5.2.5), and its inverse moments are

$$E(\hat{\alpha}^{-j}) = \Gamma(n+j-1)/[\Gamma(n-1)(n\alpha)^j], \quad j=1,2,\ldots.$$

It is clear that the UMVUE of $(2\alpha-1)^{-1} = \sum_{j=1}^{\infty}(2\alpha)^{-j}$ is given by

$$\sum_{j=1}^{\infty}\left(\frac{n}{2\hat{\alpha}}\right)^j \Gamma(n-1)/\Gamma(n+j-1)$$

provided that $\alpha > 1/2$ (for convergence). An analogous approach leads to the UMVUE of the median of the classical Pareto distribution.

Muniruzzaman (1957) discussed estimation of several measures of location for classical Pareto populations. He considered (i) the geometric mean, $\sigma e^{1/\alpha}$; (ii) the harmonic mean, $\sigma(1+\alpha^{-1})$; and (iii) the median, $\sigma 2^{1/\alpha}$. Since all three of these parametric functions are expressible as power series in σ and α^{-1}, it is not difficult to write down series expressions for the corresponding UMVUE's (recall $\hat{\sigma}$ and $\hat{\alpha}$ are independent). See also Moothathu (1986). Kern (1983) discussed the UMVUE's of the median, the k-th moment and the geometric mean of a Pareto(I) distribution with σ known.

Unfortunately, for many of the parametric functions listed as (i)–(viii) in Section 5.2.1, the corresponding UMVUE is only available in the form of a power series, frequently a hypergeometric series. For example, Moothathu (1990) provides expressions for the UMVUE's of the Lorenz curve (iii) and the Gini index (iv), in the case in which σ is known, and in the case in which both σ and α are unknown. The expressions that he provides involve generalized hypergeometric series. In Moothathu (1988), similar series expressions are provided for the UMVUE's for the coefficients of variation, skewness and kurtosis.

In the case of estimation of the survival function, $(x_0/\sigma)^{-\alpha}$ for a fixed value of x_0, less cumbersome expressions for the UMVUE are available. Since a set of complete sufficient statistics is at hand, it is only necessary to identify a convenient unbiased estimate of $(x_0/\sigma)^{-\alpha}$ and to compute its conditional expectation given the complete sufficient statistics. A convenient unbiased estimate is $I(X_1 > x_0)$. We consider 3 cases.

Case 1, α known.

Here the complete sufficient statistic for σ is $X_{1:n}$ Consequently the UMVUE of

$(x_0/\sigma)^{-\alpha}$ is $P(X_1 > x_0|X_{1:n})$, i.e.,

$$\begin{aligned} P(X \hat{>} x_0) &= \left(\frac{n-1}{n}\right)[X_{1:n}/x_0]^{\alpha}, \text{ if } x_0 > X_{1:n}, \\ &= 1, \text{ if } x_0 \leq X_{1:n}. \end{aligned} \tag{5.2.33}$$

[References: Lwin (1972), Shanmugam (1987)]

Case 2, σ known.

Here the complete sufficient statistic for α is $\sum_{i=1}^n \log X_i$ and the UMVUE of $(x_0/\sigma)^{-\alpha}$ is $P(X_1 > x_0|\sum_{i=1}^n \log X_i)$, i.e.,

$$\begin{aligned} P(X \hat{>} x_0) &= 1, \text{ if } x_0 < \sigma, \\ &= \left[1 + \frac{\log(\sigma/x_0)}{\sum_{i=1}^n \log X_i - n\log\sigma}\right]^{n-1}, \text{ if } \sigma \leq x_0 < \sigma^{-(n-1)}\prod_{i=1}^n X_i, \\ &= 0, \text{ if } x_0 \geq \sigma^{-(n-1)}\prod_{i=1}^n X_i. \end{aligned} \tag{5.2.34}$$

[References: Lwin (1972), Shanmugam (1987), Asrabadi (1990)]

For this case, Asrabadi (1990) also supplies the UMVUE of the density at x_0, i.e., $f_X(x_0) = \alpha\sigma^{\alpha}/x_0^{\alpha+1}$. It is given by

$$f(\hat{x_0}) = \frac{(n-1)[\log(\prod_{i=1}^n X_i) - \log x_0 - (n-1)\log\sigma]^{n-2}}{x_0[\log(\prod_{i=1}^n X_i) - n\log\sigma]^{n-1}}, \tag{5.2.35}$$

for $x_0 \in (\sigma, \sigma\prod_{i=1}^n (X_i/\sigma))$. Dixit and Jabbari-Nooghabi (2010) provide a series expression for the mean squared error of the estimate $f(\hat{x_0})$ and verify that it is smaller than the mean squared error of the corresponding maximum likelihood estimate.

Case 3, Both α and σ unknown.

The complete sufficient statistic in this case is $(X_{1:n}, \sum_{i=1}^n \log X_i)$ or, more conveniently since it has independent coordinates, the vector $(X_{1:n}, \sum_{i=1}^n \log(X_i/X_{1:n}))$. For simplicity, define $Z = \sum_{i=1}^n \log(X_i/X_{1:n})$. The UMVUE of $(x_0/\sigma)^{-\alpha}$ is $P(X_1 > x_0|X_{1:n}, Z)$, which can be shown to be of the form

$$\begin{aligned} P(X \hat{>} x_0) &= 1, \text{ if } x_0 < X_{1:n}, \\ &= \left(\frac{n-1}{n}\right)\left[1 + \frac{\log(X_{1:n}/x_0)}{Z}\right]^{n-2}, \text{ if } X_{1:n} \leq x_0 < X_{1:n}e^Z, \\ &= 0, \text{ if } x_0 \geq X_{1:n}e^Z. \end{aligned} \tag{5.2.36}$$

[References: Lwin (1972), Shanmugam (1987)]

Shanmugam (1987) shows that, in Case 1, α known, the UMVUE of $P(X > x_0)$ has a smaller mean squared error than the corresponding maximum likelihood estimate. This is not true in Cases 2 and 3. Ali and Woo (2003) also consider such a mean squared error comparison. Chaturvedi and Alam (2010) consider derivation of the UMVUE of $P(X > x_0)$ for a broad class of distributions which includes the Pareto (I) distribution, and indeed includes generalized Pareto distributions, but the analysis assumes that all parameters save one are known.

The UMVUE of the density function at x_0, when both parameters are unknown, was derived by Lwin (1972). See also Voinov and Nikulin (1993). It is of the form

$$f_X\hat{(x_0)} = \frac{I(x_0 = X_{1:n})}{nx_0} + \frac{(n-1)(n-2)}{nx_0 Z^{n-2}}[Z - \log(x_0/X_{1:n})]I(x_0 \geq X_{1:n}). \quad (5.2.37)$$

The mean squared error of this estimate is provided in Abusev (2004).

Likes (1985) considers unbiased estimation of a Pareto (I) quantile, i.e., $x_u = \sigma(1-u)^{-1/\alpha}$, for a fixed $u \in (0,1)$. Again there are three cases to consider.

Case 1, α known.

The complete sufficient statistic for σ is $X_{1:n}$. So the UMVUE of x_u is

$$\hat{x_u} = \frac{n\alpha - 1}{n\alpha} X_{1:n}(1-u)^{-1/\alpha}, \text{ provided that } n\alpha > 1. \quad (5.2.38)$$

Case 2, σ known.

The complete sufficient statistic for α is $W = \sum_{i=1}^{n} \log(X_i/\sigma)$ which has a $\Gamma(n, 1/\alpha)$ distribution. To identify the UMVUE of x_u in this setting, we will need to find an unbiased estimate of$(1-u)^{-1/\alpha}$. Likes defines $d = -\log(1-u)$, so that we will seek an unbiased estimate of $e^{d/\alpha} = \sum_{j=0}^{\infty} \frac{d^j}{\alpha^j j!}$. Since $W \sim \Gamma(n, 1/\alpha)$, it follows that the UMVUE of x_u is

$$\hat{x_u} = \sigma \sum_{j=0}^{\infty} \frac{\Gamma(n)}{j!\Gamma(n+j)} [-\log(1-u)]^j W^j. \quad (5.2.39)$$

Case 3, Both α and σ unknown.

The complete sufficient statistic for (σ, α) is $(X_{1:n}, Z)$ where $Z = \sum_{i=2}^{n} \log(X_{i:n}/X_{1:n})$. Moreover, $X_{1:n}$ and Z are independent with $X_{1:n} \sim P(I)(\sigma, n\alpha)$ and $Z \sim \Gamma(n-$

$1, 1/\alpha)$. Since $E(X_{1:n}) = n\alpha/(n\alpha - 1)$ and $x_u = \sigma e^{d/\alpha}$ where $d = -\log(1-u)$, it may be verified that the the UMVUE of x_u is

$$\hat{x_u} = X_{1:n} \sum_{j=0}^{\infty} \left(1 + \frac{j}{\log(1-u)}\right) \frac{\Gamma(n-1)}{j!\Gamma(n-1+j)} [-\log(1-u)]^j Z^j. \qquad (5.2.40)$$

Likes (1985) also provides expressions for the variances of these unbiased estimates of x_u in all three cases.

Remarks. *(1) Seto and Iwase (1982) also discuss construction of UMVUE's of quantiles and of the survival function of the Pareto (I) distribution.*

(2) Instead of estimating the survival function, Iwase (1986) considers the Pareto (I) cumulative hazard function, $H(x;\sigma,\alpha) = \log(1 - F(x;\sigma,\alpha)) = \alpha \log(x/\sigma)$, and derives the corresponding UMVUE.

(3) Parsian and Farsipour (1997) consider selection of a constant multiplier to be applied to the maximum likelihood estimate of α to minimize the risk associated with certain loss functions distinct from squared error loss. The two loss functions that they consider are: (i) $L_1(\alpha,t) = [(t/\alpha) - 1]^2$ and (ii) $L_2(\alpha,t) = n[(t/\alpha) - \log(t/\alpha) - 1)]$. They consider the case in which σ is known and the case in which it is unknown. As a sample of their results, they show that under entropy loss (i.e., L_2 above),with σ unknown, the optimal estimate of α is $(n-2)\hat{\alpha}/n$ where $\hat{\alpha}$ is the maximum likelihood estimate displayed in (5.2.3). But this estimate will be recognized as the unbiased modification of $\hat{\alpha}$ displayed in (5.2.13).

(4) He, Zhou and Zhang (2014), in the context of classical Pareto data with known scale parameter, discuss the relative merits of the maximum likelihood estimate and the UMVUE of α.

5.2.3 Moment and Quantile Estimates

Now we will consider other proposed estimates for the parameters of the classical Pareto distribution obtained without reference to the likelihood function. Inevitably, variations on the method of moments theme would be tried. One might equate the first and second sample moments to their expectations and solve. Such a procedure would require an assumption that $\alpha > 2$ in order to have a finite second moment. Instead of this, the approach suggested by Quandt (1966) has been more popular. He equated the sample minimum and the sample mean to their corresponding expectations and solved for α and σ. Such a procedure only involves the assumption that first moments exist (i.e., that $\alpha > 1$). The resulting estimates are of the form

$$\hat{\alpha}_M = (n\overline{X} - X_{1:n})/[n(\hat{X} - X_{1:n})] \qquad (5.2.41)$$

and

$$\hat{\sigma}_M = (n\hat{\alpha}_M - 1)X_{1:n}/(n\hat{\alpha}_M). \tag{5.2.42}$$

The estimates are readily shown to be consistent (in fact strongly consistent). Monte Carlo studies generally indicate that the method of moments estimates are slightly inferior to variants of the maximum likelihood estimates (cf. Quandt, 1966; Saksena, 1978; Cook and Mumme 1981). The use of sample fractional moments was proposed by Arnold and Laguna (1977). We will return to this idea in the context of estimating the parameters for the Pareto (IV) distribution.

A variation of the method of moments was proposed by Manas (1997). Instead of equating sample and theoretical expectations of X^k for various choices of k, the estimating equations are set up by equating sample and theoretical expectations of $[f_X(X;\underline{\theta})]^s$ for various choices of s. More details in the Pareto (I) case are supplied in Manas and Boyd (1997). In that paper they eventually set up the following two estimating equations.

$$X_{1:n} = \sigma\left(1-\frac{1}{n\alpha}\right)^{-1}, \tag{5.2.43}$$

$$\left(\prod_{i=1}^{n} X_i\right)^{1/n} = \sigma\left(1-\frac{1}{n\alpha}\right)^{-n}, \tag{5.2.44}$$

which yield the following estimate of α

$$\tilde{\alpha} = [n(1-R^{1/(n-1)})]^{-1}, \tag{5.2.45}$$

where $R = X_{1:n}/[\prod_{i=1}^{n} X_i]^{1/n}$. Manas and Boyd (1997) compare the performance of this estimate with that of the bias-corrected maximum likelihood estimate (5.2.13). For small sample sizes, the corrected maximum likelihood estimator is preferable. As sample size increases, this advantage decreases. McCune and McCune (2000) provide evidence that a jack-knife procedure can be successfully applied to the estimate (5.2.45) to reduce its bias, albeit with a small increase in the mean squared error for larger sample sizes.

Quandt (1966) also proposed quantile estimates. To this end, select two probability levels p_1 and p_2. The corresponding population quantiles x_1 and x_2 satisfy

$$(1-p_i) = (x_i/\sigma)^{-\alpha}, \quad i = 1,2. \tag{5.2.46}$$

If we substitute the corresponding sample quantiles $\{X_{[np_i]:n},\ i = 1,2\}$ for x_1, x_2 in (5.2.19) and solve for α and σ, we obtain the estimates

$$\hat{\alpha}_Q = \frac{\log[(1-p_1)/(1-p_2)]}{\log(X_{[np_2]:n}/X_{[np_1]:n})} \tag{5.2.47}$$

and

$$\hat{\sigma}_Q = X_{[np_1]:n}(1-p_1)^{1/\hat{\alpha}_Q}. \tag{5.2.48}$$

Once again, consistency is readily verified. Quandt (1966) found that the performance of the quantile estimates was not markedly inferior to that of the maximum likelihood estimates. They might, in fact, be preferable because of their resistance to outliers. A later paper (Thomas, 1976) cast doubts on such observations. The quantile estimate for α, (5.2.47), can be recognized as being of the form

$$\left(a + \sum_{i=1}^{k} b_i V_i\right)^{-1} \tag{5.2.49}$$

where the V_i's are independent exponential variables (the Sukhatme Theorem 3.4.2 is used to arrive at this representation). Thomas gave exact formulae for the moments of random variables of the form (5.2.49). With these formulae he was able to demonstrate that quantile estimates of α are decidedly inferior to other available estimates. He compared the quantile estimates to the unbiased estimate $\hat{\alpha}_U$, the maximum likelihood estimate $\hat{\alpha}$ and the method of moments estimate $\hat{\alpha}_M$ when $\alpha = 1.5$. The clear winner was the unbiased estimate $\hat{\alpha}_U$ followed in order by $\hat{\alpha}$, $\hat{\alpha}_M$ and $\hat{\alpha}_Q$.

The issue is not closed however. A Monte Carlo study reported by Koutrouvelis (1981) supports the view that quantile estimates for α are actually almost as good as the unbiased estimate, $\hat{\alpha}_U$. Koutrouvelis (1981) also discusses the optimal choice of sample quantiles to be used to obtain estimates of α and σ. In addition he considers the use of $k > 2$ quantiles that are optimally selected.

Brazauskas and Serfling (2000) proposed the use of generalized quantile estimates for the parameters of the Pareto (I) distribution. In particular they recommend use of a generalized median estimate of the shape parameter α, with σ also unknown. Two variants of this estimate are discussed in Brazauskas and Serfling (2001). The recommended variant is the median of $\binom{n}{k}$ estimates of α, each based on a subset of size k of the data set $X_1, X_2, \ldots, X_n$. The subset estimate of α based on $X_{i_1}, X_{i_2}, \ldots, X_{i_k}$ which is used is of the form

$$\widetilde{\alpha}_{subset} = C_{n,k}^{-1}\left[\frac{1}{k}\sum_{j=1}^{k}\log(X_{i_j}/X_{1:n}\right]^{-1}, \tag{5.2.50}$$

where $C_{n,k}$ is chosen so that $\widetilde{\alpha}_{subset}$ has its median equal to α. For large values of n, $C_{n,k} \approx k/[k(1-n^{-1})-(1/3)]$ provides a useful approximation. Small values of k are recommended. The strong point of such estimators is to be found in their good performance in the presence of contamination by outliers. Brazauskas and Serfling (2003) provide comparison of generalized median estimates, quantile estimates, trimmed mean estimates and maximum likelihood estimates. They suggest computing several of the available estimates and, if there is much variability in these estimates, they recommend use of the more robust ones.

Castillo and Hadi (1995) recommend use of what they call elemental quantile estimates of (σ,α). In this approach, one selects M pairs, (i_k, j_k), of indices, $k = 1,2,...,M$. For each pair of indices we obtain the elemental estimate of (σ,α) by solving the equations

$$X_{i_k:n} = \sigma\left(\frac{n+1}{n-i_k+1}\right)^{1/\alpha},$$

$$X_{j_k:n} = \sigma\left(\frac{n+1}{n-j_k+1}\right)^{1/\alpha}.$$

The medians of these M elemental estimates of σ and of α are used as our final estimates of σ and α. Several variants of this approach are discussed by Castillo and Hadi (1995).

Rahman and Pearson (2003) provide some simulation based comparisons of several competing estimates including elemental quantile estimates. One estimation strategy, that is included in this comparison study, fares quite well in terms of bias and mean squared error and merits mention. It is the maximum-product-of-spacings strategy proposed by Cheng and Amin (1983). For it, one chooses σ and α to maximize

$$\log\left\{\prod_{i=1}^{n+1}[F(X_{i:n};\sigma,\alpha) - F(X_{i-1:n};\sigma,\alpha)]\right\},$$

where $X_{0:n} = \sigma$ and $X_{n=1:n} = \infty$.

5.2.4 A Graphical Technique

Perhaps the oldest technique for estimating the parameters of a classical Pareto distribution relies on the observation that the log survival function is linear, i.e.,

$$\log\overline{F}(x) = \alpha\log\sigma - \alpha\log x. \tag{5.2.51}$$

The sample log survival function should behave in a similar manner. Fitting a straight line by least squares will provide estimates of α and σ. Note that, since $\overline{F}_n(X_{n:n}) = 0$, the largest observation must be discarded in the fitting procedure. Such estimates can be shown to be consistent, and Monte Carlo evidence suggests that they are only slightly inferior to maximum likelihood estimates. One could obtain a least squares estimate of α by treating the sample Lorenz curve in a similar fashion.

5.2.5 Bayes Estimates

Several authors have considered the problem of deriving Bayes estimates based on a Pareto (I) sample. Early contributors include Malik (1970c), Zellner (1971, pp. 34–36), Lwin (1972), Rao Tummala (1977) and Sinha and Howlader (1980). The situation is quite straightforward if σ is assumed to be known. It is convenient to define

$$U = \sum_{i=1}^{n} \log(X_i/\sigma). \tag{5.2.52}$$

We use the convention that lower case letters represent realizations of the random variables defined by the corresponding upper case letters (i.e., u is a realization of U, $\underline{x}$ of $\underline{X}$, etc.). The natural family of conjugate priors for α, in the known σ case, is the gamma family

$$f(\alpha) = [\alpha^{\gamma_1 - 1} e^{-\alpha/\gamma_2}]/[\Gamma(\gamma_1)\gamma_2^{\gamma_1}], \quad \alpha > 0, \tag{5.2.53}$$

i.e., $\alpha \sim \Gamma(\gamma_1, \gamma_2)$ where $\gamma_1, \gamma_2 > 0$ are chosen to reflect prior knowledge of the parameter α. The corresponding posterior distribution is readily obtained (using the likelihood function (5.2.1))

$$f(\alpha|\underline{x}, \sigma) = \tau(\underline{x}, \sigma)\alpha^{n+\gamma_1-1} e^{-\left(\frac{1}{\gamma_2}+u\right)\alpha}, \quad \alpha > 0 \tag{5.2.54}$$

where $\tau(\underline{x}, \sigma)$ is chosen so that the posterior density integrates to one. It is evident that the posterior density is again in the gamma family, in fact

$$\alpha|\underline{x}, \sigma \sim \Gamma(n+\gamma_1, \gamma_2/(\gamma_2 u + 1)). \tag{5.2.55}$$

With squared error loss the corresponding Bayes estimate for α will be the posterior mean. Thus we have

$$\hat{\alpha}_B = (n+\gamma_1)\gamma_2/(\gamma_2 U + 1). \tag{5.2.56}$$

In the absence of prior knowledge of α, we might use a vague or diffuse prior. The corresponding Jeffrey's invariant prior would usually be used. It is the improper prior of the form

$$f(\alpha) \propto \alpha^{-1}, \quad \alpha > 0. \tag{5.2.57}$$

With this prior the posterior distribution is again of the gamma form. One finds

$$\alpha|\underline{x}, \sigma \sim \Gamma(n, u^{-1}). \tag{5.2.58}$$

With squared error loss the resulting Bayes estimate corresponding to vague prior information is

$$\hat{\alpha}_{VB} = (U/n)^{-1} \tag{5.2.59}$$

which (not surprisingly) coincides with the maximum likelihood estimate (assuming σ is known).

Rao Tummala (1977) presented a slight variant of the above analysis. He considered the same prior densities but assumed weighted quadratic loss, of the form $c(\alpha)(\alpha - \hat{\alpha})^2$. The corresponding Bayes estimates, i.e., $E(c(\alpha)\alpha|\underline{x})/E(c(\alpha)|\underline{x})$, are readily calculated for convenient choices of the weight function $c(\alpha)$ (Rao Tummala considered $c(\alpha) = \alpha^{-2}$).

If α rather than σ is assumed to be known, then a conjugate family of prior distributions for σ is provided by the power function distributions. Thus we consider priors of the form

$$f(\sigma) = \delta\sigma^{\delta-1}\sigma_0^{-\delta}, \quad 0 < \sigma < \sigma_0 \tag{5.2.60}$$

where $\sigma_0 > 0$ and $\delta > 0$. In such a case we write $\sigma \sim PF(\delta, \sigma_0)$. With a prior of this form the posterior density may be seen to be of the form

$$f(\sigma|\underline{x}, \alpha) = \tau(\underline{x}, \alpha)\sigma^{(n\alpha+\delta)-1}, \quad 0 < \sigma < \min\{\sigma_0, x_{1:n}\}. \tag{5.2.61}$$

Clearly, the posterior density is again of the power function form. The shape and scale parameters of the posterior density are respectively $n\alpha + \delta$ and $\min\{x_{1:n}, \sigma_0\}$ (cf. Lwin, 1972, whose expression for the posterior scale parameter is slightly at variance with the present result). Assuming squared error loss, the Bayes estimate of σ will be the posterior mean, namely

$$\hat{\sigma}_B = [(n\alpha + \delta)/(n\alpha + \delta + 1)] \min\{x_{1:n}, \sigma_0\}. \tag{5.2.62}$$

If instead we assume absolute error loss, then the posterior median becomes the Bayes estimate. The power function distribution is one of the relatively few for which closed form expressions for the median are readily available. The median of the posterior distribution in this case is

$$2^{-(n\alpha+\delta)^{-1}} \min\{x_{1:n}, \sigma_0\}. \tag{5.2.63}$$

The situation most likely to be encountered in practice is that in which both of the parameters σ and α are unknown. Lwin (1972) was the first author to deal with this case. The prior density for (σ, α) used by Lwin appears to be unnaturally restrictive. The following development (abstracted from Arnold and Press, 1981) is more general, but it still involves a difficult to interpret condition which is needed to guarantee that the prior is a proper one. An independent identification of a conjugate prior family for (σ, α) was provided by Pisarewska (1982).

We assume we have a realization $x_1, x_2, \ldots, x_n$ of n independent Pareto (I) random variables and that we wish to make inferences about the unknown parameters σ and α. The likelihood function is of the form

$$L(\sigma, \alpha) = \alpha^n \sigma^{n\alpha} \left(\prod_{i=1}^{n} x_i \right)^{-(\alpha+1)} I(x_{1:n} > \sigma). \tag{5.2.64}$$

Here and in the following we use the indicator function $I(a > b)$ defined by

$$I(a > b) = \begin{cases} 1 & \text{if } a > b \\ 0 & \text{if } a \leq b. \end{cases} \tag{5.2.65}$$

A tractable family of priors is that with a kernel of the form

$$\alpha^{\delta} \sigma^{\lambda\alpha-1} \mu^{-\alpha} I(\rho > \sigma) \tag{5.2.66}$$

where δ, λ, μ and ρ are positive parameters. This kernel is integrable provided that

$$0 < \rho^{\lambda} < \mu. \tag{5.2.67}$$

This restriction is difficult to interpret, but, unless we wish to use improper priors, inevitable. Lwin (1972), using the present notation, restricted attention to the case in which $\delta = \lambda \geq 1$ and $0 < \rho = \mu < 1$. For such a choice (5.2.67) does hold, but the requirement that ρ be less than 1 implies the unlikely prior knowledge that σ must be less than 1. It is convenient to define

$$\begin{aligned} \tau(\delta,\lambda,\mu,\rho) &= \int_0^{\infty}\int_0^{\rho} \frac{\alpha^{\delta}\sigma^{\lambda\alpha-1}}{\mu^{\alpha}}\, d\sigma\, d\alpha \\ &= \lambda^{-1}\Gamma(\delta)[-\lambda\log\rho + \log\mu]^{-\delta} \end{aligned} \tag{5.2.68}$$

provided $0 < \rho^{\lambda} < \mu$. The explicit form of the prior density is

$$f(\sigma,\alpha) = \frac{\lambda[-\lambda\log\rho + \log\mu]^{\delta}\alpha^{\delta}\sigma^{\lambda\alpha-1}I(\rho > \sigma)}{\Gamma(\delta)\mu^{\alpha}}. \tag{5.2.69}$$

In this prior, α has a gamma distribution, while the conditional distribution of σ given α is of the power function form. Specifically, it is assumed that $\alpha \sim \Gamma(\delta, [\log\mu - \lambda\log\rho]^{-1})$ and $\sigma|\alpha \sim PF(\lambda\alpha,\rho)$. The joint density can be called gamma-power with parameters $(\delta,\lambda,\mu,\rho)$, for brevity $\Gamma P(\delta,\lambda,\mu,\rho)$. It is a conjugate prior family, since multiplication of the likelihood (5.2.64) and the kernel of the prior (5.5.52) indicates that the posterior density is in the same family. In fact,

$$(\sigma,\alpha)|\underline{x} \sim \Gamma P\left(n+\delta, n+\lambda, \mu\left(\prod_{i=1}^{n} x_i\right),\ \min\{\rho, x_{1:n}\}\right). \tag{5.2.70}$$

The Bayes estimate of any function of (σ,α), assuming squared error loss, is merely the posterior expectation of that function. For example, we find

$$\begin{aligned} \hat{\alpha}_B = E(\alpha|\underline{x}) &= \frac{\tau\left(n+\delta+1, n+\lambda, \mu\prod_{i=1}^{n} x_i, \min\{\rho, x_{1:n}\}\right)}{\tau\left(n+\delta, n+\lambda, \mu\prod_{i=1}^{n} x_i, \min\{\rho, x_{1:n}\}\right)} \\ &= (n+\delta)\Big/\left\{-(n+\lambda)\log[\min\{\rho, x_{1:n}\}] + \log\mu\prod_{i=1}^{n} x_i\right\}. \end{aligned} \tag{5.2.71}$$

The Bayes estimate for σ is not obtainable in closed form. There are other parametric functions which might be of interest, e.g., the survival probability $\min[(t/\sigma)^{-\alpha}, 1]$ where t is a fixed positive number, and the Gini index $(2\alpha-1)^{-1}$. Generally speaking, evaluation of the corresponding posterior median or mean will require numerical integration and/or incomplete gamma tables. The case of the survival probability is unusually tractable (Lwin (1972) is the basic reference for survival function estimation). Assuming squared error loss, the Bayes estimates for $\min[(t/\sigma)^{-\alpha},1]$ corresponding to the priors (5.2.53), (5.2.57), (5.2.60) and (5.2.69) are, respectively,

$$\min\left[\left(\frac{1}{\gamma_2}+u\right)^{n+\gamma_1}\left(\frac{1}{\gamma_2}+u+\log(t/\sigma)\right)^{-(n+\gamma_1)},1\right], \tag{5.2.72}$$

$$\min[\{u/(u+\log(t/\sigma))\}^n,1], \tag{5.2.73}$$

$$\frac{t^{-\alpha}(n\alpha+\delta)[\min\{\sigma_0,t,x_{1:n}\}]^{(n+1)\alpha+\delta}}{[(n+1)\alpha+\delta][\min\{\sigma_0,x_{1:n}\}]^{n\alpha+\delta}}+1-\left[\frac{\min\{\sigma_0,t,x_{1:n}\}}{\min\{\sigma_0,x_{1:n}\}}\right]^{n\alpha+\delta} \tag{5.2.74}$$

and

$$\left(\frac{n+\lambda}{n+\lambda+1}\right)\left[\frac{-(n+\lambda+1)\log(\min\{\rho,t,x_{1:n}\})+\log\mu t\prod_{i=1}^{n}x_i}{-(n+\lambda)\log(\min\{\rho,x_{1:n}\})+\log\mu\prod_{i=1}^{n}x_i}\right]^{-(n+\delta)}$$
$$+1-\left[\frac{-(n+\lambda)\log(\min\{\rho,t,x_{1:n}\})+\log\mu\prod_{i=1}^{n}x_i}{-(n+\lambda)\log(\min\{\rho,x_{1:n}\})+\log\mu\prod_{i=1}^{n}x_i}\right]^{-(n+\delta)}. \tag{5.2.75}$$

If we assume absolute error loss and assume that α is known (i.e., use the prior (5.2.57)), then since $\min[(t/\alpha)^{-\alpha},1]$ is a monotone function of σ, we may immediately write the Bayes estimate of $\min[(t/\sigma)^{-\alpha},1]$ (i.e., the posterior median of $\min[(t/\sigma),1]$) to be $\min[(t/\tilde{\sigma})^{-\alpha},1]$ where $\tilde{\sigma}$ is the posterior median of δ given by (5.2.63). A similar approach can be applied if δ is assumed known. If both σ,α are unknown (i.e., if the prior is (5.2.69), then the posterior median of $\min[t/\alpha)^{-\alpha},1]$ is not obtainable in closed form. Another parametric function of interest is the value of the Pareto (I) Lorenz curve evaluated at a specific point u_0, i.e., $L(u_0)=1-(1-u_0)^{1-\alpha^{-1}}$. Abdul-Sathar, Jeevanand and Muraleedharan (2010) discuss Bayes estimation of $L(u_0)$.

One should not be seduced by the mathematical tractability of generalized Lwin priors however. The assumed prior dependence between σ and α may well lead to unsurmountable assessment problems. In addition the marginal posterior density for σ turns out to be bimodal with local maxima at 0 and $\min\{\rho,x_{1:n}\}$.

No amount of accumulated data can remove the mode at 0. Arnold and Press (1981) emphasized these defects. However, as Geisser (1984) points out, the bimodal nature of the posterior density of σ is not really that troublesome. One can always focus on the reciprocal of σ which has a unimodal posterior distribution. Because of assessment problems, Arnold and Press (1981) proposed that, instead of a generalized Lwin prior, independent prior densities for α and σ be assumed, i.e., a gamma prior for α and a power-function prior for σ.

Arnold, Castillo and Sarabia (1998) suggested that a more flexible family of priors might be preferable to the generalized Lwin priors. They argued as follows. If σ were known, the natural conjugate prior for α would be a gamma distribution. If α were known, then a natural conjugate prior for σ would be a power-function distribution. Note that, in their paper, they spoke of a precision parameter $\tau = 1/\sigma$ which would have a natural Pareto prior if α were known. However, we will continue with the (σ, α) parameterization in the present discussion. In the light of these observations, it is appealing to consider, as a joint prior for (σ, α), one which has the property that for each value of σ, the conditional distribution of α given σ is of the gamma form with parameters which might depend on σ, and for each value of α, the conditional distribution of σ given α is of the power-function form with parameters which might depend on α.

The corresponding 6-parameter family of prior densities for (σ, α) is of the form

$$\begin{aligned} f(\sigma,\alpha) \propto\ & \exp[-b\log\sigma - m_{12}\log\alpha\log\sigma] \\ & \times\exp[a_1\alpha + a_2\log\alpha - m_{11}\alpha\log\sigma]I(0<\sigma<\sigma_0), \end{aligned} \tag{5.2.76}$$

where the right hand side has been broken into two factors. As will be seen below, the first factor involves hyperparameters (parameters of the prior) that are unchanged by the data. The second factor involves hyperparameters that are affected by the data.

First, it is not difficult to verify that densities of the form (5.2.76) do indeed have gamma and power-function conditionals. Specifically

1. The conditional density of α given σ is a gamma density with shape parameter $\gamma(\sigma)$ and scale parameter $\beta(\sigma)$, i.e.,

$$f(\alpha|\sigma) \propto \alpha^{\gamma(\sigma)-1}e^{-\alpha/\beta(\sigma)}I(\alpha>0), \tag{5.2.77}$$

where

$$\gamma(\sigma) = a_2 - m_{12}\log\sigma + 1, \tag{5.2.78}$$

and

$$\beta(\sigma) = (a_1 - m_{11}\log\sigma)^{-1}. \tag{5.2.79}$$

2. The conditional density of σ given α is a power-function density with shape parameter $\delta(\alpha)$ and scale parameter $\nu(\alpha)$, i.e.,

$$f(\sigma|\alpha) = \frac{\delta(\alpha)\sigma^{\delta(\alpha)-1}}{[\nu(\alpha)]^{\delta(\alpha)}}I(0<\sigma<\nu(\alpha)), \tag{5.2.80}$$

where

$$\delta(\alpha) = b + m_{12}\log\alpha + m_{11}\alpha + 1, \tag{5.2.81}$$

and

$$\nu(\alpha) = \sigma_0. \tag{5.2.82}$$

If we insist on having a proper prior density, constraints must be placed on the hyperparameters in (5.2.76) to ensure that the parameters that appear in the conditional densities are all positive. Thus we would require that $\gamma(\sigma) > 0$, $\beta(\sigma) > 0$, $\delta(\alpha) > 0$

Table 5.2.1 *Changes in the hyperparameters of the prior (5.2.76) when combined with the likelihood (5.2.83)*

Hyperparameter	Prior value	Posterior value
a_1	a_1^*	$a_1^* - \sum_{i=1}^n \log x_i$
a_2	a_2^*	$a_2^* + n$
b	b^*	b^*
m_{11}	m_{11}^*	$m_{11}^* - n$
m_{12}	m_{12}^*	m_{12}^*
σ_0	σ_0^*	$\min(x_{1:n}, \sigma_0^*)$

and $\nu(\alpha) > 0$. Even if we allow the prior to be improper, the corresponding posterior density will be proper if the sample size is sufficiently large.

The likelihood function for a sample of size n from a Pareto I (σ, α) distribution is of the form

$$L(\sigma, \alpha) = \exp[n \log \alpha + n\alpha \log \sigma - \alpha \sum_{i=1}^{n} \log x_i - \sum_{i=1}^{n} \log x_i] I(x_{1:n} > \sigma). \quad (5.2.83)$$

It is readily verified that the family of densities (5.2.76) is a conjugate family of priors for likelihoods of the form (5.2.83). If we combine the prior (5.2.76) with the likelihood (5.2.83), we obtain a posterior density that is again in the family (5.2.76). The prior and posterior hyperparameters are related as shown in Table 5.2.1. Note that the two hyperparameters b and m_{12} are unaffected by the data (they appear in the first factor in (5.2.76)).

It can be noted that the family of priors (5.2.76) includes:

1. The generalized Lwin prior (5.2.66). It corresponds to the case in which b and m_{12} are both set equal to 0.
2. The case of independent gamma and power-function priors, as suggested by Arnold and Press (1989). For this we set $m_{11} = m_{12} = 0$ in (5.2.76).

Thus the family (5.2.76) includes two of the most frequently proposed classes of joint priors for (σ, α). Moreover, it includes vague priors for which most or all of the hyperparameters are set equal to 0 or perhaps -1.

It will be noted that prior or posterior densities of the form (5.2.76) have convenient conditional densities; however the corresponding marginal densities are not of simply recognizable forms. One advantage of the generalized Lwin densities is that one of their marginal densities, that of α, is of a recognizable form, i.e., a gamma density. However, the argument in favor of using the conditionally conjugate family (5.2.76) is that it is tailor-made for the application of a Gibbs sampler to simulate realizations from the posterior density, which has gamma and power-function conditionals, just like the prior density.

It is suggested by Arnold, Castillo and Sarabia (1998) that prior hyperparameter assessment can be implemented by matching prior conditional moments and percentiles supplied by an informed expert. Alternatively, recourse may be made to the use of diffuse or partially diffuse priors in the absence of reliable prior information. Parameter estimation will be most easily implemented by use of the posterior conditional densities and a Gibbs sampler algorithm. For example, if we wish to estimate a particular parametric function, say $h(\sigma,\alpha)$, we will require a good approximation to $E(h(\sigma,\alpha)|\underline{X}=\underline{x})$. For this we successively generate $\sigma_1,\alpha_1,\sigma_2,\alpha_2,...,\sigma_N,\alpha_N$, using the posterior conditional distributions for a large value of N. Our approximation to $E(h(\sigma,\alpha)|\underline{X}=\underline{x})$ will then be $(1/N)\sum_{i=1}^{N}h(\sigma_i,\alpha_i)$, or perhaps $(1/N)\sum_{i=N'+1}^{N'+N}h(\sigma_i,\alpha_i)$, if we allow time for the sampler to stabilize.

See Arnold, Castillo and Sarabia (1998) for further discussion of conditionally conjugate priors, including a numerical example and some related extensions. Remember that they use a precision parameter $\tau=1/\sigma$ in their discussion of the Pareto (I) distribution.

One might expect a parallel development of Bayesian inference techniques for the more general Pareto distributions (i.e., Pareto II, III and IV and perhaps even for the multivariate Pareto distributions to be described in Chapter 6). Unfortunately, analytic difficulties in dealing with "natural" conjugate priors have thwarted efforts in this direction. See Section 5.2.13 below for some further discussion of this topic.

5.2.6 Bayes Estimates Based on Other Data Configurations

Several authors have considered Bayesian estimation for Pareto (I) distributions based on data sets which are not of the random sample type. Doostparast, Akbari and Balakrishnan (2011) consider record value data. Suppose that a sequence $X_1,X_2,...$ of Pareto(I) variables is observed until m lower records have been observed (see Arnold, Balakrishnan and Nagaraja (1998) for an introduction to record values). An observation is a lower record if it is smaller than all of the preceding observations in the sequence. The available data are then of the form (R_i,Δ_i), $i=1,2,...,m$, where R_i is the i'th lower record and Δ_i is the i'th inter-arrival time of the lower record values. Denoting, as usual, the observed values of the R_i's and the Δ_i's by r_i's and the δ_i's, respectively, the corresponding likelihood function will be of the form (see Doostparast, Akbari and Balakrishnan (2011) for more details)

$$\begin{aligned} L(\sigma,\alpha) &= \exp[m\log\alpha+\alpha\log(\sigma\sum_{i+1}^{m}\delta_i)-\alpha(\sum_{i=1}^{m}\delta_i\log r_i)-\sum_{i=1}^{m}\log r_i] \\ &\quad \times I(r_1>r_2>\cdots r_m)I(r_m>\sigma). \end{aligned} \tag{5.2.84}$$

This should be compared with the likelihood corresponding to a simple random sample of size n from a Pareto (I)(σ,α) distribution as found in (5.2.83). Viewed as functions of (σ,α), the kernels of these likelihoods differ only in that the data values $(n, n, \sum \log x_i, x_{1:n})$ in (5.2.83) are replaced by the values $(m, \sum \delta_i, \sum \delta_i \log r_i, r_m)$ in (5.2.84). Consequently, Bayesian analysis using either a generalized Lwin prior (5.2.66), a conditionally conjugate prior (5.2.76) or a non-informative prior will proceed exactly as in the case of an ordinary random sample from a Pareto (I)(σ,α) distribution.

If, instead of observing the sequence of lower records and the corresponding inter-record times, we only observe the values of the first m lower records, the corresponding likelihood function is somewhat more complicated. It is of the form

$$L(\sigma,\alpha) = \frac{\prod_{i=1}^{n}\left[\frac{\alpha}{\sigma}\left(\frac{r_i}{\sigma}\right)^{-(\alpha+1)}\right]}{\prod_{i=1}^{m-1}\left[1-\left(\frac{r_i}{\sigma}\right)^{-\alpha}\right]} I(r_1 > r_2 \cdots > r_m) I(r_m > \sigma). \tag{5.2.85}$$

In this case a convenient conjugate prior is lacking. If a generalized Lwin prior or a conditionally conjugate prior is used, then numerical integration of the posterior expectation of a parametric function $h(\sigma,\alpha)$ will be required to obtain the corresponding Bayes estimate.

The situation is markedly different if a set of m upper records is observed. Denote the m upper records by $S_1, S_2, \ldots S_m$ and the observed upper records by $s_1, s_2, \ldots s_m$. In this case the likelihood is less complicated. It is given by

$$\begin{aligned} L(\sigma,\alpha) &= \frac{\prod_{i=1}^{n}\left[\frac{\alpha}{\sigma}\left(\frac{r_i}{\sigma}\right)^{-(\alpha+1)}\right]}{\prod_{i=1}^{m-1}\left(\frac{r_i}{\sigma}\right)^{-\alpha}} I(\sigma < r_1 < r_2 \cdots < r_m) \qquad (5.2.86) \\ &= \exp[m\log\alpha + \alpha\log\sigma - \alpha\log r_m - \sum_{i=1}^{m}\log r_i] I(r_1 > \sigma). \end{aligned}$$

However, here too, the kernel of the likelihood as a function of (σ,α) is identical in form to that of the likelihood for a simple random sample of size n from a Pareto (I)(σ,α) distribution. They differ only in that the data values $(n, n, \sum \log x_i, x_{1:n})$ in (5.2.83) are replaced by $(m, 1, \log r_m, r_1)$. Consequently Bayesian inference for such a data configuration (upper record values) can also proceed in the same manner as it was for a simple random sample.

There are other data configurations which also have likelihoods basically of the form (5.2.83). For example, suppose that n i.i.d. observations $X_1, X_2, \ldots, X_n$ are available together with m observations that are truncated from below at points $a_1, a_2, \ldots, a_m$; i.e., we only observe $(X_j > a_j)$, $j = 1, 2, \ldots, m$. Then, provided that

$x_{1:n} < a_{1:m}$, the likelihood for this Pareto (I) data set is given by

$$\begin{aligned} L(\sigma,\alpha) &= \prod_{i=1}^{n}\left[\frac{\alpha}{\sigma}\left(\frac{x_i}{\sigma}\right)^{-(\alpha+1)}\right]\prod_{j=1}^{m}\left(\frac{a_j}{\sigma}\right)^{-\alpha} I(x_{1:n} > \sigma) \qquad (5.2.87) \\ &= \exp[n\log\alpha + (n+m)\log\sigma - \alpha(\sum_{i=1}^{n}\log x_i + \sum_{j=1}^{m}\log a_j)] \\ &\quad \times \exp[-\sum_{i=1}^{n}\log x_i] I(x_{1:n} > \sigma). \end{aligned}$$

Bayesian inference for such a data configuration will proceed in the same fashion as it does for a simple random sample. The data values $(n, n, \sum \log x_i, , x_{i:n})$ in (5.2.83) are simply replaced by $(n, n+m, \sum \log x_i + \sum \log a_j, x_{1:n})$.

Other grouped or truncated data sets lead to more complicated likelihood functions and, consequently, more complicated posterior densities corresponding to any of the prior densities discussed in this section. For details, see Arnold and Press (1986). In these situations Arnold and Press suggest a pragmatic approach which utilizes $X_{1:n}$ as an estimate of σ and then proceeds with a simplified Bayesian analysis assuming that σ is known. However, even with this approach, the posterior densities are usually troublesome.

Mention should be made of a large class of Pareto (I) data configurations that are amenable to straightforward Bayesian analysis. These correspond to what are known as generalized order statistics. These were introduced by Kamps who provides an extensive discussion of the concept in Kamps (1995). We will use the notation of that monograph to facilitate reference to it for useful distributional properties of generalized order statistics. A random vector $(X_1, X_2, ..., X_n)$ is said to correspond to a set of n generalized order statistics with a Pareto (I) base distribution if its joint density is of the form

$$\begin{aligned} f_{X_1,X_2,...,X_n}(x_1,x_2,...,x_n) &= k\left[\prod_{j=1}^{n-1}\gamma_j\left(\frac{x_j}{\sigma}\right)^{\alpha m_j}\frac{\alpha}{\sigma}\left(\frac{x_j}{\sigma}\right)^{-(\alpha+1)}\right] \qquad (5.2.88) \\ &\quad \times\left(\frac{x_n}{\sigma}\right)^{-\alpha(k-1)}\frac{\alpha}{\sigma}\left(\frac{x_n}{\sigma}\right)^{-(\alpha+1)} I(\sigma < x_1 < x_2 < \cdots < x_n), \end{aligned}$$

where $n \in \{1, 2, ...\}$, $m_1, m_2, ..., m_n \in (-\infty, \infty)$, $M_r = \sum_{j=r}^{n-1} m_j$, $\gamma_r = k + n - r + M_r \geq 1$, $\forall r \in \{1, 2, ..., n-1\}$.

This model includes as special cases, for various choices of its parameters, order statistics, upper record values, sequential order statistics, k-records, Pfeifer records and progressively censored data. Thus all such data configurations can be analyzed in the same way. Note that the Pareto (I) generalized order statistics likelihood (5.2.88) can be written as

$$\begin{aligned} L(\sigma,\alpha) &= \exp\left\{\log k+\sum_{j=1}^{n-1}\log\gamma_j+n\log\alpha\right\}I(\sigma<x_1<x_2<\cdots<x_n) \\ &\times \exp\left\{-\alpha\left[\sum_{j=1}^{n-1}(m_j+1)\log x_j+k\log x_n\right]\right\} \\ &\times \exp\left\{\alpha\log\sigma(\sum_{j=1}^{n-1}m_j+k+1)-\sum_{j=1}^{n}\log x_j\right\}. \end{aligned} \quad (5.2.89)$$

Viewed as a function of (σ,α), this likelihood is of the same form as the Pareto (I) random sample likelihood (5.2.83), but with the data values $(n,n,\sum\log x_i,x_{1:n})$ replaced by $(n,\sum_{j=1}^{n-1}m_j+k+1,\sum_{j=1}^{n-1}(m_j+1)\log x_j+k\log x_n,x_1)$. Consequently, Bayesian analysis for Pareto (I) generalized order statistics data, including all the special cases mentioned above, will proceed exactly as was the case for data corresponding to a random sample from the distribution.

A particular example of Bayesian analysis in a generalized order statistics setting is provided by Fu, Xu and Tang (2012). They discuss the particular case of progressive censoring and consider a spectrum of non-informative priors. Madi and Raqab (2008) consider Bayes estimation for Pareto record value data assuming squared error and LINEX loss functions.

5.2.7 Bayes Prediction

If we have at hand a Pareto (I) data configuration with a likelihood of the form (5.2.83), or one of similar nature, it is often of interest to predict the values of future observations, rather than to provide estimates of the parameters of the parent distribution. A strong advocate of prediction of future observables rather than estimation of unobservable parameters was Geisser. See for example Geisser (1984).

A typical prediction problem may be described as follows. Suppose that we observe the first k order statistics of a sample of size n from a Pareto (I)(σ,α) distribution where both σ and α are unknown. We wish to predict the value of the $(k+1)$'st order statistic. It is known that the conditional distribution of the $(k+1)$'st order statistic, given values of the first k order statistics, is that of the minimum of a sample of size $(n-k)$ from the parent distribution truncated below at the observed value of $X_{k:n}$ (see Arnold, Balakrishnan and Nagaraja (1992, p.23)). What we need to evaluate is, for any $t>x_{k:n}$,

$$P(X_{k+1:n}>t|X_{1:n}=x_{1:n},X_{2:n}=x_{2:n},\ldots,X_{k:n}=x_{k:n}).$$

For convenience, denote $(X_{1:n},X_{2:n},\ldots,X_{k:n})$ by $\underline{X}$, with $\underline{x}$ analogously defined. We know that

$$P(X_{k+1:n}>t|\underline{X}=\underline{x},\sigma,\alpha)=(t/x_{k:n})^{-(n-k)\alpha},\ t>x_{k:n}. \quad (5.2.90)$$

Our predictive survival function will then be

$$P(X_{k+1:n} > t|\underline{X} = \underline{x}) = \int_0^\infty (t/x_{k:n})^{-(n-k)\alpha} f(\alpha|\underline{x})d\alpha. \tag{5.2.91}$$

If we assume a generalized Lwin prior for (σ, α) then the posterior density for α will be a gamma density with parameters that depend on $\underline{x}$ and the prior hyperparameters. In this case we can evaluate (5.2.91) without difficulty, since it can be recognized as the moment generating function of $\alpha|\underline{X} = \underline{x}$ evaluated at $-(n-k)\log(t/x_{k:n})$. For details see Arnold and Press (1989). If instead a conditionally conjugate prior of the form (5.2.76) is assumed for (σ, α) then we can approximately evaluate (5.2.91) by simulating a sample $(\sigma_1, \alpha_1), (\sigma_2, \alpha_2), ..., (\sigma_N, \alpha_N)$ from the posterior density of $(\sigma, \alpha)|\underline{X} = \underline{x}$ using a Gibbs sampler. Our approximate evaluation of (5.2.91) will then be

$$\frac{1}{N}\sum_{i=1}^{N}\left(\frac{t}{x_{k:n}}\right)^{-(n-k)\alpha_i}.$$

A second prediction problem that may be of interest involves two samples. The first sample of size n, $X_1, X_2, ..., X_n$, is from a Pareto (I) population, with (σ, α) unknown, and is fully observed. A second independent sample from the same distribution, to be denoted by $Y_1, Y_2, ..., Y_m$, is to be subsequently observed. We might, for example, wish to evaluate the probability that all m of the Y_i's will lie between $X_{1:n}$ and $2X_{1:n}$, given $\underline{X} = \underline{x}$. Here $\underline{X}$ denotes $(X_1, X_2, ..., X_n)$. In this case we have, for $\sigma > x_{i:n}$,

$$\begin{aligned} P(\text{all } Y_j\text{'s} \in (X_{1:n}, 2X_{1:n})|\underline{X} = \underline{x}, \sigma, \alpha) &= \left[\left(\frac{x_{1:n}}{\sigma}\right)^{-\alpha} - \left(\frac{2x_{1:n}}{\sigma}\right)^{-\alpha}\right]^m \\ &= (x_{1:n}/\sigma)^{-m\alpha}(1-2^{-\alpha})^m. \end{aligned} \tag{5.2.92}$$

The desired quantity $P(\text{all } Y_j\text{'s} \in (X_{1:n}, 2X_{1:n})|\underline{X} = \underline{x})$ is then obtained by integrating (5.2.92) with respect to the power-gamma posterior density of $(\sigma, \alpha)|\underline{X} = \underline{x}$. Geisser (1985) provides detailed discussion of problems of this kind, assuming a generalized Lwin prior. The alternative approach would be one in which a conditionally conjugate prior is used and an approximate evaluation of the conditional expectation of (5.2.92) given $\underline{X} = \underline{x}$ is obtained by Gibbs sampler simulations of realizations from the posterior density of (σ, α) given $\underline{X} = \underline{x}$.

Several other authors have considered Bayesian prediction problems for Pareto (I) data. Nigm and Hamdy (1987) discuss the problem of evaluating $P(X_{k+s;n} > x_{k+s:n}|X_{1:n}, X_{2:n}, ...X_{k:n})$ where $\underline{X} = (X_1, X_2, ..., X_n)$ is a Pareto (I)(σ, α) sample. See also Nigm, Al-Hussaini and Jaheen (2003). Dyer (1981), using a structural probability argument, provides discussion of $P(Y > y|\underline{X} = \underline{x})$ where $\underline{X} = (X_1, X_2, ..., X_n)$ is a Pareto (I)(σ, α) sample and Y is an independent observation from the same distribution. Dunsmore and Amin (1998) discuss the predictive distribution of the remaining total time on test given observed values of the first k order statistics. Thus they define $Z = \sum_{j=k+1}^{n}(X_{j:n} - X_{k:n})$ based on a Pareto (I)(σ, α) sample, and consider $P(Z > z|X_{1:n}, X_{2:n}, ..., X_{k:n})$ assuming a generalized Lwin prior for (σ, α). Balakrishnan and Shafay (2012) discuss prediction of a future order statistic based on a hybrid censoring scheme, one in which sampling is terminated either because k failures have been observed or if the pre-fixed time of study is reached. They too assume a modified Lwin prior.

Mohie El-Din, Abdel-Aty and Shafay (2012) discuss Bayesian prediction involving Pareto (I) generalized order statistics. They consider one set of generalized order statistics, $\underline{X} = (X_1, X_2, ..., X_n)$, and a second independent set $\underline{Y} = (Y_1, Y_2, ..., Y_m)$ with the same Pareto (I) base distribution. They assume that only r of the X_i's are observed and, for a particular $s \in \{1, 2, ..., M\}$, they discuss predictive bounds for Y_s given $(X_{j_1} = x_{j_1}, X_{j_2} = x_{j_2}, ..., X_{j_r} = x_{j_r})$, and derive an expression for $P(Y_s > y|X_{j_1} = x_{j_1}, X_{j_2} = x_{j_2}, ..., X_{j_r} = x_{j_r})$.

Most of the authors who discuss Bayesian prediction for Pareto (I) data also consider the cases in which σ is known or α is known. Predictably, in these cases the analysis is somewhat simplified.

5.2.8 *Empirical Bayes Estimation*

Tiwari and Zalkikar (1990) consider an empirical Bayes estimate of the parameter σ of the Pareto (I)(σ, α) distribution when α is assumed to be known. In this setting it is assumed that we have n i.i.d. variables $(X_1, \sigma_1), (X_2, \sigma_2), ..., (X_1, \sigma_n)$ together with an additional independent variable (X^*, σ^*). It is assumed that we only observe the X's. In addition we assume that the conditional distribution of X_i given σ_i is Pareto (I) (σ_i, α) and that the σ_i's are i.i.d. with common unknown distribution function $G(\sigma)$. We wish to predict σ^* based on $X_1, X_2, ..., X_n$ and X^*. If we knew G, then the natural predictor would be $E(\sigma^*|\underline{X} = \underline{x}.X^* = x^*)$, i.e., $\phi_G(x^*)$ where

$$\begin{aligned} \phi_G(x) &= \frac{\int_0^\infty \sigma f(x|\sigma) dG(\sigma)}{\int_0^\infty f(x|\sigma) dG(\sigma)} \\ &= \frac{\alpha \int_0^x \sigma^{\alpha+1} dG(\sigma)}{x^{\alpha+1} f(x)} = \frac{A(x)}{x^{\alpha+1} f(x)}. \end{aligned}$$

Tiwari and Zalkikar (1990) and Liang (1993) assume that $G(m) = 1$ for some positive number m, and, in each paper, estimates $A_n(x)$ and $f_n(x)$ of $A(x)$ and $f(x)$ based on $X_1, X_2, ..., X_n$ are provided. The resulting empirical Bayes predictor of σ^* is then given by

$$\phi_n(x^*) = \frac{A_n(x^*)}{x^{*(\alpha+1)} f(x^*)}.$$

Tiwari and Zalkikar (1990) and Liang (1993) discuss asymptotic optimality properties of the estimates using their particular choices of $A_n(x)$ and their choices of the restricted class of possible distributions $G(\sigma)$ that are considered.

Preda and Ciumara (2007) also assumed that α is known. They investigate the performance of an empirical Bayes predictor of σ assuming a weighted squared error loss function. Asymptotic optimality of the proposed predictor is confirmed. In an earlier paper, Lwin (1974) discussed empirical Bayes prediction of σ under the assumption that, instead of just one, several X values were observed for each σ_i in the available data set, to be used to predict the value of σ associated with X^*. Lwin's presentation is somewhat less detailed than that provided by the other authors quoted in this subsection, and he does not address the optimality issue.

5.2.9 Miscellaneous Bayesian Contributions

Several authors have discussed Bayesian estimation of particular functions of (σ, α) based on Pareto(I) data. Jeevanand and Abdul-Sathar (2010) consider estimation of the total time on test transform. They also (Abdul-Sathar and Jeevanand, 2010) consider estimation of the Lorenz curve and the Gini index. Ali and Woo (2004) consider estimation of the mean, i.e., of the parametric function $\alpha\sigma/(\alpha-1)$.

Pandey, Singh and Mishra (1996) consider estimation of σ assuming a LINEX loss function (instead of squared error). They treat the known σ case and the case in which both σ and α are unknown, using a Lwin prior for (σ, α).

Podder, Roy, Bhuiyan and Karim (2004) discuss minimax estimation for Pareto (I) data. Kang, Kim and Lee (2012) consider estimation of the common shape parameter α based on samples from two Pareto (I) populations (with different scale parameters).

Finally, we mention two alternative prior distributions that have been suggested for Pareto (I) data. Trader (1985) suggests an alternative conjugate prior for (σ, α). For it, it is assumed that α has a gamma distribution, while σ given α has a normal distribution with variance c/α truncated below at 0. The posterior distribution is then of analogous form, with $\alpha|(\underline{X}=\underline{x})$ having a gamma distribution and $\sigma|(\alpha, \underline{X}=\underline{x})$ having a normal distribution truncated below at 0 and above at $x_{1:n}$. It is then possible to derive expressions for the corresponding squared error loss Bayes estimates of σ and α that involve evaluations of incomplete beta functions.

Zellner (1986) proposed the use of what he called "g-prior" densities which represent a somewhat eclectic combination of the classical and the Bayesian approaches to estimation. We will illustrate this approach in the simple and relatively non-controversial case of a sample from a Pareto (I)$(1, \alpha)$ distribution. The experimenter provides a prior "guessed " value for α (which may be data-based), say $\widetilde{\alpha}$, and also provides a "sample size" g associated with the guessed value of α (this is a measure of the precision of the guess). The prior density to be used is then of the form $\alpha \sim \Gamma(g, \widetilde{\alpha}/g)$, leading to a Bayes estimate of the form

$$E(\alpha|\underline{X}=\underline{x}) = \frac{n+g}{(g/\widetilde{\alpha}) + \sum_{i=1}^{n} \log x_i}.$$

5.2.10 Maximum Likelihood for Generalized Pareto Distributions

There is little doubt that the classical Pareto distribution provides too restrictive a family for modeling purposes and, following the discussion of Chapter 3, we are led to consider estimation problems for Pareto II, III and IV distributions. Unless we assume that some parameters are known, then, as mentioned in the introduction to the present chapter, difficulties are quickly encountered. The Pareto II differs from the classical Pareto distribution only by the introduction of a location parameter, but that is already enough to complicate matters to the extent that estimates can only be obtained by numerical means. We begin by considering maximum likelihood estimation for generalized Pareto distributions. Early work in this area was done by Silcock (1954), Harris (1968) (who considered Pareto II samples,) Harris and Singpurwalla (1969) (Pareto (IV)) and Arnold and Laguna (1977) (Feller-Pareto). We may begin with the Feller-Pareto case and specialize to study the other (simpler) models. If $X_1, X_2, \ldots, X_n$ is a sample of size n from a Feller-Pareto $(\mu, \sigma, \gamma, \gamma_1, \gamma_2)$ distribution (see (3.2.14) for the corresponding density function), the log-likelihood function is given by

$$\begin{aligned}\ell(\mu,\sigma,\gamma,\gamma_1,\gamma_2) = &\left(\frac{\gamma_2}{\gamma} - 1\right)\sum_{i=1}^{n}\log\left(\frac{X_i-\mu}{\sigma}\right)\\ &-(\gamma_1+\gamma_2)\sum_{i=1}^{n}\log\left(1+\left(\frac{X_i-\mu}{\sigma}\right)^{1/\gamma}\right)\\ &-n\log\gamma - n\log\sigma - n\log B(\gamma_1,\gamma_2).\end{aligned} \tag{5.2.93}$$

Maximization is conceptually possible, but as Harris and Singpurwalla (1969) pointed out, one must be careful with regard to multiple maxima and other anomalies of the likelihood surface. The choice $\gamma_1 = \alpha$, $\gamma_2 = 1$ leads to the log-likelihood function for a Pareto (IV) sample:

$$\begin{aligned}P(\mathrm{IV}) \qquad \ell(\mu,\sigma,\gamma,\alpha) = &\left(\frac{1}{\gamma} - 1\right)\sum_{i=1}^{n}\log\left(\frac{X_i-\mu}{\sigma}\right)\\ &-(\alpha+1)\sum_{i=1}^{n}\log\left(1+\left(\frac{X_i-\mu}{\sigma}\right)^{1/\gamma}\right)\\ &-n\log y - n\log\sigma + n\log\alpha.\end{aligned} \tag{5.2.94}$$

Harris and Singpurwalla (1969) used a slightly different parameterization, so that some care is needed in comparing (5.2.94) with their equation (4). They differentiated the log-likelihood and obtained four likelihood equations which must be solved numerically. Likelihood equations can be written down for the Pareto (III) case (i.e., when $\alpha = 1$ in (5.2.94)), and once again numerical solution is required. Life is a little simpler in the Pareto (II) situation. Here the log-likelihood is of the form

$$\begin{aligned}P(\mathrm{II}) \qquad \ell(\mu,\sigma,\alpha) = &-(\alpha+1)\sum_{i=1}^{n}\log\left(1+\frac{X_i-\mu}{\sigma}\right)\\ &-n\log\sigma + n\log\alpha.\end{aligned} \tag{5.2.95}$$

It is evident that for any α and σ, this likelihood will be maximized by setting $\hat{\mu} = X_{1:n}$. The remaining maximum likelihood estimates $\hat{\alpha}$ and $\hat{\sigma}$ must be obtained by considering the equations

$$\begin{aligned}\hat{\alpha} &= \left[\frac{1}{n}\sum_{i=1}^{n}\log\left[1+\left(\frac{X_i - X_{1:n}}{\hat{\sigma}}\right)\right]\right]^{-1} \\ \hat{\sigma} &= \frac{\hat{\alpha}+1}{n}\sum_{i=1}^{n}(X_i - X_{1:n})\left[1+\frac{X_i - X_{1:n}}{\hat{\sigma}}\right]^{-1}.\end{aligned} \tag{5.2.96}$$

Typically, (5.2.96) will have a solution (see Lucke, Myhre and Williams, 1977, for sufficient conditions). Del Castillo and Daoudi (2009), considering the Pareto (II)$(0,\sigma,\alpha)$ case (in the context of a Pickands generalized Pareto model), show that a sufficient condition for the existence of a global maximum of the likelihood based on n observations is that the empirical coefficient of variation exceeds 1. It is, unfortunately, certainly possible to encounter a data set for which no solution exists. Since $X_{1:n}$ is a consistent estimate of μ in the Pareto III, IV and the Feller-Pareto family, one may consider setting $\mu = X_{1:n}$ and solving the resulting simplified "likelihood" equations to get approximate maximum likelihood estimates. If one is willing to assume that some or most of the parameters are known, then considerable simplification naturally results. An extreme example might be a Pareto (II) sample with σ known, in which case the first equation in (5.2.96) gives an exact expression for $\hat{\alpha}$ (upon setting $\hat{\sigma}$ equal to the known value of σ).

See the discussion at the end of Section 5.6 below of a wide variety of Pareto (II) data configurations which have log-likelihood functions essentially of the form (5.2.95). These include generalized order statistics and certain truncated data scenarios.

The asymptotic distribution of the maximum likelihood estimates for the generalized Pareto distributions has received attention from several researchers. Two special cases of the Pareto II distribution have been examined (the Pareto I case was discussed following equation (5.2.12)); Kulldorff and Vannman (1973) observed that, if for a Pareto (II) sample we assume that μ and α are known, then the maximum likelihood estimate of σ is obtained by solving

$$\hat{\sigma}\sum_{i=1}^{n}(X_i - \mu + \hat{\sigma})^{-1} = n\alpha/(\alpha+1) \tag{5.2.97}$$

and for n large we have

$$\hat{\alpha} \dot{\sim} N(\sigma, \sigma^2(\alpha+2)/n\alpha). \tag{5.2.98}$$

If instead in the Pareto (II) sample we assume that σ and α are known, then the maximum likelihood estimate of μ is $X_{1:n}$, which clearly has a P(II)$(\mu,\sigma,n\alpha)$ distribution with mean $\mu + [\sigma(n\alpha-1)^{-1}]$ and variance $\sigma^2 n\alpha(n\alpha-1)^{-2}(n\alpha-2)^{-1}$ (refer to (3.3.9) and (3.3.13) or to Kulldorff and Vannman, 1973).

If we are willing to assume that μ is known (and, without loss of generality, equal to 0), it is clear that the Pareto (II)–(IV) and indeed the Feller-Pareto densities are regular families as far as maximum likelihood estimation is concerned. The corresponding maximum likelihood estimates will be asymptotically normal with variance covariance matrix $\frac{1}{n}J^{-1}$ where J is the information matrix. The exact form of the corresponding information matrices may be determined after some tedious computation. Thus for the Pareto (II) and Pareto (III) models we have

$$P(\mathrm{II})(0,\sigma,\alpha) \qquad J = \begin{pmatrix} \frac{\alpha}{\sigma^2(\alpha+2)} & -\frac{1}{(\alpha+1)\sigma} \\ -\frac{1}{(\alpha+1)\sigma} & \frac{1}{\alpha^2} \end{pmatrix} \tag{5.2.99}$$

and

$$P(\mathrm{III})(0,\sigma,\gamma) \qquad J = \begin{pmatrix} \frac{1}{3\sigma^2\gamma^2} & 0 \\ 0 & \left(\frac{3+\pi^2}{9\gamma^2}\right) \end{pmatrix}. \tag{5.2.100}$$

De Cani and Stine (1986) provided an elementary derivation of the information matrix for a two parameter logistic distribution. From this, recalling that a Pareto (III)$(0,\sigma,\gamma)$ random variable can be viewed as a log-logistic variable, it is possible to confirm the expression given in (5.2.100).

Brazauskas (2002) addressed the problem of identifying the form of the information matrix for the Feller-Pareto distribution with $\mu = 0$. The corresponding density function, obtained from (5.2.14) by substituting $\mu = 0$, is of the form

$$f_0(x) = \frac{(x/\sigma)^{(\gamma_2/\gamma)-1}}{B(\gamma_1,\gamma_2)\gamma\sigma[1+(x/\sigma)^{1/\gamma}]^{\gamma_1+\gamma_2}} I(x>0). \tag{5.2.101}$$

A typical element of the information corresponds to the negative of the expectation of a mixed partial derivative of $\log f_0(X)$. Some of the elements are not complicated. For example

$$J_{14} = -E\left(\frac{\partial^2}{\partial\sigma\partial\gamma_2}\log f_0(X)\right) = \frac{\gamma_1}{\gamma\sigma(\gamma_1+\gamma_2)}.$$

Others are more complicated. In particular the expression for

$$J_{22} = -E[(\partial^2/\partial\gamma^2)\log f_0(X)]$$

is particularly so. Brazauskas (2001) supplies the details for all of the elements of the 4×4 information matrix J for the Feller-Pareto $(0,\sigma,\gamma,\gamma_1,\gamma_2)$ distribution.

If interest is focused on the Pareto(IV) $(0,\sigma,\gamma,\alpha)$ distribution, the corresponding information matrix is of dimension 3×3 and the elements of it can be obtained from the expressions for the Feller-Pareto model by substituting $\gamma_1 = \alpha$ and $\gamma_2 = 1$. This yields the following expression for the information matrix J for the Pareto (IV) $(0,\sigma,\gamma,\alpha)$ distribution

$$\begin{pmatrix} \frac{\alpha}{(\gamma\sigma)^2(\alpha+2)} & \frac{\alpha[\psi(1)-\psi(\alpha)+1]-1}{\gamma^2\sigma(\alpha+2)} & -\frac{1}{\gamma\sigma(\alpha+1)} \\ \frac{\alpha[\psi(1)-\psi(\alpha)+1]-1}{\gamma^2\sigma(\alpha+2)} & \frac{\alpha[(\psi(\alpha)-\psi(1)-1)^2+\psi'(\alpha)+\psi'(1)]+2[\psi(\alpha)-\psi(1)]}{\gamma^2(\alpha+2)} & \frac{\psi(\alpha)-\psi(1)-1}{\gamma(\alpha+1)} \\ -\frac{1}{\gamma\sigma(\alpha+1)} & \frac{\psi(\alpha)-\psi(1)-1}{\gamma(\alpha+1)} & \frac{1}{\alpha^2} \end{pmatrix}$$

where $\psi(\alpha) = (d/d\alpha)\log\Gamma(\alpha)$ is the digamma function. An alternative derivation of the Pareto (IV) information matrix may be found in Brazauskas (2003). In an earlier contribution, Schmittlein (1983) obtained the Pareto (IV) information matrix using a slightly different parameterization. Schmittlein replaces the Pareto (IV) parameter vector (σ, γ, α) with $(\widetilde{\alpha}, \beta, r)$ where $\widetilde{\alpha} = \sigma^{1/\gamma}$, $\beta = 1/\gamma$ and $r = \alpha$. This choice of parameters results in somewhat more complicated expressions for the elements of the information matrix (see equation (2.7) in Schmittlein (1983)). Schmittlein does also provide some discussion of parameter estimation (of $\widetilde{\alpha}, \beta$ and r) when the data are grouped.

While on the topic of information matrices for generalized Pareto models, we note that a model even more complicated than the Feller-Pareto model has received some attention in this context. Zandonatti (2001) introduced a generalized Feller-Pareto distribution. A random variable with this distribution admits a representation of the form

$$X = \mu + \sigma\left[\left(\frac{Z_1}{Z_1 + Z_2}\right)^{-1/\theta} - 1\right]^{\gamma}, \tag{5.2.102}$$

where Z_1 and Z_2 are independent random variables with $Z_i \sim \Gamma(\gamma_i, 1),\ i = 1, 2$. It is clear that this model reduces to become the Feller-Pareto model when $\theta = 1$. If X has a generalized Feller-Pareto distribution then its density is of the form

$$\begin{aligned} f_X(x) &= \frac{\theta x^{\gamma^{-1}-1}}{\gamma\sigma^{1/\alpha}B(\gamma_1, \gamma_2)}\left[1 + \left(\frac{x-\mu}{\sigma}\right)^{1/\gamma}\right]^{\theta\gamma_1 - 1} \\ &\quad \times \left[1 - \left(1 + \left(\frac{x-\mu}{\sigma}\right)^{1/\gamma}\right)^{-\theta}\right]^{\gamma_2 - 1} I(x > \mu). \end{aligned} \tag{5.2.103}$$

If X has this distribution we write $X \sim GFP(\mu, \sigma, \gamma, \gamma_1, \gamma_2, \theta)$. In the case in which $\mu = 0$, a 5×5 information matrix will need to be evaluated in order to precisely identify the asymptotic distribution of the maximum likelihood estimates of the parameters. Using a slightly different notation, the task of evaluating the elements of this information matrix has been addressed by Mahmoud and Abd El-Ghafour (2014) The resulting expressions are very complicated, to the extent that their utility may be questioned. They do, however, by considering special sub-models, provide expressions for the information matrix elements corresponding to broad array of income distribution models that have been considered in the literature.

Remarks. *1. Nadarajah and Kotz (2005a) consider the information matrix for a mixture of two Pareto distributions.*

2. Chaouche and Bacro (2006) suggest an alternative strategy for identifying the maximum likelihood estimates in the Pareto (II)$(0, \sigma, \alpha)$ case which generally avoids convergence problems. Grimshaw (1993) also suggests an algorithm for this problem.

3. *Lin and Wang (2000) consider maximum likelihood estimation for Pareto (II)$(0,\sigma,\alpha)$ data that are possibly censored.*
4. *Ahsanullah (1992) considers UMVUE's and best linear estimates for the Pareto (II)$(0,\sigma,\alpha)$ distribution based on record values.*
5. *Rezaei, Tahmasbi and Mahmoodi (2010) consider maximum likelihood estimation of $P(X<Y)$ based on independent random samples with $X_i \sim P(II)(0,\sigma_1,\alpha_1)$, $i = 1,2,...,n_1$ and $Y_j \sim P(II)(0,\sigma_2,\alpha_2))$, $j = 1,2,...,n_2$. They consider the case in which $\sigma_1 = \sigma_2$ and the case in which $\sigma_1 = \sigma_2 = 1$.*
6. *Davison and Smith (1990) discuss maximum likelihood estimation in the Pickands generalized Pareto case, which includes the Pareto (II)$(0,\sigma,\alpha)$ model, in the presence of covariables.*
7. *Jalali and Watkins (2009) provide a careful discussion of maximum likelihood estimation in the case of Pareto (IV) data with $\mu = 0$ and $\sigma = 1$, assuming that some data points are observed precisely and others are censored.*
8. *Al-Hussaini, Mousa and Jaheen (1992) discuss estimation for the Pareto (IV)$(0,1,\gamma,\alpha)$ distribution. However, they assume that γ is known. Consequently the data can be transformed and then viewed as a sample from a Pareto (I)$(1,\alpha)$ distribution and the results obtained do not add much to the discussion presented in the sub-sections dealing with Pareto (I) estimation. Al-Hussaini, Mousa and Jaheen do present some discussion of empirical Bayes estimates and they provide some simulation-based comparisons of the various estimates that they consider.*

5.2.11 Estimates Using the Method of Moments and Estimating Equations for Generalized Pareto Distributions

Given the disturbing complexity of the likelihood equations for a generalized Pareto population and the possible absence of solutions, the method of moments begins to look like an appealing alternative. At the very least, it can be expected to give suitable starting values for any attempt at iterative solution of the likelihood equations. Harris (1968) suggested such a program for Pareto (II) samples (with $\mu = 0$). Harris assumed finite second moments, so his equations will be suspect in income distribution contexts in which finite second moments are not the norm. For completeness we include his estimates. Suppose $X_1, X_2, \ldots, X_n$ are independent identically distributed P(II)$(0,\sigma,\alpha)$ variables and that we wish to estimate (σ,α). The moment equations are obtained by equating first and second sample and theoretical moments. They are thus

$$M_1 = \frac{1}{n}\sum_{i=1}^{n} X_i = \sigma(\alpha-1)^{-1}$$

and

$$M_2 = \frac{1}{n}\sum_{i=1}^{n} X_i^2 = 2\sigma^2(\alpha-1)^{-1}(\alpha-2)^{-1} \tag{5.2.104}$$

with solution

$$\tilde{\alpha} = 2(M_2 - M_1^2)(M_2 - 2M_1^2)^{-1} \tag{5.2.105}$$

$$\tilde{\sigma} = M_1 M_2 (M_2 - 2M_1^2)^{-1} \tag{5.2.106}$$

(provided that $M_2 > 2M_1^2$).

We have observed that the Pareto (II) model with $\mu = 0$ can be identified with the subclass of Pickands generalized Pareto distributions with $k < 0$. Since the introduction of the Pickands model, a cornucopia of parameter estimation strategies for it have been developed and compared. de Zea Bermudez and Kotz (2010a,b) provide a detailed survey of this body of research. Only a selection of the material available in their papers will be presented here; for more coverage and more details the reader is referred to the de Zea Bermudez and Kotz papers. Note that their results are directly applicable to the Pareto (II)$(0, \sigma, \alpha)$ model provided that one identifies the parameter k in these papers as being equal to $-1/\alpha$.

Variations on the method of moments have attracted attention. Rather than equating first and second sample and population moments as Harris did, moments of other orders can be considered. Ashkar and Ouarda (1996), for example, consider moments of orders $-1, 0$ and 1. Arnold and Laguna (1977) instead urged the use of fractional moments in applying the method of moments. Harris' estimates (5.2.71–72) still might be well defined when α is less than 2, but they will fail to possess the consistency and asymptotic normality usually associated with method of moments estimates.

The method of moments estimates suggested by Arnold and Laguna will exhibit respectable behavior under much less restrictive conditions.

It is possible to describe the Arnold-Laguna procedure in the general context of Feller-Pareto distributions. We will agree to always use $X_{1:n}$ as a reasonable (consistent) estimate of μ, therefore leaving four parameters to be estimated. We thus begin with $X_1, X_2, \ldots, X_n$, a sample from a Feller-Pareto $(\mu, \sigma, \gamma, \gamma_1, \gamma_2)$ distribution. Subtracting off $X_{1:n}$ from the observations (or subtracting μ, if it is known), we can set up the following four fractional moment equations whose solution $(\tilde{\sigma}, \hat{\gamma}, \hat{\gamma}_1, \hat{\gamma}_2)$ will be the method of moments estimates of $(\sigma, \gamma, \gamma_1, \gamma_2)$ (refer to (3.3.6) for the fractional moments of a Feller-Pareto distribution):

$$\frac{1}{n-1} \sum_{i=1}^{n} (X_i - X_{1:n})^{1/k} = \sigma^{1/k} \Gamma(\gamma_1 - \gamma k^{-1}) \Gamma(\gamma_2 + \gamma k^{-1}) / \Gamma(\gamma_1) \Gamma(\gamma_2),$$
$$k = 2, 3, 4, 5. \tag{5.2.107}$$

A divisor of $n-1$ is used instead of n, since only $n-1$ of the $(X_i - X_{1:n})$'s will be non-zero. It is possible for the system of equations (5.2.107) not to have a solution. This anomaly is an event of low probability for reasonably large sample sizes. If instead of a Feller-Pareto distribution, we wish to consider a Pareto I, II, III or IV distribution, we need to make the obvious modifications in (5.2.107) using only as many equations as there are parameters to estimate and using the relevant expressions (3.3.8) – (3.3.12) for the right hand sides of the equations.

Al-Marzoug and Ahmad (1985) provide discussion of fractional moment estimation for the Pareto (IV) distribution with $\mu = 0$ and $\sigma = 1$. They compare these estimates with maximum likelihood and ordinary moment estimates in terms of bias.

Instead of using fractional moments in (5.2.107), one might consider using negative moments or moments of logarithms. Such a procedure is clearly advantageous in the classical Pareto setting, and reasonable performance for generalized Pareto samples can be expected.

Since it may well be difficult to decide which four sample and population moments to equate to estimate the parameters of the Feller-Pareto distribution with $\mu = 0$, one might consider utilizing more than four moments. For this approach we select m positive values $\{c_i : i = 1,2,...,m\}$ and then choose σ, γ, γ_1 and γ_2 to minimize the following objective function:

$$\sum_{j=1}^{m}\left[\frac{1}{n-1}\sum_{i=1}^{n}(X_i - X_{1:n})^{c_j} - \sigma^{c_j}\frac{\Gamma(\gamma_1 - \gamma c_j)\Gamma(\gamma_2 + \gamma c_j)}{\Gamma(\gamma_1)\Gamma(\gamma_2)}\right]^2. \qquad (5.2.108)$$

The use of probability weighted moments (PWM's) has also been proposed as a suitable parameter estimation strategy. The basic idea is to consider parametric functions of the form

$$E\{X^a[F_X(X)]^b[1 - F_X(X)]^c\} \qquad (5.2.109)$$

for various values of a, b and c. These parametric expressions are then equated to their corresponding sample versions in order to identify parameter estimates. In the case of the Pareto models discussed in this book, the survival function is typically less complicated than the distribution function, so that one will usually set $b = 0$ in (5.2.109). Even with that simplification the computation of the probability weighted moments will be non-trivial in the Pareto (III) and (IV) cases, and they will not be available in closed form for the Feller-Pareto distribution. In the Pareto (II) case, the probability weighted moment approach does become feasible. This is especially true if we choose $a = 1$ in (5.2.109). In this setting, we will consider $X \sim P(II)(0, \sigma, \alpha)$ and we will wish to evaluate

$$\alpha_c = E\left\{X\left[\left(1+\frac{X}{\sigma}\right)^{-\alpha}\right]^c\right\}. \qquad (5.2.110)$$

However, $X \stackrel{d}{=} \sigma(Y-1)$ where $Y \sim P(I)(1, \alpha)$. Consequently

$$\begin{aligned}\alpha_c &= \sigma E[(Y-1)Y^{-c\alpha}] \\ &= \frac{\sigma}{\alpha}(1+c)^{-1}(1+c-1/\alpha)^{-1}.\end{aligned} \qquad (5.2.111)$$

For two choices of c, this expression will be equated to its corresponding sample version, i.e.,

$$a_c = \frac{1}{n}\sum_{i=1}^{n} X_{i:n}\left(\frac{n-i+1/2}{n}\right)^c, \tag{5.2.112}$$

and the two equations are then solved to yield values of σ and α which are our PWM estimates. Note that there is a restriction on the values of c that can be used in (5.2.111). We must have $1+c > 1/\alpha$.

Singh and Ahmad (2004), based on a simulation study, suggest that PWM estimates are preferable to method of moments estimates. Hosking and Wallis (1987) recommend method of moments estimates and PWM estimates over maximum likelihood estimates, unless the sample size is very large.

Different estimating equations were suggested by Zhang (2007) for a Pareto (II)$(0,\sigma,\alpha)$ sample. They are of the form

$$\left[\frac{1}{n}\sum_{i=1}^{n}\left(1+\frac{X_i}{\sigma}\right)^r - (1+r/\alpha)^{-1}\right]/r = 0. \tag{5.2.113}$$

If we choose the two equations corresponding to $r=1$ and $r=2$ in (5.2.113), the resulting estimates are the usual method of moments estimates. If we choose $r=-1$ and $r=0$ (take the limit as $r\to 0$), then the equations are equivalent to the usual maximum likelihood equations. Zhang (2007) proposes the use of $r=0$ and $r=r^*\in(0,1/2)$ to obtain good estimates. However it appears to be difficult if not to say impossible to determine globally optimal choices for the r's to be used.

A hybrid estimation approach was suggested by Husler, Li and Raschke (2011). Suppose we have a sample from a Pareto (II)$(0,\sigma,\alpha)$ distribution. First observe that, if σ is known, the maximum likelihood estimate of α is

$$\hat{\alpha} = \left[\frac{1}{n}\sum_{i=1}^{n}\log\left(1+\frac{X_i}{\sigma}\right)\right]^{-1}.$$

This suggests that the distribution function of the X_i's can be estimated by

$$\hat{F}(X_i) = 1-\left(1+\frac{X_i}{\sigma}\right)^{-\hat{\alpha}}.$$

However, we do not know σ. For our second estimating equation we set up

$$\frac{1}{n}\sum_{i=1}^{n}\hat{F}(X_i) = 1/2.$$

The Husler, Li and Raschke estimate of (σ,α) is then the solution of the following pair of equations:

$$\hat{\alpha} = \left[\frac{1}{n}\sum_{i=1}^{n}\log\left(1+\frac{X_i}{\hat{\sigma}}\right)\right]^{-1}, \tag{5.2.114}$$

$$\sum_{i=1}^{n}\left(1+\frac{X_i}{\hat{\sigma}}\right)^{-\hat{\alpha}} = 1/2. \tag{5.2.115}$$

Husler, Li and Raschke (2011) confirm that such estimates are consistent and asymptotically normal.

Two alternative hybrid approaches to this Pareto (II)$(0,\sigma,\alpha)$ problem were proposed by Zhang (2010). Building on ideas presented in Zhang and Stephens (2009), in each of these proposals, the unknown parameter σ in (5.2.114) is estimated using a Bayesian technique with a data-dependent prior.

Relatively simple expressions are available for expectations of order statistics from Pareto I, II and III distributions (see (3.4.16) and (3.4.28)). One can use these expressions to obtain estimates of parameters by equating observed order statistics to their expectations. As usual, it seems plausible to use $X_{1:n}$ to estimate μ (or σ in the Pareto I case). For example, in the Pareto (III) case we would use $X_{1:n}$ as an estimate of μ and obtain estimates $(\tilde{\sigma},\tilde{\gamma})$ of (σ,γ) as the solution to

$$X_{k_i:n} - X_{1:n} = \sigma\Gamma(k_i+\gamma)\Gamma(n-k_i-\gamma+1)/\Gamma(k_i)\Gamma(n-k_i+1), \quad i=1,2. \tag{5.2.116}$$

One might try to choose k_1, k_2 to minimize the variability of the resulting estimates, but such a search may be non-trivial. Arnold and Laguna (1977), making no claims about optimality, suggested $(k_1,k_2) = (n/3,2n/3)$. If one chooses (k_1,k_2) in this manner and assumes that n is large, then Stirling's approximation can be invoked in solving (5.2.116) yielding as approximate solutions

$$\hat{\gamma} = (\log T)/(2\log 2) \tag{5.2.117}$$

and

$$\tilde{\sigma} = S/2^{\tilde{\gamma}}$$

where $S = X_{2n/3:n} - X_{1:n}$ and $T = [X_{2n/3:n} - X_{1:n}]/[X_{n/3:n} - X_{1:n}]$.

An alternative approach involves the use of L-moments which are particular linear combinations of expectations of order statistics, introduced by Hosking (1990). The mth L-moment is defined by

$$\lambda_m = \frac{1}{m}\sum_{i=0}^{m-1}(-1)^i\binom{m-1}{i}E(X_{m-i:m}), \tag{5.2.118}$$

$m = 1, 2, \ldots$. In the Pareto (I)–(III) cases it is possible to obtain usable expressions for such λ_m's and they can be equated to their corresponding sample versions in order to identify suitable parameter estimates. de Zea Bermudez and Kotz (2010a) provide details for the implementation of this strategy in the Pareto (II)$(0, \sigma, \alpha)$ case (as usual, with $k = -1/\alpha$).

Castillo and Hadi (1997) suggest a subsample based approach to the estimation of parameters in the Pareto (II)$(0, \sigma, \alpha)$ case. It can be utilized in the Pareto (IV)$(0, \sigma, \alpha, \gamma)$ case also. In the Pareto (IV) case we proceed as follows. Choose three indices $i_1 < i_2 < i_3$ from the set $1, 2, \ldots, n$ and set up three equations as follows:

$$\left[1 + \left(\frac{X_{i_j:n}}{\sigma}\right)^{1/\gamma}\right]^{-\alpha} = \frac{n+1-i_j}{n+1}, \quad j = 1, 2, 3. \tag{5.2.119}$$

Solve these three equations for σ, α and γ, and denote the solution by

$$(\widetilde{\sigma}(i_1, i_2, i_3), \widetilde{\alpha}(i_1, i_2, i_3), \widetilde{\gamma}(i_1, i_2, i_3))$$

to indicate the dependence on the choice of the indices i_1, i_2 and i_3.

Repeat this procedure for a large number of selections of (i_1, i_2, i_3). If computing time is not a constraint, one might do it for every possible set of three indices. The Castillo-Hadi estimates will be the medians of these three-index estimates. Thus, if all possible sets of three indices are considered, then

$$\widetilde{\sigma} = \text{median}\, \{\widetilde{\sigma}(i_1, i_2, i_3) : 1 \leq i_1 < i_2 < i_3 \leq n\}$$

with $\widetilde{\alpha}$ and $\widetilde{\gamma}$ similarly defined. In the Pareto (II) case, Castillo and Hadi provide evidence that such an estimation procedure can often be recommended over the method of moments and the PWM approaches.

5.2.12 Order Statistic Estimates for Generalized Pareto Distributions

Moore and Harter (1967, 1969), Kulldorff and Vannman (1973) and Vannman (1976) have discussed estimates based on Pareto (II) order statistics, while Ahmad (1980, 1985) has considered order statistic estimation for the Burr distribution (i.e., Pareto (IV) with μ and σ known). We summarize their results, but reference to the original papers will be necessary for certain tables required in using their procedures.

Moore and Harter (1967) considered a Pareto (II) population with μ known and $\sigma = 1$. So without loss of generality we could assume we have $\mu = \sigma = 1$ (i.e., add $1 - \mu$ to each observation) and thus have a sample from a classical Pareto distribution with $\sigma = 1$. They addressed the problem of estimating the shape parameter α based on a single order statistics $X_{k:n}$. They observed that

$$E(\log X_{k:n}) = \left[\sum_{i=0}^{k-1} \frac{1}{n-i}\right] \alpha^{-1}$$

and consequently proposed as an estimate of α, the statistic

$$\alpha^* = \left(\sum_{i=0}^{k-1} \frac{1}{n-i} \right) \Big/ [\log X_{k:n}]. \tag{5.2.120}$$

This estimate is of the form (5.2.49) (i.e., its reciprocal has the same distribution as a linear combination of independent identically distributed exponential variables). Consequently, the formulae given by Thomas (1976) could be used to determine its mean and variance. It is not an unbiased estimate of α (since its reciprocal is an unbiased estimate of α^{-1}). If one were permitted the choice of which order statistic to use (i.e., permitted to choose the k which appears in (5.2.120)), a suitable choice might be $k = (0.797)n$. This suggestion is based on two pieces of evidence. Thomas, in studying quantile estimates for classical Pareto population based on two order statistics (for the case when both σ and α were unknown), found that the best choice of the two order statistics was $X_{1:n}$ and $X_{k:n}$ where k/n was usually close to 0.797. Sarhan, Greenberg and Ogawa (1960) found that in estimating α^{-1} by expressions which were reciprocals of those appearing in (5.2.120), the mean squared error was minimized by choosing $k/n \simeq 0.7968$.

In a later paper Moore and Harter (1969) discussed the problem of estimating α based on the first m order statistics of a sample of size n from a Pareto (II) population with μ known and $\sigma = 1$. Without loss of generality we may assume that sample is from a classical Pareto distribution with $\sigma = 1$. Referring to Epstein and Sobel's (1953) work on exponential estimation or proceeding directly, we find the maximum likelihood estimate of α is given by

$$\hat{\alpha} = m \Big/ \left[\sum_{i=1}^{m-1} \log X_{i:n} + (n-m) \log X_{m:n} \right]. \tag{5.2.121}$$

It is easily verified that $\hat{\alpha}$ given in (5.2.121) is distributed as the reciprocal of a gamma random variable. The unbiased modification (i.e., $(m-1)\hat{\alpha}/m$) is the MVUE of α with relative efficiency (compared to the Cramer-Rao lower bound) of $(m-2)/m$.

Huang (2011) considers estimation for the Pareto (II)$(0, 1, \alpha)$ distribution. He proposes an estimate which is a linear combination of order statistics, but is one which puts less weight on the extreme order statistics than do the usual linear systematic estimates. Huang shows that his estimate outperforms the maximum likelihood and the minimum risk estimates of α. Extension of these results to the Pareto (II)(μ, σ, α) case with μ and σ unknown would be of interest.

There is theoretically little difficulty in extending the earlier discussion of maximum likelihood estimation for generalized Pareto populations to cover the case of censored sampling (see, e.g., Myhre and Saunders, 1976). The likelihood equations neither change much in character nor do they change much in terms of ease of solution.

If we consider the P(II)(μ, σ, α) case where α is known, we are dealing with a location and scale family. Provided that we know (as we do from (3.4.28) and (3.4.29)) the means and covariances of the order statistics, we can use the Gauss-Markov theorem to construct minimum-variance unbiased estimates of μ and σ among the class of linear systematic statistics (i.e., linear combinations of order statistics). This program was carried out by Kulldorff and Vannman (1973). The resulting estimates, called BLUE's (best linear unbiased estimates), are derived in three possible cases: (i) μ known, (ii) σ known and (iii) μ and σ both unknown. The expressions involved in the BLUE's are cumbersome, to say the least, and Kulldorff and Vannman recommended use of what they called ABLUE's (asymptotically best linear unbiased estimates). These will be based on k (a not large) prespecified number of order statistics. A set of k order statistics will have an asymptotically normal joint distribution. Based on the variance-covariance structure of the asymptotic distribution, we can choose the best linear combination of those order statistics. Then we choose which k order statistics to use in order to minimize the asymptotic variance. In this way we arrive at ABLUE's. Kulldorff and Vannman (1973) provided tables giving the ABLUE's for $1 \leq k \leq 10$ when α is known to be 0.5(0.5)5.0. In some special cases the computations yield closed form expressions, and a surprisingly high efficiency is exhibited. For example, if $\alpha = l$ and $k = 5$, the asymptotic relative efficiency of the ABLUE with respect to the maximum likelihood estimate is 0.97. In a later paper Vannman (1976) exhibited exact expressions for the BLUE's based on k selected order statistics. Use of BLUE's based on only a few order statistics (rather than on the entire sample) may be desirable, since the higher order statistics may not be stable (i.e., may not have finite means and variances). Other authors who have considered BLUE's and ABLUE's for Pareto (II) distributions include Chan and Cheng (1973), Kaminsky (1973) and Saksena (1978). In the same spirit as the above is Ahmad's (1980, 1985) work, in which he developed ABLUE's for the Burr distribution (i.e., Pareto (IV) with μ and σ known). This is, of course, not a location-scale family, but the asymptotic covariance of k order statistics may be exploited to obtain asymptotic variances for linear systematic statistics. Then one may search for the optimal selection of order statistics. Ahmad's (1980, 1985) results were not as extensive as those of Kulldorff and Vannman. In the case when α is known he determined the ABLUE based on a single order statistic for various choices of α (his notation differs from ours). In the case when γ is known, he determined the ABLUE based on a single order statistic. He found that the optimal choice of k is $(0.79681)n$. The fact that this coincides with the best choice of k when using the estimator (5.2.120) is not accidental. When both α and γ are unknown, Ahmad presented only some preliminary results on the derivation of ABLUE's. He considered the case $k = 3$ and obtained approximately optimal values by use of an ad hoc search.

Linear systematic statistics can also be used to estimate population quantiles. Ali, Umbach and Hassanein (1981) considered best linear unbiased estimates of Pareto (II) quantiles (with α known) based on two optimally chosen order statistics. They provide tables for sample sizes up to 30. In a later paper Saleh, Ali and Umbach (1985) described certain non-linear estimates of Pareto (I) quantiles based on a few selected order statistics.

The BLUE's and ABLUE's described above all share the unfortunate property that they depend on knowledge of parameters which typically will not be known.

One could bravely attack the problem of devising ABLUE's based on selected order statistics for Pareto (IV) or even Feller-Pareto families with all parameters unknown. The optimization involved may well be more troublesome than that involved in maximizing the likelihood. Lucke, Myhre and Williams (1977) reported some Monte Carlo work which suggests that BLUE's, ABLUE'S and BLUE's based on censored samples may all be inferior to competing maximum likelihood estimates in Pareto (II) samples even in the unlikely circumstance that the shape parameter α is known. These authors seem to be definitely on the maximum likelihood bandwagon.

With the burgeoning interest in the Pickands generalized Pareto distribution, the issue of identifying good unbiased estimates of its location and scale parameters has drawn the attention of several researchers. Papers dealing with this topic of course include results for estimating location and scale parameters of a Pareto (II)(μ, σ, α) distribution. The following is a representative but far from exhaustive list of available results. Cheng and Chou (2000a) discuss the BLUE of the scale parameter, and in a subsequent paper (Cheng and Chou, 2000b) they discuss the asymptotically best unbiased estimate of σ. Mahmoud, Sultan and Moshref (2005) discuss BLUE's of μ and σ based on the complete set of order statistics. Burkschat, Cramer and Kamps (2007) consider BLUE's of μ and σ based on data consisting of several independent sets of generalized order statistics.

In many instances, interest will center on estimating certain measures of inequality for generalized Pareto distributions. The formulae which relate the more common inequality measures to the underlying parameters of the distribution were listed at the end of Section 4.2. One may then substitute maximum likelihood estimates for the unknown parameters and obtain, generally, consistent asymptotically normal estimates of the inequality measures. This procedure can be expected to be preferable to using the corresponding sample measures of inequality (introduced in Section 4.3) as estimates of the population inequality measures. Even though the sample measures are consistent and asymptotically normal, they will generally be inefficient when compared with estimates which are functions of the maximum likelihood estimates of the population parameters. Kakwani (1974) showed, for example, that the sample Elteto and Frigyes inequality measures were markedly inferior to the maximum likelihood estimates, when the parent population was log-normal. The same phenomenon can be expected when the parent population is generalized Pareto.

On the other hand, the sample measures of inequality enjoy robustness properties with respect to incorrectly specified underlying distributions. In other words, the maximum likelihood estimates can be expected to be preferable only when the data exhibit a good fit to the assumed model.

Luceño (2006) suggests choosing, as parameter estimates, those values which minimize certain goodness of fit statistics. He deals with Pareto (II) data, but the approach can be applied to any model with an available closed form expression for the distribution function (for example the Pareto (IV) distribution). Thus, for example, if one uses Kolmogorov-Smirnov distance to assess goodness of fit, the estimates of the Pareto (IV) parameters $(\mu, \sigma, \gamma, \alpha)$ will be those values which minimize the following objective function

$$\sup_{x \in (-\infty,\infty)} |F_{\mu,\sigma,\gamma,\alpha}(x) - F_n(x)|,$$

where $F_{\mu,\sigma,\gamma,\alpha}$ is the distribution function of $P(IV)(\mu,\sigma,\gamma,\alpha)$ distribution and F_n is the sample distribution function. Instead of using the Kolmogorov-Smirnov statistic, Luceño also considers the Cramer-von Mises statistic and several variants of the Anderson-Darling goodness of fit statistic. Luceño provides simulation based comparisons of these estimation techniques with more standard approaches in the case of Pareto (II) data.

Remark. *Chen (2006) develops confidence intervals for the shape parameter γ based on the first k order statistics from a Pareto (III)$(0,\sigma,,\gamma)$ sample.*

5.2.13 Bayes Estimates for Generalized Pareto Distributions

If we have a sample of size n, $X_1, X_2, ..., X_n$, from a Pareto (II)(μ,σ,α) distribution, the minimal sufficient statistic is the complete set of order statistics. This precludes the identification of a convenient conjugate prior for the parameters of the model. As in earlier Sections, it is customary to adopt $X_{1:n}$ as a suitable estimate for the location parameter and then to act as if μ were known to be equal to that value. Consequently, in this sub-section, henceforth it will suffice to consider the case in which $\mu = 0$ (we will also adopt this approach when we discuss more general Pareto models below).

In the Bayesian approach we will seek a suitable joint prior density for (σ,α) that will combine with the Pareto (II)$(0,\sigma,\alpha)$ likelihood to yield a posterior density that can either be analyzed analytically or which is susceptible to sampling using a Monte Carlo Markov chain approach, usually involving a Gibbs sampler. Pickands (1993) proposed use of independent locally uniform priors for $\log\sigma$ and $\log\alpha$, i.e.,

$$f(\sigma,\alpha) \propto \frac{1}{\sigma\alpha} I(\sigma > 0, \alpha > 0). \tag{5.2.122}$$

Howlader and Hossain (2002) also use the non-informative prior (5.2.122) in developing a squared error loss Bayes estimate of $[1 + (x/\sigma)]^{-\alpha}$ based on censored data. They approximate the corresponding posterior expectation by using techniques proposed by Lindley (1980) and Tierney and Kadane (1986).

Arnold and Press (1983) proposed the use of independent prior densities for σ and α in the Pareto (I) case. Such a joint prior might also be considered in the present Pareto (II)$(0,\sigma,\alpha)$ setting. Thus we might assume that, a priori, $\alpha \sim \Gamma(c,d)$ and, independently, $\sigma \sim PF(\delta,\sigma_0)$. De Zea Bermudez and Turkman (2003), viewing the Pareto (II)$(0,\sigma,\alpha)$ model as being nested within a Pickands generalized Pareto model, suggested the use of independent priors for $1/\alpha$ and α/σ. Their suggestion is that one should take as an a priori distribution for the parameters, one for which $\alpha \sim PF(\delta,\sigma_0)$ and $\alpha/\sigma \sim \Gamma(c,d)$. Alternatively one could use a generalized Lwin prior for (σ,α) or a conditionally specified prior of the form (5.2.76). These are flexible families of priors and, by suitable choices of the hyperparameters, they may be selected to be approximately concordant with the prior beliefs of an informed expert. With any of these suggested prior densities, the final inferences will be implemented either by approximating the posterior distribution or by Monte Carlo sampling from the (complicated) posterior density.

Arnold and Press (1983) proposed an eclectic approach that they called empirical Bayes. They suggested using a method of moments estimate of σ, say $\hat{\sigma}_M$ (as in (5.2.106), and then acting as if σ were known to be equal to this value. With that assumption they suggest use of a gamma prior for α, i.e., $\alpha \sim \Gamma(c,d)$. The corresponding posterior density for α will again be of the gamma form, and the Bayes estimate of α will be

$$\hat{\alpha}_B = \frac{n+c}{d+\hat{v}} \text{ where } \hat{v} = \sum_{i=1}^{n} \log(1+\frac{X_i}{\hat{\sigma}_M}).$$

They suggest that a similar approach (combining classical and Bayesian strategies) be used for data with Pareto (III) and (IV) distributions.

Al-Hussaini and Jaheen (1992) describe an alternative procedure in the Pareto (IV) case. They assume that μ and σ are known and, without loss of generality, are equal to 0 and 1, respectively. They then propose a joint prior for (γ,α) that is equivalent to the following specification:

$$1/\gamma \sim \Gamma(\delta,\beta)$$

and

$$\alpha|\gamma \sim \Gamma(\rho+1,\gamma\tau),$$

where δ,β,ρ and τ are suitably chosen hyperparameters. Lindley's approximation of the posterior density is then used to obtain the corresponding Bayes estimates of $1/\gamma,\alpha$ and $[1+x_0^{1/\gamma}]^{-\alpha}$. In a later paper, Al-Hussaini and Jaheen (1995) consider prediction of a future observation from the Pareto (IV) distribution using the same form of joint prior for (γ,α).

5.3 Interval Estimates

Suppose that we have a sample $X_1,X_2,\ldots,X_n$ from a classical Pareto distribution (i.e., $X_i \sim P(I)(\sigma,\alpha)$ with survival function given by (3.2.1)). If σ is known, then $\log(X_i/\sigma)$ will have an exponential distribution (i.e., $\log(X_i/\sigma) \sim \Gamma(1,\alpha^{-1})$), and a suitable pivotal quantity for construction of confidence intervals is provided by

$$\alpha \sum_{i=1}^{n} \log(X_i/\sigma) \sim \Gamma(n,1). \tag{5.3.1}$$

Thus a $100(1-\gamma)\%$ confidence interval for α will be of the form

$$[T^{-1}\Gamma_{\gamma/2}(n,1), T^{-1}\Gamma_{1-\gamma/2}(n,1)] \tag{5.3.2}$$

where $T = \sum_{i=1}^{n} \log(X_i/\sigma)$ and $\Gamma_\delta(n,1)$ denotes the δth percentile of a $\Gamma(n,1)$ distribution. If n is large, the pivotal quantity (5.3.1) will be approximately normal, and normal percentiles would be used in (5.3.2). If σ rather than α is unknown, (5.3.1) remains a pivotal quantity, although a better one is available. By considering the maximum likelihood estimate of σ (for α known), we are led to the following pivotal quantity

$$n\alpha \log(X_{1:n}/\sigma) \sim \Gamma(1,1) \tag{5.3.3}$$

with a corresponding $100(1-\gamma)\%$ confidence interval given by

$$[X_{1:n}(\gamma/2)^{1/(n\alpha)}, X_{1:n}(1-\gamma/2)^{1/(n\alpha)}]. \tag{5.3.4}$$

Typically both σ and α will be unknown. The corresponding maximum likelihood estimates (given in (5.2.3)) are independent, and this may be exploited to construct a confidence region for (σ,α). We thus have two independent pivotals:

$$n\alpha\,\hat{\alpha}^{-1} \sim \Gamma(n-1,1) \tag{5.3.5}$$

$$n\alpha \log(\hat{\sigma}/\alpha) \sim \Gamma(1,1). \tag{5.3.6}$$

The resulting $100(1-\gamma)\%$ confidence region for (σ,α) is

$$\{(\sigma,\alpha) : \hat{\alpha}c_1 < n\alpha < \hat{\alpha}c_2, \hat{\sigma}e^{-d_2/(n\alpha)} < \sigma < \hat{\sigma}e^{-d_1/(n\alpha)}\} \tag{5.3.7}$$

where

$$c_1 = \Gamma_{\eta/2}(n-1,1), \quad c_2 = \Gamma_{1-\eta/2}(n-1,1)$$

$$d_1 = -\log(1-\eta/2), \quad d_2 = -\log(\eta/2)$$

in which $\eta = 1-\sqrt{1-\gamma}$.

It is, of course, also possible to derive one-sided confidence intervals and simultaneous lower (or upper) confidence bounds by using the above pivotal quantities ((5.3.1), (5.3.3), (5.3.5) and (5.3.6)). Iliescu and Viorel (1974) gave such examples (assuming α is known to be equal to 1).

If we have only a partial Pareto (I) sample (i.e., a single order statistic or a censored sample), it is still possible to construct confidence intervals and regions. If we have only a single order statistic, we must, of course, assume that one of the parameters is known. Results in this direction were reported by Moore and Harter (1967, 1969). They restricted attention to the Pareto (I) case with σ known. If one has the smallest m of the n order statistics available, then a suitable pivotal quantity for α (assuming without loss of generality that the known value of σ is 1) is given by

$$\alpha/\hat{\alpha} \sim \Gamma(m,1) \tag{5.3.8}$$

where $\hat{\alpha}$ is the maximum likelihood estimate of α based on $(X_{1:n}, X_{2:n}, \ldots, X_{m:n})$ (see (5.2.121)). If one has merely a single order statistic $X_{m:n}$ available from a

$P(I)(1,\alpha)$ population, then $\log X_{m:n}$ is distributed like the mth order statistic from an exponential distribution with scale parameter α^{-1}. Consequently, $\alpha \log X_{m:n}$ is a pivotal quantity whose distribution has been tabulated by Harter (1961).

The case in which we have a censored sample with both σ and α unknown was considered by Chen (1996). He constructed a region parallel to that in (5.3.7) based on $(X_{1:n}, X_{2:n}, ..., X_{k:n})$, utilizing the maximum likelihood estimate of (σ, α) based on the available order statistics. Subsequently Fernández (2012) discussed confidence regions based on doubly censored samples, i.e., based on $(X_{r:n}, X_{r+1:n}, ..., X_{s:n})$. He identifies a minimum area confidence region by solving a non-linear programming problem. See also Wu (2008). Parsi, Ganjali and Farsipour (2010) and independently Wu (2010) derived confidence regions for (σ, α) based on progressively censored samples. A parallel development would be possible for generalized order statistics data.

If the sample size is large, we can consider constructing approximate confidence regions for the unknown parameters of generalized Pareto distributions by using our knowledge of the asymptotic joint distribution of the corresponding maximum likelihood estimates. For the Pareto II, III and IV families with $\mu = 0$ the regularity conditions for asymptotic joint normality of the maximum likelihood estimates are satisfied, and the essential tool for constructing ellipsoidal confidence regions is a knowledge of the appropriate information matrix. This is available for the $P(II)(0, \sigma, \alpha)$, $P(III)(0, \sigma, \gamma)$ and $P(IV)(0, \sigma, \gamma, \alpha)$ cases and even for the Feller-Pareto distribution (see (5.2.99), (5.2.100) and the discussion following (5.2.101)). For example, in the $(P(III)(0, \sigma, \gamma))$ case, an approximate 95% confidence region for (σ, γ) is given by

$$\left\{(\sigma, \gamma) : \frac{(\hat{\sigma} - \sigma)^2}{3\sigma^2\gamma^2} + \frac{(3+\pi^2)(\hat{\gamma} - \gamma)^2}{9\gamma^2} < \frac{1}{n}\chi^2_{2:0.95}\right\} \tag{5.3.9}$$

where $(\hat{\sigma}, \hat{\gamma})$ is the maximum likelihood estimate of (σ, γ) obtained by numerically solving the appropriate likelihood equations.

Omar, Ibrahim and Razali (2012), assuming that σ is known, discuss asymptotic confidence intervals for α in the Pareto (II) case based on the asymptotic normality of estimates of α obtained by maximum likelihood and by the method of moments. A simulation study indicated that intervals based on the maximum likelihood estimate were preferable to those based on the method of moments estimate. This finding is consistent with the observation that maximum likelihood estimates are asymptotically efficient. Omar, Ibrahim and Razali also consider similar intervals based on sample minima, sample maxima and extreme ranked set samples.

If interest is focused on obtaining confidence intervals for inequality measures associated with the population under study, we have two avenues available for generalized Pareto samples. One technique involves writing the inequality index of interest as a function of the parameters of the given Pareto distribution (referring perhaps to the material at the end of Section 4.2) and then using available confidence regions for the unknown parameters to construct a confidence set for the parametric function of interest. Since the usual inequality measures are frequently monotone functions of the parameters of generalized Pareto distributions, the program is not as difficult to carry out as might be suspected. An extreme case is provided by the P(III)$(0,\sigma,\gamma)$ case, in which the Gini index is exactly γ, so that an approximate confidence interval can be written down using the asymptotic normality of the maximum likelihood estimate $\hat{\gamma}$. We have from (5.2.100), $\gamma \dot{\sim} N(\gamma, n^{-1}9\gamma^2/(3+\pi^2))$ so that $(\hat{\gamma}-\gamma)/\gamma \dot{\sim} N(0,n^{-1}9/(3+\pi^2))$, from which an approximate confidence interval is readily obtained. An alternative approach relies on the asymptotic normality of sample measures of inequality. General expressions for the asymptotic variances of sample measures of inequality are to be found in Section 4.3. That corresponding to the sample Gini index is given by equation (4.3.53). A glance at that equation suffices to convince one that the asymptotic variance depends in a non-trivial way on the underlying distribution. The confidence interval for the population Gini index of level $1-\alpha$ will be of the form

$$G_n \pm z_{1-\alpha/2}\sqrt{\hat{\text{var}}\, G_n} \tag{5.3.10}$$

where G_n is the sample Gini index, and the estimated variance G_n is obtained by evaluating (4.3.53) for the assumed parent distribution and estimating any unknown parameters in the resulting expression. Such estimation may well be done by maximum likelihood. If so, we find ourselves working considerably harder than, for example, the statistician who, in a $P(\text{III})(0,\sigma,\gamma)$ family, bases his confidence interval on the maximum likelihood estimate of γ, which is in this case the population Gini index. The advantage of using a confidence interval based on the sample Gini index is that we can expect such an interval to be considerably more robust with respect to violations of the underlying distributional assumptions. No matter what, G_n will be a consistent estimate of the population Gini index, and a confidence region obtained using (4.3.53) won't be too far off, even if we used the wrong distribution F in our evaluation. Of course, we are also utilizing the asymptotic distribution, and a cautionary note is included in Arnold and Laguna (1977). They reported a simulation which suggested that for certain Pareto (III) families the sample distribution of the sample Gini index may not approach its limiting form as quickly as we might hope.

Remarks. *1. Wu, Lee and Chen (2007a) assume that α is known in the Pareto (II) setting, and consider prediction intervals for a future observation based on certain order statistics from a sample of size n.*

2. *Wong (2012) considers the case of two independent samples, $X_1,X_2,\ldots,X_n$ which are i.i.d. Pareto (II)$(0,\sigma,\alpha)$, and $Y_1,Y_2,\ldots,Y_m$ which are i.i.d. Pareto (II)$(0,\sigma,\beta)$. Note that a common scale parameter is assumed. The quantity of interest is $P(Y_1 < X_1)$ which is equal to $\beta/(\alpha+\beta)$. Wong derives asymptotic confidence intervals for $\beta/(\alpha+\beta)$ based on a modified signed log-likelihood-ratio statistic. An alternative approach would be one which utilizes the pivotal quantity $[\alpha/\hat{\alpha}]/[\beta/\hat{\beta}]$, which has an $F_{n-1,m-1}$ distribution, to construct a confidence interval for α/β, from which a confidence interval for $\beta/(\alpha+\beta)$ is readily derived. Some discussion of point estimation of $P(Y_1 < X_1)$ may be found at the end of Section 5.6.*

3. Ghosh and Wackerly (1986) discuss sequential fixed width confidence intervals for σ in the Pareto (I) case with α unknown.

5.4 Parametric Hypotheses

Suppose that we have available a sample $X_1, \ldots, X_n$ from some parametric family of generalized Pareto distributions. To test hypotheses in these settings generalized likelihood ratio tests are often the only available tool. Muniruzzaman's (1957) early paper and Iliescu and Viorel's (1974) more recent note are the only publications which specifically address the problem. Both these contributions dealt with the case of a classical Pareto distribution (Iliescu and Viorel assumed a unit scale parameter). Naturally, their results are parallel to well-known results for testing hypotheses regarding the scale parameter of an exponential, since the problem can be put in an exponential setting by taking logarithms. We may summarize Muniruzzaman's results as follows.

(i) $H : \alpha = \alpha_0$. Assume $X_1, X_2, \ldots, X_n$ are independent identically distributed $P(I)(\sigma, \alpha)$ variables, and we wish to test $\alpha = \alpha_0$ where α_0 is a specified real value. When H is true, we find

$$S_1 = 2\alpha_0 n / \hat{\alpha} \sim X^2_{2n-2} \tag{5.4.1}$$

where $\hat{\alpha}$ is the maximum likelihood estimate of α (see (5.2.3)). The hypothesis $\alpha = \alpha_0$ will be rejected, if S_1 is too large or too small (assuming an alternative of the form $\alpha \neq \alpha_0$).

(ii) $H : \alpha_1 = \alpha_2$. Assume $X_1, X_2, \ldots, X_{n_1}$ is a sample from a $P(I)(\sigma_1, \alpha_1)$ population while $Y_1, Y_2, \ldots, Y_{n_2}$ is an independent sample from a $P(I)(\sigma_2, \alpha_2)$ population. We wish to test $H : \alpha_1 = \alpha_2$ against $A : \alpha_1 \neq \alpha_2$. If we let $\hat{\alpha}_1$ and $\hat{\alpha}_2$ be the corresponding maximum likelihood estimates of α_1 and α_2 (obtained from (5.2.3)), we find that

$$S_2 = [n_1(n_2 - 1)\hat{\alpha}_2] / [n_2(n_1 - 1)\hat{\alpha}_1] \sim F_{2n_1-2, 2n_2-2}, \tag{5.4.2}$$

and we reject H if S_2 is too large or too small.

(iii) $H : \alpha_1 = \alpha_2 = \ldots = \alpha_K$. Here a generalized likelihood ratio test is readily derived. The generalized likelihood ratio statistic is of the form

$$\lambda = \prod_{j=1}^{k} (\hat{\alpha}_p / \hat{\alpha}_j)^{(n_j - 1)} \tag{5.4.3}$$

where $\hat{\alpha}_p$ is the maximum likelihood estimate of α, if all α_i's are assumed to be equal to α. Note that the parameters $\sigma_1, \sigma_2, \ldots, \sigma_k$ for the k populations need not be the same. The test proposed by Muniruzzaman in this setting is not quite the generalized likelihood ratio test derived from (5.4.3), but it is asymptotically equivalent. The problem at hand is essentially the same as that faced when testing for homogeneity of variances in an analysis of variance setting. Several modifications (e.g., Bartlett's test) of the corresponding generalized likelihood ratio test have been proposed. Analogous modifications can be considered for the present problem.

Chaouche and Bacro (2004) provide a suggested approach to the problem of testing $H : \alpha = \alpha_0, \sigma > 0$ based on a sample of size n from a Pareto (II)$(0, \sigma, \alpha)$ distribution. Let $T = T(X_1, X_2, ..., X_n)$ be a scale invariant statistic based on the sample, and let t be the observed value of T. The significance level of the observed value relative to H is then approximated by simulating a large number of realizations of T where the X_i's have a Pareto (II)$(0, 1, \alpha_0)$ distribution, and observing the relative position of t among these simulated values. Chaouche and Bacro provide suggestions for selecting the form of the scale invariant statistic T to be used in this process. To test $H : \alpha = \alpha_0, \mu \in (0, \infty), \sigma > 0$ based on a sample of size n from a Pareto (II)(μ, σ, α) distribution, an analogous procedure could be followed utilizing a location and scale invariant statistic T.

For testing parametric hypotheses based on samples from generalized Pareto distributions (i.e., P(II), P(III), P(IV) and Feller-Pareto), the lack of a simple sufficient statistic has hindered the development of tests. Essentially two routes are open. Both are asymptotic, and, in fact, when both techniques are available, they usually lead to equivalent approximate tests. The first is the generalized likelihood ratio test. For it, numerical methods are usually required to obtain maximum likelihood estimates of the unknown parameters when H is true and when the parameter space is unrestricted. If our hypothesis is that a certain parameter (or group of parameters) has a specified value, we may consider a test of size α which rejects the hypothesis if a $100(1-\alpha)\%$ confidence interval (or region) fails to cover the hypothesized parametric value(s). Confidence intervals and regions, which can be used in such procedures, were described in Section 5.3. It will be recalled that for generalized Pareto distributions, only asymptotic approximate confidence intervals are usually available.

5.5 Tests to Aid in Model Selection

A basic problem is that of deciding which, if any, of the available generalized Pareto families of distributions adequately describe a given data set. Various standard procedures are certainly available. If we wish to test whether, for example, a Pareto (IV) model is adequate against an unspecified alternative, we could use a Kolmogorov-Smirnov test with four parameters numerically estimated from the data, or we could use a chi-squared test with parameters estimated from the data. The usual caveats regarding the power of such procedures are of course in order. Alternatively, a Cramer-von Mises test or an Anderson-Darling test with estimated parameters can be used to test goodness of fit for generalized Pareto models. Choulakian and Stephens (2001) consider such tests for the Pareto (II)$(0, \sigma, \alpha)$ model. Several authors have considered the development of Neyman smooth tests for the Pareto (II)$(0, \sigma, \alpha)$ model (usually in the Pickands generalized Pareto context). In this approach, the Pareto model is embedded in a suitably chosen multiparameter exponential family of densities. Details of this procedure, including some discussion of simulation based results on the power of such tests against various alternatives, may be found, for example, in Radouane and Cretois (2002), Falk, Guillou and Toulemonde (2008) and De Boeck, Thas, Rayner and Best (2011).

Gulati and Shapiro (2008) consider goodness of fit for the Pareto(I) and Pareto (II) models. The approach that they suggest involves first obtaining numerically the maximum likelihood estimates of some or all of the parameters of these models, and then, acting as if these were the true values, transforming the data to conform with well-known distributions such as exponential and uniform. Several specialized goodness of fit tests are then used to assess the fit of the transformed data. Simulation based evaluations of the power of these tests are provided for several alternative distributions.

If we wish to test whether, for example, a Pareto (II) model is adequate, having accepted that a Pareto (IV) model is appropriate, we can use a generalized likelihood ratio test.

Kalbfleisch and Prentice (1980, pp. 63–67) describe such generalized likelihood ratio tests appropriate for their class of generalized F distributions (i.e., Feller-Pareto distributions with $\mu = 0$).

In addition to the above general techniques several tests, which focus on qualitative aspects of the data set, e.g., heavy tails, have been proposed for use in determining whether a Pareto model is appropriate. Sometimes a specific alternative, such as lognormal, is visualized, while in other cases the qualitative test is proposed to take on all comers. Before turning to this list of specialized tests, we will mention one omnibus procedure proposed by Quandt (1966) as an alternative to Kolmogorov-Smirnov and chi-squared. We will describe it for testing the adequacy of a $P(IV)(0, \sigma, \alpha, \gamma)$ model, although the procedure is quite general.

Suppose $X_{1:n}, X_{2:n}, \ldots, X_{n:n}$ represents the order statistics based on a sample of size n and that we wish to test the hypothesis that the parent distribution is $P(IV)(0, \sigma, \alpha, \gamma)$ where σ, α and γ are unspecified. Denote the $P(IV)(0, \sigma, \alpha, \gamma)$ distribution function by $F(x, \sigma, \alpha, \gamma)$. Quandt proposes considering the statistic

$$S = \inf_{\sigma,\alpha,\gamma} \sum_{i=1}^{n+1} [F(X_{i:n}; \sigma, \alpha, \gamma) - F(X_{i-1:n}; \sigma, \alpha, \gamma) - (n+1)^{-1}]^2 \tag{5.5.1}$$

where $X_{0:n} = 0$ and $X_{n+1:n} = \infty$. A small value of S will render the hypothesis plausible that indeed the parent distribution is $P(IV)(0, \sigma, \alpha, \gamma)$. A small Monte Carlo study was conducted to obtain estimated critical values of the S statistic when testing whether the parent distribution is classical Pareto. The critical values for S when the parent distribution is Pareto (IV) have not been determined. It is not superficially evident that a procedure such as Quandt's is superior to, say, Kolmogorov-Smirnov, but then neither is it evidently inferior. The test is recognizable as a "spacings" test in the sense of Pyke (1965) with parameters estimated from the data. The corresponding test when parameters are known and not estimated from the data was attributed by Pyke to Irwin and Kimball.

Gastwirth and Smith (1972) proposed a test for model adequacy based on Gastwirth's (1972) Gini index bounds. Suppose that the data are grouped into intervals, as is often the case. Using the results summarized in Theorems 4.2.1 and 4.2.3, it is possible to derive upper and lower bounds for the ungrouped sample Gini index. The bounds in question involve the group means which might not be available. To test the adequacy of, say, a $P(III)(0, \sigma, \gamma)$ model, we would estimate γ by maximum likelihood or least squares from the grouped data assuming the Pareto (III) model and compare this value (which is the estimated Gini index) with the upper and lower bounds computed from the grouped data and group means. If the resulting estimated Gini Index is far outside the bounds, we will have reason to doubt the adequacy of the Pareto (III) model. Gastwirth and Smith (1972) illustrated this technique using Internal Revenue Service data. As mentioned in Chapter 1, the results of this analysis were not supportive of either a log-normal or a $P(III)(0,\sigma,\gamma)$ (or sech^2) model.

Bryson (1974) addressed the problem of testing for an exponential distribution when the alternative distribution is $P(II)(0,\sigma,\alpha)$. He was unable to derive a uniformly best test, but he did propose, on heuristic grounds, a statistic designed to be sensitive to the heavy tail of the Pareto (II) distribution. In fact, he proposed it as a test to be used when the alternative was any distribution with increasing mean residual life. It may be recalled that the Pareto (II) distribution has linearly increasing mean residual life (see (3.5.4)). Bryson's statistic based on a sample $X_1, X_2, \ldots, X_n$ is of the form

$$T = X_{1:n}\left(\sum_{i=1}^{n} X_i\right)\left[\prod_{i=1}^{n}(X_i + \hat{A})\right]^{-2/n} / [n(n-1)] \tag{5.5.2}$$

where $\hat{A} = X_{n:n}(n^{1/\alpha_0} - 1)^{-1}$ in which α_0 is chosen by the experimenter. The test is designed to be an approximation to the most powerful test of exponentiality against the alternative that the distribution is $P(II)(0,\sigma,\alpha_0)$. The hypothesis of exponentiality is to be rejected if T is too large. Since the exponential distribution is the limiting distribution encountered if in a $P(II)(0,\sigma,\alpha)$ distribution we let $\alpha \to \infty$, another potential use of Bryson's statistic becomes apparent. It could be used to test $H : \alpha = \alpha_0$ (or $\alpha \le \alpha_0$) against $A : \alpha > \alpha_0$. Here one would reject for small values of T. Bryson simulated the distribution of T when the underlying distribution is exponential, to determine appropriate critical values. If it is to be used for testing $\alpha = \alpha_0$ against $\alpha > \alpha_0$ in a $P(II)(0,\sigma,\alpha)$ setting, the distribution of T when $\alpha = \alpha_0$ will need to be simulated. There are other available tests for exponentiality against increasing mean residual life alternatives. They too could be used to test one-sided hypotheses about the parameter α in a Pareto (II) population.

One of the common defenses for studying characterizations is that they may prove useful in providing tests of model acceptability. As we saw in Section 3.7, most of the available Pareto characterizations deal with the Pareto I and II families. As a consequence, the associated tests can often be translated into tests of exponentiality based on exponential characterizations. A convenient starting point in tracking down available results of that type is provided by Galambos and Kotz's (1978) monograph on characterizations. In the present section attention will be focused on tests based on characterizations of other generalized Pareto families. Very little work has been done here, undoubtedly because of the paucity of characterizations available. Arnold and Laguna (1976, 1977) report on a preliminary study of a test based on the Pareto (III) characterization involving geometric minimization.

The result in question (described at the end of Section 3.7) states that, under a mild regularity condition ($\lim_{x\to 0} x^{-\lambda}F(x) = \eta > 0$), if a geometric minimum of i.i.d. X_i's has, up to a change of scale, the same distribution as the individual X_i's, then the common distribution of the X_i's must be $\text{P(III)}(0,\sigma,\gamma)$ (or log-logistic). To test whether a $\text{P(III)}(0,\sigma,\gamma)$ model is acceptable, we divide the data into two parts. The first contains $n/3$ observations. The remaining $2n/3$ observations are used to construct by (simulating) tossing a coin another sample of approximately $n/3$ geometric minima. The two samples of size $n/3$ should have, under the null hypothesis, distributions which differ only by scale. If we take logarithms and then center at medians, a standard non-parametric test such as Kolmogorov-Smirnov can be used to determine whether, indeed, the identity of distributions up to scale is plausible. If the Kolmogorov-Smirnov test results in rejection, then we reject the original hypothesis that the data came from a $\text{P(III)}(0,\sigma,\gamma)$ population. A small simulation of such a test procedure was reported in Arnold and Laguna (1976, 1977). The test does seem to be approximately unbiased, but its power (against uniform and exponential alternatives) is not as high as one might hope.

Arnold and Laguna (1977) discussed another more ad hoc test of the hypothesis that a data set is from a $\text{P(III)}(0,\sigma,\gamma)$ population. The test was based on the observation that in such a population the parameter γ has a dual role. It is both a shape parameter and, at the same time, the Gini index. Suppose that $X_1,X_2,\ldots,X_n$ are i.i.d. $\text{P(III)}(0,\sigma,\gamma)$, then $X_1^\delta,X_2^\delta,\ldots,X_n^\delta$ will be i.i.d. $\text{P(III)}(0,\sigma,\delta\gamma)$. The parameter γ can be consistently estimated by the sample Gini index G_n (4.3.42), provided $\gamma<1$. Consequently, if we define

$$Y_i = X_i^{1/2G_n}, \quad i=1,2,\ldots,n, \tag{5.5.3}$$

we can expect that the Y_i's will behave much like a sample of size n from a $\text{P(III)}(0,\sigma,1/2)$ population. So the Y_i's should have a corresponding sample Gini index denoted by $G_n^*(1/2)$, computed using (4.3.42), that is quite close to $1/2$. Discrepancy between $G_n^*(1/2)$ and $1/2$ is evidence against a P(III) model.

Of course, there is no a priori reason to use the value $1/2$ in the above discussion. For any $\eta \in (0,1)$ we can define

$$Y_i(\eta) = X_i^{\eta/G_n}, \quad i=1,2,\ldots,n, \tag{5.5.4}$$

and then base our test on the distance between $G_n^*(\eta)$, the sample Gini index computed using the $Y_i(\eta)$'s, and the value η. Some Monte Carlo results described in Arnold and Laguna (1977) suggest that even for samples of size 100 from Pareto (III) populations, the values of $G_n^*(1/2)$ and $G_n^*(3/4)$ may be surprisingly far from 1/2 and 3/4, respectively. Further investigation of the operating characteristics of tests based on $|G_n^*(\eta)-\eta|$ for different values of η might identify a better choice for η.

Sometimes a test which arises in an ad hoc fashion by focusing on some special feature of the hypothesized family will turn out to be quite good. An example is provided by the following test for the classical Pareto distribution. Suppose that $X_1,X_2,\ldots,X_n$ are i.i.d. $P(I)(\sigma,\alpha)$. To simplify matters, assume that σ is known (we would estimate it by $X_{1:n}$ in any event) and without loss of generality $\sigma=1$. Let $Z_i=\log X_i$, then the Z_i's are exponentially distributed. An exponential distribution has the unusual property that its variance is always the square of its mean. If we denote the sample mean and variance of the Z_i's by $\overline{Z}$ and S^2, then a plausible test of the exponentiality of the Z_i's (and thus a test of the Paretian model for the X_i's) would reject for large or small values of

$$U = [1 - (S^2/\overline{Z}^2)]. \tag{5.5.5}$$

DuMouchel and Olshen (1975) showed that a one-sided test (rejecting for large values) based on U is asymptotically locally most powerful for testing for the exponential distribution against alternatives with "normal" tail behavior, i.e., distributions of the form

$$F(x) = 1 - e^{-\alpha x - \tau x^2} \tag{5.5.6}$$

where $\tau > 0$. The case $\tau = 0$ corresponds to the exponential distribution. If this test is applied to the Z_i's regarded as logarithms of X_i's, then we are testing a classical Pareto hypothesized distribution for the X_i's against an alternative which mimics the log-normal distribution in tail behavior. DuMouchel and Olshen developed their test in the context of insurance claims distributions. The family of alternatives they considered is perhaps a little contrived, but it is interesting that a test based on such a simple statistic as (5.5.5) can be reasonably expected to be quite good against log-normal alternatives. Perhaps similar claims can be made for other tests, such as those based on characterizations which focus on specific properties of the classical Pareto distribution.

Recently, several interesting alternative approaches have been suggested for the problem of testing goodness of fit to either the Pareto (I) or Pareto (II) models. Meintanis (2009) considers Pareto (I)(σ, α) goodness of fit. He observes that if X has a Pareto (I) distribution, then $E(X^{-t})$ is finite for every $t > 0$. This is the Laplace transform of $\log X$. He then defines

$$M(t) = E(X^{-t}) = \sigma^{-t}\left(1 + \frac{t}{\alpha}\right)^{-1} \tag{5.5.7}$$

and a corresponding empirical function

$$M_n(t) = \frac{1}{n}\sum_{i=1}^{n} X_i^{-t}. \tag{5.5.8}$$

Consequently the following related function is always equal to 0, i.e.,

$$D(t) = (\alpha + t)\sigma^t M(t) - \alpha = 0, \quad t > 0. \tag{5.5.9}$$

If α is fixed, say equal to α_0, and we wish to test $H : X_i \sim P(I)(\sigma, \alpha_0)$ based on a sample $X_1, X_2, ..., X_n$, Meintanis proposes that we estimate σ using $X_{1:n}$ and define

$$D_n(t, \alpha_0) = (\alpha_0 + t)\widetilde{M}_n(t) - \alpha_0, \tag{5.5.10}$$

where

$$\widetilde{M}_n(t) = \frac{1}{n}\sum_{i=1}^{n}\left(\frac{X_i}{X_{1:n}}\right)^{-t}.$$

The test for H will then be of the form: Reject H for large values of

$$P_{n,w}(\alpha_0) = n\int_0^\infty D_n^2(t,\alpha_0)w(t)dt, \qquad (5.5.11)$$

where $w(t)$ is a suitable non-negative weight function. Meintanis suggests use of a weight function of the form $w(t) = e^{-at}$, for some positive a.

If we wish to test $H : X_i \sim P(I)(\sigma, \alpha)$ where both σ and α are unspecified, it will be necessary to estimate α in (5.5.11).

Rizzo (2009) suggested a selection of goodness-of-fit tests based on the following result.

Lemma 5.5.1. *If $\underline{X}$ and $\underline{Y}$ are independent k-dimensional random variables with $E(\|\underline{X}\|^\beta + \|\underline{Y}\|^\beta) < \infty$ for some $\beta \in (0,2)$, then*

$$Q = 2E(\|\underline{X}-\underline{Y}\|^\beta) - E(\|\underline{X}-\underline{X}'\|^\beta) - E(\|\underline{Y}-\underline{Y}'\|^\beta) \geq 0,$$

with equality if and only if $\underline{X} \stackrel{d}{=} \underline{Y}$. Here $\underline{X}'$ is an independent copy of $\underline{X}$ and $\underline{Y}'$ is an independent copy of $\underline{Y}$.

To test $H : X_i \sim P(I)(\sigma_0, \alpha_0)$, we consider an empirical version $\widetilde{Q}$ of the one-dimensional form of Q (i.e., with $k = 1$), in which the distribution of Y is taken to be the empirical distribution function corresponding to the sample $X_1, X_2, ..., X_n$. The hypothesis is rejected if $\widetilde{Q}$ is large. To test goodness of fit to a Pareto (I) model with σ and α unspecified, a version of $\widetilde{Q}$ involving estimated values of σ and α is used. Modifications of this procedure can be used to test goodness of fit to more general Pareto models.

Saldaña-Zepeda, Vaquera-Huerta and Arnold (2010) discuss a Pareto (I) goodness of fit test based on samples that are subject to Type II right censoring, i.e., only the first r order statistics, $X_{1:n}, X_{2:n}, ..., X_{r:n}$, are available. They transform to a shifted exponential model by taking logarithms. Then they define $Z_{i:r-1} = \log X_{i+1:n} - \log X_{1:n}$, $i = 1,2,...,r-1$. The cumulative hazard function of an exponential distribution is a linear function (i.e., $H(x) = \lambda x$). The proposed test statistic is the sample correlation between the vector of $Z_{i:r-1}$'s and estimated values of the hazard function at the points $Z_{i:r-1}$, $i = 1,2,...,r-1$, i.e., the $\widetilde{H}(Z_{i:r-1})$'s, where $\widetilde{H}$ is a Nelson-Aalen estimate of H. Some simulation evidence regarding the power of this test is provided. In particular, it is shown to have good power against log-normal alternatives.

Villaseñor and Gonzalez (2009) address the question of goodness of fit to a Pareto (II)$(0, \sigma, \alpha)$ model. They observe that, for this model, $[\overline{F}(x)]^{-1/\alpha}$ is a linear function of x. To test the goodness of fit, they consider the sample correlation between $X_{1:n}, X_{2:n}, ..., X_{n:n}$ and $[\overline{F}_n(X_{1:n})]^{-1/\widetilde{\alpha}}, [\overline{F}_n(X_{2:n})]^{-1/\widetilde{\alpha}}, ..., [\overline{F}_n(X_{n:n})]^{-1/\widetilde{\alpha}}$ where $\widetilde{\alpha}$ is a suitable estimate of α, and reject the hypothesized Pareto (II) model if this sample correlation deviates considerably from 1.

5.6 Specialized Techniques for Various Data Configurations

This section is a brief potpourri of inferential procedures related to Pareto distributions. The only qualification for inclusion is that the material in question do not seem to fit in any other section of the chapter. In many cases the motivation for developing the procedure was not related to income distributions, and the utility of the procedures in such a context is not always apparent.

Mukhopadhyay and Ekwo (1987) consider sequential estimation of the parameters of a Pareto (I)(σ, α) model. Both scale and shape parameters are considered, assuming a loss function which incorporates a cost for each observation taken. An alternative approach involving a three stage procedure is described in Hamdy and Pallotta (1987). Recently there has been a revival of interest in an alternative multi-stage estimation strategy, one that incorporates preliminary hypothesis test(s). Early work on preliminary test estimation was due to Bancroft (1944). In this approach, the estimation strategy to be used depends on the result of a hypothesis test that is carried out using the same data set that is subsequently used to estimate the parameter(s). A typical example begins with a guessed value of the parameter of interest. For example, with data from a Pareto (I)(σ, α) distribution with both parameters unknown, we may be interested only in the value of the shape parameter α. A prior guessed value of α, say α_0, is utilized. We first test the hypothesis that $\alpha = \alpha_0$. If it is accepted, then we naturally use α_0 as our estimate of α. If the hypothesis is rejected, then the maximum likelihood estimate of α is used. Let $\widetilde{\alpha}$ denote the preliminary test estimate and let $\hat{\alpha}$ denote the maximum likelihood estimate. The relative efficiency of the preliminary test estimate is then given by

$$R(\alpha) = \frac{E_{\sigma,\alpha}[(\hat{\alpha} - \alpha)^2]}{E_{\sigma,\alpha}[(\widetilde{\alpha} - \alpha)^2]}.$$

In a few cases, preliminary test estimates can be preferred to maximum likelihood estimates in the sense that their relative efficiency is uniformly greater than 1. For example, such was shown to be the case when estimating a normal variance with unknown mean (Davis and J. C. Arnold, 1970). However, more typically the relative efficiency of the preliminary test estimate only exceeds 1 in a neighborhood of the guessed value of the parameter, i.e., if the guess was a good one. In the Pareto (I)(σ, α) context, Singh, Prakash and Singh (2007) consider a preliminary test estimate of α using a prior guessed value of α. They utilize a LINEX loss function rather than squared error loss, but it is still true that the preliminary test estimate has smaller risk than the maximum likelihood estimate only for values of α in a neighborhood, albeit a large neighborhood, of the guessed value α_0. Baklizi (2008) considers squared error loss in this example and makes a recommendation about the size of the preliminary test to be used. In the Pareto (I) setting, Prakash, Singh and Singh (2007) discuss estimation of σ using a prior guessed value α_0.. Here, too, a reduced mean squared error is only encountered if α/α_0 is in a neighborhood of 1. Kibria and Saleh (2010) consider estimation of α based on a doubly censored Pareto (I)(σ, α) sample, $X_{r+1:n}, X_{r+2:n}, \ldots, X_{n-s:n}$, utilizing a preliminary test of $H : \alpha = \alpha_0$. They also consider estimation of σ. If one is considering using such a preliminary test estimate, perhaps because its mean squared error is well behaved in an interval $(\alpha_0 - \delta, \alpha_0 + \delta)$, it is recommended that the preliminary test estimate be compared with a more standard estimate which is obtained assuming that indeed the parameter α is restricted to the interval $(\alpha_0 - \delta, \alpha_0 + \delta)$. If the preliminary test estimate still outperforms the restricted parameter space estimate uniformly on $(\alpha_0 - \delta, \alpha_0 + \delta)$, then its use can be recommended, provided that you are convinced that only values of α satisfying $|\alpha - \alpha_0| < \delta$ will be encountered in practice.

Kaminsky and Nelson (1975) considered best linear unbiased prediction of $X_{k:n}$ based on the first i order statistics $(i < k)$ corresponding to a sample from a $P(II)(\mu, \sigma, \alpha)$, in which α and μ are assumed to be known. Without loss of generality, we assume $\mu = 0$. The coefficients for the best linear predictor are computable in terms of the variances and covariances of the order statistics. If i, k and n are large, then considerably simpler asymptotic approximations were provided by Kaminsky and Nelson. In such a case the best linear unbiased predictor of $X_{k:n}$, denoted by $\hat{X}_{k:n}$, is well approximated by

$$\hat{X}_{k:n} \doteq qX_{i:n} - (l - q)\hat{\sigma} \tag{5.6.1}$$

where $q = [(n-1)/(n-k)]^{1/\alpha}$ and $\hat{\sigma}$ is the best linear unbiased estimate of σ based on $(X_{1:n}, X_{2:n}, \ldots, X_{i:n})$. The exact form of $\hat{\sigma}$ is not simple. One finds

$$\hat{\sigma} = \{(\alpha+1)\sum_{j=1}^{i-1} B_j X_{j:n} + [(n-i+1)\alpha - 1]B_i X_{i:n}\}/K_i \tag{5.6.2}$$

in which

$$K_i = n\alpha - 2 - (\alpha+1)\sum_{j=1}^{i-1} B_j - [(n-i+1)\alpha - 1]B_i \tag{5.6.3}$$

$$\text{and } B_j = \frac{\Gamma(n-j+1)\Gamma(n+1-2\alpha^{-1})}{\Gamma(n-j+1-2\alpha^{-1})\Gamma(n+1)}, \quad j = 1, 2, \ldots, n. \tag{5.6.4}$$

Kaminsky and Nelson pointed out that the above analysis extends readily in principle to the problem of predicting several or a linear combination of several subsequent order statistics based on a subset of the first i order statistics. No change in strategy is required either, if we wish to consider the cases (i) σ known, μ unknown and (ii) μ and σ both unknown.

In a later contribution, Kaminsky and Rhodin (1978) discussed the loss of information when all but the latest failure is discarded in a Pareto (IV) setting (they considered other distributions, as well). Thus, we assume we have order statistics $X_{1:n}, X_{2:n}, \ldots, X_{i:n}$ available from a $P(IV)(\mu, \sigma, \alpha, \gamma)$ population where we assume μ,α and γ are known. The task at hand is to predict $X_{k:n}$ using an asymptotically best unbiased predictor based either on $X_{i:n}$ alone or the full set of observations up until the ith, i.e., $X_{1:n}, X_{2:n}, \ldots, X_{i:n}$. Obviously the estimates using more information will be more efficient. The question addressed by Kaminsky and Rhodin was whether one can put a bound on the loss of efficiency associated with the use of the latest failure ($X_{i:n}$) only. They showed that the estimate based on the latest failure ($X_{i:n}$) is always at least 96.58% efficient (asymptotically) relative to the estimate based on all of the first i order statistics. Since we know α (which is Kaminsky and Rhodin's γ), we can actually state a little more. If $\alpha = 2$, then the asymptotic efficiency is at least 98.01%. If $\alpha = 0.5$, then the lower bound on the asymptotic efficiency becomes 98.70%. Kaminsky and Rhodin (1978) provide a short table of such results. Results such as these are comforting when sample sizes are large and storage of all past failure times becomes a problem.

Doubly censored samples from a Pareto (II)$(0, \sigma, \alpha)$ distribution are sometimes encountered. In this formulation we observe only $X_{r+1:n}, X_{r+2:n}, \ldots, X_{n-s:n}$ because the r smallest and the s largest observations are unavailable. The relevant likelihood function is of the form

$$\left[1-\left(1+\frac{x_{r+1:n}}{\sigma}\right)^{-\alpha}\right]^{r}\left(1+\frac{x_{n-s:n}}{\sigma}\right)^{-\alpha s}\left(\frac{\alpha}{\sigma}\right)^{n-s-r}\prod_{j=r+1}^{n-s}\left(1+\frac{x_{j:n}}{\sigma}\right)^{-(\alpha+1)},$$

which is somewhat more complicated than the likelihood would be if $r = 0$. In the case in which $r = 0$, the likelihood is analogous in form to the likelihood of a random sample of size $n - s$ from the Pareto (II) population and the analysis is no more difficult than it was in the simple random sample case. Wu, Lee and Chen (2006, 2007b) have addressed the problem of estimating σ for such censored data sets assuming that α is known. They derive weighted moment estimates and also consider prediction of $X_{n-s+k:n}$. A more complicated model was investigated by Abdel-Ghaly, Attia and Aly (1998). They assume that several censored samples are collected at different stress levels. An accelerated failure model is assumed relating the corresponding shape parameters (the α's) to the various stress levels via an inverse power law. Maximum likelihood estimates of the model parameters are obtained numerically.

W. T. Huang (1974) discussed selection procedures for classical Pareto populations. He assumed observations are available from k different classical Pareto populations $\pi_1, \pi_2, \ldots, \pi_k$. The distribution associated with population π_i is assumed to be $P(I)(\sigma, \alpha_i)$. Observe that a common unknown value of σ is assumed for all k populations. We wish to select the population which has the largest (smallest) value of α associated with it. A similar problem is that of selecting a subset of the k populations such that the probability that the subset includes the population with the largest (smallest) α_i is at least as large as some prescribed number. Suppose that for some $s < 1$, $\alpha_{[1]} \leq s\alpha_{[2]}$ where $(\alpha_{[1]}, \ldots, \alpha_{[k]})$ denotes the α_i's listed in increasing order. This is known as an indifference zone formulation. A correct selection occurs, if $\pi_{[1]}$ is selected. Selection problems of this type for gamma populations were discussed by Bechofer, Kiefer and Sobel (1968). Their results can be utilized, if we make the usual logarithmic transformation which takes our $P(I)(\sigma, \alpha_i)$ variables into translated exponential (α_i) variables. If we denote the order statistics from the jth population by $X_{i:n}^{(j)}$, then the appropriate selection rule would be:

$$\text{Select } \pi_p \text{ if and only if } \prod_{i=1}^{n}\left(\frac{X_{i:n}^{(p)}}{X_{1:n}^{(p)}}\right) = \max_{1\leq r\leq k}\prod_{i=1}^{n}\left(\frac{X_{i:n}^{(r)}}{X_{1:n}^{(r)}}\right). \tag{5.6.5}$$

W. T. Huang assumed that the π_i's share a common value of σ. The rule (5.6.5) is actually appropriate even when different values of σ are associated with the various populations. A better rule when σ is common to all π_i's is provided by

$$\text{Select } \pi_p \text{ if and only if } \prod_{i=1}^{n} X_{i:n}^{(p)} = \max_{1 \le r \le k} \prod_{i=1}^{n} X_{i:n}^{(r)}. \tag{5.6.6}$$

The rule (5.6.6) increases the effective sample size by 1 from that associated with (5.6.5). W. T. Huang (1974) presented analogous material for the subset selection formulation of the problem.

In the setting described by W. T. Huang (1974), we may consider the problem of estimating the common scale parameter, usually but not always assuming that the shape parameters are unknown. The data then consist of $\sum_{i=1}^{k} n_i = n$ independent observations $\{X_{ij}\}_{i=1,j=1}^{k,n_i}$ where $X_{ij} \sim P(I)(\sigma, \alpha_i)$, $i = 1,2,...,k$, $j = 1,2,...,n_i$. Dey and Liu (1992) using weighted squared error loss, i.e., $L(\hat{\sigma}, \sigma) = (\hat{\sigma} - \sigma)^2/\sigma^2$, show that a modified maximum likelihood estimator of σ defined by

$$\tilde{\sigma} = \hat{\sigma}[1 - (\sum_{i=1}^{k} c_i \hat{\sigma}_i)^{-1}] \tag{5.6.7}$$

has smaller risk than the maximum likelihood estimate $\hat{\sigma}$, provided that $c_i \ge n_i/2$, $i = 1,2,...,k$. Dey and Liu also develop an estimate of σ which dominates the maximum likelihood estimate in terms of Pitman closeness. Subsequently, Elfessi and Jin (1996) consider a slightly more general loss function $L(\hat{\sigma}, \sigma) = (\hat{\sigma} - \sigma)^{2r}/\sigma^{2r}$ where r is a positive integer, and identify improved estimates of the form

$$\tilde{\tilde{\sigma}} = \hat{\sigma}\left[1 - \frac{c}{g(\sum_{i=1}^{k} c_i \hat{\alpha}_i)}\right], \tag{5.6.8}$$

for suitable choices of c, the function g and the constants c_i, $i = 1,2,...,k$. The same problem, but this time involving Type II censored samples, was considered by Jin and Elfessi (2001) using the loss function $L(\hat{\sigma}, \sigma) = (\hat{\sigma} - \sigma)^2/\sigma^2$. In this case the data consistof order statistics $X_{1:n_i}, X_{2:n_i}, ..., X_{r_i:n_i}$ from a Pareto (I)(σ, α_i) population, $i = 1,2,...,k$. A class of improved estimates of the form (5.6.8) is identified in this setting also (note that the maximum likelihood estimates now are based on the censored samples).

Kumar and Gangopadhyay (2005) considered a related estimation-after-selection problem for Pareto (I) populations using a simple selection rule. They consider k different Pareto (I) populations $\pi_1, \pi_2, \ldots, \pi_k$. The distribution associated with population π_i is Pareto (I)(σ_i, α). Random samples of size n are taken from each population and the corresponding sample minima are denoted by $X_i = \min\{X_{i1}, X_{i2}, ..., X_{in}\}$, $i = 1,2,...,k$. The population corresponding to the largest X_i is selected and its corresponding scale parameter is to be estimated. Kumar and Gangopadhyay derive the mean squared error of the UMVUE of the selected scale parameter under the assumption that α is known.

Garren, Hume and Leppert (2007) consider data which consists of $n_1 + n_2$ independent X_{ij}'s with $X_{ij} \sim P(I)(\sigma_i, \alpha_{ij})$, $i = 1,2;\ j = 1,2,...,n_i$. It is assumed that the α_{ij}'s are known and that σ_1, σ_2 are unknown except that

$$\sigma_1 \leq k\sigma_2 \quad \text{where} \quad k = \left(\frac{\alpha_1 + \alpha_2 - 2}{\alpha_1 + \alpha_2 - 1}\right)\left(\frac{\alpha_1 - 1}{\alpha_2 - 2}\right), \tag{5.6.9}$$

in which

$$\alpha_i = \sum_{j=1}^{n_i} \alpha_{ij}. \ i = 1,2.$$

The minimal sufficient statistics for (σ_1, σ_2) are $T_i = \min\{X_{i_1}, X_{i,2}, ..., X_{i,n_i}\}$, $i = 1,2$. We have $T_i \sim P(I)(\sigma_i, \alpha_i)$, $i = 1,2$. The focus is on estimation of σ_1. The minimum mean squared error estimate of the form cT_1 for estimating σ_1 is $\widetilde{\sigma} = (\alpha_1 - 2)T_1/(\alpha_1 - 1)$. Garren, Hume and Leppert verify that this estimate is dominated by an isotonic regression estimate of the form

$$\widetilde{\sigma}_I = \min\left\{\frac{\alpha_2 - 2}{\alpha_2 - 1}T_1, \frac{\alpha_1 + \alpha_2 - 2}{\alpha_1 + \alpha_2 - 1}T_2\right\}, \tag{5.6.10}$$

under the restriction (5.6.9). The order restriction (5.6.9) is contrived to ensure the mean squared error domination of (5.6.10) over the estimate $\widetilde{\sigma}$. However, it should be noted that if $\sigma_1 \leq \sigma_2$ (a more natural condition), then (5.6.9) will hold. A parallel analysis involving less assumed knowledge of the α_{ij}'s would be interesting.

Drane, Owen and Seibert (1978) pointed out the possibility of using the Burr distribution (i.e., Pareto (IV)) as a response function in quantal response experiments. They assumed that α and γ are known and suggested use of approximate minimum chi-squared estimates of μ and σ. The approach is parallel to that introduced by Berkson (1944) with "Burrits" playing the role of Berkson's well-known logits.

Several authors have discussed estimation and other inference procedures for Pareto distributions based on data configurations distinct from random samples from the distribution in question. However, as was remarked on in the discussion of Bayesian inference, the likelihood functions in these non-standard situations are only minimally different from the random sample likelihood and inference generally proceeds mutatis mutandis. Researchers have generally focused on Pareto (I) and (II) models. Our discussion will be of the Pareto (II) model. The corresponding Pareto (I) results correspond to a special case and are often somewhat simpler to deal with, as we have seen in earlier sections of this chapter.

We begin with a quite general case. Suppose that the available data consist of a set of generalized order statistics (in the sense of Kamps (1995)) $X_1, X_2, ..., X_n$ corresponding to a Pareto (II) parent distribution. The corresponding likelihood function is of the form

$$k\left(\prod_{j=1}^{n-1}\gamma_j\right)\left(\prod_{j=1}^{n-1}[\overline{F}(x_j)]^{m_j}f(x_j)\right)[\overline{F}(x_n)]^{k-1}f(x_n)I(x_1<x_2<\cdots<x_n)$$

where

$$\overline{F}(x)=\left[1+\left(\frac{x-\mu}{\sigma}\right)\right]^{-\alpha},\quad x>\mu,$$

and

$$f(x)=\frac{\alpha}{\sigma}\left[1+\left(\frac{x-\mu}{\sigma}\right)\right]^{-(\alpha+1)}I(x>\mu).$$

Comparison of this with the likelihood for a random sample of size n from a Pareto (II)(μ,σ,α) distribution will confirm that both likelihoods are of the same general form

$$L(\mu,\sigma,\alpha)\propto(\alpha/\sigma)^c\prod_{i=1}^{n}\left(1+\frac{x_i-\mu}{\sigma}\right)^{-d_i\alpha}\prod_{i=1}^{n}\left(1+\frac{x_i-\mu}{\sigma}\right)^{-1}I(x_1>\mu),$$

where c and the d_i's are known constants. As a consequence, inference for Pareto (II) generalized order statistics data sets will proceed in a manner parallel to that of inference based on a random sample from a Pareto (II) distribution. Record values and progressive censoring are two of the most commonly discussed variants of the generalized order statistics paradigm. Specific details of estimation, confidence intervals and related inference in such settings may be found in several papers in the literature. A selection of relevant references will provide an opportunity to see the necessary similarities between the results obtained for random samples, record values and progressive censoring cases. Wu (2003) and Kus and Kaya (2007) consider progressive censoring in the Pareto (I) case. Wu and Chang (2003) consider progressive censoring with random removals in a Pareto (I)$(1,\alpha)$ setting. Raqab, Asgharzadeh and Valiollahi (2010) deal with progressive censoring in the Pareto (II)(μ,σ,α) case with α known. Inference based on Pareto (I) record values is discussed in Ahsanullah and Houchens (1989), Raqab, Ahmadi, and Doostparast (2007) and Asgharzadeh, Abdi and Kus (2011). Ahsanullah (1992) considers Pareto (II)(μ,σ,α) record values, as do Abd-El-Hakim and Sultan (2004). Habibullah and Ahsanullah (2000) consider parameter estimates for a Pareto (II)(μ,σ,α) distribution based on generalized order statistics. As indicated above, Habibullah and Ahsanullah's discussion actually covers record values and progressive censoring as special cases. Burkschat (2010) discusses linear estimation and prediction based on Pareto (II) generalized order statistics.

Van Zyl (2012) suggests an alternative approach to parameter estimation for the Pareto (II)(μ,σ,α) distribution. First he observes that for this distribution we have

$$1-[\overline{F}(x)]^{-1/\alpha}=\beta_0+\beta_1 x, \tag{5.6.11}$$

where $\beta_0 = -\mu/\sigma$ and $\beta_1 = 1/\sigma$. For each order statistic, $X_{i:n}$, we then can write (using $\hat{\overline{F}}(X_{i:n}) = i/(n+1)$),

$$1 - \left(\frac{i}{n+1}\right)^{-1/\alpha} = \beta_0 + \beta_1 X_{i:n}. \tag{5.6.12}$$

For a series of different values of α, we estimate β_0 and β_1 by minimizing the sum of absolute deviations between the left and right sides of (5.6.12). Denote these estimates by $\hat{\beta}_0(\alpha)$ and $\hat{\beta}_1(\alpha)$. The value of α is then estimated by maximizing the likelihood $L(\alpha, \hat{\beta}_0(\alpha), \hat{\beta}_1(\alpha))$ by a search over the series of values of α for which the median regression estimates were obtained. Van Zyl recommends this procedure for relatively small sample sizes.

Castillo, Hadi and Sarabia (1998) suggest an approach to parameter estimation that is effective in cases in which an analytic expression is available for the Lorenz curve corresponding to the distribution in question (such as is the case, for example, with a Pareto(I) distribution). Suppose that we have a sample of size n from a distribution with Lorenz curve $L(u; \underline{\theta})$ which depends on a p-dimensional parameter $\underline{\theta}$. Choose p indices, say $1 \leq i_1 < i_2 < \cdots < i_p \leq n$, and simultaneously solve the following system of p equations:

$$\frac{\sum_{j=1}^{i_\ell} X_{j:n}}{\sum_{j=1}^{n} X_{j:n}} = L(i_\ell/n, \underline{\theta}), \;\; \ell = 1, 2, ..., p. \tag{5.6.13}$$

Perform this operation for every (or for many) set of p indices. The coordinatewise median of the $\binom{n}{p}$ estimates of $\underline{\theta}$ obtained in this way will be the final estimate of $\underline{\theta}$. If this approach is applied to a sample from a Pareto (I)(σ, α) distribution, the resulting estimate of α will be

$$\widetilde{\alpha} = \text{median}\{\widetilde{\alpha}_i : i = 1, 2, ..., n\}, \tag{5.6.14}$$

where

$$\widetilde{\alpha}_i = \left[1 - \frac{\log\left(\sum_{j=i+1}^{n} X_{j:n} / \sum_{j=1}^{n} X_{j:n}\right)}{\log\left(\frac{n-i}{n}\right)}\right]^{-1}.$$

Castillo, Hadi and Sarabia provide simulation based evidence that such estimates compare favorably with more traditional alternatives. Note that any estimation strategy based on the Lorenz curve will fail to provide an estimate of the scale parameter of the distribution, since the Lorenz curve is scale invariant.

Chotikapanich and Griffiths (2002) describe an alternative approach to parameter estimation based on analytic Lorenz curves. They assume that the available data consists of m points on the sample Lorenz curve where m exceeds p, the dimension of the unknown parameter vector in the Lorenz curve $L(u; \underline{\theta})$. If the available points on the sample Lorenz curve are

$$\left(\frac{i_\ell}{n}, \frac{\sum_{j=1}^{i_\ell} X_{j:n}}{\sum_{j=1}^{n} X_{j:n}}\right), \quad \ell = 1,2,...,m, \tag{5.6.15}$$

then Chotikapanich and Griffiths define

$$q_\ell = \frac{\sum_{j=i_{\ell-1}+1}^{i_\ell} X_{j:n}}{\sum_{j=1}^{n} X_{j:n}}, \quad \ell = 1,2,...,m,$$

and assume that $(q_1, q_2, ..., q_m)$ has a Dirichlet$(\alpha_1, \alpha_2, ..., \alpha_{m+1})$ distribution with

$$\alpha_\ell \propto L(i_\ell/n; \underline{\theta}) - L(i_{\ell-1}/n; \underline{\theta}), \quad \ell = 1,2,...m+1.$$

Here $\ell_0 = 1$ and $\ell_{m+1} = n$. Under this distributional assumption, the parameters $\underline{\theta}$ can be estimated using maximum likelihood. Since this approach involves an arbitrary and perhaps questionable assumption (involving a Dirichlet distribution), it will require more extensive investigation of its operating characteristics than is provided by Chotikapanich and Griffiths before it can be recommended.

The problem of outliers with Pareto data is not easy to resolve. One formulation, suggested by Dixit and Jabbari-Nooghabi (2011), is as follows. A set of n random variables $(X_1, X_2, ..., X_n)$ are such that k of them (the outliers) have a Pareto (I)$(\beta\sigma, \alpha)$ distribution, where $\beta > 1$. The remaining $n-k$ observations have a Pareto (I)(σ, α) distribution. If the outliers are identified (i.e., we know which ones have a Pareto (I)$(\beta\sigma, \alpha)$ distribution), then estimation of the parameters α, β and σ will be relatively routine. For example, maximum likelihood or the method of moments could be utilized. The problem is more complicated if the outliers are not identified. And, of course, it will be further complicated if k is not known in advance (as would usually be the case). Dixit and Jabbari-Nooghabi consider the perhaps unrealistic case in which σ, β and k are known. In such a situation, the likelihood equation for α is of the form

$$(n/\alpha) + n\log\sigma + k\log\beta - \sum_{i=1}^{n} \log x_i = 0,$$

which is readily solved. Dixit and Jabbari-Nooghabi discuss the distribution of this maximum likelihood estimate and compare its performance, via simulation, with that of the UMVUE of α. The extension to consider cases in which σ, β and/or k are unknown would be interesting if it can be achieved.

If we consider two independent Pareto (I) variables, $X \sim P(I)(\sigma, \alpha)$ and $Y \sim P(I)(\sigma, \beta)$ (with common scale parameter σ), it may be of interest to estimate $P(X < Y)$. However as remarked earlier, this parametric quantity admits a simple description. Using the fact that $\log(X/\sigma)$ (respectively $\log(Y/\sigma)$) has an exponential(α) (respectively exponential(β)) distribution, we may verify that

$$P(X < Y) = \frac{\alpha}{\alpha + \beta}.$$

Suppose that we have available a sample of n i.i.d. copies of (X,Y), i.e., (X_i,Y_i), $i=1,2,...,n$. If σ is known, the complete sufficient statistics for (α,β) are $(Z_X,Z_Y)=(\sum_{i=1}^n \log(X_i/\sigma),\sum_{i=1}^n \log(Y_i/\sigma))$. The random variables Z_X and Z_Y are independent with $Z_X \sim \Gamma(n,1/\alpha)$ and $Z_Y \sim \Gamma(n,1/\beta)$. In this case a series expansion for the UMVUE of $P(X<Y)$ can be obtained (cf., Tong (1974)). If σ is unknown then the complete sufficient statistic for (α,β,σ) is (U,W_X,W_Y) with independent coordinates $U=\min\{X_{1:n},Y_{1:n}\}$, $W_X=\sum_{i=1}^n \log(X_i/U)$ and $W_Y=\sum_{i=1}^n \log(Y_i/U)$. See Rohatgi and Saleh (1987) who consider unbiased estimation of σ in this setting with possibly censored samples. See Remark 2 at the end of Section 5.3 for discussion of related interval estimation of $P(X<Y)$.

5.7 Grouped Data

Typically, in income modeling, the available data are grouped. One approach is to ignore the grouping. If we act as if the original data are ungrouped and apply the inferential techniques discussed in Sections 5.2–5.6, we probably will not make any serious errors. However, it is often not much more difficult to generate specific procedures which take direct account of the discrete nature of our data.

When Champernowne (1952) introduced his family of models for income (see equation (2.11.3)), he proposed seven methods of fitting the distribution to data sets which were clearly grouped. His model included the Pareto (III) family. He eschewed maximum likelihood, preferring to choose a parametric model which would agree with the observed distribution in certain economically important matters such as total income, average income and slope of Pareto line for high incomes. The grouped nature of the data was not really exploited in this analysis. For example, the median income, which is one of the economic attributes to be matched by the fitted distribution, is estimated by "rough interpolation in the table."

Fisk (1961b) discussed maximum likelihood estimation for his sech square distribution (or log-logistic) based on grouped and possibly truncated data sets. He also treated the classical Pareto model from the same viewpoint. For completeness, we will describe the procedure in the context of estimating the parameters of a Pareto (IV) population. We will assume that μ, the location parameter, is known and, without loss of generality, is zero. Therefore, we assume that we have n observations $X_1,X_2,\ldots,X_n$ which are independent and identically distributed according to a $P(IV)$ $(0,\sigma,\alpha,\gamma)$ distribution (see (3.2.4)). It is assumed that the data are grouped. Thus there exist k numbers $x_1<x_2<\ldots<x_k$, and all that is reported about a particular X_i is the interval $(x_{j-1},x_j]$ into which it falls. We use the notation $x_0=0$ and $x_{k+1}=\infty$. It is not assumed that the intervals $(x_{j-1},x_j]$; $j=1,2,\ldots,k$ are of equal length. Let N_j denote the number of observations falling into the jth interval. The probability that an observation falls into the jth interval denoted p_j is given by

$$p_j=\left[1+\left(\frac{x_{j-1}}{\sigma}\right)^{1/\gamma}\right]^{-\alpha}-\left[1+\left(\frac{x_j}{\sigma}\right)^{1/\gamma}\right]^{-\alpha}. \tag{5.7.1}$$

If truncation exists at the extremities, then we set $p_0=p_{k+1}=0$ and replace p_j by $p_j/(\sum_{i=1}^k p_i)$, $j=1,2,\ldots,k$. No other changes are necessary. We will continue the analysis, assuming for notational simplicity that no truncation has occurred. The likelihood of observing a particular outcome $\underline{N} = \underline{n}$ is of the form

$$L(\sigma,\gamma,\alpha) = n! \prod_{j=1}^{k+1} \frac{p_j^{n_j}}{(n_j)!} \tag{5.7.2}$$

where the p_j's are given by (5.7.1). The corresponding likelihood equations must be solved by numerical techniques. Fisk (1961b) gave a detailed algorithm for use when $\alpha = 1$, while Fisk (1961b) and Aigner and Goldberger (1970) discussed the classical Pareto case with known scale parameter σ (i.e., $P(I)(\sigma,\alpha)$). Both of these references include expressions for the asymptotic variances and covariances of the maximum likelihood estimates. The covariance structure for maximum likelihood estimates obtained in the three parameter case (from (5.7.2)) has not been explicitly determined. There is one special case in which a closed form expression is available for the maximum likelihood estimate. If we are dealing with a classical Pareto population with $\sigma = x_0$ and if, by chance or by design, $x_i = cx_{i-1}, i = 1,2,\ldots,k$, then our grouped observations have a truncated log-geometric distribution. The corresponding likelihood function will be

$$L(\alpha) = n! \left[\prod_{j=1}^{k} (1-c^{-\alpha})^{n_j} c^{-(n-1)\alpha n_j} / n_j! \right] [c^{-k\alpha n_{k+1}} / n_{k+1}!]. \tag{5.7.3}$$

It is convenient to reparameterize setting $p = c^{-\alpha}$. The maximum likelihood estimate of p is then readily found to be

$$p = \left[\sum_{j=1}^{k+1} (j-1)N_j \right] \Big/ \left[n - N_{k+1} + \sum_{j=1}^{k+1} (j-1)N_j \right]. \tag{5.7.4}$$

Ismail (2004) also considered grouped Pareto(I)(σ,α) data with σ known. He makes use of the fact that the Pareto (I) hazard function is of the form $f(x)/\overline{F}(x) = \alpha/x$. He then obtains estimates of the hazard rate within each of the grouping intervals (x_i, x_{i+1}), $i = 0,1,\ldots,k-1$, and then estimates α by

$$\widetilde{\alpha} = \frac{1}{k} \hat{h}(x_{mi}) x_{mi}, \tag{5.7.5}$$

where x_{mi} is the midpoint of the interval (x_i, x_{i+1}). He also considers the use of the geometric mean rather than the arithmetic mean in (5.7.5). If one uses Kimball's (1960) estimate of the hazard rate in each interval, the estimate (5.7.5) (or its cousin, using the geometric mean) has an advantage of simplicity in computation. A simulation based investigation by Ismail suggests that these hazard rate based estimates may outperform the maximum likelihood estimate of α, if α is small.

Returning to the Pareto (IV) setting, there are several alternative ways in which we can measure deviations between observed frequencies $f_j = N_j/n$ and their corresponding expectations (the p_j's given by (5.7.1)). Each such measure of discrepancy can be used to define a method of estimation; the general policy being to use as estimates of the parameters those values which minimize the discrepancy between the vectors $\underline{f}$ and $\underline{p}$. McDonald and Ransom (1979) provided an empirical comparison of the three most common techniques of this type using U.S. family income data for the years 1960 and 1969–1975. They identified the Pareto $(0,\sigma,\gamma,\alpha)$ distribution as the Singh-Maddala distribution and used notation consistent with that used by Singh and Maddala (1976). The Pareto (IV) or Singh-Maddala distribution fared quite well in their comparisons. The three techniques in question were, following McDonald and Ransom's nomenclature: (i) Pearson estimates, (ii) Neyman-Wald estimates and (iii) least squares estimates. They also mentioned the possibility of using what they called least lines estimates. The Pearson estimates are those values of σ, α and γ which minimize the quantity

$$X_1^2 = \sum_{j=1}^{k+1} (f_j - p_j)^2 / p_j, \tag{5.7.6}$$

while the Neyman-Wald estimates are those which minimize

$$X_2^2 = \sum_{j=1}^{k+1} (f_j - p_j)^2 / f_j, \tag{5.7.7}$$

where the p_j's (given by (5.7.1)) and the f_j's are, respectively, the theoretical and observed frequencies in the intervals. Although both techniques are asymptotically equivalent to the use of maximum likelihood, their small sample behavior is different, and the choice of which technique to use is, perhaps, subjective. The computational simplicity associated with the Neyman-Wald estimates is appealing, but it is not known whether any serious small sample loss of efficiency is encountered in the Pareto (IV) setting. That the chi-squared estimates and the maximum likelihood estimates are different even for quite large samples is evident from inspecting McDonald and Ransom's (1979) tables.

The least squares technique is predictably less efficient. The least squares estimates of σ,γ and α are those values which minimize the sum of squared deviations:

$$\mathrm{SSD} = \sum_{j=1}^{k+1} (f_j - p_j)^2. \tag{5.7.8}$$

Analogously, McDonald and Ransom defined least lines estimates of σ,γ and α to be those values which minimize the sum of absolute deviations:

$$\mathrm{SAD} = \sum_{j=1}^{k+1} |f_j - p_j|. \tag{5.7.9}$$

Aigner and Goldberger (1970) discussed least squares estimation of α in the classical Pareto setting. Since the basic data were multinomial, it was also possible to use weighted least squares, i.e., to minimize

$$(\underline{f} - \underline{p})' \, \Sigma^{-1} \, (\underline{f} - \underline{p}) \tag{5.7.10}$$

where Σ was the variance covariance matrix of $\underline{f}$, and in the general Pareto (IV) setting $\underline{p}$ was given by (5.7.1). A hybrid scheme involving using least squares estimates to estimate Σ in (5.7.10) and then minimizing was also suggested. Such techniques may still involve non-linear minimization and cannot be said to be markedly simpler than maximum likelihood. Since they also cannot be expected to be superior to maximum likelihood, they may be of limited utility. Aigner and Goldberger (1970) also discussed estimates for the classical Pareto based on regression of cumulative and log-cumulative frequencies. If the fitting is done by least squares, then new consistent estimates are obtained in this fashion. If the fitting is done by weighted least squares, the resulting estimates will coincide with those obtained by minimizing (5.7.10). The fitting by unweighted least squares of a line to the log-cumulative-frequencies as functions of log income is (ignoring the grouping) equivalent to constructing a Pareto chart for the data (see Section 4.3). Aigner and Goldberger included expressions for the asymptotic variance of the maximum likelihood and generalized least squares estimates of α in the classical Pareto situation and Aigner (1970) discussed the asymptotic variance of related estimates of the moments of the classical Pareto population.

Other estimation techniques have been proposed. Singh and Maddala (1976) estimated σ, γ and α by minimizing

$$\sum_{j=1}^{k+1}\left[\left(\log\sum_{i=j}^{k+1} p_i\right)-\left(\log\sum_{i=j}^{k+1} f_i\right)\right]^2. \tag{5.7.11}$$

McDonald and Ransom (1979) pointed out that this estimation technique is apparently sensitive to algorithmic inaccuracies. The values they obtained as estimates are markedly different from the original values published by Singh and Maddala. In an earlier note, Singh and Maddala (1975) proposed estimation by minimizing the sum of squared deviations between the sample and theoretical Lorenz curve (evaluated at the points $f_1, f_1+f_2, \ldots$). Their technique would prove most useful in situations where a closed form is available for the Lorenz curve but not for the corresponding distribution function (see Singh and Maddala, 1975, and Rasche, Gaffney, Koo and Obst, 1980, for examples). Kakwani and Podder (1973) discussed estimates obtained by fitting the Lorenz curve by weighted least squares. Actually they fitted a curve closely related to the Lorenz curve. Instead of $L(u)$ vs u, they considered $\log L(u)$ vs $\log u$. In Kakwani and Podder (1976) they reparameterized the Lorenz curve before applying least squares or weighted least squares. They did not specifically consider Pareto populations. In fact their new coordinate system was not particularly suited to describing Paretian Lorenz curves. The issue is somewhat clouded by the observation of Rasche, Gaffney, Koo and Obst (1980) that the parametric family of Lorenz curves actually used by Kakwani and Podder in their fitting procedure included many curves which were not Lorenz curves. Kakwani (1980b) in his rejoinder asserted that this lacuna does not vitiate the utility of his technique of fitting. It would seem, however, to play havoc with the philosophy behind the technique and casts a shadow over the desirability of using the Kakwani and Podder coordinate system.

If, indeed, the aim is to estimate the Lorenz curve and, perhaps related inequality measures, then one should not overlook the work of Gastwirth and his colleagues on bounding the Lorenz curve and measures of inequality based on grouped data and minimal distributional assumptions (see Section 4.2). As mentioned in Section 5.5, violation of Gastwirth bounds may be used as an indication of model inadequacy. It would be a shame to announce a laboriously fitted distribution which had a Lorenz curve, which could not possibly be associated with the given set of grouped data.

Finally, we turn to the question of Bayesian analysis of grouped data. The horrendous form of the likelihood (5.7.2) is a warning that friendly conjugate priors will not be encountered. It is probably, then, not coincidental that little attention has been paid to the problem. Zellner (1971, pp. 36–38) did present a small example in which he assumed a classical Pareto model with known σ and vague prior information about α (i.e., $f(\alpha) \propto \alpha^{-1}$). Numerical integration yielded the correct normalization for the posterior density and values of the posterior mean and variance. In principle, the same approach can be applied to samples from generalized Pareto distributions. There is one (contrived?) situation in which an informative prior can be used without numerical integration. Consider a classical Pareto setting with $\sigma = x_0$ and a grouping such that $x_i = cx_{i-1}$, $i = 1, 2, \ldots, k$. The corresponding likelihood is given by (5.7.3), and it is evident that a Beta prior can be used for $p = c^{-\alpha}$. The detailed analysis, which is exactly that for a truncated geometric sample, may be found in Arnold and Press (1986).

In that paper, Arnold and Press also consider data from Pareto populations that include both precise observations and observations that are only known to belong to some interval. A typical data set then consists of n observations $X_1, X_2, \ldots, X_n$ which are precisely observed and m additional observations $Y_1, Y_2, \ldots, Y_m$ for which it is only known that $a_j \leq Y_j \leq b_j$, $j = 1, 2, \ldots, m$. If we assume a Pareto(I)(σ, α) model, then the corresponding likelihood is of the form

$$
\begin{aligned}
L(\sigma,\alpha) &= \prod_{i=1}^{n} \frac{\alpha}{\sigma}\left(\frac{x_i}{\sigma}\right)^{-(\alpha+1)} I(x_{1:n} > \sigma) \\
&\quad \times \prod_{j=1}^{m}\left[\min\left\{\left(\frac{\sigma}{a_j}\right)^{\alpha}, 1\right\} - \min\left\{\left(\frac{\sigma}{b_j}\right)^{\alpha}, 1\right\}\right].
\end{aligned} \tag{5.7.12}
$$

If both σ and α are unknown, then no matter what joint prior density is assumed for them, the corresponding posterior density will be complicated and the Bayes estimates will need to be obtained by numerical integration. However, we can adopt an eclectic approach in which we estimate σ by $X_{1:n}$ and then act as if σ is known. It is then reasonable to assume that $a_j \geq \sigma$ for every j, which simplifies the likelihood considerably. In this known σ case, following the procedure used for samples which include only precisely observed values, it is natural to consider a gamma prior for α, i.e., $\alpha \sim \Gamma(\gamma_1, \gamma_2)$ (as in (5.2.53)). The corresponding posterior density of α corresponding to the likelihood (5.7.12) will be

$$
\begin{aligned}
f(\alpha|\underline{x},\underline{a},\underline{b},\sigma) &\propto \alpha^{\gamma_1+n-1} \prod_{j=1}^{m}\left[1 - \left(\frac{a_j}{b_j}\right)^{\alpha}\right] \\
&\quad \times \exp\left\{-\left[\frac{1}{\gamma_2} + \sum_{i=1}^{n} \log(x_i/\sigma) + \sum_{j=1}^{m} \log(a_i/\sigma)\right]\alpha\right\}.
\end{aligned} \tag{5.7.13}
$$

This posterior density is of the form

$$f(\alpha|\underline{x},\underline{a},\underline{b},\sigma) \propto \alpha^{\gamma_1+n-1} \sum_{j=0}^{m} (-1)^j e^{-d_j\alpha},$$

so that it can be represented as a linear combination of gamma densities (with the same shape parameter $\gamma_1 + n$, but differing scale parameters). The posterior mean of α can then be evaluated without difficulty.

If instead we consider similar samples, including precise and imprecise observations, from more general Pareto models (i.e., P(II),P(III),P(IV) or FP), and we wish to obtain Bayes estimates of the parameters, then Arnold and Press (1986) have little to offer us. For the Pareto (II)–(IV) cases we can write down a likelihood function analogous to (5.7.12) but somewhat more complicated. No matter what prior density that we assume for the parameters, the only recourse for obtaining Bayes estimates will be numerical integration. An alternative ("empirical Bayes") approach would involve estimating all parameters save one, say α, by frequentist methods and then, acting as if those parameters were known, adopting a suitable prior for α. Finally we remark that, in the Pareto (I) case, if all the b_j's in (5.7.12) are equal to ∞, then we are dealing with Type I censored observations and the likelihood (5.7.12) simplifies considerably to become as in (5.2.87. For this a generalized Lwin density will be a suitable conjugate prior, or a conditionally conjugate prior for (σ, α) can be used.

5.8 Inference for Related Distributions

In Section 3.11 discrete Pareto or Zipf distributions were introduced. These together with the zeta distributions and the Simon distributions (see Sections 2.6 and 3.11) are the natural analogs of the generalized Pareto distributions, when the support set of the distribution is assumed to be the non-negative integers. Even though, as Zipf so convincingly demonstrated, there are innumerable instances of socio-economic phenomena which exhibit discrete Paretian or Zipf behavior, surprisingly little attention has been devoted to the statistical analysis of samples from such distributions. In part this appearance is illusory. The generalized Zipf distributions (Zipf II, III and IV defined in (3.11.3–3.11.5)) can be analyzed as grouped generalized Pareto distributions, and the techniques of Section 5.7 may be appropriately used. This very feature can be used to explain the paucity of references on discrete Pareto variables. The "nice" things that work for the continuous Pareto distributions usually lose their aesthetic appeal and even sometimes their feasibility when we move over to consider grouped data. Maximum likelihood estimation can however be implemented numerically. Krishna and Singh Pundir (2009) discuss maximum likelihood estimation of the parameters of the Zipf (IV)$(0, 1, \gamma, \alpha)$ distribution (which they call a discrete Burr distribution). The corresponding likelihood equations are solved numerically. They also provide an example in which the model is fitted to some dental data. In addition, they consider maximum likelihood estimation of the parameters of a Feller-Zipf distribution.

The zeta and Simon distributions are not discretized Pareto variables, and specialized inferential techniques need to be developed for them.

5.8.1 Zeta Distribution

Seal (1952) considered maximum likelihood estimation for the zeta distribution. Suppose that n observations are available from a zeta(α) distribution (see (3.11.1)). For $j = 1, 2, \ldots$, let N_j denote the number of observations among the n that are equal to j. The corresponding likelihood function is given by

$$L(\alpha) = \left[\prod_{j=1}^{\infty} j^{N_j}\right]^{-(\alpha+1)} \Big/ [\zeta(\alpha+1)]^n. \tag{5.8.1}$$

If we differentiate the log likelihood, we obtain the corresponding likelihood equation

$$-[\zeta'(\alpha+1)/\zeta(\alpha+1)] = \frac{1}{n}\sum_{j=1}^{\infty} N_j \log j. \tag{5.8.2}$$

This can be solved by interpolation in a table of values of the function $\zeta'(y)/\zeta(y)$ (Seal supplied a reference to Walther (1926) as a source of such a table). Seal also discussed maximum likelihood estimation for zeta random variables whose parameter α is assumed to be a linear function of some background variable x (age in Seal's example). Again interpolation in Walther's table is needed. Two other estimation techniques were mentioned by Seal. The mean of a zeta(α) distribution is $\zeta(\alpha)/\zeta(\alpha+1)$ (from (3.11.2)). This can be equated to the sample mean, and a method of moments estimate of α can be obtained by trial and error in a table of values of the zeta function. A second approach is to equate the ratio of N_1 and N_2 to the ratio of their expectations and solve for α. The resulting estimate is

$$\tilde{\alpha} = [\log (N_1/N_2) - \log 2]/\log 2. \tag{5.8.3}$$

All the proposed estimation techniques are consistent. It is to be expected that maximum likelihood will be the most efficient technique, while use of (5.8.3), since it utilizes so little information, might be the least efficient. Seal presented an example in which all four estimation techniques were applied to 12 populations. The four estimation techniques led to remarkably dissimilar values in many of the populations. This casts some doubt on the appropriateness of the zeta distribution model for the given examples. In order to compare the small sample efficiencies of the techniques, simulation will probably be required.

5.8.2 Simon Distributions

The Simon distributions as natural extensions of the Yule and Waring distributions were introduced in Section 3.11. They are most easily defined in terms of the ratio of successive terms in the discrete density. Repeating (3.11.12), we have for such random variables (whose support is $0, 1, 2, \ldots$)

$$P(X = k)/P(X = k-1) = (k+\alpha)/(\gamma k+\beta) \tag{5.8.4}$$

for $k = 1, 2, \ldots$. Tripathi and Gurland (1976) described estimation techniques for such distributions. They arrived at these distributions by a slightly different route, used a different parameterization, and spoke of distributions in the extended Katz family. Katz (1963) had considered distributions which correspond to the choice $\beta = 0$ in (5.8.4). Actually Tripathi and Gurland's extended Katz family is both a little more and a little less general than the Simon family. It is more general in that they permit the parameters α, β and γ to assume values for which eventually the ratio in (5.8.4) becomes negative. The support of such distributions is then restricted, as it must be, to those values of k for which the ratio is non-negative. We will assume the parametric restrictions related to those introduced following (3.11.12), namely $\alpha > -1, \beta > -\gamma$ and $\gamma > 1$. For such values of the parameters the support of X can be taken to be the full set of non-negative integers. The discrete density corresponding to the Simon distribution (5.8.4) involves a hypergeometric function:

$$P(X=k) = \frac{\gamma^{-k}\Gamma(\alpha+k+1)\Gamma\left(\frac{\beta}{\gamma}+1\right)}{\Gamma\left(\frac{\beta}{\gamma}+k+1\right)\Gamma(\alpha+1)} \frac{1}{{}_2F_1\left(\alpha+1,1,\frac{\beta}{\gamma}+1,\gamma^{-1}\right)}$$

$$k = 0,1,2,\dots . \tag{5.8.5}$$

The expression (5.8.5) remains valid when $\gamma = 1$ provided that $\beta > \alpha + 1$. Such distributions are not included in the extended Katz family as analyzed by Tripathi and Gurland. They were included in the family of Simon distributions introduced in Section 3.11. The hypergeometric function appearing in (5.8.5) can, when $\gamma = 1$, be expressed in terms of gamma functions. After a little manipulation, we recognize the retrospectively obvious fact that a Simon distribution with $\gamma = 1$ is precisely a Waring distribution (translated by 1) whose discrete density, albeit in an alternative parameterization, was displayed in (3.11.10). The resulting simplification renders the Waring ($\gamma = 1$) distribution more amenable to analysis than the Simon distributions ($\gamma > 1$, i.e., extended Katz). For the remainder of this section we will treat the two situations separately using the names Simon and Waring to distinguish them.

Tripathi and Gurland (1976) supplied a detailed investigation of alternative estimation schemes for the Simon distribution. It is appropriate to begin with a brief review of the distributional properties of the Simon distribution. If X has a Simon (α, β, γ) distribution given by (5.8.4), the corresponding probability generating function is of the form

$$E(s^X) = {}_2F_1\left(\alpha+1,1,\frac{\beta}{\gamma}+1,\gamma^{-1}s\right) / {}_2F_1\left(\alpha+1,1,\frac{\beta}{\gamma}+1,\gamma^{-1}\right). \tag{5.8.6}$$

The defining relation (5.8.4) may be rewritten yielding

$$(\gamma k + \beta)p_k = (k+\alpha)p_{k-1} \tag{5.8.7}$$

where here and henceforth $p_k = P(X=k)$. If we sum (5.8.7) over $k = 1,2,\dots$ we obtain the relationship

$$(\gamma - 1)E(X) = \beta(1-p_0) + \alpha + 1. \tag{5.8.8}$$

If one multiplies (5.8.7) by k^m on each side and then sums over k, one obtains an expression relating the first m moments of X. For example, with $m = 2$ and 3 we find:

$$(\gamma - 1)E(X^2) = (2 + \alpha - \beta)E(X) + \alpha + 1, \tag{5.8.9}$$

$$(\gamma - 1)E(X^3) = (3 + \alpha - \beta)E(X^2) + (3 + 2\alpha)E(X) + \alpha + 1. \tag{5.8.10}$$

Alternatively, one may derive a second order difference equation relating the factorial moments (Tripathi and Gurland's (1976) equation (2.2)).

The form of the discrete density (5.8.5) is sufficiently complicated to deter all but enthusiastic programmers from using maximum likelihood. Even the corresponding information matrix must be evaluated numerically, since it involves partial derivatives of the hypergeometric function. Tripathi and Gurland verify that, although it is not the most efficient way, one can differentiate termwise to evaluate such partial derivatives. The estimation procedure recommended by these authors is a minimum chi-square technique introduced by Hinz and Gurland (1970). It is, as will be recognized, a close relative of one of the techniques proposed by Aigner and Goldberger (1970) for the analysis of grouped Pareto data. We will merely sketch the procedure. More detail (considerably more) is available in Tripathi and Gurland (1976). We use the relations (5.8.7)–(5.8.10) to determine s functions $\tau_1, \tau_2, \ldots, \tau_s$ of moments μ_k and/or probabilities p_k, which are all linear functions of the parameters α, β and γ with known coefficients. Thus,

$$\underline{\tau} = W\underline{\theta} \tag{5.8.11}$$

with W known and $\underline{\theta} = (\alpha, \beta, \gamma)'$. Let $\underline{t}$ be the corresponding sample version of τ (obtained by substituting sample moments for the μ_k's and observed frequencies for the p_k's). Determine the variance matrix Σ_t of the random vector $\underline{t}$ as a function of α,β,γ. Let $\hat{\Sigma}_{\underline{t}}$ be a consistent estimate of $\Sigma_{\underline{t}}$, probably most easily obtained by using simple method of moments estimates for α, β and γ in $\Sigma_{\underline{t}}$. The generalized minimum chi-squared estimates of (α, β, γ) are then obtained by minimizing

$$Q = n[\underline{t} - W\underline{\theta}]'\hat{\Sigma}_{\underline{t}}^{-1}[\underline{t} - W\underline{\theta}]. \tag{5.8.12}$$

Thus the desired estimates are given by

$$\hat{\theta} = (W'\hat{\Sigma}_{\underline{t}}^{-1}W)^{-1}W'\hat{\Sigma}_{\underline{t}}^{-1}\underline{t}. \tag{5.8.13}$$

The linear form of the resulting estimate is quite appealing. Hidden in the development was the need to solve possibly non-linear equations to obtain the method of moments estimates of α, β and γ, which were used to get $\hat{\Sigma}_{\underline{t}}$. Nevertheless, the procedure is relatively straightforward and much easier than maximum likelihood. The choice of how many and which τ_i's to use is subjective. The more we use, the better we can expect our estimates to be, but at an increasing cost in the complexity of evaluating (5.8.13). Tripathi and Gurland recommended using six functions $\tau_1, \ldots, \tau_6$ which are linear functions of (α, β, σ) obtained from (5.8.7) with $k = 1, 2, 3$, (5.8.8), (5.8.9) and (5.8.10). Other choices were discussed, but this selection appears most efficient over much of the parameter space.

5.8.3 *Waring Distribution*

Turning to the Waring distribution (i.e., $\gamma = 1$ in (5.8.5)), we find that the likelihood function is no longer as formidable. Chen (1979a) described two estimation procedures for data from such a distribution. There is, of course, no reason why the generalized minimum chi-squared technique described above for the Simon distribution could not be equally well applied to Waring data. We will use the parameterization used in (3.11.10) where the Waring distribution was developed as a Beta mixture of geometric distributions. Thus

$$P(X = k) = B(\alpha + 1, \beta + k)/B(\alpha, \beta) \tag{5.8.14}$$

where $\alpha > 0$, $\beta > 0$. The log-likelihood of a sample of size n from this distribution is

$$\begin{aligned} \ell(\alpha, \beta) &= n[\log\ \alpha + \log\ \Gamma(\alpha + \beta) - \log\ \Gamma(\beta)] \\ &\quad + \sum_{i=1}^{n} [\log\ \Gamma(\beta + X_i) - \log\ \Gamma(\beta + \alpha + X_i + 1)]. \end{aligned} \tag{5.8.15}$$

Since $\frac{d}{dz} \log\ \Gamma(z) = -\gamma - \frac{1}{z} + \sum_{j=1}^{\infty}(\frac{1}{j} - \frac{1}{j+z})$ (where γ is Euler's γ), we find a relatively simple set of likelihood equations

$$\begin{aligned} &\frac{n}{\alpha} - \frac{n}{\alpha + \beta} + \sum_{i=1}^{n} \frac{1}{\alpha + \beta + X_i + 1} - \sum_{i=1}^{n} \sum_{j=1}^{X_i+1} \frac{1}{\alpha + \beta + j} = 0, \\ &-\frac{n}{\alpha} + \frac{n}{\beta} - \sum_{i=1}^{n} \frac{1}{\beta + X_i} + \sum_{i=1}^{n} \sum_{j=1}^{X_i} \frac{1}{\beta + j} = 0, \end{aligned} \tag{5.8.16}$$

where $\sum_{j=1}^{0}$ is zero by convention. Numerical or iterative techniques will be needed to determine the solution to (5.8.16). It may turn out that a direct search for the location of the maximum of the likelihood function is quite efficient (Ord, personal communication).

Chen (1979a) suggested an alternative estimation technique which yields what he calls the transformed data estimates of α and β. This is a somewhat ad hoc procedure yet, remarkably, Chen reported that it performs better than maximum likelihood when applied to Willis' (1922) much analyzed chrysomelidae and lizard data. The two stage procedure relies on the fact that the Waring distribution exhibits Paretian tail behavior. That is, for k large,

$$P(X > k) \sim Ck^{-\alpha}. \tag{5.8.17}$$

Chen estimated α by regressing $\log P(X > k)$ on $\log k$. Denoting the resulting estimate of α by α_t, he then estimated β by maximizing the likelihood (5.8.15), assuming that α is equal to α_t. The resulting transformed data estimate (α_t, β_t) clearly did not, in general, maximize the likelihood (5.8.15) globally, but in Chen's examples chi-squared goodness-of-fit statistics yielded smaller values for (α_t, β_t) than for the maximum likelihood estimate $(\hat{\alpha}, \hat{\beta})$. It is conceivable, but unlikely, that the transformed data estimates are more efficient in small and medium sized samples than the maximum likelihood estimates.

5.8.4 *Under-reported Income Distributions*

Krishnaji's (1970) Pareto characterization based on under-reported incomes (described in Section 3.9.4) stimulated interest in models where Paretian true income is under-reported and, consequently, reported income may have a non-Paretian distribution. We particularly focus on a Krishnaji type model as analyzed first by Hartley and Revankar (1974) and subsequently by Hinkley and Revankar (1977). We will follow their practice of relating actual and reported income in an additive fashion as contrasted to Krishnaji's (1970) multiplicative formulation (see (3.9.26)). The change is cosmetic. We assume that actual income X has a classical Pareto distribution (i.e., $X \sim P(I)(\sigma, \alpha)$). We assume that $\alpha > 2$, in order to have finite second moments. The observable reported income Y is assumed to be related to X by the equation

$$Y = X - U \tag{5.8.18}$$

where $U \leq X$. It is assumed that the proportion of under-reporting is distributed independently of X. More specifically, it is assumed that for any value of X, the variable

$$W = U/X = 1 - (Y/X) \tag{5.8.19}$$

has a particular form of the Beta distribution, i.e., for some $\lambda \in (0, \infty)$

$$f_W(w) = \lambda(1-w)^{\lambda-1}, \quad 0 < w < 1. \tag{5.8.20}$$

More transparently, it is assumed that Y/X has a power function distribution. This is precisely the distribution for Y/X that was assumed by Krishnaji. The implications of other distributional assumptions for Y/X were discussed in the characterization context in Section 3.9.4. If the distribution of W is as given in (5.8.20), we may verify that the density corresponding to reported income Y is of the form

$$f_Y(y) = \begin{cases} \frac{\lambda\alpha}{\sigma(\lambda+\alpha)}\left(\frac{y}{\sigma}\right)^{\lambda-1}, & 0 \leq y \leq \sigma \\ \frac{\lambda\alpha}{\sigma(\lambda+\alpha)}\left(\frac{\sigma}{y}\right)^{\alpha+1}, & \sigma < y < \infty \\ 0, & \text{otherwise.} \end{cases} \tag{5.8.21}$$

From (5.8.21) we readily observe that Y truncated at σ has the same $P(I)(\sigma, \alpha)$ distribution that X had originally. It will be recalled that this observation was the basis for Krishnaji's (1970) characterization. Hartley and Revankar (1974) observed that with this model the ratio

$$q = E(Y)/E(X) = \lambda/(\lambda+1) \tag{5.8.22}$$

is independent of the parameters σ and α of the distribution of X. They suggested that q may be regarded as an index of dishonesty. Actually this quantity q does not depend on the distribution of X in any way. This is a simple consequence of the assumption in (5.8.20) that W has a distribution which does not depend on X.

The case where $X \sim P(I)$ and W defined in (5.8.19) has a general Beta distribution has been discussed by Ratnaparkhi (1981). He gave expressions for the joint density of X and Y and the marginal density of Y including (5.8.21) as a special case. See also McDonald (1981).

Suppose that we have a sample $Y_1, Y_2, \ldots, Y_n$ from the reported income density (5.8.21). How should we estimate the three parameters σ, α and λ? It is possible to apply the method of moments, since routine computations yield the following expression for the γ'th moment of Y

$$E(Y^\gamma) = \left(\frac{\lambda}{\lambda+\gamma}\right)\left(\frac{\alpha}{\alpha-\gamma}\right)\sigma^\gamma, \quad \alpha > \gamma. \tag{5.8.23}$$

This expression is actually most easily derived using Krishnaji's multiplicative formulation of the model. Alternatively, we can use maximum likelihood estimates. To this end, Hartley and Revankar considered two cases, depending on whether or not σ is assumed to be known.

If σ is known, then we may, without loss of generality, assume it to be equal to one. The log-likelihood function of a sample of size n from the density (5.8.21) with $\sigma = 1$ then assumes the form:

$$\ell(\alpha, \lambda) = n\left[\log\left(\frac{\lambda\alpha}{\lambda+\alpha}\right) - (\lambda-1)S_1 - (\alpha+1)S_2\right] \tag{5.8.24}$$

where

$$S_1 = \frac{1}{n}\sum_{i=1}^{n}(\log Y_i)^-$$

and

$$S_2 = \frac{1}{n}\sum_{i=1}^{n}(\log Y_i)^+. \tag{5.8.25}$$

In (5.8.25) we have used the notation $(z)^+ = \max(0,z)$ and $(z)^- = -\min(0,z)$. It is not difficult, provided that we assume $S_1 \neq 0$ and $S_2 \neq 0$, to solve the corresponding likelihood equations to obtain

$$\begin{aligned} \hat{\alpha} &= [\sqrt{S_2}(\sqrt{S_1}+\sqrt{S_2})]^{-1} \\ \hat{\lambda} &= [\sqrt{S_1}(\sqrt{S_1}+\sqrt{S_2})]^{-1}. \end{aligned} \tag{5.8.26}$$

By considering second partial derivatives of $\ell(\alpha,\lambda)$, one may verify that the values $(\hat{\alpha}, \hat{\lambda})$ in (5.8.26) do correspond to a global maximum. In so doing, one will do the necessary computations, in order to evaluate the information matrix and, consequently, to determine the asymptotic distribution of $(\hat{\alpha}, \hat{\lambda})$. For large n one finds

$$(\hat{\alpha}, \hat{\lambda}) \dot{\sim} N\left((\alpha,\lambda), \frac{1}{n}\Sigma\right) \tag{5.8.27}$$

where

$$\Sigma = \begin{bmatrix} \alpha^2\left(1+\frac{\alpha}{2\lambda}\right) & \frac{\lambda\alpha}{2} \\ \frac{\lambda\alpha}{2} & \lambda^2\left(1+\frac{\lambda}{2\alpha}\right) \end{bmatrix}. \tag{5.8.28}$$

As a consequence of (5.8.27), we may conclude that $\hat{\alpha}$ and $\hat{\lambda}$ are consistent estimates of α and λ, respectively.

The situation is more challenging when σ is unknown. Hartley and Revankar (1974) described a technique for determining the corresponding maximum likelihood estimates but since the situation is non-regular, they did not resolve the problem of determining the asymptotic properties of the estimates. This gap was subsequently filled by Hinkley and Revankar (1977) using some results of Huber (1967) on maximum likelihood estimation under non-standard conditions. In their analysis Hinkley and Revankar transformed the data by taking logarithms. If Y has density (5.8.21), then $\log Y$ has an asymmetric Laplace distribution. This suggests how to determine the maximum likelihood estimates by analogy to the technique used for the symmetric Laplace distribution. If $Y_1, \ldots, Y_n$ have common density (5.8.21), then the log-likelihood function is expressible as

$$\begin{aligned}\ell(\alpha,\lambda,\sigma) = {} & n \log \frac{\lambda\alpha}{\lambda+\alpha} + (\lambda-1)T_1 - (\alpha+1)T_2 \\ & + [(\alpha+1)(n-R) - (\lambda-1)R] \log \sigma \end{aligned} \tag{5.8.29}$$

where

$$R = \sum_{i=1}^{n} I(Y_i \leq \sigma), \tag{5.8.30}$$

$$T_1 = \sum_{i=1}^{R} \log Y_{i:n} \tag{5.8.31}$$

and

$$T_2 = \sum_{i=R+1}^{n} \log Y_{i:n}. \tag{5.8.32}$$

For fixed values of α and λ this likelihood is maximized by choosing $\hat{\sigma} = Y_{r+1:n}$ where r is the integer part of $n(\alpha+1)/(\lambda+\alpha)$. With this value of σ the maximizing choices for α and λ may then be found exactly as in the case when σ is known. With small sample sizes the issue is complicated somewhat if $\hat{\sigma}$ is $Y_{1:n}$ or $Y_{n:n}$. For large sample sizes this complication can be ignored.

It will be observed that the log-likelihood function (5.8.29) as a function of σ has discontinuities at $\sigma = Y_{i:n}$, $(i = 1, 2, \ldots, n)$. Despite this non-standard feature the information matrix remains well defined and asymptotic normality may be verified using Huber's (1967) results. Hinkley and Revankar (1977) carefully spelled out the details. For large n, we have

$$(\hat{\alpha}, \hat{\lambda}, \hat{\sigma}) \sim N\left((\alpha, \lambda, \sigma), \frac{1}{n}\Sigma\right) \tag{5.8.33}$$

where

$$\Sigma = \begin{pmatrix} \frac{\alpha^2(\lambda+\alpha)}{\lambda} & 0 & \frac{\alpha\sigma}{\lambda} \\ 0 & \frac{\lambda^2(\lambda+\alpha)}{\alpha} & -\frac{\lambda\sigma}{\alpha} \\ \frac{\alpha\sigma}{\lambda} & -\frac{\sigma\lambda}{\alpha} & \frac{2\sigma^2}{\lambda\alpha} \end{pmatrix}. \tag{5.8.34}$$

To see how the precision of our estimates of α and λ depend on the knowledge (or lack of knowledge) of σ, we may compare the asymptotic variances of $\hat{\alpha}$ and $\hat{\lambda}$ in (5.8.27) and (5.8.33). The ratio

$$\frac{(\text{var } \hat{\alpha} \text{ when } \sigma \text{ known})}{(\text{var } \hat{\alpha} \text{ when } \sigma \text{ unknown})} \simeq \frac{2\lambda+\alpha}{2(\lambda+\alpha)} \tag{5.8.35}$$

can be significantly less than 1. When α and λ are approximately equal, this ratio is nearly 3/4. If, however, λ is quite small relative to α, then the ratio is approximately 1/2. Small values of λ are associated with extreme cases of under-reporting (cf. (5.8.20)). If instead we are interested in estimating λ, the effect on the variance of not knowing σ is most pronounced when λ is very large (i.e., when little under-reporting is occurring).

McDonald (1981) discussed the form of the reported income distribution when a gamma distribution is assumed for actual income X and a scaled Beta distribution is assumed for the misreporting factor U (in (5.8.36)). The resulting distribution is distressingly complex. This is not surprising. It is not frequently the case that scale mixtures of nice distributions are again nice. One exception, of course, is the case of an inverse gamma scale mixture of exponentials which yields the Pareto (II) distribution (see Section 3.2 or Maguire, Pearson, and Wynn, 1952).

5.8.5 *Inference for Flexible Extensions of Pareto Models*

Several authors have addressed the problem of inference on the parameters of flexible alternative models that subsume and extend Pareto models. A selection of such models was described in Section 3.10.

Mahmoudi (2011) discusses maximum likelihood estimation of the parameters of the beta generalized Pareto (II)$(0,\sigma,\alpha)$ distribution (as displayed in (3.10.3) with $\gamma = 1$). Shawky and Abu-Zinadah (2009) consider inference for the exponentiated Pareto (II) model (as in (3.10.5), but with $\gamma = \sigma = 1$). They consider a variety of estimation strategies for this model including maximum likelihood, percentile estimates and certain weighted and unweighted least squares estimates. A simulation study presented in their paper suggests that maximum likelihood estimates can be recommended. Afify (2010) also considers the exponentiated Pareto model (with $\gamma = \sigma = 1$). He derives maximum likelihood estimates and also Bayes estimates obtained using independent non-informative priors for the two parameters in the model. Both squared error and LINEX loss are considered.

Estimation for the parameters of hidden truncation Pareto (II) models (as in

(3.10.22) with $\gamma = 1$), is discussed in Arnold and Ghosh (2011). Maximum likelihood, method of moments and percentile estimates are considered in these papers, together with diffuse prior Bayes estimates.

The truncated Pareto (II) distribution was introduced in (3.2.4). Its density is of the form

$$f(x;\mu,\sigma,\alpha,\tau) = \frac{\frac{\alpha}{\sigma}\left(1+\frac{x-\mu}{\sigma}\right)^{-(\alpha+1)}}{1-\left(1+\frac{\tau-\mu}{\sigma}\right)^{-\alpha}} I(\mu < x < \tau). \tag{5.8.36}$$

Parameter estimation for samples from this distribution is most simply implemented by setting $\hat{\mu} = X_{1:n}$ and $\hat{\tau} = X_{n:n}$, and then numerically solving two likelihood equations in α and σ (after substituting $X_{i:n}$ for μ and $X_{n:n}$ for τ). If we consider instead the simpler truncated Pareto (I) model (obtained by setting $\mu = \sigma$ in (5.8.36)),we find that several authors have provided relevant discussion. An early contribution was that of Beg (1981). He derived the UMVUE of $P(X > t_0)$ for this distribution. Subsequently, in Beg (1983), the UMVUE's of σ, α and τ were provided. Cohen (1991) discussed maximum likelihood estimation of μ and σ when τ is known. Aban, Meerschaert and Panorska (2006) verified that the maximum likelihood estimates of the parameters of this distribution are such that $\hat{\sigma} = X_{i:n}$, $\hat{\tau} = X_{n:n}$ and $\hat{\alpha}$ is the solution to the equation

$$\frac{\partial}{\partial \alpha} \log L(\hat{\sigma}, \alpha, \hat{\tau}) = 0,$$

which must be solved numerically. Be aware that Aban, Meerschaert and Panorska use $X_{(n)}$ to denote $X_{1:n}$ and $X_{(1)}$ to denote $X_{n:n}$ in their paper. They provide a simulation based comparison of the maximum likelihood estimates, Beg's estimates and a competing estimate of α due to Hill (1975). Huang and Zhao (2010) suggest instead the use of weighted estimates for the parameters of truncated Pareto models.

Picard and Tribouley (2002) discuss parameter estimation based on samples from a distribution that they call an evolutionary Pareto distribution. This distribution is of the form

$$\begin{aligned} F(x;\sigma_0,\sigma_1,\alpha,\beta) &= \left[1-\left(\frac{x}{\sigma_0}\right)^{-\alpha}\right] I(\sigma_0 < x < \sigma_1) \\ &\quad + \left[1-\left(\frac{x}{\sigma_1}\right)^{-\beta}\left(\frac{\sigma_0}{\sigma_1}\right)^{\alpha}\right] I(\sigma_1 \leq x < \infty), \end{aligned} \tag{5.8.37}$$

where $0 < \sigma_0 < \sigma_1 < \infty$ and $\alpha, \beta \geq 1$. The parameter σ_1 is a "change-point" in this distribution and consequently will present more estimation difficulty than the other parameters. Maximum likelihood estimation of all the parameters is discussed by Picard and Tribouley. The four maximum likelihood estimates are shown to be independent. The maximum likelihood estimate of σ_0 is $\hat{\sigma_0} = X_{1:n}$, which has an asymptotic exponential distribution. The estimates $\hat{\alpha}$ and $\hat{\beta}$ are consistent and asymptotically normal. Determination of the asymptotic behavior of $\hat{\sigma_1}$ requires more complicated analysis; see Picard and Tribouley (2002) for details.

5.8.6 Back to Pareto

The last entries to be discussed in this section on Pareto-related distributions will bring us back full-circle to Vito Pareto's book. It will be recalled that a variant model (1.2.2) was proposed by Pareto to describe the distribution of wealth. Although several early economists considered such a model as a plausible alternative when the Pareto charts exhibited a nonlinear character, for many years it was generally ignored in the analytic literature on income models. The model was however eventually resuscitated (Davis and Felstein, 1979). They viewed the model as one having potential in life testing situations. They considered the density to have support $(0,\infty)$ and discussed the nature of maximum likelihood estimates based on possibly censored data sets assuming that the scale parameter is known.

Kagan and Schoenberg (2001) consider Pareto's second model (1.2.2), which they called a tapered Pareto distribution. The survival function in question is of the form

$$\overline{F}(x) = \left(\frac{x}{\sigma}\right)^{-\alpha} e^{-\beta(x-\sigma)}, \; x \geq \sigma.$$

It is assumed that σ is known (if not, it would most likely be estimated by $X_{1:n}$), and interest centers on estimating α and β. Kagan and Schoenberg discuss maximum likelihood estimation of α and β based on a sample of size n from the tapered distribution. The parameter $\theta = 1/\beta$ is called the cutoff parameter. Assuming that α is known (as well as σ), the authors concentrate on estimating the cutoff parameter. Simulation based evidence suggests that the maximum likelihood estimate of θ is noticeably biased. This prompts consideration of several alternative estimation strategies. A method of moments estimator has an advantage of simplicity, but no final recommendation is made. Kagan and Schoenberg suggest use of this model in a context of seismic hazard quantification, but its potential in an income distribution setting is clearly apparent (as indeed it was to Pareto).

Hill (1975) provided an analysis of the upper tail of a distribution that is very faithful to the original enunciation of Pareto's income distribution model. In Hill (1975) it was only assumed that the distribution generating the data had a Paretian survival function $\overline{F}(y)$ for values of y greater than some unknown value D. Below D, the functional form of $\overline{F}$ was unspecified. A conditional likelihood analysis was proposed. Appropriate Bayesian modifications were also discussed.

Chapter 6

Multivariate Pareto distributions

6.0 Introduction

In this chapter we will discuss several k-dimensional distributions which qualify as multivariate Pareto distributions by virtue of having Pareto marginals. Sometimes conditional and other related distributions are also Paretian, but this is frequently not the case. Distributional properties of these models are discussed. Later sections of the chapter are concerned with models with Pareto conditional distributions and with flexible extensions of multivariate Pareto models. Inference issues are briefly addressed. The final section deals with discrete versions of multivariate Pareto models.

6.1 A Hierarchy of Multivariate Pareto Models

6.1.1 Mardia's First Multivariate Pareto Model

The first author to systematically study k-dimensional Pareto distributions was Mardia (1962). Mardia's type I multivariate Pareto distribution is the classic instance in which both marginals and conditional distributions are Paretian in nature. We will say that a k-dimensional random vector $\underline{X}$ has a type I multivariate Pareto distribution, if the joint survival function is of the form

$$\overline{F}_{\underline{X}}(\underline{x}) = \left[\sum_{i=1}^{k}(x_i/\sigma_i) - k + 1\right]^{-\alpha}, \quad x_i > \sigma_i \qquad (6.1.1)$$

$$i = 1, 2, \dots, k,$$

and we write $\underline{X} \sim MP^{(k)}(\mathrm{I})(\underline{\sigma}, \alpha)$. The σ_i's are non-negative marginal scale parameters. The non-negative parameter α is an inequality parameter (common to all marginals). It is immediately obvious from (6.1.1) that the one-dimensional marginals are classical Pareto distributions. Thus $X_i \sim P(\mathrm{I})(\sigma_i, \alpha)$, $i = 1, 2, \dots, k$. By setting selected x_i's equal to σ_i in (6.1.1), it is apparent that, for any $k_1 < k$, all k_1 dimensional marginals are again multivariate Pareto. If we use the notational device $\underline{X} = (\dot{\underline{X}}, \ddot{\underline{X}})$ where $\dot{\underline{X}}$ is k_1 dimensional with an analogous partition of the vector $\underline{\sigma} = (\dot{\underline{\sigma}}, \ddot{\underline{\sigma}})$, we may write

$$\dot{\underline{X}} \sim MP^{(k_1)}(\mathrm{I})(\dot{\underline{\sigma}}, \alpha). \qquad (6.1.2)$$

Conditional distributions are of the form (6.1.1), but with a change of location. One may verify (again use the dot/double dot notation to partition vectors) that

$$\underline{\dot{X}}|\underline{\ddot{X}} = \underline{\ddot{x}} \sim c(\underline{\ddot{x}})MP^{(k_1)}(\mathrm{I})(\underline{\dot{\sigma}}, \alpha + k - k_1) + [1 - c(\underline{\ddot{x}})]\underline{\dot{\sigma}} \tag{6.1.3}$$

where $$c(\underline{\ddot{x}}) = \left[\sum_{i=k_1+1}^{k} \frac{x_i}{\sigma_i} - k + k_1 + 1\right]. \tag{6.1.4}$$

The associated correlation structure can be deduced from the bivariate densities. One finds, provided $\alpha > 2$, that

$$E(X_i) = \alpha\sigma_i/(\alpha - 1), \tag{6.1.5}$$

$$\mathrm{var}(X_i) = \alpha\sigma_i^2/[(\alpha-1)^2(\alpha-2)], \tag{6.1.6}$$

$$\mathrm{cov}(X_i, X_j) = \sigma_i\sigma_j/[(\alpha-1)^2(\alpha-2)], \tag{6.1.7}$$

$$\mathrm{corr}(X_i, X_j) = \alpha^{-1}. \tag{6.1.8}$$

6.1.2 A Hierarchy of Generalizations

The expression (6.1.1) used by Mardia for the multivariate Pareto is, in fact, unfortunate in that it obscures the dual role that the σ_i's play. They are, in fact, both location and scale parameters. A better representation would be

$$\overline{F}_{\underline{X}}(\underline{x}) = \left[1 + \sum_{i=1}^{k}\left(\frac{x_i - \sigma_i}{\sigma_i}\right)\right]^{-\alpha}, \quad x_i > \sigma_i \quad i = 1, 2, \ldots, k. \tag{6.1.9}$$

Additional flexibility at essentially no cost in terms of computational difficulty can be obtained by allowing the location parameter to be possibly different from the scale parameter. The resulting multivariate Pareto distribution of type II (not to be confused with Mardia's (1962) Type 2 distribution, which will be discussed further in Section 6.2.3) has joint survival function of the form

$$\overline{F}_{\underline{X}}(\underline{x}) = \left[1 + \sum_{i=1}^{k}\left(\frac{x_i - \mu_i}{\sigma_i}\right)\right]^{-\alpha}, \quad x_i > \mu_i \quad i = 1, 2, \ldots, k. \tag{6.1.10}$$

In such a case we write $\underline{X} \sim MP^{(k)}(\mathrm{II})(\underline{\mu}, \underline{\sigma}, \alpha)$. The corresponding marginals are, again, multivariate Pareto of type II; the univariate marginals being Pareto II (cf.(3.2.2)). Conditional distribution are, again, multivariate Pareto of type II; in fact using our dot/double dot notation we find

$$\underline{\dot{X}} \sim MP^{(k_1)}(\text{II})(\underline{\dot{\mu}}, \underline{\dot{\sigma}}, \alpha) \tag{6.1.11}$$

and

$$\underline{\dot{X}}|\underline{\ddot{X}} = \ddot{x} \sim MP^{(k_1)}(\text{II})(\underline{\dot{\mu}}, \ddot{c}(\underline{x})\underline{\dot{\sigma}}, \alpha + k - k_1) \tag{6.1.12}$$

$$\text{where} \quad c(\underline{\ddot{x}}) = \left[1 + \sum_{i=k_1+1}^{k} \left(\frac{x_i - \mu_i}{\sigma_i}\right)\right]. \tag{6.1.13}$$

The marginal means are given by

$$E(X_i) = \mu_i + [\sigma_i/(\alpha - 1)], \tag{6.1.14}$$

while the covariance expressions (6.1.6)–(6.1.8) remain valid for the type II multivariate Pareto.

The regression functions in this multivariate Pareto distribution are linear. In fact, using (6.1.12), we find

$$E(X_i|X_j) = \mu_i + \sigma_i \alpha^{-1}\left(1 + \frac{X_j - \mu_j}{\sigma_j}\right) \tag{6.1.15}$$

and

$$\text{var}(X_i|X_j) = \sigma_i^2(\alpha+1)\alpha^{-2}(\alpha-1)^{-1}\left(1 + \frac{X_j - \mu_j}{\sigma_j}\right)^2. \tag{6.1.16}$$

Cook and Johnson (1981, 1986) discuss a variety of models which share the $MP^{(k)}(II)$ copula, i.e., which are marginal transformations of $MP^{(k)}(II)$ distributions.

It is, of course, possible to repeat these calculations for a type III multivariate Pareto distribution of the form

$$\overline{F}_{\underline{X}}(\underline{x}) = \left[1 - \sum_{i=1}^{k} \left(\frac{x_i - \mu_i}{\sigma_i}\right)^{1/\gamma_i}\right]^{-1}, \quad x_i > \mu_i \quad i = 1, 2, \ldots, k \tag{6.1.17}$$

which clearly has Pareto III marginals. However, this class is not closed with respect to conditional distributions.

Malik and Abraham (1973) discuss a multivariate logistic distribution which is a marginal transformation of an $MP^{(k)}(III)$ distribution.

The more general multivariate Pareto IV distribution does have the conditional closure property. We will say that $\underline{X}$ has a k-dimensional Pareto distribution of type IV, if its joint survival function is of the form

$$\overline{F}_{\underline{X}}(\underline{x}) = \left[1+\sum_{i=1}^{k}\left(\frac{x_i-\mu_i}{\sigma_i}\right)^{1/\gamma_i}\right]^{-\alpha}, \quad x_i > \mu_i, \quad i=1,2,\ldots,k \tag{6.1.18}$$

and we write $\underline{X} \sim MP^{(k)}(\text{IV})(\underline{\mu},\underline{\sigma},\underline{\gamma},\alpha)$. It is clear that the marginal distributions are again multivariate Pareto (IV) and, after a little algebraic manipulation, that the conditional distributions are also multivariate Pareto (IV). Specifically, using the dot/double dot notation, we have

$$\dot{\underline{X}} \sim MP^{(k_1)}(\text{IV})(\dot{\underline{\mu}},\dot{\underline{\sigma}},\dot{\underline{\gamma}},\alpha) \tag{6.1.19}$$

and

$$\dot{\underline{X}}|\ddot{\underline{X}} = \ddot{\underline{x}} \sim MP^{(k_1)}(\text{IV})(\dot{\underline{\mu}},\dot{\underline{\tau}},\dot{\underline{\gamma}},\alpha+k-k_1), \tag{6.1.20}$$

where

$$\tau_i = \sigma_i\left[1+\sum_{j=k_1+1}^{k}\left(\frac{x_j-\mu_j}{\sigma_j}\right)^{1/\gamma_j}\right]^{\gamma_i}, \quad i=1,2,\ldots,k_1. \tag{6.1.21}$$

The distribution (6.1.18) with $\underline{\mu}=\underline{0}$ and $\underline{\sigma}=\underline{1}$ was introduced by Takahasi (1965) and called a multivariate Burr distribution. Takahasi observed that this multivariate distribution has marginals and conditionals of the same type as shown above in the more general case in which $\underline{\mu}$ and $\underline{\sigma}$ are not constrained to be $\underline{0}$ and $\underline{1}$, respectively. Johnson, Kotz and Balakrishnan (2000, pp. 606–611) provided some additional material on Takahasi's distribution.

Using (6.1.20) and our knowledge of moments of the univariate Pareto IV distribution (recall (3.3.11)), we may easily deduce that

$$E(X_1|X_2,X_3,\ldots,X_k) = \mu_1+\sigma_1\left[1+\sum_{j=2}^{k}\left(\frac{X_j-\mu_j}{\sigma_j}\right)^{1/\gamma_j}\right]^{\gamma_1}\frac{\Gamma(\alpha+k-1-\gamma_1)\Gamma(1+\gamma_1)}{\Gamma(\alpha+k-1)} \tag{6.1.22}$$

and

$$E(X_i|X_j) = \mu_i+\sigma_i\left[1+\left(\frac{X_j-\mu_j}{\sigma_j}\right)^{1/\gamma_j}\right]^{\gamma_i}\frac{\Gamma(\alpha+1-\gamma_i)\Gamma(1+\gamma_i)}{\Gamma(\alpha+1)}. \tag{6.1.23}$$

To obtain expressions for the variances and covariances, it is convenient to take advantage of a representation of the multivariate Pareto (IV) distribution as a gamma mixture of independent Weibull variables. In fact, Takahasi (1965) used such a representation to introduce his multivariate Burr distribution. Suppose that $Z \sim \Gamma(\alpha,1)$

and that, given $Z = z$, $X_1, X_2, \ldots, X_k$ are independent Weibull random variables with

$$P(X_i > x_i) = e^{-z\left(\frac{x_i - \mu_i}{\sigma_i}\right)^{1/\gamma_i}}, \quad x_i > \mu_i \quad i = 1, 2, \ldots k. \tag{6.1.24}$$

An elementary computation shows that unconditionally it is the case that $\underline{X} \sim MP^{(k)}(\text{IV})(\underline{\mu}, \underline{\sigma}, \underline{\gamma}, \alpha)$. Thus, if $\underline{X}$ has a k-dimensional Pareto distribution of type IV, we may act as if the X_i's have the representation

$$X_i = \mu_i + \sigma_i (W_i/Z)^{\gamma_i}, \quad i = 1, 2, \ldots, k \tag{6.1.25}$$

where the W_i's are independent identically distributed $\Gamma(1,1)$ variables (i.e., standard exponential variables) and Z, independent of the W_i's, has a $\Gamma(\alpha, 1)$ distribution. Using (6.1.25) we find

$$E(X_i | Z) = \mu_i + \sigma_i \Gamma(\gamma_i + 1) Z^{-\gamma_i} \tag{6.1.26}$$

and thus

$$E(X_i) = \mu_i + \sigma_i \Gamma(\alpha - \gamma_i)\Gamma(\gamma_i + 1)/\Gamma(\alpha) \tag{6.1.27}$$

(a result earlier derived for the univariate Pareto IV, see (3.3.11)). By the same kind of conditioning argument (or, if one prefers, by use of the representation (6.1.25)) one finds

$$\text{var}(X_i) = \sigma_i^2 \left[\frac{\Gamma(\alpha - 2\gamma_i)\Gamma(2\gamma_i + 1)}{\Gamma(\alpha)} \left(\frac{\Gamma(\alpha - \gamma_i)\Gamma(\gamma_i + 1)}{\Gamma(\alpha)} \right)^2 \right], \tag{6.1.28}$$

and

$$\text{cov}(X_i, X_j) = \frac{\sigma_i \sigma_j \Gamma(1 + \gamma_i)\Gamma(1 + \gamma_j)\{\Gamma(\alpha)\Gamma(\alpha - \gamma_i - \gamma_j) - \Gamma(\alpha - \gamma_i)\Gamma(\alpha - \gamma_j)\}}{[\Gamma(\alpha)]^2}. \tag{6.1.29}$$

An expression for the correlation between X_1 and X_2 was displayed in Takahasi (1965). Johnson, Kotz and Balakrishnan (2000) provided approximations for the covariances when the γ_i's are small.

Remark. *The $MP^{(k)}(II)(\underline{0}, \underline{\sigma}, \alpha)$ distribution was used by Lindley and Singpurwalla (1986) as an example of a model for the distribution of lifelengths of components sharing a common environment. A representation of the form (6.1.25) with $\underline{\mu} = \underline{0}$ and $\underline{\gamma} = \underline{1}$ was used in their example.*

A natural generalization of the representation (6.1.25) suggests itself. Instead of requiring the W_i's to be exponential variables, we can permit them to be gamma variables. The resulting distribution will be called k-dimensional Feller-Pareto, since its marginals are of the Feller-Pareto form. Thus $\underline{X} \sim FP^{(k)}(\underline{\mu}, \underline{\sigma}, \underline{\gamma}, \alpha, \underline{\beta})$ if

$$X_i = \mu_i + \sigma_i (W_i/Z)^{\gamma_i}, \quad i = 1, 2, \ldots, k \tag{6.1.30}$$

where the W_i's and Z are independent random variables with $W_i \sim \Gamma(\beta_i, 1)$, $(i = 1, 2, \ldots, k)$, and $Z \sim \Gamma(\alpha, 1)$. The marginal and conditional distributions of the multivariate Feller-Pareto distribution are again multivariate Feller-Pareto. The covariance structure can be readily obtained from the representation (6.1.30).

An enormously flexible extension of this model which still has Feller-Pareto marginals but which has, for most applications, far too many parameters is one of the form

$$X_i = \mu_i + \sigma_i (W_i/Z_i)^{\gamma_i}, \quad i = 1, 2, \ldots, k,$$

where $\underline{W}$ and $\underline{Z}$ are independent random vectors with k-variate gamma distributions with unit scale parameters. There are many k-variate gamma distributions that might be used in this construction.

6.1.3 *Distributional Properties of the Generalized Multivariate Pareto Models*

The multivariate Pareto II distribution has many interesting structural properties related to its representation in terms of independent exponential and gamma variables. Referring to (6.1.25), a representation of the $MP^{(k)}(\text{II})$ distribution may be written in the form (for the $MP^{(k)}(\text{II}), \underline{\gamma} = \underline{1}$)

$$X_i = \mu_i + \sigma_i (W_i/Z) \tag{6.1.31}$$

where the W_i's are independent standard exponential variables and $Z \sim \Gamma(\alpha, 1)$ (independent of the W_i's). To further simplify the analysis, assume that $\underline{\mu} = \underline{0}$. Thus our vector $\underline{X}$ has survival function

$$\overline{F}_{\underline{X}}(\underline{x}) = \left(1 + \sum_{i=1}^{k} \frac{x_i}{\sigma_i}\right)^{-\alpha}, \quad x_i > 0, \quad i = 1, 2, \ldots, k. \tag{6.1.32}$$

Much of the subsequent analysis can be extended to the $MP^{(k)}(\text{IV})$ case under the restriction of homogeneous γ_i's (i.e., $\gamma_i = \gamma$, $i = 1, 2, \ldots, k$) and $\underline{\mu} = \underline{0}$. Such a distribution is obtainable from the $MP^{(k)}(\text{II})$ distribution by applying the same power transformation to all coordinates.

Suppose $\underline{X} \sim MP^{(k)}(\text{II})(\underline{0}, \underline{\sigma}, \alpha)$. What can we say about the distribution of X_i/X_j? Using the representation (6.1.31), we see that the random scale parameter Z cancels out in such expressions, and from known results for independent exponential variables we find

$$\begin{aligned} P((X_i/X_j) > u) &= P\left((W_i/W_j) > \tfrac{\sigma_j}{\sigma_i} u\right) = P\left(\tfrac{W_j}{W_i + W_j} < \left(1 + \tfrac{\sigma_j}{\sigma_i} u\right)^{-1}\right) \\ &= \left(1 + \tfrac{\sigma_j}{\sigma_i} u\right)^{-1} \end{aligned} \tag{6.1.33}$$

since $W_j/(W_i + W_j) \sim \text{Uniform}(0, 1)$. Thus $(X_i/X_j) \sim P(\text{II})(0, (\sigma_i/\sigma_j), 1)$. Nadarajah (2009) derives the density of $X_i/(X_i + X_j)$ in this setting, from which (6.1.33) could be readily derived. He also provides expressions for the densities of X_i+X_j and X_iX_j. For example

$$f_{X_i+X_j}(z) = \frac{\alpha}{\sigma_i - \sigma_j)}\left[\left(1+\frac{z}{\sigma_i}\right)^{-\alpha-1} - \left(1+\frac{z}{\sigma_j}\right)^{-\alpha-1}\right], \quad z>0,$$

provided that $\sigma_i \neq \sigma_j$.

See Malik and Trudel (1985) for an expression for the density of X_i/X_j when $\underline{X} \sim MP^{(k)}(\text{I})(\underline{\sigma}, \alpha)$.

The above analysis can be carried even further in the case of the standardized multivariate Pareto II distribution (i.e., $\underline{\mu} = \underline{0}$, $\underline{\sigma} = \underline{1}$). Here the joint survival function is of the form

$$\overline{F}_{\underline{X}}(\underline{x}) = \left(1+\sum_{i=1}^{k} x_i\right)^{-\alpha}, \quad x_i > 0, \quad i = 1,2,\ldots,k, \tag{6.1.34}$$

and the corresponding density is

$$f_{\underline{X}}(\underline{x}) = \Gamma(\alpha+k)\bigg/\left[\Gamma(\alpha)\left(1+\sum_{i=1}^{k} x_i\right)^{\alpha+k}\right], \quad x_i > 0, \quad i = 1,2,\ldots,k. \tag{6.1.35}$$

If we let $Y = \sum_{i=1}^{k} X_i$, it follows, using Jacobians, that Y is a Feller-Pareto variable. One finds $Y \sim FP(0,1,1,\alpha,k)$. Clearly, the sum of any subgroup of the X_i's has a Feller-Pareto distribution, i.e., $\sum_{i=1}^{k_1} X_i \sim FP(0,1,1,\alpha,k_1)$. We already know that $(X_i/X_j) \sim P(\text{II})(0,1,1)$. Another useful property of the standardized $MP^{(k)}(\text{II})$ distribution is that for any $j \leq k$, we have

$$Y_j = X_j\bigg/\left(1+\sum_{i=1}^{j-1} X_i\right) \sim P(\text{II})(0,1,\alpha+j-1). \tag{6.1.36}$$

Moreover, one may verify that the Y_j's $(j = 1,2,\ldots,k)$ are independent random variables (this can be verified using the representation (6.1.31)). It is, thus, possible to construct a multivariate Pareto II random vector in a straightforward manner using independent univariate Pareto II variables.

Theorem 6.1.1. *Let $Y_1, Y_2, \ldots, Y_k$ be independent random variables with $Y_j \sim P(\text{II})(0,1,\alpha+j-1)$, $j = 1,2,\ldots,k$. Define $X_1 = Y_1$ and for $1 < j \leq k$ define $X_j = Y_j \prod_{i<j}(1+Y_i)$. It follows that $\underline{X} \sim MP^{(k)}(\text{II})(\underline{0},\underline{1},\alpha)$.*

Proof. Use Jacobians. □

The representation given by Theorem 6.1.1 can be used in computing the covariance and correlation structure of an $MP^{(k)}(\text{II})$ distribution. It also might be helpful when one wishes to generate pseudo-random samples from an $MP^{(k)}(\text{II})$ distribution. Of course, for these purposes, it is hard to improve upon the advantages already inherent in the representation (6.1.31).

Closely related to the above representation is one involving a Dirichlet distribution. If we begin with $(V_1, V_2, \ldots, V_{k+1}) \sim \text{Dirichlet}(\beta_1, \ldots, \beta_k, \alpha)$ and define $X_i = V_i/V_{k+1}$, $i = 1, 2, \ldots, k$, we find that $\underline{X} \sim FP^{(k)}(\underline{0}, \underline{1}, \underline{1}, \alpha, \underline{\beta})$. The particular choice $\beta_i = 1$, for $i = 1, 2, \ldots, k$, leads to the standardized $MP^{(k)}(\text{II})$ distribution.

Even though the coordinates of an $MP^{(k)}(\text{II})$ random vector are not independent they still can be ordered, and it makes sense to discuss the order statistics $X_{1:k}, X_{2:k}, \ldots, X_{k:k}$ and the spacings, $Y_{i:k} = X_{i:k} - X_{i-1:k}$. If we assume $X \sim MP^{(k)}(\text{II})(\underline{0}, \underline{\sigma}, \alpha)$ then we find

$$\begin{aligned} P(X_{1:k} > y) = P(X_1 > y, \ldots, X_k > y) &= \left(1 + \sum_{i=1}^{k} \frac{y}{\sigma_i}\right)^{-\alpha} \\ &= \left[1 + y\left(\sum_{i=1}^{k} \frac{1}{\sigma_i}\right)\right]^{-\alpha}. \end{aligned}$$

Thus $$X_{1:k} \sim P(\text{II})\left(0, \left(\sum_{i=1}^{k} \frac{1}{\sigma}\right)^{-1}, \alpha\right). \tag{6.1.37}$$

The distributions associated with higher order statistics can be evaluated by use of the inclusion-exclusion formula, but relatively simple expressions are only obtainable when one assumes scale homogeneity, i.e., all σ_i's equal. For convenience assume all σ_i's are 1 so that we are again dealing with the standardized $MP^{(k)}(\text{II})$ distribution (see (6.1.34)). Such a distribution is a scale mixture of independent exponentials. Given a particular value of the mixing variable, the scaled spacings are independent identically distributed exponential variables (using the Sukhatme result, Theorem 3.4.2). To remove the conditioning, we again integrate out the random scale and once more arrive at a standardized $MP^{(k)}(\text{II})$ distribution. We may summarize these observations in:

Theorem 6.1.2. *Let $\underline{X} \sim MP^{(k)}(\text{II})(\underline{0}, \underline{1}, \alpha)$ and let $\underline{S}$ represent the corresponding vector of scaled spacings, i.e., $S_i = (k - i + 1)(X_{i:k} - X_{i-1:k})$, $i = 1, 2, \ldots, k$. It follows that $\underline{S} \sim MP^{(k)}(II)(\underline{0}, \underline{1}, \alpha)$.*

The conclusion of the theorem can be stated in the form $\underline{S} \overset{d}{=} \underline{X}$. The argument used to justify this conclusion clearly extends to any case where $\underline{X}$ is a scale mixture of independent exponential variables.

The technique used to obtain Theorem 6.1.2 can be used to obtain the distribution of the ith order statistic and the $(k_2 - k_1)$ generalized spacing of a standardized $MP^{(k)}(\text{II})$ random vector. These can be represented as linear functions of the spacings. For example, we can write

$$X_{i:k} = \sum_{j=1}^{i} (X_{j:n} - X_{j-1:n}) \overset{d}{=} \left[\sum_{j=1}^{i} (W_{j:n} - W_{j-1:n})\right] \Big/ Z \tag{6.1.38}$$

where the $W_{j:n}$'s are standard exponential order statistics and $Z \sim \Gamma(\alpha, 1)$. Given $Z = z$, it follows that $X_{i:n}$ is distributed as a sum of i independent exponential variables with means $[(k-j+1)z]^{-1}$, $j=1,2,\ldots,i$. Using the observation in Feller (1971, p. 40), we can write

$$P(X_{i:k} > u|Z=z) = \sum_{j=1}^{i} e^{-z(k-j+1)u} \prod_{\substack{\ell=1 \\ \ell\neq j}}^{i} \left(\frac{k-\ell+1}{\ell-j}\right). \tag{6.1.39}$$

Recalling that $Z \sim \Gamma(\alpha,1)$, this yields immediately

$$P(X_{i:k} > u) = \sum_{j=1}^{i} [1+(k-j+1)u]^{-\alpha} \prod_{\substack{\ell=1 \\ \ell\neq j}}^{i} \left(\frac{k-\ell+1}{\ell-j}\right). \tag{6.1.40}$$

This should be compared with (3.4.10), which gives the corresponding result for the case in which the X_i's are independent P(II) variables. It is clearly not difficult to write an expression analogous to (6.1.40) for the distribution of the (k_2-k_1) generalized spacing $(X_{k_2:k}-X_{k_1:k})$ of a standardized $MP^{(k)}$(II) random vector.

In a financial setting, a multivariate Pareto distribution might be used to model the dependent risks associated with lines of business. Vernic (2011) discusses such scenarios in some detail. In that context we might have $\underline{X} \sim MP^{(k)}(\text{II})(\mu\underline{1},\sigma\underline{1},\alpha)$, and we will be interested in the variables $X_{1:k}, X_{k:k}$ and $T=\sum_{i=1}^{k} X_i$. Note that we have assumed common values of μ and σ for each coordinate variable. In this case we may use a mixture representation of the $MP^{k}(II)$ distribution, as in (6.1.24) with $\gamma_i = 1$ for each i. It is then possible to express the distributions of $X_{1:k}, X_{k:k}$ and T as finite scale mixtures of $P(II)(\mu,\sigma,\alpha)$ distributions. See Vernic (2011) for more detail on this claim. He also treats a more general problem with heterogeneous μ_i's and σ_i's.

Remark. *Goodman and Kotz (1981) discuss the $MP^{(k)}$(I) distribution in the context of hazard rates based on isoprobability contours.*

Next, we will discuss certain distributional properties of the $MP^{(k)}(IV)$ distribution with survival function (6.1.18). Of course these results will also apply to $MP^{(k)}(I)$, $MP^{(k)}(II)$ and $MP^{(k)}(III)$ distributions. Before discussing the general $MP^{(k)}(IV)$ distribution, we note that a homogeneous version of the $MP^{(k)}(IV)$ distribution, in which all the inequality parameters (the γ_i's) are assumed to be equal to γ say, is in general not much more difficult to deal with than is the $MP^{(k)}(II)$ model. Many of the nice properties of the $MP^{(k)}(II)$ model discussed above require only minor modification if we are dealing with the homogeneous $MP^{(k)}(IV)(\underline{\mu},\underline{\sigma},\gamma\underline{1},\alpha)$ distribution. For details, see Yeh (1994).

A salient feature of the $MP^{(k)}(IV)$ distribution, that we will utilize subsequently, is that it admits a representation as a scale mixture of independent Weibull variables. This is evident from (6.1.24) in which it is clear that, given $Z=z$, the X_i's are indeed independent Weibull variables. Among the coordinates of a single $MP^{(k)}(IV)$ variable there can be closure under minimization. It will occur if $\underline{\mu}=\underline{0}$, $\underline{\sigma}=\sigma\underline{1}$ and $\underline{\gamma}=\gamma\underline{1}$. Thus for $\underline{X} \sim MP^{(k)}(IV)(\underline{0},\sigma\underline{1},\gamma\underline{1},\alpha)$ with coordinate random variables $X_1,X_2,\ldots,X_k$, if we consider $X_{1:k} = min\{X_1,X_2,\ldots,X_k\}$ it may be verified that

$X_{1:k} \sim P(IV)(0, \sigma k^{-\gamma}, \gamma, \alpha)$. See Yeh (2009) for discussion of weighted minima of the coordinates of an $MP^{(k)}(IV)$ variable.

Mardia (1964) observed that coordinatewise minima of random samples from a multivariate Pareto distribution of type I are themselves multivariate Pareto I. This result is generally true for multivariate Pareto distributions of types I, II, III and IV. The result for type IV may be stated as follows. Let $\underline{X}^{(1)}, \ldots, \underline{X}^{(n)}$ be independent identically distributed $MP^{(k)}(\mathrm{IV})(\underline{\mu}, \underline{\sigma}, \underline{\gamma}, \alpha)$ variables and define $\underline{Y} = \min_j \underline{X}^{(j)}$ (the coordinatewise minimum). It follows that $\underline{Y} \sim MP^{(k)}(\mathrm{IV})(\underline{\mu}, \underline{\sigma}, \underline{\gamma}, n\alpha)$. This result is transparent from (6.1.18). Other order statistics are less tractable. A slight generalization of this result is one in which the $\underline{X}^{(j)}$'s are independent with $\underline{X}^{(j)} \sim MP^{(k)}(IV)((\underline{\mu}, \underline{\sigma}, \underline{\gamma}, \alpha_j)$. In this case we have

$$\min_j \underline{X}^{(j)} \sim MP^{(k)}(IV)((\underline{\mu}, \underline{\sigma}, \underline{\gamma}, \sum_{i=1}^{n} \alpha_j).$$

Mardia (1964) did propose a technique for evaluating the joint density of the coordinatewise ranges (i.e., $\underline{R} = \max_j \underline{X}^{(j)} - \min_j \underline{X}^{(j)}$). For $k = 2$ and α an integer, the required integration can be effected by using partial fractions.

The observation that a minimum of n i.i.d. $MP^{(k)}(IV)$ variables is again an $MP^{(k)}(IV)$ variable allows us to conclude that an $MP^{(k)}(IV)$ distribution is min-infinitely divisible (as defined by Alzaid and Proschan (1994)). A random vector $\underline{X}$ is min-infinitely divisible if, for every integer n, the n'th root of its joint survival function is a valid joint survival function. This property of the $MP^{(k)}(IV)$ distribution becomes transparently true upon reference to (6.1.18).

Closure under geometric minimization and maximization, which was shown in Chapters 2 and 3 to occur in one dimension for $P(III)$ variables, continues to hold in k-dimensions for $MP^{(k)}(III)$ variables. Thus if $\underline{X}^{(1)}, \underline{X}^{(2)}, \ldots$ is a sequence of i.i.d. $MP^{(k)}(III)(\underline{\mu}, \underline{\sigma}, \underline{\gamma})$ variables and if $N(p)$, independent of the $\underline{X}^{(i)}$'s has a geometric (p) distribution, then

$$\underline{Y} = \min_{1=1,2,\ldots,N(p)} \underline{X}^{(i)} \sim MP^{(k)}(III)(\underline{\mu}, p\underline{\sigma}, \underline{\gamma}) \tag{6.1.41}$$

and

$$\underline{Z} = \max_{1=1,2,\ldots,N(p)} \underline{X}^{(i)} \sim MP^{(k)}(III)(\underline{\mu}, p^{-1}\underline{\sigma}, \underline{\gamma}). \tag{6.1.42}$$

As a consequence of these observations, it can be stated that an $MP^{(k)}(III)$ distribution is min-geometric stable and max-geometric stable. Since the results in (6.1.41) and (6.1.42) hold for every $p \in (0,1)$, it follows that $MP^{(k)}(III)$ distributions are max-geometric-infinitely divisible and min-geometric-infinitely divisible (for further discussion of these concepts and their consequences, see Rachev and Resnick (1991)).

Remark. *Zografos and Nadarajah (2005) supply expressions for the Renyi and Shannon entropies of* $MP^{(k)}(IV)$ *distributions.*

6.1.4 Some Characterizations of Multivariate Pareto Models

Jupp and Mardia (1982) discussed certain characteristic properties of the $MP^{(k)}(I)(\underline{\sigma},\alpha)$ distribution. In one dimension, it is well known that a linear mean residual life function is a characteristic property of the Pickands generalized Pareto distribution. Jupp and Mardia consider a k-dimensional version of this result. To avoid complications due to the fact that the marginal densities of an $MP^{(k)}(I)$ distribution have support sets which depend on $\underline{\sigma}$, we will instead consider $MP^{(k)}(II)(0,\underline{\sigma},\alpha)$ variables whose coordinate random variables are positive (i.e., have support $(0,\infty)$). For a k-dimensional random vector with positive coordinate random variables, we define its mean residual life function to be

$$m_{\underline{X}}(\underline{t}) = E(\underline{X}-\underline{t}|\underline{X}>\underline{t}), \ \ \underline{t}\in(0,\infty)^k. \tag{6.1.43}$$

In the case of an $MP^{(k)}(II)(0,\underline{\sigma},\alpha)$ distribution one finds

$$m_{\underline{X}}(\underline{t}) = c\underline{\sigma}\left[1+\sum_{i=1}^{k}\frac{t_i}{\sigma_i}\right], \ \ \underline{t}\in(0,\infty)^k, \tag{6.1.44}$$

i.e., it is a linear function of the form $A\underline{t}+\underline{b}$ where $\underline{b}\geq\underline{0}$ and all elements of the matrix A are positive. Jupp and Mardia's results show that it is only a $MP^{(k)}(II)(0,\underline{\sigma},\alpha)$ distribution that has such a linear mean residual life function (paralleling the one dimensional result discussed in Chapter 3). See Pusz (1989) for a more general version of this result.

Yeh (2004a) discusses characterizations of the $MP^{(k)}(III)$ distribution based on its closure under geometric minimization (or maximization). Suppose that $\underline{X}^{(1)},\underline{X}^{(2)},\ldots$ is a sequence of i.i.d. non-negative k-dimensional random vectors and that $N(p)$, independent of the $\underline{X}^{(i)}$'s, has a geometric (p) distribution. Define $\underline{Y(p)} = \min_{1=1,2,\ldots,N(p)}\underline{X}^{(i)}$ and assume that

$$\underline{Y(p)} \stackrel{d}{=} c(p)\underline{X}^{(1)} \tag{6.1.45}$$

for some $c(p)<1$. Provided that the distribution of $\underline{X}^{(1)}$ satisfies the following regularity condition

$$P(\underline{X}^{(1)}>\underline{x}) \sim 1-\sum_{i=1}^{k}\left(\frac{x_i}{\sigma_i}\right)^{1/\gamma_i}, \ \ \underline{x}\to\underline{0}, \tag{6.1.46}$$

then $\underline{X}^{(1)} \sim MP^{(k)}(III)(\underline{0},\underline{\sigma},\underline{\gamma})$. If (6.1.45) is assumed to hold for every $p\in(0,1)$, then the regularity condition (6.1.46) is not required to characterize the $MP^{(k)}(III)(\underline{0},\underline{\sigma},\underline{\gamma})$ distribution. Yeh also shows that the $MP^{(k)}(III)(\underline{0},\underline{\sigma},\underline{\gamma})$ distribution can arise as a limit distribution under repeated geometric minimization (paralleling the one dimensional result discussed in Chapter 3).

The homogeneous $MP^{(k)}(IV)$ distribution admits a characterization involving weighted minima summarized by the following theorem due to Yeh (2004a):

Theorem 6.1.3. *Let $\underline{X}$ be a non-negative k-dimensional random vector. Let $\gamma > 0$ and $\underline{\sigma} > \underline{0}$. The following two statements are equivalent.*

- *For every $\underline{a}$ such that $(\sum_{i=1}^{k} a_i^{1/\gamma})^{\gamma} = 1$ the weighted minimum of the coordinates of $\underline{X}$, denoted by $M(\underline{a})$, has a Pareto(IV) distribution, i.e., $M(\underline{a}) = \min_{1 \le i \le k}\{X_i/(a_i\sigma_i)\} \sim P(IV)(0,1,\gamma,\alpha)$.*
- *$\underline{X} \sim MP^{(k)}(IV)(\underline{0},\underline{\sigma},\gamma\underline{1},\alpha)$.*

6.2 Alternative Multivariate Pareto Distributions

Mardia's original paper on multivariate Pareto distributions discussed two classes of multivariate Pareto distributions. He was concerned with obtaining distributions with Pareto I (i.e., classical Pareto) marginals. There is no difficulty extending his analysis to obtain multivariate distributions with Pareto IV marginals. In fact, it is possible to generate a plethora of such distributions. A convenient reference which discusses methods of generating bivariate distributions with specified marginals is Mardia (1970). Three basic techniques are available for generating multivariate Pareto (IV) distributions. Several other techniques which generate multivariate Pareto III distributions will also be described. The step from bivariate to multivariate in these examples is fraught with notational rather than conceptual difficulties. Rather than risk submersion in a sea of subscripts and parameters, attention will generally be restricted to bivariate distributions in the remainder of this section. The reader will have no difficulty in visualizing the appropriate extensions to the multivariate case. In some cases, indications of the analogous multivariate models will be provided.

We begin by discussing three tried and true methods for generating bivariate Pareto IV distributions. They may be labeled with the terms: (i) mixtures of bivariate Weibulls, (ii) transformations of bivariate exponentials and (iii) trivariate reduction. Following discussion of these three methods, a selection of alternative approaches will be described.

6.2.1 Mixtures of Weibull Variables

Method (i) is motivated by Takahasi's (1965) representation of the $MP^{(2)}$(IV) distribution (discussed in Section 6.1) as a gamma mixture of conditionally independent Weibull variables. The key to this result is that if, given $Z = z$, X has a Weibull survival function of the form $\exp\{-z(\frac{x-\mu}{\sigma})^{1/\gamma}\}$ and if $Z \sim \Gamma(\alpha,1)$, then, unconditionally, $X \sim P(\text{IV})(\mu,\sigma,\gamma,\alpha)$. Takahasi chose to consider independent Weibull variables, but the marginals will still be P(IV), even if we take scale mixtures of dependent Weibulls. Many bivariate Weibull distributions are available for such scale mixing. They can be constructed by taking powers of random vectors with distributions selected from the available wide variety of bivariate exponential distributions. Arnold (1975) described many such bivariate exponential distributions. As an example, suppose that (U_1,U_2) has a bivariate exponential distribution of the Marshall-Olkin type with parameters 1, 1 and λ. Thus, for $u_1,u_2 > 0$,

$$P(U_1 > u_1, U_2 > u_2) = e^{-u_1-u_2-\lambda\max(u_1,u_2)}. \tag{6.2.1}$$

Now suppose $Z \sim \Gamma(\alpha, 1)$ is independent of (U_1, U_2) and define (X_1, X_2) by

$$X_i = \mu_i + \sigma_i (U_i/Z)^{\gamma_i}, \quad i = 1, 2. \tag{6.2.2}$$

It is not difficult to verify that $X_i \sim P(\text{IV})(\mu_i, \sigma_i, \gamma_i, \alpha)$, $i = 1, 2$, so that we do have a bivariate Pareto IV distribution. The joint survival function is of the form

$$\begin{aligned} \overline{F}_{X_1,X_2}(x_1,x_2) = & \left[1 + \left(\frac{x_1 - \mu_1}{\sigma_1} \right)^{1/\gamma_1} + \left(\frac{x_2 - \mu_2}{\sigma_2} \right)^{1/\gamma_2} \right. \\ & \left. + \lambda \max \left\{ \left(\frac{x_1 - \mu_1}{\sigma_1} \right)^{1/\gamma_1}, \left(\frac{x_2 - \mu_2}{\sigma_2} \right)^{1/\gamma_2} \right\} \right]^{-\alpha}, \\ & x_1 > \mu_1, x_2 > \mu_2. \end{aligned} \tag{6.2.3}$$

Observe that in any bivariate Pareto (IV) distribution generated by method (i), the marginals share a common value of α. This will be seen to not be the case for bivariate Pareto (IV) distributions obtained using methods (ii) and (iii). It is not difficult to envision k-dimensional versions of the model (6.2.3).

Chiragiev and Landsman (2009) propose an alternative multivariate Pareto (IV) distribution which also involves a Marshall-Olkin dependency structure. A model with independent Pareto (IV) marginals can be of the following form:

$$X_i = \mu_i + \sigma_i (u_i/Z_i)^{\gamma_i}, \; i = 1, 2, ..., k, \tag{6.2.4}$$

where the U_i's are i.i.d. *exponential*(1) and the Z_i's are independent $\Gamma(\alpha_i, 1)$ random variables, independent of the U_i's. The model (6.2.3) was obtained by assuming a Marshall-Olkin joint density for the U_i's and assuming that $Z_1 = Z_2$. Chiragiev and Landsman (2009) instead assume a Marshall-Olkin joint distribution for the Z_i's and assume that the U_i's are independent. In particular, they assume a simplified Marshall-Olkin type of k-dimensional gamma distribution for the Z_i's involving just one dependence parameter. Thus they assume that $V_0, V_1, ..., V_k$ are independent random variables with $V_i \sim \Gamma(\alpha_i, 1), \; i = 0, 1, ..., k$, and define

$$Z_i = V_i + V_0, \; i = 1, 2, ..., k.$$

If we then define $\underline{X}$ as in (6.2.4), we obtain the following expression for the joint survival function of $\underline{X}$.

$$\overline{F}_{\underline{X}}(\underline{x}) = \left[1 + \sum_{i=1}^{k} \left(\frac{x_i - \mu_i}{\sigma_i} \right)^{1/\gamma_i} \right]^{-\alpha_0} \prod_{i=1}^{k} \left[1 + \left(\frac{x_i - \mu_i}{\sigma_i} \right)^{1/\gamma_i} \right]^{-\alpha_i}, \tag{6.2.5}$$

where $\underline{x} > \underline{\mu}$.

Chiragiev and Landsman (2009) focus on the case in which $\underline{\mu} = \underline{0}$ and $\underline{\sigma} = \underline{1}$. They observe that the model (6.2.5) includes the case with independent marginals, corresponding to the choice of $\alpha_0 = 0$. They also consider an alternative model built by using a dependence structure suggested by Mathai and Moschopoulos (1991). It utilizes a k-variate gamma distribution of the form:

$$Z_i = \sum_{j=1}^{i} Y_j, \quad i = 1,2,...,k, \tag{6.2.6}$$

where the Y_j's are independent gamma variables. Using this construction results in the following joint survival function:

$$\overline{F}_{\underline{X}}(\underline{x}) = \prod_{i=1}^{k} \left[1 + \sum_{j=i}^{k} \left(\frac{x_i - \mu_i}{\sigma_i} \right)^{1/\gamma_i} \right]^{-\alpha_i}, \tag{6.2.7}$$

where $\underline{x} > \underline{\mu}$. Both (6.2.5) and (6.2.7) can be viewed as special cases of a more general model in which the Z_i's are sums of subsets of a collection of $2^k - 1$ independent gamma variables. This results in the following model:

$$\begin{aligned} \overline{F}_{\underline{X}}(\underline{x}) &= \prod_{i=1}^{k} \left[1 + \left(\frac{x_i - \mu_i}{\sigma_i} \right)^{1/\gamma_i} \right]^{-\alpha_i} \\ &\times \prod_{i<j} \left[1 + \left(\frac{x_i - \mu_i}{\sigma_i} \right)^{1/\gamma_i} + \left(\frac{x_j - \mu_j}{\sigma_j} \right)^{1/\gamma_j} \right]^{-\alpha_{ij}} \\ &\times \cdots \times \left[1 + \sum_{i=1}^{k} \left(\frac{x_i - \mu_i}{\sigma_i} \right)^{1/\gamma_i} \right]^{-\alpha_{12\ldots k}}, \end{aligned} \tag{6.2.8}$$

where $\underline{x} > \underline{\mu}$. Parsimonious submodels in which many of the α's are set equal to 0 are most likely to be useful.

6.2.2 *Transformed Exponential Variables*

Method (ii) relies on the fact that if $U \sim \Gamma(1,1)$ (i.e., standard exponential), then

$$\mu + \sigma(e^{U/\alpha} - 1)^{\gamma} \sim P(\text{IV})(\mu, \sigma, \gamma, \alpha) \tag{6.2.9}$$

(compare with (3.2.26) or verify directly). Thus, if (U_1, U_2) is a bivariate random variable having an arbitrary distribution subject only to the constraint that it have standard exponential marginals and if we define

$$X_i = \mu_i + \sigma_i(e^{U_i/\alpha_i} - 1)^{\gamma_i}, \quad i = 1,2, \tag{6.2.10}$$

then (X_1, X_2) has a bivariate Pareto (IV) distribution. As mentioned earlier, there are many bivariate exponential distributions which can be utilized in this manner. For example, consider the bivariate exponential distribution with joint survival function

$$\overline{F}_{\underline{U}}(\underline{u}) = (e^{u_1} + e^{u_2} - 1)^{-1}, \quad u_1 > 0, u_2 > 0. \tag{6.2.11}$$

Application of (6.2.10) to this distribution yields the $MP^{(2)}(\text{IV})$ distribution described in Section 6.1 (which includes Mardia's first type of bivariate Pareto). Mardia's second type of bivariate Pareto is obtained by applying the transformation (6.2.10) to a bivariate exponential distribution of the Wicksell (1933)-Kibble (1941) type. It is possible to write down the density of a Wicksell-Kibble random vector using modified Bessel functions (see, e.g., Mardia, 1962). However, a simple stochastic model, which leads to such a distribution, may be easier to grasp. Fix $p \in (0,1)$ and consider two sequences $\{U_{1j}\}$, $\{U_{2j}\}$ of independent identically distributed exponential random variables with mean p. Let N, independent of the U_{ij}'s, be a geometric random variable with $P(N=n) = p(1-p)^{n-1}$, $n = 1,2,\ldots$. Define

$$U_i = \sum_{j=1}^{N} U_{ij}, \quad i = 1,2. \tag{6.2.12}$$

The random vector (U_1, U_2) defined in this manner may be shown to have joint characteristic function

$$\varphi_{U_1,U_2}(t_1,t_2) = (1 - it_1 - it_2 - pt_1t_2)^{-1}. \tag{6.2.13}$$

That this is the characteristic function of the Wicksell-Kibble bivariate exponential may be verified by reference to Kibble (1941). The representation (6.2.12) as a geometric sum of independent exponentials permits us to recognize the Wicksell-Kibble distribution as a member of one of the classes $(\xi_2^{(2)})$ of bivariate exponential distributions based on hierarchical successive damage introduced by Arnold (1975). Applying (6.2.10) to the density with characteristic function (6.2.13), we are led to a bivariate Pareto IV family which includes Mardia's second type of bivariate Pareto distribution. Mardia (1962) supplied expressions for the joint density, conditional densities, conditional means and variances and the covariance structure applicable when $\gamma_1 = \gamma_2 = 1$. For example, the expression for the correlation between X_1 and X_2 is (when $\gamma_1 = \gamma_2 = 1$)

$$\text{corr}(X_1, X_2) = \frac{(1-p)\sqrt{\alpha_1\alpha_2(\alpha_1-2)(\alpha_2-2)}}{(\alpha_1-1)(\alpha_2-1)-(1-p)^2}, \quad \alpha_1, \alpha_2 > 2. \tag{6.2.14}$$

As an alternative, one might apply the transformation (6.2.10) to a bivariate exponential model of the Marshall-Olkin (1967) type. Muliere and Scarsini (1987) discuss the $MP^{(2)}(I)$ version of this construction. Nadarajah and Kotz (2005b) derive the distributions of sums, products and ratios of the coordinate variables of the Muliere-Scarsini bivariate Pareto distribution. An analogous distribution with $P(II)$ marginals was considered by Al-Ruzaiza and El-Gohary (2007). Further discussion of the k-dimensional version of this construction will be found at the end of the following subsection.

6.2.3 *Trivariate Reduction*

The third technique for generating bivariate Pareto IV is to utilize what Mardia (1970) calls the method of trivariate reduction (cf., Arnold, 1967). This technique is applicable, since the class of univariate Pareto IV distributions is closed under minimization in the sense that if X_i $(i = 1,2)$ are independent with $X_i \sim P(\text{IV})(\mu, \sigma, \gamma, \alpha_i)$, then $\min(X_1, X_2) \sim P(\text{IV})(\mu, \sigma, \gamma, \alpha_1 + \alpha_2)$. With this in mind, consider three independent random variables U_1, U_2, U_3 with $U_i \sim P(\text{IV})(\mu, \sigma, \gamma, \alpha_i)$ and define

$$\begin{aligned} X_1 &= \min(U_1, U_3), \\ X_2 &= \min(U_2, U_3). \end{aligned} \tag{6.2.15}$$

It follows that (X_1, X_2) has a bivariate distribution with Pareto IV marginals. The corresponding joint survival function is of the form

$$\begin{aligned} \overline{F}_{\underline{X}}(\underline{x}) = &\left(1 + \left(\frac{\max(x_1, x_2) - \mu}{\sigma}\right)^{1/\gamma}\right)^{-\alpha_3} \\ &\cdot \left(1 + \left(\frac{x_1 - \mu}{\sigma}\right)^{1/\gamma}\right)^{-\alpha_1} \left(1 + \left(\frac{x_2 - \mu}{\sigma}\right)^{1/\gamma}\right)^{-\alpha_2}, \\ &x_1, x_2 > \mu. \end{aligned} \tag{6.2.16}$$

Besides having Pareto IV marginals, it is clear that the distribution described by (6.2.16) has the property that $\min(X_1, X_2) \sim P(\text{IV})(\mu, \sigma, \gamma, \alpha_1 + \alpha_2 + \alpha_3)$. It will be observed that this distribution, obtained by trivariate reduction, has the perhaps undesirable property that the marginals share common values of μ, σ and γ's The family (6.2.16) can be embedded in a larger class of bivariate Pareto IV distributions which may have differing μ's, σ's and γ's. Such a class is obtainable by applying the transformation (6.2.10) to a bivariate exponential distribution of the Marshall-Olkin type (6.2.1). The resulting family includes (6.2.16), obtainable by setting $\mu_1 = \mu_2 = \mu$, $\sigma_1 = \sigma_2 = \sigma$ and $\gamma_1 = \gamma_2 = \gamma$. The property of having a Pareto IV distribution for $\min(X_1, X_2)$ is only encountered when the marginals share common values for μ, σ and γ. The property of having Pareto distributed minima is not common. Recall that it was encountered earlier with the $MP^{(2)}(\text{II})$ family (see (6.1.37)).

Yeh (2004b) discusses a multivariate Pareto model obtained via marginal transformations applied to a k-dimensional Marshall-Olkin exponential model. The bivariate version of this model has a joint survival function given by

$$\begin{aligned} \overline{F}_{\underline{X}}(\underline{x}) = &\left[1 + \left(\frac{x_1 - \mu_1}{\sigma_1}\right)^{1/\gamma_1}\right]^{-\alpha_1 \lambda_1} \left[1 + \left(\frac{x_2 - \mu_2}{\sigma_2}\right)^{1/\gamma_2}\right]^{-\alpha_2 \lambda_2} \\ &\times \left[\max_{i=1,2} \left\{\left[1 + \left(\frac{x_i - \mu_i}{\sigma_i}\right)^{1/\gamma_i}\right]^{-\alpha_i}\right\}\right]^{\lambda_{12}}, \end{aligned} \tag{6.2.17}$$

where $\underline{x} > \underline{\mu}$. Here the non-negative parameters λ_1, λ_2 and λ_{12} are inherited from the Marshall-Olkin exponential model. The corresponding k-dimensional model is replete with parameters. It is of the form:

$$\begin{aligned}
\overline{F}_{\underline{X}}(\underline{x}) &= \prod_{i=1}^{k}\left[1+\left(\frac{x_i-\mu_i}{\sigma_i}\right)^{1/\gamma_i}\right]^{-\alpha_i\lambda_i} \qquad (6.2.18)\\
&\times \prod_{i<j}\left(\max\left\{\left[1+\left(\frac{x_i-\mu_i}{\sigma_i}\right)^{1/\gamma_i}\right]^{-\alpha_i},\left[1+\left(\frac{x_j-\mu_j}{\sigma_j}\right)^{1/\gamma_j}\right]^{-\alpha_j}\right\}\right)^{\lambda_{ij}}\\
&\times \cdots\times\left[\max_{i=1,2,\ldots,k}\left\{1+\left(\frac{x_i-\mu_i}{\sigma_i}\right)^{1/\gamma_i}\right\}^{-\alpha_i}\right]^{\lambda_{12\ldots k}},
\end{aligned}$$

where $\underline{x} > \underline{\mu}$. Here also the λ parameters are inherited from the Marshall-Olkin exponential model. This distribution does have Pareto (IV) marginals, i.e., $X_i \sim P(IV)(\mu_i,\sigma_i,\gamma_i,\alpha_i\tau_i)$ where τ_i is the sum of all the λ's in (6.2.18) whose subscript includes i. Note that Yeh (2004b) only considers the case in which all the λ's in (6.2.18) are equal to 1. The general model (6.2.18) has a property parallel to the univariate result, that truncation from below is equivalent to marginal rescaling.

A simplified version of (6.2.18) may be of interest. After setting most of the λ parameters equal to zero and relabeling the parameters in the model, it becomes

$$\begin{aligned}
\overline{F}_{\underline{X}}(\underline{x}) &= \prod_{i=1}^{k}\left[1+\left(\frac{x_i-\mu_i}{\sigma_i}\right)^{1/\gamma_i}\right]^{-\beta_i} \qquad (6.2.19)\\
&\times \left[\max_{1=1,2,\ldots,k}\left\{1+\left(\frac{x_i-\mu_i}{\sigma_i}\right)^{1/\gamma_i}\right\}\right]^{-\beta_0},
\end{aligned}$$

where $\underline{x} > \underline{\mu}$. This is recognizable as a k-variate version of (6.2.17). Asimit, Furman and Vernic (2010) consider this distribution in the special case in which $\underline{\gamma} = \underline{1}$. Hanagal (1996) investigates a much simpler version of this distribution, obtained by setting $\underline{\gamma} = \underline{1}$ and $\underline{\mu} = \underline{\sigma} = \sigma\underline{1}$. This distribution has classical Pareto marginal distributions and a joint survival function of the form

$$\overline{F}_{\underline{X}}(\underline{x}) = \left[\frac{\max\{x_1,x_2,\ldots,x_k\}}{\sigma}\right]^{-\beta_0}\prod_{i=1}^{k}\left(\frac{x_i}{\sigma}\right)^{-\beta_i}, \qquad (6.2.20)$$

where $\underline{x} > \sigma\underline{1}$. The bivariate version of this model was discussed by Muliere and Scarsini (1987). Padamadan and Muraleedharan Nair (1994) provide a characterization of this bivariate model involving income misreporting.

6.2.4 *Geometric Minimization and Maximization*

Next we turn to a technique for generating bivariate Pareto III distributions. The key idea in this development is the observation that geometric minima of Pareto III variables are themselves Pareto III variables (see the discussion preceding (3.9.93)).

There are several ways in which such an idea can be used to generate bivariate Pareto III variables.

Suppose $\{U_{1j}\}_{j=1}^{\infty}$ and $\{U_{2j}\}_{j=1}^{\infty}$ are two independent sequences of independent random variables. Assume that $U_{1j} \sim P(\mathrm{III})(\mu_1, \sigma_1, \gamma_1), j = 1,2,\ldots,$ while $U_{2j} \sim P(\mathrm{III})(\mu_2, \sigma_2, \gamma_2), j = 1,2,\ldots$. Let N be a geometric (p) random variable, i.e., $[P(N=k) = pq^{k-1}, k = 1,2,\ldots]$, that is independent of the U_{ij}'s and define

$$X_i = \min\{U_{i1}, U_{i2}, \ldots, U_{iN}\}, \quad i = 1,2. \tag{6.2.21}$$

It is not difficult to determine that (X_1, X_2) has a bivariate distribution with Pareto III marginals. In fact, the resulting joint survival function takes the form

$$\begin{aligned} \overline{F}_{\underline{X}}(\underline{x}) = \Bigg\{ 1 + \frac{1}{p}\Bigg[\left(\frac{x_1-\mu_1}{\sigma_1}\right)^{1/\gamma_1} + \left(\frac{x_2-\mu_2}{\sigma_2}\right)^{1/\gamma_2} \\ + \left(\frac{x_1-\mu_1}{\sigma_1}\right)^{1/\gamma_1} \left(\frac{x_2-\mu_2}{\sigma_2}\right)^{1/\gamma_2} \Bigg] \Bigg\}^{-1} \\ x_1 > \mu_1, \quad x_2 > \mu_2. \end{aligned} \tag{6.2.22}$$

Note that the limiting case, as $p \to 1$, has independent P(III) marginals. Also note that if one takes the α'th power of the survival function in (6.2.22), one obtains a survival function for a bivariate Pareto (IV) distribution. This is a special case of the bivariate Burr distribution introduced by Durling, Owen and Drane (1970). See also Durling (1975).

Instead of taking a univariate geometric minimum of bivariate independent Pareto (III) variables, we can consider a bivariate geometric minimum of univariate Pareto (III) variables. To this end consider $U_1, U_2, \ldots,$ a sequence of independent identically distributed $P(\mathrm{III})(\mu, \sigma, \gamma)$ random variables and let (N_1, N_2) be a bivariate geometric random variable with survival function

$$\begin{aligned} P(N_1 > n_1, N_2 > n_2) &= p_{00}^{n_1}(p_{10} + p_{00})^{n_2 - n_1}, \quad n_1 \leq n_2, \\ &= p_{00}^{n_2}(p_{01} + p_{00})^{n_1 - n_2}, \quad n_2 < n_1. \end{aligned} \tag{6.2.23}$$

This is a fairly standard bivariate geometric distribution. Block (1975) describes several physical models leading to such a distribution and provides an extensive bibliography. Using the bivariate geometric distribution defined by (6.2.23), we may define

$$X_i = \min(U_1, U_2, \ldots, U_{N_i}), \quad i = 1,2. \tag{6.2.24}$$

The random vector (X_1, X_2) has Pareto III marginals (because the marginal distributions of the N_i's are geometric). The survival function of the random vector defined by (6.2.24) is obtained by considering the following stochastic mechanism which generates a bivariate geometric distribution of the type defined by (6.2.23).

An experiment with four possible outcomes (0,0), (0,1), (1,0) and (1,1) is run repeatedly. The probabilities associated with the outcomes on a particular trial are p_{00}, p_{01}, p_{10} and p_{11} (summing to 1). Trials are assumed to be independent. Let N_1

be the number of the first trial on which for the first time either (1,0) or (1,1) occurs and let N_2 be the number of the trial on which for the first time either (0,1) or (1,1) occurs. It is easy to verify that with this definition, (N_1, N_2) has the survival function (6.2.23). To evaluate the survival function of $\underline{X}$ (defined using (6.2.23)), we condition on the outcome of the first of the sequence of trials which generate $\underline{N}$. In this manner we find $\overline{F}_{\underline{X}}(\underline{x})$ must satisfy

$$\overline{F}_{\underline{X}}(\underline{x}) = \left[1 + \left(\frac{\max(x_1, x_2) - \mu}{\sigma}\right)^{1/\gamma}\right]^{-1}$$

$$\times [p_{00}\overline{F}_{\underline{X}}(\underline{x}) + p_{10}\overline{F}_{\underline{X}}(\mu, x_2) + p_{01}\overline{F}_{\underline{X}}(x_1, \mu) + p_{11}], \tag{6.2.25}$$

$$x_1 > \mu, x_2 > \mu.$$

Functional equations analogous to (6.2.25) were discussed by Paulson (1973) (see also Hawkes, 1972, and Block, 1975). In the present case, (6.2.25) is readily solved, since we already know the marginal distributions, i.e., we know

$$\overline{F}_{\underline{X}}(x_1, \mu) = F_{X_1}(x_1) = \left[1 + \frac{1}{p_{10} + p_{11}}\left(\frac{x_1 - \mu}{\sigma}\right)^{1/\gamma}\right]^{-1}, \tag{6.2.26}$$

$$x_1 > \mu,$$

because X_1 is a (univariate) geometric$(p_{10} + p_{11})$ minimum of independent $P(\text{III})(\mu, \sigma, \gamma)$ variables. Similarly, one can write an expression $\overline{F}_{\underline{X}}(\mu, x_2)$. The resulting survival function for (X_1, X_2) is

$$\overline{F}_{\underline{X}}(\underline{x}) = \left[1 - p_{00} + \left(\frac{\max(x_1, x_2) - \mu}{\sigma}\right)^{1/\gamma}\right]^{-1}$$

$$\times \left\{\frac{p_{01}(p_{10} + p_{11})}{p_{10} + p_{11} + \left(\frac{x_1 - \mu}{\sigma}\right)^{1/\gamma}} + \frac{p_{10}(p_{01} + p_{11})}{p_{01} + p_{11} + \left(\frac{x_2 - \mu}{\sigma}\right)^{1/\gamma}} + p_{11}\right\}, \tag{6.2.27}$$

$$x_1 > \mu, x_2 > \mu.$$

Both (6.2.22) and (6.2.27) are bivariate distributions with Pareto III marginals. They can be embedded in a hierarchy of more and more complicated bivariate Pareto III distributions. The manner of constructing these is completely analogous to that used by Arnold (1975) to construct a hierarchy of bivariate exponential distributions. The only basic change is that geometric compounding is replaced by geometric minimization. The essential ingredient of the scheme is as follows. Let $\{(U_{1i}, U_{2i})\}_{i=1}^{\infty}$ be a sequence of independent identically distributed bivariate random variables with $U_{1i} \sim P(\mathrm{III})(\mu_1, \sigma_1, \gamma_1)$ and $U_{2i} \sim P(\mathrm{III})(\mu_2, \sigma_2, \gamma_2)$ and let $\underline{N}$ be a bivariate geometric random variable with survival function (6.2.23) (other bivariate geometric distributions could of course be used). Then if we define

$$X_1 = \min\{U_{11}, U_{12}, \ldots, U_{1N_1}\},$$
$$X_2 = \min\{U_{21}, U_{22}, \ldots, U_{2N_2}\}, \tag{6.2.28}$$

it is clear that $\underline{X}$ is a bivariate Pareto III variable. If we let $\overline{G}$ be the common survival function of the (U_{1i}, U_{2i})'s and let $\overline{F}$ be the survival function of $\underline{X}$, then conditioning on the outcome of the first of the series of trials associated with $\underline{N}$, we find (parallel to (6.2.25)) that

$$\overline{F}(x_1,x_2) = \overline{G}(x_1,x_2)[p_{00}\overline{F}(x_1,x_2) + p_{10}\overline{F}(\mu_1,x_2) + p_{01}\overline{F}(x_1,\mu_2) + p_{11}]. \tag{6.2.29}$$

Repeated application of (6.2.29) yields the desired hierarchy of bivariate Pareto III distributions. Distributions obtained by use of (6.2.29) do not generally have the attractive features of the $MP^{(2)}$(III) described in Section 6.1. Thus, they may fail to have Paretian conditional distributions, tractable moments, etc. They will be of interest if some plausible stochastic mechanism involving a bivariate geometric distribution can be identified as a factor in generating observed bivariate income statistics. Two natural settings in which one might search for such mechanisms are (i) when (X_1, X_2) represents the income of husband and wife pairs and (ii) when (X_1, X_2) represents the income of one individual measured at two different time points. In both cases the corresponding distributions might be expected to have Paretian marginals and to exhibit some degree of dependence.

Since the set of univariate Pareto (III) distributions is closed under both geometric minimization and geometric maximization, it is possible to derive analogs to (6.2.22) and (6.2.27) involving geometric maximization. Details are provided in Yeh (1990). For example, the model analogous to (6.2.22) obtained via geometric maximization has a joint distribution function (not the joint survival function) given by

$$F_{\underline{X}}(\underline{x}) = \left\{1 + \frac{1}{p}\left[\left(\frac{x_1-\mu_1}{\sigma_1}\right)^{-1/\gamma_1} + \left(\frac{x_2-\mu_2}{\sigma_2}\right)^{-1/\gamma_2} + \left(\frac{x_1-\mu_1}{\sigma_1}\right)^{-1/\gamma_1}\left(\frac{x_2-\mu_2}{\sigma_2}\right)^{-1/\gamma_2}\right]\right\}^{-1}, \quad x_1 > \mu_1, \quad x_2 > \mu_2. \tag{6.2.30}$$

Remark. *Another flexible multivariate distribution constructed via geometric minimization will be discussed in Section 6.2.8.*

6.2.5 *Building Multivariate Pareto Models Using Independent Gamma Distributed Components*

Recall that a Feller-Pareto random variable X admits a representation of the form

$$X = \mu + \sigma(Z_2/Z_1)^{\gamma},$$

where $Z_i \sim \Gamma(\gamma_i, 1)$, $i = 1,2$ are independent random variables and $\mu \in (-\infty,\infty)$, $\sigma, \gamma_1, \gamma_2, \gamma \in (0,\infty)$. Equivalently

$$X = \mu + \sigma V^{\gamma}$$

where V has a Beta distribution of the second kind with parameters γ_2 and γ_1. A natural k-dimensional extension of this model will be one in which

$$X_i = \mu_i + \sigma V_i^{\gamma_i}, \quad i = 1,2,...,k$$

where $\underline{V} = (V_1, V_2, ...V_k)$ has a suitable k-dimensional Beta distribution of the second kind. We will focus the discussion on the bivariate case initially and later indicate possible multivariate constructions.

One possible bivariate Beta distribution of the second kind may be represented in the form

$$V_1 = \frac{U_1}{U_0}, \qquad V_2 = \frac{U_2}{U_0}, \tag{6.2.31}$$

where $U_i \sim \Gamma(\alpha_i, 1)$, $i = 0,1,2$, are independent variables. If (V_1, V_2) has the structure indicated in (6.2.31), then $\underline{W}$, defined by $W_1 = V_1/(1+V_1)$ and $W_2 = V_2/(1+V_2)$, has the bivariate Beta distribution (of the first kind) introduced by Olkin and Liu (2003). A positive feature of this bivariate distribution is that it does have a closed form expression for its joint density. See (6.2.33) below. A drawback associated with the model is that, in it, the variables V_1 and V_2 are necessarily positively dependent. A related model which has negative dependence is one in which

$$\widetilde{V}_1 = V_1, \qquad \widetilde{V}_2 = 1/V_1. \tag{6.2.32}$$

Clearly both (V_1, V_2) and $(\widetilde{V}_1, \widetilde{V}_2)$ qualify as having bivariate Beta distributions of the second kind, since their marginal distributions are indeed Beta distributions of the second kind. (V_1, V_2) has positive dependence while $(\widetilde{V}_1, \widetilde{V}_2)$ has negative dependence. Both bivariate random variables have relatively simple joint densities which can be obtained via Jacobians from the Olkin-Liu density given in Equation (1.4) of Olkin and Liu (2003). Thus

$$f_{V_1,V_2}(v_1, v_2) = \frac{v_1^{\alpha_1-1} v_2^{\alpha_2-1} \Gamma(\alpha_0+\alpha_1+\alpha_2) I(0 < v_1 < 1, 0 < v_2 < 1)}{(1+v_1+v_2)^{\alpha_0+\alpha_1+\alpha_2} \Gamma(\alpha_0)\Gamma(\alpha_1)\Gamma(\alpha_2)} \tag{6.2.33}$$

and

$$f_{\widetilde{V}_1,\widetilde{V}_2}(\widetilde{v}_1, \widetilde{v}_2) = \frac{\widetilde{v}_1^{\alpha_1-1} \widetilde{v}_2^{-\alpha_2-1} \Gamma(\alpha_0+\alpha_1+\alpha_2) I(0 < \widetilde{v}_1 < 1, 0 < \widetilde{v}_2 < 1)}{(1+\widetilde{v}_1+\widetilde{v}_2^{-1})^{\alpha_0+\alpha_1+\alpha_2} \Gamma(\alpha_0)\Gamma(\alpha_1)\Gamma(\alpha_2)}. \tag{6.2.34}$$

From these densities the corresponding joint densities of the bivariate Feller-Pareto variables $\underline{X}$ and $\underline{\widetilde{X}}$ where $X_i = \mu_i + \sigma_i V_i^{\gamma_i}$ and $\widetilde{X}_i = \mu_i + \sigma_i \widetilde{V}_i^{\gamma_i}$ are readily obtained.

Arnold and Ng (2011) considered the task of building bivariate Beta distributions of the second kind, using independent gamma distributed components, that will exhibit both positive and negative dependence. They introduced a 5-parameter model and a highly flexible 8-parameter model. To construct the 5-parameter model, we begin with independent gamma variables $U_i \sim \Gamma(\alpha_i, 1),\ \ i = 1,2,...,5$ and define

$$V_1 = \frac{U_1 + U_3}{U_4 + U_5}, \qquad V_2 = \frac{U_2 + U_4}{U_3 + U_5}. \tag{6.2.35}$$

The corresponding $FP^{(2)}$ random vector is then $\underline{X} = (X_1, X_2)$ where

$$X_i = \mu_i + \sigma_i V_i^{\gamma_i}, \ \ i = 1,2. \tag{6.2.36}$$

In general, this model will not have a joint density that is available in a closed form. It does have Feller-Pareto marginals so that an estimation strategy involving marginal maximum likelihood and the method of moments can yield reasonable parameter estimates. For details in the bivariate Beta of the second kind case, see Arnold and Ng (2011).

An even more flexible eight parameter bivariate Beta model can be considered. For it, let $U_1, U_2, ..., U_8$ be independent random variables with $U_i \sim \Gamma(\alpha_i, 1)$ $i = 1,2,...,8$, and define

$$V_1 = \frac{U_1 + U_5 + U_7}{U_3 + U_6 + U_8}, \qquad V_2 = \frac{U_2 + U_5 + U_8}{U_4 + U_6 + U_7}, \tag{6.2.37}$$

and then define $\underline{X} = (X_1, X_2)$ as in (6.2.36).

Multivariate extensions of these concepts are readily envisioned, although they may involve an inordinate number of parameters in the most general cases. An example of a k-dimensional Feller-Pareto model with relatively few parameters but capable of exhibiting both positive and negative dependence among its coordinate random variables may be constructed as follows. Begin with $2k+1$ independent random variables U_0 and $\{U_{i,j}\}_{i=1,j=1}^{2,k}$ where $U_0 \sim \Gamma(\alpha_0, 1)$ and $U_{i,j} \sim \Gamma(\alpha_{i,j}, 1)$. Then define, for $i = 1,2,...,k$,

$$V_i = \frac{U_{1,i} + U_0}{U_{2,i} + \sum_{j=1, j \neq i}^{k} U_{1,j}}, \tag{6.2.38}$$

and then, once more, define $\underline{X}$ using (6.2.36).

Needless to say, an enormous spectrum of k-variate Feller-Pareto distributions can be constructed in this fashion. The only requirement is that we begin with a large collection of independent random variables $\{U_\ell : \ell \in \Lambda\}$ with $U_\ell \sim \Gamma(\alpha_\ell, 1)$ for each $\ell \in \Lambda$. Then identify k pairs of disjoint subsets of Λ, say $(\Lambda_i^{(1)}, \Lambda_i^{(2)}),\ \ i = 1,2,...,k$. The corresponding k-dimensional Feller-Pareto variable is then defined by applying the marginal transformation (6.2.36) to the random vector $\underline{V}$ defined as follows.

$$V_i = \frac{\sum_{\ell \in \Lambda_i^{(1)}} U_\ell}{\sum_{\ell \in \Lambda_i^{(2)}} U_\ell}, \ \ i = 1,2,...,k. \tag{6.2.39}$$

Note that it is only assumed that $\Lambda_i^{(1)} \cap \Lambda_i^{(2)} = \phi$ for each i. Except for this requirement, there can be considerable quite arbitrary non-empty intersections among the $2k$ subsets $\{\Lambda_i^{(1)}\}_{i=1}^k$ and $\{\Lambda_i^{(2)}\}_{i=1}^k$. Only in special cases will the resulting joint density of $\underline{V}$ be available in closed form. However, since the structure of $\underline{V}$ involves simple functions of independent gamma variables, parameter estimation and selection among hierarchically ordered submodels will be feasible.

There are other available strategies for building multivariate Feller-Pareto models using gamma distributed components. It will suffice to mention just two of them, and, for simplicity, focus on the bivariate case for each of them. It will be easy to visualize how to construct analogous higher dimensional models. We will make use of some properties of the Beta distribution. We will write $X \sim B(\alpha, \beta)$ to indicate that X has a Beta distribution (of the first kind) and $X \sim B^{(2)}(\alpha, \beta)$ to indicate that X has a Beta distribution of the second kind. First recall the well-known fact that if $U_1 \sim B(\alpha_1, \alpha_2)$ and $U_2 \sim B(\alpha_1 + \alpha_2, \alpha_3)$ are independent random variables then

$$U_1 U_2 \sim B(\alpha_1, \alpha_2 + \alpha_3). \tag{6.2.40}$$

A less well-known fact is that if $U_1 \sim B(\frac{\alpha}{2}, \frac{\beta}{2})$ and $U_2 \sim B(\frac{\alpha+1}{2}, \frac{\beta}{2})$ are independent random variables then

$$\sqrt{U_1 U_2} \sim B(\alpha, \beta). \tag{6.2.41}$$

See Krysicki (1999) for discussion of and extensions of this result.

Using result (6.2.40) and defining $V_1 = U_1/(1 - U_1)$ and $V_2 = U_1 U_2/(1 - U_1 U_2)$ yields a bivariate Beta variable of the second kind of the form

$$(V_1, V_2) \stackrel{d}{=} \left(\frac{W_1}{W_2}, \frac{W_1}{W_2 + W_3} \right), \tag{6.2.42}$$

where $W_i \sim \Gamma(\alpha_i, 1),\ i = 1, 2, 3$ are independent random variables.

Using result (6.2.41) leads to

$$(V_1, V_2) \stackrel{d}{=} \left(\frac{W_1}{W_2}, \frac{\sqrt{\frac{W_1}{W_1+W_2} \frac{W_3}{W_3+W_4}}}{1 - \sqrt{\frac{W_1}{W_1+W_2} \frac{W_3}{W_3+W_4}}} \right), \tag{6.2.43}$$

where the W_i's are independent random variables with $W_1 \sim \Gamma(\alpha/2, 1)$, $W_2 \sim \Gamma(\beta/2, 1)$, $W_3 \sim \Gamma((\alpha+1)/2, 1)$ and $W_4 \sim \Gamma(\beta/2, 1)$.

Application of the transformation (6.2.36) will then yield bivariate Feller-Pareto models. Note that (6.2.42) is a special case of the general model (6.2.39), while (6.2.43) is an, albeit complicated, alternative model.

6.2.6 *Other Bivariate and Multivariate Pareto Models*

One popular approach to the construction of bivariate and multivariate distributions with given marginals is usually described as being one based on copulas. A multivariate copula is simply a k-dimensional distribution function with $uniform(0, 1)$

marginals. Marginal transformations of any copula will yield a multivariate distribution with desired marginals. The dependence structure will be inherited from that of the particular copula used in the construction. Equivalently, we could begin with any analytically tractable multivariate distribution for a random vector $(Z_1, Z_2, ..., Z_k)$ and via marginal transformations obtain a multivariate distribution with Pareto (IV) marginals. Thus if we denote the distribution function of a $P(IV)(\mu, \sigma, \gamma, \alpha)$ variable by $F_{\mu,\sigma,\gamma,\alpha}$ and denote the marginal distribution function of each Z_i by G_i, then a random vector $\underline{X} = (X_1, X_2, ..., X_k)$ with Pareto (IV) marginals can be defined by

$$X_i = F^{-1}_{\mu_i,\sigma_i,\gamma_i,\alpha_i}(G_i(Z_i)). \tag{6.2.44}$$

The copula involved in this construction is the joint distribution function of the random vector $(G_1(Z_1), G_2(Z_2), ..., G_k(Z_k))$.

A popular choice for the distribution of $\underline{Z}$ is a k-variate normal distribution with zero means, unit variances and correlation matrix R. In this case $G_i(z) = \Phi(z),\ i = 1, 2, ..., k$. Another possible choice for the distribution of $\underline{Z}$ is a multivariate exponential distribution of one of the many available kinds. The model obtained by using a Marshall-Olkin multivariate exponential model in this approach has been discussed in Section 6.2.3 (see equations (6.2.17) and (6.2.18)).

It is interesting to note that, since for a $uniform(0,1)$ random variable U we have $1 - U \stackrel{d}{=} U$, a minor modification of the copula approach involving the transformation (6.2.36) will yield different k-dimensional models, sometimes with less complicated structure. For it, instead of (6.2.44), we apply marginal transformations of the form

$$X_i = F^{-1}_{\mu_i,\sigma_i,\gamma_i,\alpha_i}(1 - G_i(Z_i)). \tag{6.2.45}$$

As an illustration of this phenomenon, consider the bivariate case in which (Z_1, Z_2) has a bivariate uniform distribution of the Cuadras and Augé (1981) form, i.e.,

$$P(Z_1 \leq z_1, Z_2 \leq z_2) = z_1^{1-\theta} z_2^{1-\theta} (min\{z_1, z_2\})^{\theta}, \tag{6.2.46}$$

where $0 \leq z_1, z_2 \leq 1$ and $\theta \in (0,1)$. If we apply the transformation (6.2.44) to this distribution (which in fact is a copula), the resulting distribution is quite complicated. Whereas, if we use the transformation (6.2.45), the resulting bivariate survival function is a familiar one, i.e., the transformed Marshall-Olkin model (6.2.17).

In the context of multivariate exceedences over high thresholds, another class of multivariate Pareto distributions has a clear role. They are usually described as distributions with marginals of the Pickands generalized Pareto type. The key property of such multivariate Pareto models is that, when marginally transformed, the maximum of the coordinates has a Pareto distribution. For example, we can begin with an arbitrary k-dimensional random variable $\underline{W}$ with non-negative coordinate random variables $W_1, W_2, ..., W_k$ which have positive expectations and are such that $\max_{i=1,2,...,k}\{W_i\} = 1$ with probability 1. Now consider an independent random variable Z with a $P(I)(1,1)$ distribution and define

$$\underline{Y} = Z\underline{W}. \tag{6.2.47}$$

Note that $\max_{i=1,2,\ldots,k}\{Y_i\} \sim P(I)(1,1)$. Such a random variable $\underline{Y}$ can be said to have a standard k-dimensional Pareto distribution parameterized by $F_{\underline{W}}$, the distribution of $\underline{W}$. Marginal transformations of the form $X_i = \mu_i + \sigma_i Y_i^{\gamma_i}$ then yield a more general model. For more details on such distributions, especially with regard to their role in extreme value theory, see Rootzen and Tajvidi (2006), Rakonczai and Zempléni (2011) and Ferreira and de Haan (2013).

As a final entry in this list of other multivariate Pareto models, we will consider the bilateral bivariate Pareto distribution introduced by de Groot (1970). For this model there are three parameters, $r_1 < r_2$ and $\alpha > 0$. The joint density of (X_1, X_2) is of the form

$$f_{X_1,X_2}(x_1,x_2) = \frac{\alpha(\alpha+1)(r_2-r_1)^{\alpha}}{(x_2-x_1)^{\alpha+2}} I(x_1 < r_1, x_2 > r_2). \tag{6.2.48}$$

For this model, it was noted that $r_1 - X_1$ and $X_2 - r_2$ each have a Pareto distribution, actually they each have Pareto (II) distributions. However, more can be said. If we define $(Y_1, Y_2) = (r_1 - X_1, X_2 - r_2)$, then it may be verified that $\underline{Y} \sim MP^{(2)}(\underline{0}, (r_2 - r_1)\underline{1}, \alpha)$. So, in a sense, although the bilateral bivariate Pareto appears to be a new distribution, it is in fact just obtained via a linear transformation applied to a well-known model.

6.2.7 *General Classes of Bivariate Pareto Distributions*

If we specify that a random vector (X_1, X_2) is to have Paretian marginals, let us say for definiteness that $X_i \sim P(\text{IV})(\mu_i, \sigma_i, \gamma_i, \alpha_i)$, then we do know something about the joint survival function. The Frechet-Hoeffding bounds on $\overline{F}_{\underline{X}}(\underline{x})$ are

$$\max\{0, \overline{F}_1(x_1) + \overline{F}_2(x_2) - 1\} \leq \overline{F}_X(x) \leq \min\{\overline{F}_1(x_1), \overline{F}_2(x_2)\} \tag{6.2.49}$$

where $\overline{F}_i$ is the survival function of X_i. We will use the notation:

$$\overline{H}_1(\underline{x}) = \max\{0, \overline{F}_1(x_1) + \overline{F}_2(x_2) - 1\} \tag{6.2.50}$$

and

$$\overline{H}_2(x) = \min\{\overline{F}_1(x_1), \overline{F}_2(x_2)\}. \tag{6.2.51}$$

Useful references for the subsequent comments are Whitt (1976) and Conway (1979). It may be verified that $\overline{H}_1$ is the joint survival function with minimal correlation among those whose marginals are $\overline{F}_1$ and $\overline{F}_2$, assumed to have finite second moments. Analogously, $\overline{H}_2$ corresponds to the case of maximal correlation. This is obvious for non-negative random variables since

$$E(X_1, X_2) = \int_0^{\infty} \int_0^{\infty} \overline{F}_{\underline{X}}(\underline{x}) dx_1 \, dx_2. \tag{6.2.52}$$

In the case of Pareto II marginals, i.e., $X_i \sim P(\mathrm{II})(\mu_i, \sigma_i, \alpha_i)$, $i = 1,2$, Conway (1979) supplied the following bounds for the correlation

$$\frac{B(1-\alpha_1^{-1}, 1-\alpha_2^{-1}) - \xi_1}{\sqrt{\xi_2}} \leq \rho(X_1, X_2) \leq \frac{\frac{\alpha_1\alpha_2}{\alpha_1\alpha_2-\alpha_1-\alpha_2} - \xi_1}{\sqrt{\xi_2}} \tag{6.2.53}$$

where

$$\xi_1 = \alpha_1\alpha_2(\alpha_1 - 1)^{-1}(\alpha_2 - 1)^{-1}$$

and

$$\xi_2 = \alpha_1\alpha_2(\alpha_1 - 1)^{-2}(\alpha_2 - 1)^{-2}(\alpha_1 - 2)^{-1}(\alpha_2 - 2)^{-1}.$$

When $\alpha_1 = \alpha_2$, the upper bound in (6.2.53) becomes 1.

It is not difficult to verify that $\overline{H}_2$ is not only the joint distribution with marginals $\overline{F}_1, \overline{F}_2$ with maximal correlation, but also is the joint distribution for which $E[\max(X_1, X_2)]$ is maximal. For classical Pareto marginals, i.e., $X_i \sim P(\mathrm{I})(\sigma_i, \alpha)$ one may verify that

$$E[\max(X_1, X_2)] \leq \frac{\alpha}{\alpha - 1}(\sigma_1^\alpha + \sigma_2^\alpha)^{1/\alpha}. \tag{6.2.54}$$

If, additionally, one assumes $\sigma_i = \sigma$, $(i = 1,2)$, (i.e., identical classical Pareto marginals), then one finds

$$E[\max(X_1, X_2)] \leq 2^{1/\alpha} E(X_1). \tag{6.2.55}$$

Several authors have constructed one or two parameter families of distributions with specified marginals $\overline{F}_1$, $\overline{F}_2$ which include the Frechet-Hoeffding distributions $\overline{H}_1$ and $\overline{H}_2$ as limiting cases. We mention the two best known such families (Conway, 1979, provided a useful summary of related material).

(i) The Frechet family

$$\overline{F}_\alpha(\underline{x}) = (1 - \alpha^2)\overline{F}_1(x_1)\overline{F}_2(x_2) + \frac{\alpha^2 - \alpha^3}{2}\overline{H}_1(\underline{x}) + \frac{\alpha^2 + \alpha^3}{2}\overline{H}_2(\underline{x}) \tag{6.2.56}$$

where $\alpha \in [-1, 1]$. This is just a subclass of the set of all possible convex combinations of $\overline{H}_1$, $\overline{H}_2$ and the joint survival function with independent marginals $\overline{F}_1(x_1)\overline{F}_2(x_2)$.

(ii) The Plackett family

Let F_ψ be the unique solution to

$$\psi = \frac{F_\psi(x_1, x_2)[1 - F_1(x_1) - F_2(x_2) + F_\psi(x_1, x_2)]}{[F_1(x_1) - F_\psi(x_1, x_2)][F_2(x_2) - F_\psi(x_1, x_2)]}, \tag{6.2.57}$$

where $\psi \in [0, \infty)$.

A popular family of bivariate distributions with marginals $\overline{F}_1(x_1)$ and $\overline{F}_2(x_2)$

which does not include the limiting cases $\overline{H}_1$ and $\overline{H}_2$ is the Farley-Gumbel-Morgenstern family:

$$\overline{F}_\alpha(\underline{x}) = \overline{F}_1(x_1)\overline{F}_2(x_2)[1+\alpha\varphi_1(x_1)\varphi_2(x_2)] \tag{6.2.58}$$

where $\alpha \in [-1,1]$ and φ_1 and φ_2 are suitably chosen functions.

Selecting Pareto (IV) survival functions to play the roles of $\overline{F}_1$ and $\overline{F}_2$ in (6.2.56)–(6.2.58) gives us three additional families of bivariate Pareto (IV) distributions.

Conway (1979) discussed these in the case of classical Pareto marginals. Several other parametric families with specified marginals were included in her discussion. These general purpose families do not yield natural bivariate Pareto distributions, in the sense that it does not seem possible to identify a stochastic mechanism which might account for such distributions. They do provide convenient spectrums of bivariate Pareto distributions with a wide range of levels of dependence. In this sense they are viable competitors to all the other bivariate and multivariate Pareto distributions introduced in this and the preceding section. All provide flexible families with Paretian marginals but, as of yet, with no identifiable model to relate them to actual multivariate income distributions.

6.2.8 A Flexible Multivariate Pareto Model

Arnold (1990) introduced a flexible family of multivariate Pareto distributions that is constructed by considering multivariate geometric minima of univariate Pareto variables. The construction makes use of the fact that a Pareto (III)(μ,σ,γ) random variable X can be viewed as a transformed standard Pareto variable. Thus, if Z has as its survival function $P(Z>z)=(1+z)^{-1}$, $z>0$, and if $X=\mu+\sigma Z^\gamma$ then $X \sim P(III)(\mu,\sigma,\gamma)$. The final step utilized to obtain a multivariate Pareto (IV) distribution is to observe that if $\overline{F}_{\underline{X}}(\underline{x})$ is a k-dimensional survival function then, for many values of $\alpha>0$, $[\overline{F}_{\underline{X}}(\underline{x})]^\alpha$ is also a k-dimensional survival function.

Begin by considering a sequence of independent trials with $k+1$ possible outcomes labeled $0,1,2,...,k$ with associated probabilities $p_0,p_1,p_2,...,p_k$ where $\sum_{i=0}^k p_i = 1$. Define a random vector $\underline{N}=(N_1,N_2,...,N_k)$ where, for each $j=1,2,...,k$, N_j denotes the number of outcomes of type$j$ that precede the first outcome of type 0 in the sequence of trials. Then for any $\underline{n}=(n_1,n_2,...,n_k)$ where each n_j is a non-negative integer we have

$$P(\underline{N}=\underline{n}) = \binom{\sum_{i=1}^k n_i}{n_1,n_2,\ldots,n_k} p_1^{n_1}p_2^{n_2}\ldots p_k^{n_k}p_0. \tag{6.2.59}$$

To construct the k-dimensional standard Pareto distribution, we begin with k independent sequences of independent standard Pareto variables $Z_i^{(j)}$, $j=1,2,...,k$, $i=1,2,...$, together with an independent multivariate geometric variable $\underline{N}$ with density (6.2.59). Define $\underline{U}=(U_1,U_2,...,U_k)$ by

$$U_j = \min_{i\le N_j+1}\{Z_i^{(j)}\}, \quad j=1,2,...,k. \tag{6.2.60}$$

By conditioning on $\underline{N}$ we obtain

$$P(\underline{U} \geq \underline{u}) = \frac{p_0}{\left(1 - \sum_{j=1}^{k} \frac{p_j}{1+u_j}\right)\left[\prod_{j=1}^{k}(1+u_j)\right]}.$$

Define $\delta_j = p_j/p_0$ and $V_j = u_j/(1+\delta_j)$ to obtain a k-dimensional standard Pareto variable $\underline{V}$ with joint survival function of the form

$$\begin{aligned} P(\underline{V} \geq \underline{v}) = & \left[1 + \sum_{j=1}^{k} v_j + \sum_{j_1}\sum_{j_2>j_1} \eta_{j_1,j_2} v_{j_1} v_{j_2} \right. \\ & + \sum_{j_1}\sum_{j_2>j_1}\sum_{j_3>j_2} \eta_{j_1,j_2,j_3} v_{j_1} v_{j_2} v_{j_3} \\ & \left. + \cdots + \eta_{1,2,\ldots,k} v_1 v_2 \ldots v_k \right]^{-1}, \quad \underline{v} > \underline{0}, \end{aligned} \tag{6.2.61}$$

where

$$\eta_{j_1,j_2,\ldots j_\ell} = \frac{[1 + \delta_{j_1} + \delta_{j_2} + \cdots + \delta_{j_\ell}]}{\prod_{i=1}^{\ell}(1+\delta_{j_i})}.$$

Up to this point the expression in (6.2.61) is linked to the geometric minima construction. However it continues to be a valid joint survival function for a quite extensive array of choices of values for the η's in (6.2.61). There are some restrictions however on the η's. For example when $k = 2$, we must have $\eta_{1,2} \in [0,2]$. When $k = 3$ the η's must satisfy $\eta_{1,2}, \eta_{1,3}, \eta_{2,3} \in [0,2]$ and $\eta_{1,2,3} \geq 2(\eta_{1,2} + \eta_{1,3} + \eta_{2,3}) - 6$.

If we denote the right hand side of (6.2.61) by $G_{\underline{\eta}}(\underline{v})$ then a corresponding $MP^{(k)}(IV)$ survival function will be of the form

$$\left[G_{\underline{\eta}}\left(\left(\frac{v_1 - \mu_1}{\sigma_1}\right)^{1/\gamma_1}, \left(\frac{v_2 - \mu_2}{\sigma_2}\right)^{1/\gamma_2}, \ldots, \left(\frac{v_k - \mu_k}{\sigma_k}\right)^{1/\gamma_k}\right)\right]^{\alpha}, \quad \underline{v} > \underline{\mu}, \tag{6.2.62}$$

where $\underline{\mu} \in (-\infty, \infty)^k$, $\underline{\sigma}, \underline{\gamma} \in (0,\infty)^k$ and $\alpha \in (0,\infty)$.

For this to be a valid joint survival function, the constraints on the η's will now depend on the value of α. For example, it will now be required that $0 \leq \eta_{j_1,j_2} \leq \alpha + 1$ for each pair $j_1 < j_2$. Durling (1975) discussed the bivariate version of (6.2.62), i.e., when $k = 2$. See also Sankaran and Unnikrishnan Nair (1993a)

Nadarajah and Kotz (2006b) consider the special case of (6.2.62) when $k = 2$ and $\underline{\mu} = \underline{0}$. They obtain an expression for $P(V_1 < V_2)$ involving hypergeometric functions. Nadarajah and Espejo (2006) derive expressions for the distributions of $V_1 + V_2$, $V_1 V_2$ and $V_1/(V_1 + V_2)$ when $\underline{V}$ has a bivariate distribution closely related to (6.2.62). Here, too, the expressions involve hypergeometric functions. Gupta and Sankaran (1998) characterize the distribution of the form (6.2.62) with $k = 2$, $\underline{\mu} = \underline{0}$, $\underline{\gamma} = \underline{1}$ and $\alpha = n$ in terms of its bivariate mean residual life function (defined in the sense proposed by Zahedi (1985)). Sankaran and Unnikrishnan Nair (1993b) discuss a characterization of the distribution (6.2.62) with $k = 2$ based on the form of the corresponding residual life distributions.

Higher dimensional versions of (6.2.62), i.e., $k > 2$, have not received much attention in the literature.

Remark. *It is possible to develop an analogous (but distinct) family of k-dimensional Pareto (IV) distributions based on geometric maxima rather than minima.*

6.2.9 *Matrix-variate Pareto Distributions*

Jupp (1986) goes one step beyond consideration of multivariate Pareto variables. Rather than considering vector valued random variables, he considers matrix valued random variables.

Following Jupp's notation, let $S(p)$ denote the class of all symmetric $p \times p$ real matrices. Next let $S^+(p)$ be the set of all positive definite matrices in $S(p)$. Jupp denotes this class by P, but we continue, as much as possible, to reserve the letter P to denote probability. For a random matrix X taking values in $S(p)$, we define its survival function $\overline{F}_X$ by

$$\overline{F}_X(C) = P(X > C) \text{ for } C \in S(p). \tag{6.2.63}$$

Here we write $A > B$ iff $A - B \in S^+(p)$, i.e., the usual partial order on $S(p)$. We make use of the standard inner product defined on $S(p)$, i.e., $< A, B >= \text{tr}(AB)$. With this notation in place, we may define our matrix-variate Pareto distribution as follows:

A random matrix X taking values in $S(p)$ has a matrix-variate Pareto distribution with parameters $A \in S^+(p)$, $B \in S(p)$ and $\alpha > 0$, if its survival function is of the form

$$\overline{F}_X(C) = P(X > C) = [1 + < A, C - B >]^{-\alpha} \text{ for } C > B. \tag{6.2.64}$$

Paralleling the result in the multivariate case, a matrix-variate Pareto random variable X admits a representation of the form

$$X = \frac{1}{Z} Y, \tag{6.2.65}$$

where Z and Y are independent with $Z \sim \Gamma(\alpha, 1)$ and Y having a matrix-variate exponential distribution.

Jupp (1986) shows that, under regularity conditions, a matrix valued random variable taking values in $S(p)$ is characterized by its mean residual life function, $E(X - C|X > C)$, as in the real valued case ($p = 1$). Moreover the matrix-variate Pareto distribution is essentially characterized by the property that its mean residual life function is a non-constant affine function, a result that might be predicted from knowledge of what occurs when $p = 1$.

The survival function (6.2.64) can be recognized as a matrix-variate extension of the univariate Pareto (II)(μ, σ, α) distribution. A matrix-variate analog of the Feller-Pareto distribution with $\gamma = 1$, i.e., $FP(\mu, \sigma, 1, \gamma_1, \gamma_2)$, can be constructed in the form

$$X = \frac{1}{Z} W, \tag{6.2.66}$$

where Z and W are independent with $Z \sim \Gamma(\gamma_1, 1)$ and W having a matrix-variate gamma distribution.

6.3 Related Multivariate Models

In this Section we will discuss a variety of models closely related to multivariate Pareto models. These models can often be viewed as parametric extensions of multivariate Pareto models and often include Pareto distributions as special cases.

6.3.1 *Conditionally Specified Models*

Rather than insist on having models with univariate Pareto marginals, it might at times be more appropriate to consider models with Pareto conditional distributions. A convenient introduction to conditionally specified models in general is Arnold, Castillo and Sarabia (1999). Arnold and Strauss (1991) were able to completely characterize all bivariate distributions with conditionals in given exponential families. The Pareto models discussed in this book do not comprise exponential families; nevertheless it is possible to identify flexible families of bivariate and multivariate distributions that do have Pareto conditionals.

Initially we will focus on the problem of identifying bivariate distributions with Pareto (II) conditional distributions. It turns out that the extension to the k-dimensional case and the extension to consideration of Pareto (IV) conditionals will be relatively straightforward.

Thus we wish to identify all bivariate densities $f(x,y)$ defined on $(0,\infty)^2$ with the property that all of its conditional densities are of the Pareto (II)$(0,\sigma,\alpha)$ form, i.e., of the form

$$f(x) = \frac{\alpha}{\sigma}\left(1+\frac{x}{\sigma}\right)^{-(\alpha+1)} I(x>0).$$

Specifically we must have, for each $y>0$,

$$f(x|y) = \frac{\alpha}{\sigma(y)}\left(1+\frac{x}{\sigma(y)}\right)^{-(\alpha+1)} I(x>0), \tag{6.3.1}$$

and, for each $x>0$,

$$f(y|x) = \frac{\alpha}{\tau(x)}\left(1+\frac{y}{\tau(x)}\right)^{-(\alpha+1)} I(y>0), \tag{6.3.2}$$

for some functions $\sigma(y)$ and $\tau(x)$. If we denote the marginal densities by $g(x)$ and $h(y)$ then, writing

$$f(x,y) = g(x)f(y|x) = h(y)f(x|y),$$

we arrive at the following equation

$$\frac{\alpha h(y)[\sigma(y)]^\alpha}{(\sigma(y)+x)^{\alpha+1}} = \frac{\alpha g(x)[\tau(x)]^\alpha}{(\tau(x)+y)^{\alpha+1}} \tag{6.3.3}$$

which must be solved for $g(x)$, $h(y)$, $\sigma(y)$ and $\tau(x)$.

It is convenient to define

$$\delta(x) = \{\alpha h(y)[\sigma(y)]^{\alpha}\}^{1/(\alpha+1)} \text{ and } \rho(y) = \{\alpha g(x))[\tau(x)]^{\alpha}\}^{1/(\alpha+1)},$$

so that (6.3.3) becomes

$$\frac{\rho(y)}{\sigma(y)+x} = \frac{\delta(x)}{\tau(x)+y},$$

or, equivalently,

$$\rho(y)\tau(x) + y\rho(y) = \delta(x)\sigma(y) + x\delta(x). \tag{6.3.4}$$

The functional equation (6.3.4) can be readily solved by differencing with respect to x and y, to yield the following general solution:

$$\begin{aligned} \tau(x) &= \frac{\lambda_{00}+\lambda_{01}y}{\lambda_{01}+\lambda_{11}x}, \\ \sigma(y) &= \frac{\lambda_{00}+\lambda_{10}x}{\lambda_{10}+\lambda_{11}y}, \\ \delta(x) &= [\lambda_{01}+\lambda_{11}x]^{-1}, \\ \rho(y) &= [\lambda_{10}+\lambda_{11}y]^{-1}. \end{aligned} \tag{6.3.5}$$

From this we obtain the joint density in the form

$$f(x,y) \propto (\lambda_{00}+\lambda_{10}x+\lambda_{01}y+\lambda_{11}xy)^{-(\alpha+1)} I(x>0.y>0), \tag{6.3.6}$$

with corresponding marginal densities

$$f_X(x) \propto (\lambda_{01}+\lambda_{11}x)^{-1}(\lambda_{00}+\lambda_{10}x)^{-\alpha}, \tag{6.3.7}$$

$$f_Y(y) \propto (\lambda_{10}+\lambda_{11}y)^{-1}(\lambda_{00}+\lambda_{01}y)^{-\alpha}. \tag{6.3.8}$$

The conditional distributions of this density are indeed of the Pareto (II) form, thus

$$X|Y=y \sim P(II)(0, \frac{\lambda_{00}+\lambda_{01}y}{\lambda_{10}+\lambda_{11}y}, \alpha) \tag{6.3.9}$$

and

$$Y|X=x \sim P(II)(0, \frac{\lambda_{00}+\lambda_{10}x}{\lambda_{01}+\lambda_{11}x}, \alpha). \tag{6.3.10}$$

It is necessary to impose some constraints on the λ_{ij}'s in this model to ensure that the expressions in (6.3.6), (6.3.7) and (6.3.8) are non-negative and integrable. The necessary constraints depend on the value of the parameter α as follows.

- If $0<\alpha<1$, then it is necessary that $\lambda_{10}, \lambda_{01}$ and λ_{11} be strictly positive while λ_{00} must be non-negative (it can take the value 0).

- If $\alpha = 1$, all four λ_{ij}'s must be strictly positive.
- If $\alpha > 1$, then $\lambda_{00}, \lambda_{10}$ and λ_{01} must be strictly positive while λ_{11} must be non-negative (it can take the value 0).

Note that if $\alpha > 1$ and if $\lambda_{11} = 0$, the density of (X, Y) has Pareto (II) marginals and conditionals. It is recognizable as the Mardia bivariate Pareto density corresponding to the $MP^{(2)}(II)(\underline{0}, \underline{\sigma}, \alpha - 1)$ distribution (recall that $\alpha > 1$) with survival function (6.1.10). It will be noted from the expressions in (6.3.7) and (6.3.8) that it is only in the case in which $\alpha > 1$ and $\lambda_{11} = 0$ that the marginal densities will be of the Pareto (II) type. The required normalizing constant for the density (6.3.6) is expressible in terms of hypergeometric functions. See Arnold, Castillo and Sarabia (1999) for details. Needless to say, this provides a strong impetus for the use of conditional likelihood in estimating parameters for these models.

Having identified all possible bivariate densities with Pareto (II)$(0, \sigma, \alpha)$ conditionals for a fixed value of α, i.e., densities of the form (6.3.6), the obvious marginal transformations will lead us to bivariate densities with Pareto (IV) conditionals. Thus if $\underline{X}$ has a density of the form (6.3.6), and if $\underline{Y}$ is defined by

$$Y_i = \mu_i + \sigma_i X_i^{\gamma_i}, \quad i = 1, 2,$$

then $\underline{Y}$ will have a bivariate distribution with Pareto (IV) conditional densities. The density of $\underline{Y}$ will be

$$f_{Y_1,Y_2}(y_1, y_2) \propto \tag{6.3.11}$$

$$\frac{\frac{1}{\sigma_1\sigma_2\gamma_1\gamma_2}\left(\frac{y_1-\mu_1}{\sigma_1}\right)^{(1/\gamma_1)-1}\left(\frac{y_2-\mu_2}{\sigma_2}\right)^{(1/\gamma_2)-1} I(y_1 > \mu_1, y_2 > \mu_2)}{\left[\lambda_{00} + \lambda_{10}\left(\frac{y_1-\mu_1}{\sigma_1}\right)^{1/\gamma_1} + \lambda_{01}\left(\frac{y_2-\mu_2}{\sigma_2}\right)^{1/\gamma_2} + \lambda_{11}\left(\frac{y_1-\mu_1}{\sigma_1}\right)^{1/\gamma_1}\left(\frac{y_2-\mu_2}{\sigma_2}\right)^{1/\gamma_2}\right]^{\alpha+1}}.$$

Higher dimensional versions of the models (6.3.6) and (6.3.11) are readily envisioned as follows. For such k-dimensional models we will require that $\underline{X}$ have conditional distributions of each X_i given $\underline{X}_{(i)} = \underline{x}_{(i)}$ that are of the Pareto (II) or Pareto (IV) form (recall that $\underline{X}_{(i)}$ is the vector $\underline{X}$ with the i-th coordinate variable deleted and $\underline{x}_{(i)}$ is analogously defined). Such densities with $P(II)(0, \sigma, \alpha)$ conditionals will be of the form

$$f_{\underline{X}}(\underline{x}) \propto \left(\sum_{\underline{s}\in\xi_k} \delta_{\underline{s}}\left(\prod_{i=1}^{k} x_i^{s_i}\right)\right)^{-(\alpha+1)}, \tag{6.3.12}$$

where ξ_k is the set of all vectors of 0's and 1's of dimension k. This model includes the k-variate Mardia model as a special case (when $\delta_{\underline{s}} = 0$ for every $\underline{s}$ with more than one coordinate equal to 1). See Arnold, Castillo and Sarabia (1999, pp.192-3) for discussion of characterization of the Mardia model within the class of distributions with Pareto (II) conditionals. Marginal transformations applied to the density (6.3.12) yield k-variate distributions with Pareto (IV) conditionals.

A more general class of distributions with Pareto (IV) conditionals can be considered. For simplicity, we will consider the bivariate case first. We might thus try to identify the class of all bivariate random variables (X_1, X_2) such that, for every $x_2 > 0$,

$$X_1|X_2 = x_2 \sim P(IV)(0, \sigma(x_2), \gamma(x_2), \alpha(x_2)),$$

and, for every $x_1 > 0$,

$$X_2|X_1 = x_1 \sim P(IV)(0, \tau(x_1), \delta(x_1), \beta(x_1)).$$

Denote the marginal densities of X_1 and X_2 by $f_1(x_1)$ and $f_2(x_2)$, respectively. Then, since we must have

$$f_{X_1}(x_1) f_{X_2|X_1}(x_2|x_1) = f_{X_2}(x_2) f_{X_1|X_2}(x_1|x_2),$$

the following functional equation must hold:

$$\frac{\frac{f_1(x_1)\beta(x_1)}{\delta(x_1)} x_2^{(1/\delta(x_1))-1}}{[\tau(x_1)]^{(1/\delta(x_1)} \left[1 + \left(\frac{x_2}{\tau(x_1)}\right)^{(1/\delta(x_1))}\right]^{\beta(x_1)+1}} = \frac{\frac{f_2(x_2)\alpha(x_2)}{\gamma(x_2)} x_1^{(1/\gamma(x_2))-1}}{[\sigma(x_2)]^{(1/\gamma(x_2)} \left[1 + \left(\frac{x_1}{\sigma(x_2)}\right)^{(1/\gamma(x_2))}\right]^{\alpha(x_2)+1}}. \tag{6.3.13}$$

This functional equation must be solved for $f_1(x_1), f_2(x_2), \tau(x_1), \sigma(x_2), \delta(x_1), \gamma(x_2), \beta(x_1)$ and $\alpha(x_2)$. Efforts to identify a general solution to this equation are discussed in Arnold, Castillo and Sarabia (1993). Besides the densities of the form (6.3.11), a second family of bivariate densities with Pareto (IV) conditionals (with $\underline{\mu} = \underline{0}$) is identified in that paper. They are of the form

$$\begin{aligned} f_{X_1,X_2}(x_1,x_2) &= x_1^{(1/\delta)-1} x_2^{(1/\gamma)-1} \\ &\times \exp\{\theta_1 + \theta_2 \log(\theta_5 + x_1^{1/\delta}) + \theta_3 \log(\theta_6 + x_2^{1/\gamma})\} \\ &\times \exp\{\theta_4 \log(\theta_5 + x_1^{1/\delta}) \log(\theta_6 + x_2^{1/\gamma})\} I(x_1 > 0, x_2 > 0). \end{aligned} \tag{6.3.14}$$

Whereas in the model (6.3.11), both negative and positive correlation is possible, in (6.3.14) only non-negative correlation will be encountered. Note that in both of the models (6.3.11) and (6.3.14) the functions $\gamma(x_2)$ and $\delta(x_1)$ are constant functions, while in (6.3.11) $\alpha(x_2)$ and $\beta(x_1)$ share a common constant value and in (6.3.14) $\sigma(x_2)$ and $\tau(x_1)$ share a common constant value. Arnold, Castillo and Sarabia (1993) conjecture that there will not exist solutions to (6.3.13) if more than one of the functions $\tau(x_1), \delta(x_1)$ and $\beta(x_1)$ is non-constant. If this conjecture can be shown to be true, then the models (6.3.11) and (6.3.14) will comprise the totality of all bivariate models with Pareto (IV)$(0,\sigma,\gamma,\alpha)$ conditionals. Of course, a k-variate version of the model (6.3.14) is available, i.e.,

$$f_{\underline{X}}(\underline{x}) = \left[\prod_{i=1}^{k} x_j^{(1/\delta_j)-1}\right] \exp\left\{\sum_{\underline{s}\in\xi_k} \lambda_{\underline{s}} \left[\prod_{j=1}^{k} [\log(\theta_j + x_j^{(1/\delta_j)}]^{s_j}\right]\right\} I(\underline{x} > \underline{0}), \quad (6.3.15)$$

where ξ_k is as defined following equation (6.3.12).

Wesolowski (1995) provided further insight into the structure of the density (6.3.6). Suppose that (X,Y) is such that

$$X|Y=y \sim P(II)(0, \frac{a+by}{1+cy}, \alpha).$$

He shows that, in such a case, the joint density of (X,Y) is uniquely determined by $E(Y|X)$. So that if $E(Y|X) = (a+X)(\alpha-1)^{-1}(b+cX)^{-1}$ then (X,Y) has a Pareto (II) conditionals density of the form (6.3.6). Moreover, if $E(Y|X)$ is linear, i.e., if $c=0$, then (X,Y) has the Mardia bivariate Pareto distribution (6.1.10). In Wesolowski and Ahsanullah (1995) a characterization of an exchangeable Mardia multivariate Pareto model is provided. The bivariate case was considered by Arnold and Pouramahdi (1988). In the k-dimensional setting, we assume that a random vector $\underline{X}$ of dimension k is such that

$$X_k|\underline{X}_{(k)} = \underline{x}_{(k)} \sim P(II)(\alpha+k-1, 1+\sum_{i=1}^{k-1} x_i),$$

and that $\underline{X}_{(1)} \stackrel{d}{=} \underline{X}_{(k)}$. It can then be verified that $\underline{X}$ has a k-variate Mardia distribution, specifically that $\underline{X} \sim MP^{(k)}(II)(\underline{0},\underline{1},\alpha)$.

Remark. *Gupta (2001) provides details on the failure rate functions and the mean residual life functions of the marginal and conditional distributions corresponding to the bivariate density (6.3.6) with Pareto (II) conditionals. For example, it is shown that the marginal distributions have decreasing failure rates, while the conditional densities will have either increasing or decreasing failure rates depending on the sign of the correlation between the coordinate variables.*

6.3.2 *Multivariate Hidden Truncation Models*

Paralleling the discussion in Chapter 3, we may consider multivariate hidden truncation scenarios. For this, we begin with a k-dimensional $MP^{(k)}(IV)(\underline{\mu},\underline{\sigma},\underline{\gamma},\alpha)$ random variable $\underline{X}$ which is partitioned into $(\underline{\dot{X}},\underline{\ddot{X}})$ where $\underline{\dot{X}}$ is of dimension k_1 and $\underline{\ddot{X}}$ is of dimension $k_2 = k-k_1$. Likewise we partition the vectors $\underline{\mu},\underline{\sigma},\underline{\gamma}$ and $\underline{x}$. For a given vector $\underline{\ddot{x}}^{(0)}$ with $\underline{\ddot{x}}^{(0)} > \underline{\ddot{\mu}}$, we consider the situation in which $\underline{\dot{X}}$ can only be observed if $\underline{\ddot{X}} > \underline{\ddot{x}}^{(0)}$. The resulting distribution will be called the hidden truncation distribution of $\underline{\dot{X}}$. It initially involves an additional parameter vector $\underline{\ddot{x}}^{(0)}$. The assumption in this model is that such truncation has occurred but that we do not know the value of $\underline{\ddot{x}}^{(0)}$. For this truncation from below, as in the one-dimensional case, the resulting hidden truncation density is readily identified.

We have

$$P(\underline{\dot{X}} > \underline{\dot{x}} | \underline{\ddot{X}} > \underline{\ddot{x}}^{(0)}) = \frac{\left[1+\sum_{i=1}^{k_1}\left(\frac{x_i-\mu_i}{\sigma_i}\right)^{1/\gamma_i}+\sum_{j=k_1+1}^{k}\left(\frac{x_j-\mu_j}{\sigma_j}\right)^{1/\gamma_j}\right]^{-\alpha}}{\left[1+\sum_{j=k_1+1}^{k}\left(\frac{x_j-\mu_j}{\sigma_j}\right)^{1/\gamma_j}\right]^{-\alpha}}$$

$$= \left[1+\sum_{i=1}^{k_1}\left(\frac{x_i-\mu_i}{\tau_i}\right)^{1/\gamma_i}\right]^{-\alpha}, \quad \underline{\dot{x}} > \underline{\dot{\mu}}, \tag{6.3.16}$$

where

$$\tau_i = \sigma_i\left[1+\sum_{j=k_1+1}^{k}\left(\frac{x_j^{(0)}-\mu_j}{\sigma_j}\right)^{1/\gamma_j}\right]^{\gamma_i}, \quad i=1,2,...,k_1.$$

Consequently, the hidden truncation distribution of $\underline{\dot{X}}$ is again of the $MP^{(k_1)}(IV)$ form.

As in one dimension, the case of hidden truncation from above is somewhat more interesting. In such a setting, we consider

$$P(\underline{\dot{X}} > \underline{\dot{x}} | \underline{\ddot{X}} \le \underline{\ddot{x}}^{(0)}) \propto [P(\underline{\dot{X}} > \underline{\dot{x}}) - P(\underline{\dot{X}} > \underline{\dot{x}}, \underline{\ddot{X}} > \underline{\ddot{x}}^{(0)})]$$

$$= \left[1+\sum_{i=1}^{k_1}\left(\frac{x_i-\mu_i}{\sigma_i}\right)^{1/\gamma_i}\right]^{-\alpha} - \left[1+\sum_{i=1}^{k_1}\left(\frac{x_i-\mu_i}{\sigma_i}\right)^{1/\gamma_i}+\sum_{j=k_1+1}^{k}\left(\frac{x_j^{(0)}-\mu_j}{\sigma_j}\right)^{1/\gamma_j}\right]^{-\alpha}$$

$$= \left[1+\sum_{i=1}^{k_1}\left(\frac{x_i-\mu_i}{\sigma_i}\right)^{1/\gamma_i}\right]^{-\alpha} - \left[1+\sum_{i=1}^{k_1}\left(\frac{x_i-\mu_i}{\sigma_i}\right)^{1/\gamma_i}+\theta\right]^{-\alpha},$$

where

$$\theta = \sum_{j=k_1+1}^{k}\left(\frac{x_j^{(0)}-\mu_j}{\sigma_j}\right)^{1/\gamma_j}.$$

The required normalizing constant can be identified making use of the fact that $P(\underline{\dot{X}} > \underline{\dot{\mu}} | \underline{\ddot{X}} \le \underline{\ddot{x}}^{(0)}) = 1$. It is thus equal to $1-(1+\theta)^{-\alpha}$.

The density of this hidden truncation distribution is a linear combination of two $MP^{(k_1)}(IV)$ densities, one with parameters $\underline{\dot{\mu}}, \underline{\dot{\sigma}}, \underline{\dot{\gamma}}$ and α, and the other with parameters $\underline{\dot{\mu}}, \underline{\dot{\tau}}, \underline{\dot{\gamma}}$ and α, where

$$\tau_i = (1+\theta)^{\gamma_i}\sigma_i, \quad i=1,2,...,k_1.$$

Two-sided truncation in which $\dot{\underline{X}}$ is observed only if $\underline{\ddot{a}} < \underline{\ddot{X}} \leq \underline{\ddot{b}}$ can be shown to also correspond to a density which is a linear combination of $MP^{(k_1)}$ densities with different scale parameters. Thus it would be impossible to know from the data whether two-sided truncation had occurred, or whether the density had resulted from truncation from above alone.

Analyses of hidden truncation models in which $\dot{\underline{X}}$ is observed only when $\underline{\ddot{X}} \in M \subset (0,\infty)^{k_2}$, where the definition of M can be quite general, will in most cases be predictably more complicated.

6.3.3 *Beta Extensions*

In Chapter 3, in the one dimensional case, extended Pareto models were constructed by considering the composition of a Beta distribution function with a Pareto (IV) distribution function. Thus we could begin with $F_{\mu,\sigma,\gamma,\alpha}$, the distribution function of a $P(IV)(\mu,\sigma,\gamma,\alpha)$ variable, and compose it with G_{λ_1,λ_2}, the distribution function of a Beta random variable (of the first kind) with parameters λ_1 and λ_2. The resulting Beta generalized Pareto distribution is then of the form $G_{\lambda_1,\lambda_2}(F_{\mu,\sigma,\gamma,\alpha}(x))$ with density

$$f_X(x;\mu,\sigma,\gamma,\alpha,\lambda_1,\lambda_2) = \frac{\left[F_{\mu,\sigma,\gamma,\alpha}(x)\right]^{\lambda_1-1}\left[1-F_{\mu,\sigma,\gamma,\alpha}(x)\right]^{\lambda_2-1} f_{\mu,\sigma,\gamma,\alpha}(x)}{B(\lambda_1,\lambda_2)}.$$

A random variable with this density can viewed as having a representation of the form $F^{-1}_{\mu,\sigma,\gamma,\alpha}(B_{(\lambda_1,\lambda_2)})$ where $B_{(\lambda_1,\lambda_2)}$ has a Beta (λ_1,λ_2) distribution. A k-dimensional version of such a model would begin with $\underline{B} = (B_1,B_2,...,B_k)$ having a suitable k-dimensional Beta distribution (perhaps one of those discussed in Section 6.2.5), and then define $\underline{X} = (X_1,X_2,...,X_k)$ by $X_i = F^{-1}_{\mu_i,\sigma_i,\gamma_i,\alpha_i}(B_i),\ i = 1,2,...,k$.

6.3.4 *Kumaraswamy Extensions*

Recall that a Kumaraswamy (λ_1,λ_2) distribution function is of the form

$$H_{\lambda_1,\lambda_2}(x) = 1-(1-x^{\lambda_1})^{\lambda_2} I(0<x<1).$$

A one dimensional random variable X has a Kumaraswamy generalized Pareto distribution if it is representable in the form

$$X = F^{-1}_{\mu,\sigma,\gamma,\alpha}(K_{(\lambda_1,\lambda_2)}),$$

where $K_{(\lambda_1,\lambda_2)}$ has a Kumaraswamy (λ_1,λ_2) distribution and, as usual, $F^{-1}_{\mu,\sigma,\gamma,\alpha}$ is the quantile function of a Pareto (IV)$(\mu,\sigma,\gamma,\alpha)$ distribution.

For a k-dimensional version of this construction, a k-dimensional Kumaraswamy distribution will be required. One such distribution was suggested by Nadarajah, Cordeiro and Ortega (2012). If $\underline{K} = (K_1,K_2,...,K_k)$ has a k-variate Kumaraswamy distribution, we then define $X_i = F^{-1}_{\mu_i,\sigma_i,\gamma_i,\alpha_i}(K_i),\ i = 1,2,...,k$. Such a random vector $\underline{X}$ would then be said to have a k-variate Kumaraswamy Pareto (IV) distribution.

6.3.5 *Multivariate Semi-Pareto Distributions*

Univariate semi-Pareto distributions were discussed in Chapter 3. They were motivated by the observation that a geometric (p) minimum of i.i.d. Pareto (III)$(0,\sigma,\alpha)$ variables was again a Pareto (III) variable with just a change of scale. This property is true for every $p \in (0,1)$. The semi-Pareto distributions were defined to be those for which this geometric minimum property holds for one particular value of p. It is natural to consider a k-dimensional extension of this semi-Pareto concept.

In k dimensions, consider a sequence $\underline{X}^{(1)}, \underline{X}^{(2)}, \ldots$ of i.i.d. $MP^{(k)}(\underline{0},\underline{\sigma},\underline{\gamma})$ variables and an independent geometric (p) random variable $N(p)$. It may be then verified that

$$\min_{j \le N(p)} \underline{X}^{(j)} \sim MP^{(k)}(0,\underline{\tau},\underline{\gamma}), \tag{6.3.17}$$

where $\tau_i = p^{\gamma_i}\sigma_i,\ i = 1,2,\ldots,k$. Consequently, if we define $\underline{Y}(p)$ by

$$Y_i(p) = p^{-\gamma_i} \min_{j \le N(p)} \underline{X}_i^{(j)}, \quad i = 1,2,\ldots,k, \tag{6.3.18}$$

we have

$$\underline{Y}(p) \stackrel{d}{=} \underline{X}^{(1)}, \tag{6.3.19}$$

a result that is in accord with the situation in one dimension. The $MP^{(k)}(\underline{0},\underline{\sigma},\underline{\gamma})$ distribution satisfies (6.3.19) for every $p \in (0,1)$. If we only insist on (6.3.19) being true for one particular value of p, then a broader class of distributions can be expected to be characterized.

If for a fixed choice of value for p, (6.3.19) holds, then the joint survival function of $\underline{X}^{(1)}$, denoted by $\overline{F}_{\underline{X}^{(1)}}(\underline{x})$, must satisfy

$$\overline{F}_{\underline{X}^{(1)}}(\underline{x}) = \frac{p\overline{F}_{\underline{X}^{(1)}}(p^{\gamma_1}x_1, p^{\gamma_2}x_2, \ldots, p^{\gamma_k}x_k)}{1-(1-p)\overline{F}_{\underline{X}^{(1)}}(p^{\gamma_1}x_1, p^{\gamma_2}x_2, \ldots, p^{\gamma_k}x_k)}. \tag{6.3.20}$$

If we define

$$\varphi(\underline{x}) = \frac{1-\overline{F}_{\underline{X}^{(1)}}(\underline{x})}{\overline{F}_{\underline{X}^{(1)}}(\underline{x})},$$

then φ must satisfy

$$\varphi(\underline{x}) = \frac{1}{p}\varphi(p^{\gamma_1}x_1, p^{\gamma_2}x_2, \ldots, p^{\gamma_k}x_k). \tag{6.3.21}$$

Consequently, the general solution to (6.3.20) is of the form

$$\overline{F}_{\underline{X}^{(1)}}(\underline{x}) = \frac{1}{1+\varphi(\underline{x})}, \tag{6.3.22}$$

where $\varphi(\underline{x})$ satisfies the functional equation (6.3.21). See Yeh (2007) for further details in the k-dimensional case. Earlier, Balakrishna and Jayakumar (1997) discussed the bivariate case, i.e., $k = 2$.

All joint survival functions of the form (6.3.22) will be called k-dimensional semi-Pareto survival functions. A particular solution to (6.3.21) is of the form

$$\varphi^*(\underline{x}) = \sum_{i=1}^{k} \left(\frac{x_i}{\sigma_i}\right)^{1/\gamma_i},$$

which corresponds to an $MP^{(k)}(\underline{0}, \underline{\sigma}, \underline{\gamma})$ distribution. Unfortunately, the class of solutions to (6.3.21) is extremely broad. Yeh (2007) discusses solutions to (6.3.21), assuming that it holds for every $p \in (0,1)$. Balakrishna and Jayakumar (1997), in the bivariate case with $\sigma_1 = \sigma_2 = 1$, provide a general solution assuming that (6.3.21) holds for just one value of p. The general solution of (6.3.21) for one fixed value of p, when $k > 2$, is not easily determined.

Yeh (2007) also discusses limit theorems under repeated geometric minimization of k-variate semi-Pareto variables, analogous to the one dimensional theorem for Pareto variable presented in Chapter 3 (Theorem 3.8.1).

6.4 Pareto and Semi-Pareto Processes

Many socio-economic time series can be expected to have Pareto distributed marginals and perhaps Paretian limiting distributions. One possible approach to the construction of such processes is to begin with a normal process $\{X_n\}$ (perhaps a moving average and/or auto-regressive process) and suitably transform the X_n's to become Pareto variables. An alternative approach, the one to be described in this Section, is to consider an autoregressive scheme in which minima replace sums. Such processes are frequently called minification processes. The presentation will focus on one-dimensional processes but k-dimensional analogs are readily envisioned.

Let $\{Y_n\}_{n=1}^{\infty}$ be a sequence of i.i.d. non-negative extended real valued random variables and let $c > 1$. A corresponding minification process $\{X_n\}$ is defined by

$$X_n = c\min\{X_{n-1}, Y_n\}, \quad n = 1, 2, \ldots \tag{6.4.1}$$

where X_0 has survival function $\overline{F}$. If this process has a stationary distribution with survival function $\overline{F}$, it must be the case that

$$\overline{F}(x) = \overline{F}\left(\frac{x}{c}\right)\overline{F}_Y\left(\frac{x}{c}\right), \tag{6.4.2}$$

and so, for $x > 0$,

$$\overline{F}(x) = \overline{F}(0)\prod_{i=1}^{\infty}\overline{F}\left(\frac{x}{c^i}\right). \tag{6.4.3}$$

If we assume that the initial distribution is of the form (6.4.3) with $\overline{F}(0) = 1$, then provided that the infinite product does not diverge to 0, the process will be completely stationary, with

$$\overline{F}_{X_n}(x) = \prod \overline{F}_Y\left(\frac{x}{c^i}\right), \quad n = 0, 1, 2, \ldots. \tag{6.4.4}$$

For this to be a Pareto (III) process, we may choose $X_0 \sim P(III)(0,\sigma,\gamma)$ and for some $p \in (0,1)$ and for $n = 1,2,...,$ define

$$X_n = \begin{cases} p^{-\gamma}X_{n-1} & \text{with probability } p, \\ \min\{p^{-\gamma}X_{n-1}, \varepsilon_n\} & \text{with probability } 1-p, \end{cases} \tag{6.4.5}$$

where the ε_n's are i.i.d. $P(III)(0,\sigma,\gamma)$ variables.

Note first that this is a minification process of the form (6.4.1). To see this, define Y_n as follows

$$Y_n = \begin{cases} \infty & \text{with probability } p, \\ p^{\gamma}\varepsilon_n & \text{with probability } 1-p. \end{cases}$$

Next we confirm that the stationary distribution of the process (6.4.5) is of the Pareto (III)$(0,\sigma,\gamma)$ form. First note that

$$\begin{aligned} P(X_1 > x) &= pP(p^{-\gamma}X_0 > x) + (1-p)P(p^{-\gamma}X_0 > x)P(\varepsilon_1 > x) \\ &= P(p^{-\gamma}X_0 > x)[p + (1-p)P(\varepsilon_1 > x)] \\ &= \left[1 + \left(\frac{p^{\gamma}x}{\sigma}\right)^{1/\gamma}\right]^{-1} \left[p + (1-p)\left[1 + \left(\frac{x}{\sigma}\right)^{1/\gamma}\right]^{-1}\right] \\ &= \left[1 + \left(\frac{x}{\sigma}\right)^{1/\gamma}\right]^{-1}, \end{aligned}$$

so that $X_1 \stackrel{d}{=} X_0$, and an induction argument confirms that $X_n \stackrel{d}{=} X_0$, $\forall n$. The sample paths of this stochastic process exhibit stretches of geometric growth followed by abrupt drops. Note that the probability of a drop at time n is $P(X_n < X_{n-1}) = (1-p)/2$. See Yeh, Arnold and Robertson (1988) for further details.

Higher dimensional versions of this construction are possible. For them, consider $\underline{X}^{(0)} \sim MP^{(k)}(III)(\underline{0},\underline{\sigma},\underline{\gamma})$ and $\{\underline{\varepsilon}^{(n)}\}_{n=1}^{\infty}$ i.i.d. also with an $MP^{(k)}(III)(\underline{0},\underline{\sigma},\underline{\gamma})$ distribution. Then, for some $p \in (0,1)$ we can define

$$\underline{X}^{(n)} = \begin{cases} p^{-\gamma}\underline{X}^{(n-1)} & \text{with probability } p, \\ \min\{p^{-\gamma}\underline{X}^{(n-1)}, \underline{\varepsilon}^{(n)}\} & \text{with probability } 1-p. \end{cases}$$

In this way, a completely stationary $MP^{(k)}(III)$ process is obtained.

Pillai (1991) considers the semi-Pareto version of the process defined in (6.4.5). The only change required is the assumption that X_0 and the ε_n's are i.i.d. semi-Pareto variables rather than Pareto (III) variables. Bivariate semi-Pareto processes are discussed in Balakrishna and Jayakumar (1997) and Thomas and Jose (2004).

Remarks. *(1) Since the Pareto (III) distribution is identifiable as a log-logistic distribution, a logarithmic transformation of a Pareto (III) or semi-Pareto process will result in a logistic or semi-logistic process. Useful information about Pareto processes can thus often be found by searching in the literature for logistic processes. See for example Arnold (1992).*

(2) Arnold and Hallett (1989) characterize the Pareto (III) process (6.4.5) as the only minification process for which the level crossing processes $\{Z_n(t) = I(X_n > t)\}$ are Markovian for every t.

(3) Higher order autoregressive and moving average versions of Pareto (III) processes can be considered. See, for example, Yeh (1983), Jayakumar (2004) and Thomas and Jose (2003).

(4) See Dziubdziela (1997) for further discussion of the special role played by the Pareto (III) distribution in the construction of minification processes of the form (6.4.1).

(5) Ferreira (2012) provides detailed discussion of the extremal behavior of the Pareto process (6.4.5).

An alternative construction of a stationary Pareto (III) process was proposed by Arnold (1988a). The construction relies on the fact that geometric minima of Pareto (III) variables are again of the Pareto (III) form. The construction involves dependent geometric minima of independent Pareto (III) variables.

Begin with a sequence $\{U_n\}_{n=0}^{\infty}$ of i.i.d. Bernoulli (p) random variables; thus

$$P(U_n = 0) = 1 - p \text{ and } P(U_n = 1) = p, \tag{6.4.6}$$

where $p \in (0,1)$. In addition, let $\{V_n\}$ be a sequence of i.i.d. $P(III)(0,\sigma,\gamma)$ random variables independent of the U_n's. Thus

$$P(V_n > v) = \left[1 + \left(\frac{v}{\sigma}\right)^{1/\gamma}\right]^{-1}, \quad 0, v, \infty.$$

Now, if N, independent of the V_n's, has a geometric (p) distribution (i.e., $P(N = n) = p(1-p)^{n-1},\ n = 1,2,...$) and if we define

$$V^* = \min_{i \le N} V_i,$$

then, as shown in Chapter 3 and used many times in the present chapter, $V^* \sim P(III)(0, p^{\gamma}\sigma, \gamma)$.

Now there is a natural sequence of geometric (p) random variables $\{N_n\}\}_{n=0}^{\infty}$

associated with the Bernoulli (p) sequence $\{U_n\}_{n=0}^{\infty}$, defined as follows

$$\{N_n = 1\} = \{U_n = 1\},$$

$$\text{while for } i = 2, 3, ..., \tag{6.4.7}$$

$$\{N_n = i\} = \{U_n = 0, U_{n=1} = 0, ..., U_{n+i-2} = 0, U_{n+i-1} = 1\}.$$

Thus N_n is the waiting time for "success" among the "trials", $n, n+1, n+2, ...$ Clearly the N_n's are geometric (p) random variables, but they are not independent of each other. They are, however, all independent of the V_n's , since the U_n's are independent of the V_n's.

A geometric-Pareto (III) process $\{X_n\}$ may then be defined as a function of the sequences $\{N_n\}$ and $\{V_n\}$, as follows. For each $n = 0, 1, 2, ...$ define

$$X_n = p^{-\gamma} \min_{i \leq N_n} V_{n+i-1}. \tag{6.4.8}$$

It is clear that $\{X_n\}$ is a stationary process with $X_n \sim P(III)(0, \sigma, \gamma)$ for each n. For many computations related to this process, it is convenient to consider the standardized process $\{Z_n\}$, where

$$Z_n = (X_n/\sigma)^{1/\gamma}. \tag{6.4.9}$$

Each Z_n has a Pareto(III)$(0, 1, 1)$ distribution, i.e.,

$$P(Z_n > z) = (1+z)^{-1}, \ z > 0.$$

The joint survival function of (Z_n, Z_{n+1}) can be evaluated by conditioning on U_n. In this manner one finds

$$P(Z_n > z_n, Z_{n+1} > z_{n+1}) = \frac{[p(1+z_{n+1})^{-1} + (1-p)(1+\max\{z_n, z_{n+1}\})^{-1}]}{(1+pz_n)}.$$

The fluctuation probabilities associated with this process can help one visualize the nature of typical sample paths and can be used to develop estimates of p. It may be verified that

$$P(Z_n = Z_{n+1}) = 1 + \frac{p}{1-p} \log p,$$

$$P(Z_n > Z_{n+1}) = \frac{p}{1-p}\left[1 + \frac{p}{1-p} \log p\right],$$

$$P(Z_n < Z_{n+1}) = \frac{p}{(1-p)^2}[p - \log p - 1].$$

Note that, if the model is correct, then the ratio of the observed frequency of steps down to the observed frequency of ties in the series should be approximately equal to $p/(1-p)$.

A related model can be considered in which the sequence $\{U_n\}$, instead of consisting of i.i.d. Bernoulli random variables, is a Markov chain with state space $\{0,1\}$. See Arnold (1993) for some discussion of such a model. One could also assume that $\{U_n\}$ is a higher order Markov chain in order to obtain a more flexible model. Arnold (1993) provides some discussion of the case in which $\{U_n\}$ is a second order Markov chain.

Higher dimensional versions of these processes can be readily developed. The change that is necessary is to replace the sequence $\{V_n\}$ of i.i.d. Pareto (III) variables by a sequence $\{\underline{V}^{(n)}\}$ of i.i.d. standard k-variate Pareto (III) variables, i.e., $\underline{V}^{(n)} \sim MP^{(k)}(\underline{0},\underline{1},\underline{1}),\ n=1,2,\ldots$. Then, parallel to (6.4.8), define

$$\underline{Z}^{(n)} = p^{-1} \min_{i \le N_n} \underline{V}^{(n+i-1)}.$$

Then $\underline{X}^{(n)}$ is defined by

$$X_j^{(n)} = \sigma_j (Z_j^{(n)})^{\gamma_j}, \quad j=1,2,\ldots,k.$$

In this k-variate case, $\{U_n\}$ can be an i.i.d. Bernoulli sequence or it can be a Markov chain with state space $\{0,1\}$. It is also possible to consider semi-Pareto versions of these processes, both univariate and k-variate. Finally, for the fearless, it is possible to define matrix-variate versions.

6.5 Inference for Multivariate Pareto Distributions

Historically a paucity of multivariate data, a lack of models which predict multivariate Paretian behavior and the relative recency of the introduction of the majority of the multivariate Pareto distributions all combined to inhibit the development of inferential techniques. Marginal inferences can of course be made using the methods of Chapter 5, but inference procedures which capitalize on the multivariate structure of the data remain poorly developed. Naturally, the relatively ancient Mardia multivariate Pareto distributions are the best explored. In fact, Mardia (1962) devoted a considerable proportion of his pioneering paper to the development of a variety of estimates for the parameters of the distributions he introduced.

6.5.1 *Estimation for Mardia's Multivariate Pareto Families*

Mardia concentrated on the analysis of his second type of bivariate Pareto distribution. This was obtained by beginning with (U_1,U_2) having a bivariate Wicksell-Kibble exponential distribution (the characteristic function of which is given in (6.2.13)) and applying the transformation

$$X_i = \sigma_i e^{U_i/\alpha_i}, \quad i=1,2 \tag{6.5.1}$$

(a special case of (6.2.10) yielding classical Pareto marginals). Not only do we get the time honored classical Pareto marginals using such a transformation, but, of course, the transformation is invertible. In fact, if we define

$$Y_i = \log X_i, \quad i = 1,2 \tag{6.5.2}$$

then the vector $\underline{Y}$ has a bivariate Wicksell-Kibble exponential distribution with location parameters log σ_i, $(i = 1,2)$, and scale parameters α_1^{-1}, α_2^{-1}. The parameter p which appears in the characteristic function (6.2.13) is actually related to the correlation between Y_1 and Y_2 by the relation

$$1 - p = [\text{corr}(Y_1, Y_2)]^2. \tag{6.5.3}$$

Thus, by taking logarithms the problem reduces to that of analyzing a sample from a reasonably well-known bivariate exponential distribution. If the transformation (6.2.10) had been used in its full generality to obtain Pareto IV marginals, then the simple logarithmic transformation would not bring us back to exponential marginals. This was precisely the reason why univariate Pareto I analysis was considerably simpler than univariate Pareto IV analysis (in Chapter 5). We can summarize Mardia's results, as follows, where we introduce the notation $WKP^{(2)}(\text{I})$ to denote Mardia's second bivariate Pareto. The notation emphasizes its genesis from the Wicksell-Kibble distribution and the fact that it is a bivariate Pareto I distribution.

Theorem 6.5.1. *Let $\underline{X}^{(1)}, \underline{X}^{(2)}, \ldots, \underline{X}^{(n)}$ be a random sample of size n from a $WKP^{(2)}(\text{I})(\underline{\sigma}, \underline{\alpha}, p)$ distribution. Define $Y_j^{(i)} = \log X_j^{(i)}$, $i = 1,2,\ldots,n$; $j = 1,2$, and let $\overline{Y}_j = [\sum_{i=1}^n Y_j^{(i)}]/n$, $j = 1,2$.*

(i) Consistent estimates of σ_1 and σ_2 are provided by $\hat{\sigma}_1 = \min_i X_1^{(i)}$ and $\hat{\sigma}_2 = \min_i X_2^{(i)}$, respectively.

(ii) Consistent estimates of α_1 and α_2 are provided by $\hat{\alpha}_1 = \hat{\sigma}_1/\overline{Y}_1$ and $\hat{\alpha}_2 = \hat{\sigma}_2/\overline{Y}_2$, respectively.

(iii) A consistent estimate of p is provided by $\hat{p} = 1 - \sqrt{R}$ where R is the sample correlation of the $Y_j^{(i)}$'s.

The estimates $\hat{\alpha}_1$, $\hat{\alpha}_2$ and $\hat{p}$ are, of course, consistent asymptotically normal. The estimates $\hat{\sigma}_1$, $\hat{\sigma}_2$, $\hat{\alpha}_1$ and $\hat{\alpha}_2$ described in the theorem are maximum likelihood estimates. It is possible to write down an equation involving modified Bessel functions whose solution is the maximum likelihood estimate of p (see Mardia, 1962, p. 1012, or Johnson, Kotz and Balakrishnan, 1972, p. 585).

Mardia also considered estimation of the mean of the $Y^{(i)}$'s, introduced in the above theorem. We may use the notation

$$\gamma_j = E(Y_j^{(i)}), \quad j = 1,2. \tag{6.5.4}$$

Clearly $\overline{Y}_2$ is a consistent unbiased asymptotically normal estimate of γ_2. If γ_1 happens to be known, then one can consider a ratio estimate of e^{γ_2} which leads to an estimate of γ_2 of the form

$$\tilde{\gamma}_2 = \overline{Y}_2 - (\overline{Y}_1 - \gamma_1). \tag{6.5.5}$$

Mardia showed that $\tilde{\gamma}_2$ is consistent, unbiased, asymptotically normal and has smaller variance than $\overline{Y}_2$, when the correlation is high enough, i.e., when $(1-p) > \alpha_2/2\alpha_1$ (recall equation (6.5.3) which relates p to the correlation). One may also fit a straight line regression to the points $(Y_1^{(i)}, Y_2^{(i)})$ and use the estimated slope $\hat{\beta}$ to construct a regression estimate of γ_2 of the form

$$\tilde{\tilde{\gamma}} = \overline{Y}_2 - \hat{\beta}(\overline{Y}_1 - \gamma_1) \tag{6.5.6}$$

(assuming γ_1 is known). $\tilde{\tilde{\gamma}}_2$ is asymptotically more efficient than $\overline{Y}_2$ as an estimate of γ_2. Mardia also considered a double sampling version of (6.5.6) where we only observe or only utilize some of the $Y_1^{(i)}$'s to improve our estimate $\overline{Y}_2$.

In addition Mardia discussed estimation for his type I bivariate distribution, i.e., an $MP^{(2)}(\text{I})(\underline{\sigma}, \alpha)$ distribution as defined by (6.1.1). If $\underline{X}^{(i)}, i = 1, 2, \ldots, n$, is a sample from such a distribution, it is easy to obtain the maximum likelihood estimates of the parameters. One finds (Mardia, 1962)

$$\hat{\sigma}_j = \min_i X_j^{(i)}, \quad j = 1, 2 \tag{6.5.7}$$

$$\text{and} \quad \hat{\alpha} = (C^{-1} - 1/2) + (C^{-2} + 1/4)^{1/2} \tag{6.5.8}$$

$$\text{where} \quad C = \frac{1}{n}\sum_{i=1}^{n} \log\,(\hat{\sigma}_1^{-1} X_1^{(i)} + \hat{\sigma}_2^{-1} X_2^{(i)} - 1). \tag{6.5.9}$$

Actually, referring to (6.1.8), an appealing consistent alternative to $\hat{\alpha}$ (defined by (6.5.8)) is provided by the reciprocal of the sample correlation coefficient. If we try to extend these results to the k-dimensional case (see (6.1.1)), we find that the maximum likelihood estimates of the σ_j's are again of the form $\hat{\sigma}_j = \min_i X_j^{(i)}$. However, obtaining the maximum likelihood estimate of α involves solution of a kth degree equation (instead of a quadratic). As in the bivariate case, it is tempting to forego maximum likelihood and construct a consistent estimate using the pairwise sample correlations, since all of the population correlations are equal to α^{-1} (6.1.8).

Remark. *Chacko and Thomas (2007) consider estimation of a parameter in the bivariate Mardia Type I distribution, i.e., $MP^{(2)}(I)(\underline{\sigma}, \alpha)$, based on a ranked set sample. Thus they consider data consisting of n independent observations based on n independent samples of size n from the $MP^{(2)}(I)$ distribution. In the r'th sample, the r'th largest X value, $X_{r:n}$, and its concomitant Y value, $Y_{[r:n]}$, are measured. Using this data set consisting of n independent order statistics and their concomitants, they discuss estimation of σ_2 assuming that α is known.*

6.5.2 Estimation for More General Multivariate Pareto Families

If we turn to the multivariate Pareto distribution of types II, III, (IV) ((6.1.10), (6.1.17) and (6.1.18)) and the multivariate Feller-Pareto distribution (6.1.30), life becomes more complicated (predictably in the light of the corresponding univariate results described in Chapter 5). In the case of samples from a $MP^{(k)}(\text{II})(\underline{\mu}, \underline{\sigma}, \alpha)$ distribution, we can identify the maximum likelihood estimates of the μ_j's as $\hat{\mu}_j =$

$\min_i X_j^{(i)}$. But, in general, we must be content with numerical solutions of formidable systems of likelihood equations. Life is simplified a little (as in the univariate case), if we, as we shall, routinely use the estimates of the form $\min_i X_j^{(i)}$ for the μ_j's, and then act as though the μ_j's were known. In the following discussion, the multivariate Feller-Pareto distribution will be considered, since the multivariate Pareto (II)-(IV) models will be submodels, covered by the more general presentation.

Suppose that a sample $\underline{X}^{(1)}, \underline{X}^{(2)}, ..., \underline{X}^{(n)}$ is available from a k-dimensional Feller-Pareto distribution with $\underline{\mu} = \underline{0}$. Recall that $\underline{X} \sim FP^{(k)}(\underline{0}, \underline{\sigma}, \underline{\gamma}, \alpha, \underline{\beta})$ if

$$X_i \stackrel{d}{=} \sigma_j \left(\frac{W_j}{Z}\right)^{\gamma_j}, \quad j = 1, 2, ..., k,$$

where the W_j's and Z are independent random variables with $W_j \sim \Gamma(\beta_j, 1)$ and $Z \sim \Gamma(\alpha, 1)$. Consequently the joint density of $\underline{X}$ can be written as

$$f_{\underline{X}}(\underline{x}; \underline{\sigma}, \underline{\gamma}, \alpha, \underline{\beta}) = \frac{\Gamma(\alpha + \sum_{j=1}^k \beta_j)}{\Gamma(\alpha) \prod_{j=1}^k \Gamma(\beta_j)} \times \frac{\prod_{j=1}^k \left[\frac{x_j^{(\beta_j/\gamma_j)-1}}{\gamma_j \sigma_j^{(\beta_j/\gamma_j)}}\right]}{\left[1 + \sum_{j=1}^k \left(\frac{x_j}{\sigma_j}\right)^{1/\gamma_j}\right]^{\alpha + \sum_{j=1}^k \beta_j}}, \quad \underline{x} > \underline{0}. \tag{6.5.10}$$

The likelihood function of the sample will then be a product of n factors of the form (6.5.10), i.e.,

$$L(\underline{\sigma}, \underline{\gamma}, \alpha, \underline{\beta}) = \prod_{i=1}^n f_{\underline{X}}(\underline{X}^{(i)}; \underline{\sigma}, \underline{\gamma}, \alpha, \underline{\beta}).$$

Although this is a rather formidable expression, it is a regular likelihood and the corresponding maximum likelihood estimates of the parameters will have an asymptotic multivariate normal distribution and will be consistent. Some simplification will be encountered in the $MP^{(k)}(II)$, $MP^{(k)}(III)$ and $MP^{(k)}(IV)$ cases. In very special cases it is possible to obtain expressions for the elements in the corresponding information matrix. For example, Gupta and Nadarajah (2007) provide the relevant computations in the bivariate Lomax distribution case, i.e., when $\underline{X} \sim FP^{(2)}(\underline{0}, \underline{\sigma}, \underline{1}, \alpha, \underline{1})$.

Maximum likelihood estimation can also be considered for the flexible Pareto model with survival function given in (6.2.61) and (6.2.62)), from which the corresponding density may be obtained. In high dimensions, it would be natural to set many of the parameters in the model equal to zero. In two dimensions (assuming, as usual, that $\underline{\mu} = \underline{0}$), the density is of the form

$$f_{X_1,X_2}(x_1,x_2) = \frac{(\alpha^2 + \alpha - \alpha\eta) + \eta\alpha^2 \left[\frac{x_1}{\sigma_1} + \frac{x_2}{\sigma_2} + \eta \frac{x_1 x_2}{\sigma_1 \sigma_2}\right]}{\sigma_1 \sigma_2 \left[1 + \frac{x_1}{\sigma_1} + \frac{x_2}{\sigma_2} + \eta \frac{x_1 x_2}{\sigma_1 \sigma_2}\right]^{\alpha+2}}, \quad \underline{x} > \underline{0}. \tag{6.5.11}$$

Arnold and Ganeshalingam (1988) provide simulation based evidence of the performance of maximum likelihood estimates of the parameters $\sigma_1, \sigma_2, \alpha$ and η in this setting. They also discuss a hybrid estimation strategy involving the marginal likelihood functions.

If we have data from a conditionally specified multivariate Pareto model such as (6.3.12), then conditional likelihood estimates might be considered as alternatives to the more difficult to identify maximum likelihood estimates. Thus, instead of choosing parameter values to maximize

$$\prod_{i=1}^{n} f_{\underline{X}}(\underline{X}^{(i)}|\underline{\theta}),$$

where $\underline{\theta}$ denotes the full vector of parameters of the model (with $\underline{\mu} = \underline{0}$), we would maximize

$$\prod_{i=1}^{n}\prod_{j=1}^{k} f_{X_j|\underline{X}_{(j)}}(X_j^{(i)}|\underline{X}_{(j)}^{(i)};\underline{\theta}).$$

An example of such computations in the bivariate case ($k = 2$) may be found in Arnold (1988b).

Hanagal (1996) considers estimation in the case of a Pareto model obtained by transforming a Marshall-Olkin multivariate exponential model. He assumes that observations are available from the simplified distribution (6.2.20), which has a single dependence parameter. However, since such data can be readily transformed back to become Marshall-Olkin multivariate exponential data, a variety of estimation and hypothesis testing results are already available in the literature (see, e.g., Proschan and Sullo (1976)). In the bivariate case, Gupta and Nadarajah (2007) provide the corresponding information matrix, but again this is already available in the Marshall-Olkin multivariate exponential context.

Castillo, Sarabia and Hadi (1997) consider two alternative approaches to fitting continuous bivariate distributions to data. The approaches are motivated by consideration of two well-known estimation strategies useful in one dimension. If $X_1, X_2, ..., X_n$ are i.i.d. with common distribution function $F(x;\underline{\theta})$ where $\underline{\theta}$ is ℓ-dimensional, then one could seek, as an estimate of $\underline{\theta}$, that value of $\underline{\theta}$ which minimizes one of the following two possible objective functions:

$$\sum_{i=1}^{n}\left(F(X_{i:n};\underline{\theta}) - \frac{i}{n}\right)^2 \tag{6.5.12}$$

or

$$\sum_{i=1}^{n}\left(X_{i:n} - F^{-1}\left(\frac{i}{n};\underline{\theta}\right)\right)^2. \tag{6.5.13}$$

Often i/n will be replaced by an alternative "plotting position" $(i-\gamma)/(n+\delta)$ for some $\gamma, \delta \in (0,1)$.

Instead of (6.5.12), when it is more convenient to deal with survival functions, one might make use of the following objective function

$$\sum_{i=1}^{n} \left(\overline{F}(X_{i:n}; \underline{\theta}) - \frac{n-i}{n} \right)^2 . \tag{6.5.14}$$

Use of this objective function could be recommended for parameter estimation based on a univariate Pareto (IV) sample.

For a k-variate Pareto (IV) sample $\underline{X}^{(1)}, \underline{X}^{(2)}, ..., \underline{X}^{(n)}$, it is a k-dimensional version of the last objective function (i.e., (6.5.14)) that is convenient to use to estimate the parameter vector $\underline{\theta} = (\underline{\mu}, \underline{\sigma}, \underline{\gamma}, \alpha)$. For it, we may, for each observation $\underline{X}^{(i)}$, define

$$N(\underline{X}^{(i)}) = \sum_{\ell=1}^{n} I(\underline{X}^{(\ell)} > \underline{X}^{(i)})$$

and then choose $\underline{\theta}$ to minimize

$$\sum_{i=1}^{n} \left[\overline{F}(\underline{X}^{(i)}; \underline{\theta}) - \frac{N(\underline{X}^{(i)})}{n} \right]^2 .$$

Rather than use this approach, Castillo, Sarabia and Hadi (1997) focus on an alternative method which uses an objective function that can be viewed as a multivariate extension of (6.5.13). They describe the bivariate case in detail and provide an outline of how one might proceed in higher dimensions. The specific case in which the $\underline{X}^{(i)}$'s have an $MP^{(2)}(IV)(\underline{0}, \underline{\sigma}, \alpha)$ distribution is worked out explicitly as an example of application of the method.

It is natural to consider a Bayesian approach to the estimation of the parameters in multivariate Pareto models. A full Bayesian approach would require a joint prior for all the many parameters in the model. For example, if $\underline{x}^{(1)}, \underline{x}^{(2)}, ..., \underline{x}^{(n)}$ are the observed values of a sample of size n from an $MP^{(k)}(IV)(\underline{\mu}, \underline{\sigma}, \underline{\gamma}, \alpha)$ distribution, then the corresponding likelihood will be

$$L(\underline{\mu}, \underline{\sigma}, \underline{\gamma}, \alpha) = \prod_{i=1}^{n} \left\{ \frac{\prod_{j=1}^{k} \frac{1}{\gamma_j \sigma_j} \left(\frac{x_j^{(i)} - \mu_j}{\sigma_j} \right)^{(1/\gamma_j)-1}}{\left[1 + \sum_{j=1}^{k} \left(\frac{x_j^{(i)} - \mu_j}{\sigma_j} \right)^{1/\gamma_j} \right]^{\alpha+k}} I(\underline{x}^{(i)} > \underline{\mu}) \right\} .$$

A joint prior for the $(3k+1)$-dimensional parameter vector will be required, i.e., $\pi(\underline{\mu}, \underline{\sigma}, \underline{\gamma}, \alpha)$ leading to a posterior density of the form

$$\pi(\underline{\mu}, \underline{\sigma}, \underline{\gamma}, \alpha | \underline{x}^{(1)}, \underline{x}^{(2)}, ..., \underline{x}^{(n)}) \propto L(\underline{\mu}, \underline{\sigma}, \underline{\gamma}, \alpha) \pi(\underline{\mu}, \underline{\sigma}, \underline{\gamma}, \alpha).$$

Even if, to simplify matters, a locally uniform prior is chosen for the parameters, i.e.,

$$\pi(\underline{\mu}, \underline{\sigma}, \underline{\gamma}, \alpha) \propto 1,$$

the resulting posterior density will not be easy to deal with.

Arnold and Press (1983, 1986) focus on the $MP^{(k)}(II)$ model. Their approach is a hybrid one, involving classical and Bayesian estimation strategies. They consider a sample $\underline{X}^{(1)}, \underline{X}^{(2)}, ..., \underline{X}^{(n)}$ from an $MP^{(k)}(II)(\underline{\mu}, \underline{\sigma}, \alpha) = MP^{(k)}(IV)(\underline{\mu}, \underline{\sigma}, \underline{\gamma}, \alpha)$ distribution. However, they assume that marginal minima of the data values are used to estimate $\underline{\mu}$ and that marginal moment estimates are used for $\underline{\sigma}$. They then proceed to act as if $\underline{\mu}$ and $\underline{\sigma}$ were known and transform the data to obtain

$$\underline{Y}^{(i)} \sim MP^{(k)}(II)(\underline{0}, \underline{1}, \alpha), \quad i = 1, 2, ..., n.$$

The corresponding likelihood will then be

$$L(\alpha) = [\alpha(\alpha+1)\cdots(\alpha+k-1)]^n \prod_{i=1}^{n} \left(1 + \sum_{j=1}^{k} y_j^{(i)}\right)^{-(\alpha+k)}.$$

If an informative prior for α of the gamma form is adopted in this situation, it will be of the form

$$\pi(\alpha) \propto \alpha^{c-1} e^{-d\alpha} I(\alpha > 0),$$

and the corresponding posterior density will be

$$\pi(\alpha | \underline{y}^{(1)}, \underline{y}^{(2)}, ..., \underline{y}^{(n)}) \propto \alpha^{n+c-1} \left[\prod_{\ell=1}^{k-1} (\alpha + \ell)\right]^n e^{-(d+z))\alpha} I(\alpha > 0),$$

where

$$z = \sum_{i=1}^{n} \log \left(1 + \sum_{j=1}^{k} z_j^{(i)}\right).$$

This posterior density is sufficiently complicated to preclude the possibility of identifying analytic expressions for its mean or even for its mode (when $k > 2$), but it is relatively simple to obtain numerical values using suitable optimization and/or integration algorithms. Arnold and Press (1983) also consider a more complicated conjugate prior for use in this situation. In Arnold and Press (1986) they also consider data configurations which include some censored observations.

Papadakis and Tsionas (2010) present a more ambitious Bayesian analysis of $MP^{(k)}(II)(\underline{\mu}, \underline{\sigma}, \alpha)$ data. They assume independent exponential priors for α and for each μ_j, together with independent gamma priors for each $1/\sigma_j$. Even with this very simple choice for the joint prior for the parameters, the corresponding posterior density is complicated. Papadakis and Tsionas utilize a Gibbs sampler with data augmentation to simulate draws from this posterior density and thus to obtain approximate Bayes estimates of the parameters.

More generally, if a sample of size n is available from an $MP^{(k)}(IV)(\underline{\mu}, \underline{\sigma}, \underline{\gamma}, \alpha)$ distribution, then a reasonable Bayesian approach to parameter estimation might begin with the usual simplification of estimating $\underline{\mu}$ by the vector of marginal minima and then, assuming that, after subtraction, $\underline{\mu} = \underline{0}$, i.e., that we have a sample from an $MP^{(k)}(IV)(\underline{0}, \underline{\sigma}, \underline{\gamma}, \alpha)$ distribution. It will simplify computations if we reparameterize, setting $\tau_j = 1/\sigma_j$ and $\delta_j = 1/\gamma_j$, $j = 1, 2, ..., k$. The likelihood of the sample $\underline{x}^{(1)}, \underline{x}^{(2)}, ..., \underline{x}^{(n)}$ will be

$$L(\alpha, \underline{\tau}, \underline{\delta}) = \prod_{i=1}^{n} \left\{ \frac{\alpha(\alpha+1)\cdots(\alpha+k-1)\prod_{j=1}^{k}[\delta_j \tau_j^{\delta_j}(x_j^{(i)})^{\delta_j - 1}]}{\left[1 + \sum_{j=1}^{k}(\tau_j x_j^{(i)})^{\delta_j}\right]^{\alpha+k}} \right\},$$

a rather formidable expression. However, if we assume independent gamma priors for all of the $(2k+1)$ parameters, it is possible to sample from the posterior conditional distributions using a rejection method, and consequently a Gibbs sampler approach can be used to simulate draws from the joint posterior distribution. If the dimension k is large, improved more efficient algorithms will undoubtedly be required.

Jeevanand and Padamadan (1996) address the problem of Bayesian estimation of the parameters of a bivariate Pareto (I) distribution of the Marshall-Olkin type. The survival function that they consider is of the form

$$\overline{F}_{\underline{X}}(\underline{x}) = \left(\frac{x_1}{\beta}\right)^{-\lambda_1} \left(\frac{x_2}{\beta}\right)^{-\lambda_2} \left(\frac{\max\{x_1, x_2\}}{\beta}\right)^{-\lambda_3} \tag{6.5.15}$$

where $x_1, x_2 > \beta$. If β is known, then the minimal sufficient statistic based on a sample of size n from this distribution will consist of a gamma distributed random variable and an independent multinomial random variable. Conjugate priors for $\lambda = \lambda_1 + \lambda_2 + \lambda_3$ and $\underline{\delta} = (\lambda_1/\lambda, \lambda_2/\lambda, \lambda_3/\lambda)$ are of the gamma and Dirichlet form, respectively. The posterior density has independent gamma and Dirichlet marginals for λ and $\underline{\delta}$ and posterior means are readily evaluated. If, in addition, β is unknown, a joint prior for $(\beta, \lambda, \underline{\delta})$ will be required. However the form of the joint likelihood (obtained from (6.5.15)) suggests the form of an appropriate conjugate prior in this case also. For details, see Jeevanand and Padamadan (1996).

6.5.3 A Confidence Interval Based on a Multivariate Pareto Sample

Targhetta (1979) discussed a technique for obtaining a confidence interval for α in the case of a multivariate Pareto (I) distribution with known scale parameters. Targhetta argued in terms of generalizing the following univariate result, which might be called the universal pivotal technique. If $X_1, \ldots, X_n$ are independent identically distributed with common univariate distribution $F(x; \theta)$, then for every θ

$$-\sum_{i=1}^{n} \log F(X_i; \theta) \sim \Gamma(n, 1). \tag{6.5.16}$$

Thus, if we let c_1, c_2 represent the $(\gamma/2)$ and $1 - (\gamma/2)$ percentiles of a $\Gamma(n, 1)$ distribution, then

$$\left\{ \theta : -\sum_{i=1}^{n} \log F(X_i; \theta) \varepsilon (c_1, c_2) \right\} \tag{6.5.17}$$

describes a $(1-\gamma)$ level confidence region for θ. If we consider a sample from a k-dimensional random vector for which suitable conditional distributions are well defined, we can, following Targhetta, let

$$U_{i1}(\theta) = F_{X_1}(X_1^{(i)};\theta), \quad i = 1,2,\ldots,n \tag{6.5.18}$$

and

$$\begin{aligned} U_{ij}(\theta) = F_{X_j|X_1,\ldots,X_{j-1}}(X_j^{(i)}|X_1^{(i)},X_2^{(i)},\ldots,X_{j-1}^{(i)};\theta),& \\ i = 1,2,\ldots,n,& \\ j = 2,3,\ldots,k,& \end{aligned} \tag{6.5.19}$$

and use the quantity

$$Z_n(\theta) = -\sum_{i=1}^{n}\sum_{j=1}^{k} \log U_{ij}(\theta) \sim \Gamma(kn,1) \tag{6.5.20}$$

to construct a confidence region for θ. The observation that these $U_{ij}(\theta)$'s are independent $uniform(0,1)$ random variables can be traced back, at least, to Rosenblatt (1952).

In (6.5.16), (6.5.18) and (6.5.19) it is possible, and sometimes convenient, to use (marginal and conditional) survival functions instead of distribution functions. In the case of a bivariate Pareto (I) population with known scale parameters, Targhetta derived the following expression to be used to construct a confidence interval for α:

$$\begin{aligned} Z_n(\alpha) = &-\sum_{i=1}^{n} \log\,[1-(X_1^{(i)})^{-\alpha}] \\ &-\sum_{i=1}^{n} \log\left[1-\left\{\frac{X_1^{(i)}}{X_2^{(i)}+X_1^{(i)}-1}\right\}^{\alpha+1}\right]. \end{aligned} \tag{6.5.21}$$

In (6.5.21) we have assumed, without loss of generality, that the known scale parameters are all equal to 1 (i.e., that $\underline{\sigma} = 1$). We will continue to make this simplifying assumption in the subsequent discussion. Actually use of survival functions instead of the distributions in the example considered by Targhetta results in an expression considerably more tractable than (6.5.21). Using a bar to remind us of the use of survival functions, we find

$$\overline{Z}_n(\alpha) = \alpha T_1 + (\alpha+1)T_2 \sim \Gamma(2n,1) \tag{6.5.22}$$

$$\begin{aligned} \text{where} \quad T_1 &= \sum_{i=1}^{n} \log X_1^{(i)} \\ T_2 &= \sum_{i=1}^{n} \log\,[(X_2^{(i)}+X_1^{(i)}-1)/X_1^{(i)}]. \end{aligned} \tag{6.5.23}$$

From (6.5.22) we obtain as a confidence interval for α

$$\left(\frac{c_1-T_2}{T_1+T_2}, \frac{c_2-T_2}{T_1+T_2}\right) \tag{6.5.24}$$

where c_1, c_2 are appropriate $\Gamma(2n,1)$ percentiles.

The explanation for this pleasing simplicity lies in the possibility of representing a standard multivariate Pareto II distribution in terms of independent univariate Pareto II variables (i.e., Theorem 6.1.1 and the discussion preceding that theorem). If we have a sample from a multivariate Pareto I distribution with unit scale parameters, then upon subtracting 1 from all observations, we have a sample from a standard multivariate Pareto II distribution (i.e., $MP^{(k)}(\text{II})(\underline{0},\underline{1},\alpha)$). Let $\underline{U}^{(i)}$, $i = 1,2,\dots,n$, denote such a sample. Now, if we define (cf. (6.1.36))

$$Y_j^{(i)} = U_j^{(i)} \Big/ \left(1 + \sum_{\ell=1}^{j-1} U_\ell^{(i)}\right), \quad i = 1,2,\dots,n, \quad j = 1,2,\dots,k, \tag{6.5.25}$$

then all the $Y_j^{(i)}$'s are independent random variables with

$$Y_j^{(i)} \sim P(\text{II})(0,1,\alpha + j - 1). \tag{6.5.26}$$

The construction of a confidence interval analogous to (6.5.24) is, then, a routine matter. The representation (6.5.25) may also be used to obtain the maximum likelihood estimate of α based on a k-dimensional Pareto I sample.

6.5.4 Remarks

There is very little additional discussion of multivariate Pareto inference in the literature. General techniques such as construction of generalized likelihood ratio tests can, of course, be applied, but even this does not seem to have been pursued in the literature. There is clearly scope for additional work on Bayesian inference strategies for multivariate Pareto data.

Perhaps, as mentioned in the introduction of this section, things are as they should be. If, indeed, little multivariate Pareto data is encountered, and if, indeed, no plausible stochastic mechanisms exist which lead to data which appear to be well described by multivariate Pareto distributions, then development of inferential techniques especially geared to multivariate Pareto distributions becomes a sterile exercise. Some data do exist and more will become available in an era in which "big" data sets are not to be feared. Hutchinson (1979) lists four practical settings in which the bivariate Pareto might be a plausible component of a stochastic model explaining the data. Moreover, a cynic might observe that statisticians do not have a consistent record of eschewing sterile exercises, especially when lured by aesthetic considerations. The possibilities of the structural representation described in Theorem 6.1.1 might prove to be an enticing siren call to workers now focusing on multivariate normal inference and its ramifications.

Remark. *Inference for the parameters of, and goodness-of-fit tests for, multivariate generalized Pareto models of the type described in (6.2.47) are of considerable* *interest in extreme value scenarios. Michel (2009) discusses maximum likelihood estimation for such distributions. Falk and Michel (2009) address the goodness-of-fit issue. A specific example of inference for such models involving bivariate data may be found in Rakonczai and Zempléni (2012). Vigorous development in this area of inference for multivariate extremes can be expected in the near future.*

6.6 Multivariate Discrete Pareto (Zipf) Distributions

The basic reference for details on multivariate Zipf distributions is Yeh (2002). The models are appropriate for k-dimensional random vectors whose coordinate variables assume integer values. A k-dimensional integer valued random variable $\underline{X} = (X_1, X_2, ..., X_k)$ is said to have a k-variate Zipf (IV) distribution if

$$P(\underline{X} \geq \underline{m}) = \left[1 + \sum_{j=1}^{k} \left(\frac{m_j - m_j^{(0)}}{\sigma_j} \right)^{1/\gamma_j} \right]^{-\alpha}, \tag{6.6.1}$$

where $\underline{m}^{(0)}$ is a vector of integers (the "location" parameters), $\underline{m} - \underline{m}^{(0)}$ is a vector of non-negative integers and the σ_j's, γ_j's and α are positive parameters. If $\underline{X}$ has such a survival function then we write

$$\underline{X} \sim MZipf^{(k)}(IV)(\underline{m}^{(0)}, \underline{\sigma}, \underline{\gamma}, \alpha).$$

Analogous definitions of $MZipf^{(k)}(II)$ and $MZipf^{(k)}(III)$ distributions can be written. The $MZipf^{(k)}(II)$ model has $\underline{\gamma} = \underline{1}$ while the $MZipf^{(k)}(III)$ model has $\alpha = 1$. Evidently these k-variate Zipf models have univariate marginals of the corresponding Zipf form. More generally, the $MZipf^{(k)}(IV)$ distribution has $MZipf^{(\ell)}(IV)$ marginal distributions for every ℓ-dimensional subset of its coordinate variables, $1 \leq \ell \leq k-1$.

These models are identifiable as discretized versions of k-variate Pareto models. Thus, for example, if $\underline{X} \sim MP^{(k)}(IV)(\underline{\mu}, \underline{\sigma}, \underline{\gamma}, \alpha)$ where $\underline{\mu}$ has integer valued coordinates and can be denoted by $\underline{m}^{(0)}$ then, if we define

$$\underline{Y} = (Y_1, Y_2, ..., Y_k)$$

by $Y_j = [X_j], \ \ j = 1, 2, ..., k$, in which $[\cdot]$ denotes the integer part, then

$$\underline{Y} \sim MZipf^{(k)}(IV)(\underline{m}^{(0)}, \underline{\sigma}, \underline{\gamma}, \alpha).$$

This "integer part" approach can be applied to any of the multivariate Pareto models discussed in this Chapter to define parallel multivariate Zipf models. Yeh (2002) discusses two such integer part models in addition to the $MZipf^{(k)}(II), (III)$ and (IV) described above. Yeh (2002) also introduces a standard k-dimensional Zipf distribution. It is identifiable with the $MZipf^{(k)}(IV)(\underline{0}, \underline{1}, \underline{1}, 1)$ distribution. Some care must be used in discussions involving such a standard model. In the k-variate Pareto setting, it is the case that if $\underline{X} \sim MP^{(k)}(IV)(\underline{\mu}, \underline{\sigma}, \underline{\gamma}, 1)$, then one could define $\underline{U}$ by

$$U_j = \left(\frac{X_j - \mu_j}{\sigma_j} \right)^{1/\gamma_j}, \quad j = 1, 2, ..., k,$$

and have $\underline{U} \sim MP^{(k)}(IV)(\underline{0}, \underline{1}, \underline{1}, 1)$, i.e., a standard k-variate Pareto distribution. It is however not possible to similarly transform an $MZipf^{(k)}(IV)(\underline{m}^{(0)}, \underline{\sigma}, \underline{\gamma}, 1)$ variable to a standard k-dimensional Zipf variable. This is true since if $\underline{X}$ has integer valued coordinate random variables then $[(X_j - m_j^{(0)})/\sigma_j]^{-1/\gamma_j}$ will not assume integer values.

In the $MZipf^{(k)}(III)$ case there is closure under geometric minimization. Thus if $\underline{X}^{(1)}, \underline{X}^{(2)}, ...$ are i.i.d. $MZipf^{(k)}(III)(\underline{m}^{(0)}, \underline{\sigma}, \underline{\gamma})$ variables and if N is a geometric random variable, independent of the $\underline{X}^{(i)}$'s (with $P(N = n) = p(1-p)^{n-1}$), then $\underline{Y} = \min_{i \leq N} \underline{X}^{(i)}$ will also have a k-variate Zipf distribution. Specifically

$$\underline{Y} = \min_{i \leq N} \underline{X}^{(i)} \sim MZipf^{(k)}(III)(\underline{m}^{(0)}, \underline{\tau}, \underline{\gamma}),$$

where

$$\tau_j = p^{\gamma_j} \sigma_j, \quad j = 1, 2, ..., k.$$

Based on experience with geometric maxima of Pareto variables, one might conjecture that the random variable $\widetilde{\underline{Y}} = \max_{i \leq N} \underline{X}^{(i)}$ would also have a k-variate Zipf distribution. Unfortunately, as Yeh observes, this is not true.

Appendix A

Historical income data sources

The present list includes income data sources published prior to 1980. It provides the location of partial and complete income data sets to be found in the pre-1980 references listed in the bibliography. In many instances only a Lorenz curve or perhaps a list of relevant Gini indices is provided. Frequently the same data set has been "worked over" by several researchers. Such comparative analyses are of historical as well as technical interest. In the original discussion provided by Pareto (1897) in the "Cours," the range of populations studied was already remarkably diverse, both in geographic and temporal terms. It is to be emphasized that the references are culled only from the bibliography of this monograph. No priority claims are made. In many instances the data were earlier published and/or analyzed and appropriate references were given.

Table A.1 *Data in the form of a distribution function, Lorenz curve or probit diagram (usually grouped data)*

Provenance and/or nature of data	Referred to by	Reported Gini index (G)	Reported Pareto index (α)
Augusta (1471)	Pareto (1897)		1.43
Augusta (1498)	Pareto (1897)		1.40
Augusta (1512)	Pareto (1897)		1.20
Australia (1950–1)	Rutherford (1955)		
Australia (1967–8)	Kakwani and Podder (1976)		
Basel (1454)	Pareto (1897)		
Basel (1887)	Pareto (1897)		1.24
	Hagstroem (1925)		
Bohemia (1933)	Champernowne (1952)		1.94
	Rutherford (1955)		
	Fisk (1961a)		

Table A.1 *(continued)*

Provenance and/or nature of data	Referred to by	Reported Gini index (G)	Reported Pareto index (α)
Botswana (1974)	Republic of Botswana (1976)		
Canada (1947)	Rutherford (1955)		
Canton de Vaud (1892) (by source of income)	Pareto (1897) Hayakawa (1951)		
Ceylon (1950)	Rutherford (1955)		
Ceylon (1952–3)	Ranadive (1965)	0.43	
England (1843)	Pareto (1897)		1.50
England (1879–80)	Pareto (1897)		1.35
England (1893–4)	Pareto (1897)		1.50
England and Wales (1951–2)	Ranadive (1965)	0.36	
England and Wales (1956–7) (by occupation)	Fisk (1961a)		
Finland (1967)	Vartia and Vartia (1980)		
Florence (1887)	Pareto (1897)		1.41
Germany (1965)	Peterson (1979)		
India (1917–8)	Shirras (1935)		1.39
India (1923–4)	Shirras (1935)		1.13
India (1926–7)	Shirras (1935)		1.22
India (1929–30)	Shirras (1935)		1.48
India (1950)	Ranadive (1965)	0.41	
India (1953–7)	Ranadive (1965)	0.31	
India (1955–6)	Ranadive (1965)	0.34	
Ireland (1893–4)	Pareto (1897)		
Italian Cities (1887)	Pareto (1897)		1.45
Japan (1926)	Hayakawa (1951)		
Japan (1930)	Hayakawa (1951)		
Japan (1934)	Hayakawa (1951)		
Japan (1938)	Hayakawa (1951)		
Mexico (1957)	Ranadive (1965)	0.50	
Norway (1930)	Champernowne (1952)		2.00
Oldenburg (1890)	Pareto (1897)		1.47
	Hagstroem (1925)		1.60
Peru (1800)	Pareto (1897)		1.79
Prussia (1852)	Pareto (1897) Hagstroem (1925)		1.89
Prussia (1876)	Pareto (1897) Hagstroem (1925)		1.72

Table A.1 *(continued)*

Provenance and/or nature of data	Referred to by	Reported Gini index (G)	Reported Pareto index (α)
Prussia (1881)	Pareto (1897)		1.73
	Hagstroem (1925)		
Prussia (1886)	Pareto (1897)		1.68
	Hagstroem (1925)		
Prussia (1892)	Lorenz (1905)		
Prussia (1893–4)	Pareto (1897)		
	Hagstroem (1925)		
Prussia (1894–5)	Pareto (1897)		1.60
	Hagstroem (1925)		
Prussia (1901)	Lorenz (1905)		
Santiago and Valparaiso (1958)	Blitz and Brittain (1964)		
Saxony (1880)	Pareto (1897)		1.58
Saxony (1886)	Pareto (1897)		1.51
Sweden (1912)	Hagstroem (1925)		
Sweden (1919)	Hagstroem (1925)		
Sweden (1931)	Ericson (1945)		
	Champernowne (1952)		1.80
Sweden (1943)	Ericson (1945)		
U.K. (1918–9)	Macgregor (1936)		1.47
U.K. (1938–9)	Kalecki (1945)		
	Champernowne (1952)		1.60
	Rutherford (1955)		
U.K. (1949)	Rutherford (1955)		
U.S. (1914–33)	Johnson (1937)		1.34–1.90
U.S. (1918)	Yntema (1933)		1.80
	Champernowne (1952)		1.59
	Rutherford (1955)		
	Fisk (1961a)		
	Fisk (1961b)		
U.S. (1929) (family Income)	Champernowne (1952)		1.48
U.S. (1935–6)	Bowman (1945)		
	Schutz (1951)		
U.S. (1947)	Champernowne (1952)		1.96
	Rutherford (1955)		
	Fisk (1961a)		
U.S. (1954) (by occupation)	Fisk (1961a)		

Table A.1 *(continued)*

Provenance and/or nature of data	Referred to by	Reported Gini index (G)	Reported Pareto index (α)
U.S. (1955)	Gastwirth and Smith (1972)		
	Singh and Maddala (1975)		
U.S. (1959)	Verway (1966)		
U.S. (1960)	Salem and Mount (1974)		
	Singh and Maddala (1976)		
U.S. (1961)	Aigner and Goldberger (1970)		
	Zellner (1971)		
U.S. (1968)	Gastwirth (1972)	0.40	
U.S. (1969)	Salem and Mount (1974)		
	Singh and Maddala (1976)		
W. Germany (1950)	Ranadive (1965)	0.41	
Zurich (1848)	Lorenz (1905)		
Zurich (1888)	Lorenz (1905)		

Table A.2 *Bivariate distributional data*

Provenance and/or nature of data	Referred to by	Reported Gini index (G)	Reported Pareto index (α)
U.K. (1951–2)	Champernowne (1953)		
U.K. (1963, 1966, 1970)	Hart (1976)		

Table A.3 *Data in the form of inequality indices only*

Provenance and/or nature of data	Referred to by	Reported Gini index (G)	Reported Pareto index (α)
Australia (1915)	Champernowne (1952)		1.69
Australia (1915) (by sex)	Yntema (1933)		1.67–1.98
Australia (1920)	Yntema (1933)		1.55
Australia (1945–6)	Champernowne (1952)		2.32

Table A.3 *(continued)*

Provenance and/or nature of data	Referred to by	Reported Gini index (G)	Reported Pareto index (α)
Barbados (1951–2)	Ranadive (1965)	0.44	
Czechoslovakia (1933)	Champernowne (1952)		2.01
Denmark (1945–50)	Champernowne (1952)		2.13–2.21
Denmark (1952)	Ranadive (1965)	0.40	
Italy (1948)	Ranadive (1965)	0.38	
Moravia (1933)	Champernowne (1952)		2.06
Netherlands (1950)	Ranadive (1965)	0.41	
Norway (1929)	Champernowne (1952)		2.15
Prussia (1910)	Yntema (1933)		1.53
Puerto Rico (1953)	Ranadive (1965)	0.39	
Sweden (1912–55)	Hagstroem (1960)		1.24–2.54
Sweden (1946–7)	Champernowne (1952)		2.16–2.40
Sweden (1948)	Ranadive (1965)	0.41	
U.K. (1919–20)	Yntema (1933)		1.78
U.K. (1945–9)	Champernowne (1952)		1.74–1.87
U.S. (1910) (family income)	Yntema (1933)		1.99
U.S. (1914–9)	Hagstroem (1925)		1.1–1.7
U.S. (1944–68)	Hagerbaumer (1977)	0.41–0.44	
U.S. (1946)	Champernowne (1952)		2.00
U.S. (1950)	Ranadive (1965)	0.37	
U.S. (1955–67)	Gastwirth (1972)		1.7–2.7

Appendix B

Two representative data sets

In order to illustrate some of the inferential techniques described in this monograph (especially in Chapter 5), two small data sets have been selected. It is perhaps appropriate to describe the selection process. First it was decided that the data sets should represent income and, second, the corresponding Pareto chart ($\log \overline{F}(x)$ vs $\log x$) should be approximately linear. According to the income "folk-lore" it should not matter much whether we deal with income accruing to individuals or to groups and it should not matter much whether we deal with income over a short or a long period of time. One selected data set involves observations on total annual income accruing to large groups of individuals. The other data set deals with long term (in fact lifetime) earnings accruing to certain individuals. Such data sets are sufficiently unusual to make it not at all obvious that they will be well fitted by Pareto or generalized Pareto distributions. Nevertheless preliminary Pareto charts looked respectably linear and, in fact, as will be seen, the Pareto models provide, in general, satisfactory fits.

Data set number 1, to be referred to subsequently as the "golfer data," consists of 50 observations. They represent the lifetime tournament earnings through 1980 of all those professional golfers whose lifetime tournament earnings exceed \$700,000 during that period. The data set is displayed in Table B.1. A preliminary Pareto chart drawn through the sample deciles yielded a slope of approximately -2.5 and acceptable linearity.

Data set number 2, to be referred to subsequently as the "Texas counties data," consists of 157 observations. Each observation represents the total personal income accruing to the population of some one of the 254 counties in Texas in 1969. The 157 included in the present data set represent all the Texas counties in which total personal income exceeds \$20,000,000. The data set is displayed in Table B.2. A preliminary Pareto chart yielded a slope of -0.914 and acceptable linearity.

The slope of the Pareto chart gives us valuable information about the existence of moments of the underlying distribution. The Texas county distribution will not have a finite first moment and clearly exhibits a great deal of inequality. The golfer distribution can be expected to have finite first and second moments. It exhibits a considerably smaller degree of inequality. The golfer Lorenz curve, as a consequence, is only slightly bent with a corresponding low value of the sample Gini index ($G_{50} = 0.217$, using equation (4.3.48)). It is interesting to note that if we accept the

Table B.1 *Golfer data (lifetime earnings in thousands of dollars). (Source: Golf magazine, 1981 yearbook)*

3581	1433	1066	883	778
2474	1410	1056	878	778
2202	1374	1051	871	771
1858	1338	1031	849	769
1829	1208	1016	844	759
1690	1184	1005	841	753
1684	1171	1001	825	746
1627	1109	965	820	729
1537	1095	944	816	712
1519	1092	912	814	708

Table B.2 *Texas county data (total personal income in 1969 in millions of dollars). (Source: County and City Data Book, 1972; Bureau of the Census)*

6007.1	206.4	78.1	52.6	38.8	28.5	21.9
4860.5	198.1	77.8	52.4	38.2	28.4	21.8
2367.8	194.5	75.9	51.3	37.9	28.1	21.5
2135.6	188.2	75.1	50.9	36.4	27.8	21.4
890.1	181.4	74.8	50.5	36.1	27.8	21.2
840.4	179.8	74.6	50.4	35.7	27.4	20.6
710.2	150.1	73.5	46.3	35.2	27.1	20.6
601.4	130.5	72.5	45.7	35.0	27.0	20.6
513.9	125.9	71.0	44.1	35.0	27.0	20.4
486.9	125.1	71.0	43.6	34.3	26.6	20.4
379.0	122.3	70.9	43.2	34.2	26.4	20.3
336.0	117.8	68.1	42.7	34.1	26.3	20.2
313.5	115.8	64.4	42.4	34.1	25.8	20.2
301.8	115.6	63.3	42.2	34.0	25.5	
281.7	112.9	62.5	41.6	32.8	25.1	
269.7	104.6	60.5	41.4	32.8	24.9	
269.6	99.5	60.0	41.3	32.7	24.3	
266.8	95.8	58.9	40.7	32.4	24.1	
250.5	94.1	58.4	40.2	31.3	23.3	
248.3	83.2	56.2	39.7	31.0	23.1	
226.0	83.0	54.0	39.4	31.0	22.9	
220.8	82.7	53.3	39.2	30.1	22.7	
219.8	82.3	53.3	39.1	29.4	22.5	
216.1	81.5	53.1	39.0	29.1	21.9	

fitted slope (-2.5) of the Pareto chart as an estimate of $-\alpha$ where α is the inequality parameter in the classical Pareto distribution (3.2.1), we arrive at an estimate of $(2\alpha-1)^{-1} = 0.25$ for the Gini index. This is respectably close to the observed value of 0.217.

Since the support sets of the distributions were predetermined, i.e., $(700,\infty)$ and $(20,\infty)$, we will always use these as known values of μ when fitting generalized Pareto models to the data sets (or as known values of σ when the distribution being fitted is classical Pareto). It is convenient to subtract 700 from each of the 50 golfer data points and to subtract 20 from the 157 Texas county data points and thus arrive at distributions supported on $(0,\infty)$.

We address the question of whether any or all of the generalized Pareto models (3.2.1), (3.2.2), (3.2.5) and (3.2.8) fit the two data sets. All four models can be thought of as specializations of the Pareto (IV) model. In what follows recall that by subtracting 700 (respectively 20) from the golfer (respectively Texas county) data we have obtained a distribution with support $(0,\infty)$ so $\mu = 0$ in all cases.

First consider the golfer data. The full P(IV) $(0,\sigma,\gamma,\alpha)$ model has three free parameters. The corresponding likelihood equations take the form

$$\alpha = \left(\frac{1}{n}\sum_{i=1}^{n}\log\left(1+\left(\frac{x_i}{\sigma}\right)^{1/\gamma}\right)\right)^{-1},$$

$$\gamma = \frac{1}{n}\sum_{i=1}^{n}\left\{\frac{\left[\alpha\left(\frac{x_i}{\sigma}\right)^{\frac{1}{\gamma}}-1\right]}{\left[\left(\frac{x_i}{\sigma}\right)^{\frac{1}{\gamma}}+1\right]}\log\left(\frac{x_i}{\sigma}\right)\right\},$$

$$\sigma = \frac{1+\alpha}{n}\sum_{i=1}^{n}\left\{\frac{\left(\frac{x_i}{\sigma}\right)^{\frac{1}{\gamma}-1}x_i}{\left[1+\left(\frac{x_i}{\sigma}\right)^{1/\gamma}\right]}\right\}. \tag{B1}$$

These equations may be solved readily by an iterative re-substitution technique. An examination of the likelihood surface in a neighborhood of the solution to equation (B1) is necessary to be sure that a maximum and indeed a global maximum has been found. The P(II) $(0,\sigma,\alpha)$ model involves setting $\gamma = 1$ in the full Pareto (IV) model. The corresponding likelihood equations are the first and third equations in (B1) with $\gamma = 1$ whenever it appears. Only a minor modification of the program used to obtain the unrestricted maximum likelihood estimates is required. For the Pareto (III) $(0,\sigma,\gamma)$ model we set $\alpha = 1$. For the standard Pareto model we set $\alpha = 1$ and $\gamma = 1$ retaining only a scale parameter α. Finally, the classical Pareto model (recall we translated to get support $(0,\infty)$) corresponds to a P(IV) $(0,700,1,\alpha)$ model. The maximum likelihood estimates of the parameters for these five models are displayed in Table B.3. In addition, for each model, the resulting value of (-2) times the log of the likelihood is displayed, i.e., $(-2\log L)$.

Table B.3 tells us about the relative merits of the competing models. It does not tell us about the goodness-of-fit. The models are all quite good. The Kolmogorov-Smirnov statistic fails to reach significance for any of the models. If instead one uses

Table B.3 *Maximum likelihood estimates for generalized Pareto models for golfer data*

Model	Estimated parameters (fixed parameters in parentheses)			
	α	γ	σ	$-2\log L$
Classical Pareto	2.275	(1)	(700)	716.855
Standard Pareto	(1)	(1)	268.139	725.362
Pareto (II)	8.463	(1)	3,496.277	714.192
Pareto (III)	(1)	0.694	273.381	716.923
Pareto (IV)	4.676	0.927	1,633.566	714.043

a chi-squared test with 8 cells (cell boundaries 0, 60, 125, 200, 300, 425, 610, 950) only the standard Pareto model exhibits significant lack-of-fit. The inappropriateness of the standard Pareto model was already evident from the last column of Table B.3. Generalized likelihood ratio tests would suggest that the Pareto (II) model is the appropriate model. The classical Pareto and the Pareto (III) models are only slightly inferior although in both cases the reduction in likelihood from the full Pareto (IV) model approaches significance. It is remarkable that such large changes in the parameters lead to such small changes in the likelihood. For the present data set, the extra parameters in the Pareto (IV) model do not avail us much in terms of model flexibility. The values of the goodness-of-fit measures for the five maximum likelihood models are displayed in Table B.4. The chi-squared statistics smile a little more kindly on the Pareto (III) model but the Pareto (II) model remains a clear winner. In Figure B.1, the sample distribution and the fitted Pareto (II) distribution are compared. It appears that the fit is satisfactory.

Several alternative estimation techniques were described in Chapter 5. It is of interest to compare their messages with those obtained using maximum likelihood.

Since, based on the golfer Pareto chart, the second moment of the underlying distribution is finite, we can consider using the moment estimates for a Pareto (II) model using equations (5.2.105) and (5.2.106). The resulting estimates are

$$\alpha_M = 9.92 \text{ and } \sigma_M = 4171.303.$$

Table B.4 *Goodness-of-fit statistics for fitted maximum likelihood models*

Model	Number of parameters estimated	Kolmogorov-Smirnov $(\sup_x \lvert F_n(x) - \hat{F}(x)\rvert)$	Chi-squared (8 cells)
Classical Pareto	1	.077	8.425
Standard Pareto	1	.128	13.603
Pareto (II)	2	.052	6.547
Pareto (III)	2	.059	7.272
Pareto (IV)	3	.056	6.369

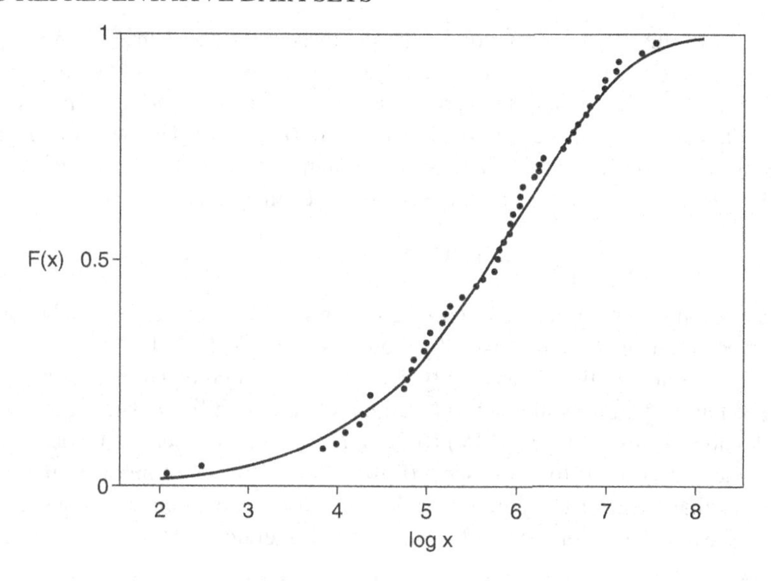

Figure B.1 *Pareto (II) model fitted to golfer data.*

Although they differ considerably from the maximum likelihood estimates (namely 8.463 and 3496.277) there is little difference in the corresponding likelihood. Remarkably, if one uses a chi-squared measure of goodness-of-fit, the distribution using the moment estimates fits the data slightly better than does the distribution using the maximum likelihood estimates (see Table B.5).

If we assume a Pareto (III) model we may estimate γ and σ by setting two sample quantiles equal to their expectations and solving for γ and σ. Using the 16th and 34th order statistics for this purpose the corresponding estimates obtained were (cf. (5.2.116))

$$Y_{EQ} = 0.78611 \text{ and } \sigma_{EQ} = 262.4504.$$

These are in general agreement with the maximum likelihood estimates (0.69389 and 273.38148) but, as measured by likelihood and by goodness-of-fit statistics, these expected quantile estimates appear to be inferior (see Table B.5).

Table B.5 *Alternative estimates for the golfer data*

Model	Equation used	Estimated parameters (fixed parameters in parentheses)			$-2\log L$	Kolmogorov-Smirnov	Chi-squared (8 cells)
		σ	γ	σ			
Pareto II	(5.2.105–6)	9.92	(1)	4,171.303	714.204	.055	6.531
Pareto III	(5.2.107)	(1)	0.642	257.340	717.531	.081	8.211
Pareto IV	(5.2.116)	(1)	0.786	262.450	718.037	.071	8.145

A final alternative estimation technique to be considered involves setting fractional moments equal to their expectations and solving for unknown parameters (cf. (5.2.107)). The equations were set up for $\frac{1}{2}$, $\frac{1}{4}$ and $\frac{1}{8}$ fractional moments. No solution was found for the Pareto (II) and Pareto (IV) cases. This was somewhat surprising since 50 appears to be a reasonable sample size. In the Pareto (III) case, use of this method of (fractional) moments yielded estimates

$$\gamma_{FM} = 0.64223 \text{ and } \sigma_{FM} = 257.3404.$$

These are in general agreement with the maximum likelihood estimates but are inferior when compared on the basis of goodness-of-fit (see Table B.5).

We turn now to the Texas county data. The likelihood equations are again given by (B1) and again a re-substitution technique may be utilized. The classical Pareto model now corresponds to a P(IV) $(0,20,1,\alpha)$ distribution (recall 20 was subtracted from the observations to get support $(0,\infty)$). The hierarchy of models fitted was the same as that used for the golfer data. The resulting maximum likelihood estimates and corresponding values of (-2) times the log-likelihood are displayed in Table B.

The models are all remarkably good. From Table B.7, it can be seen that the Kolmogorov-Smirnov statistic fails to reach significance for any of the models. In addition the models all prove acceptable when subjected to a chi-squared test with 12 cells (cell boundaries 0, 2, 5, 9, 13, 19, 27, 38, 54, 80, 135, 300).

On grounds of parsimony the appropriate model for the Texas data would appear to be a standard Pareto with $\sigma = 27.694$. The introduction of further shape parameters (i.e., use of a Pareto (II), (III) or (IV) model) does not result in a worthwhile improvement of fit. The classical Pareto model might well be a viable competitor if some theoretical considerations point in that direction. As can be seen from Figure B.2, both the standard Pareto model and the classical Pareto model fit the data remarkably well.

Turning to alternative estimation techniques, we first consider one which is predictably bad. Moment estimates for the Pareto (II) model using (5.2.71–2) assume

Table B.6 *Maximum likelihood estimates for generalized Pareto models for Texas counties data*

	Estimated parameters (fixed parameters in parentheses)			
Model	α	γ	σ	$-2\log L$
Classical Pareto	0.848	(1)	(20)	1677.006
Standard Pareto	(1)	(1)	27.694	1676.020
Pareto (II)	0.969	(1)	26.401	1675.979
Pareto (III)	(1)	1.024	27.692	1675.897
Pareto (IV)	1.050	1.041	29.880	1675.873

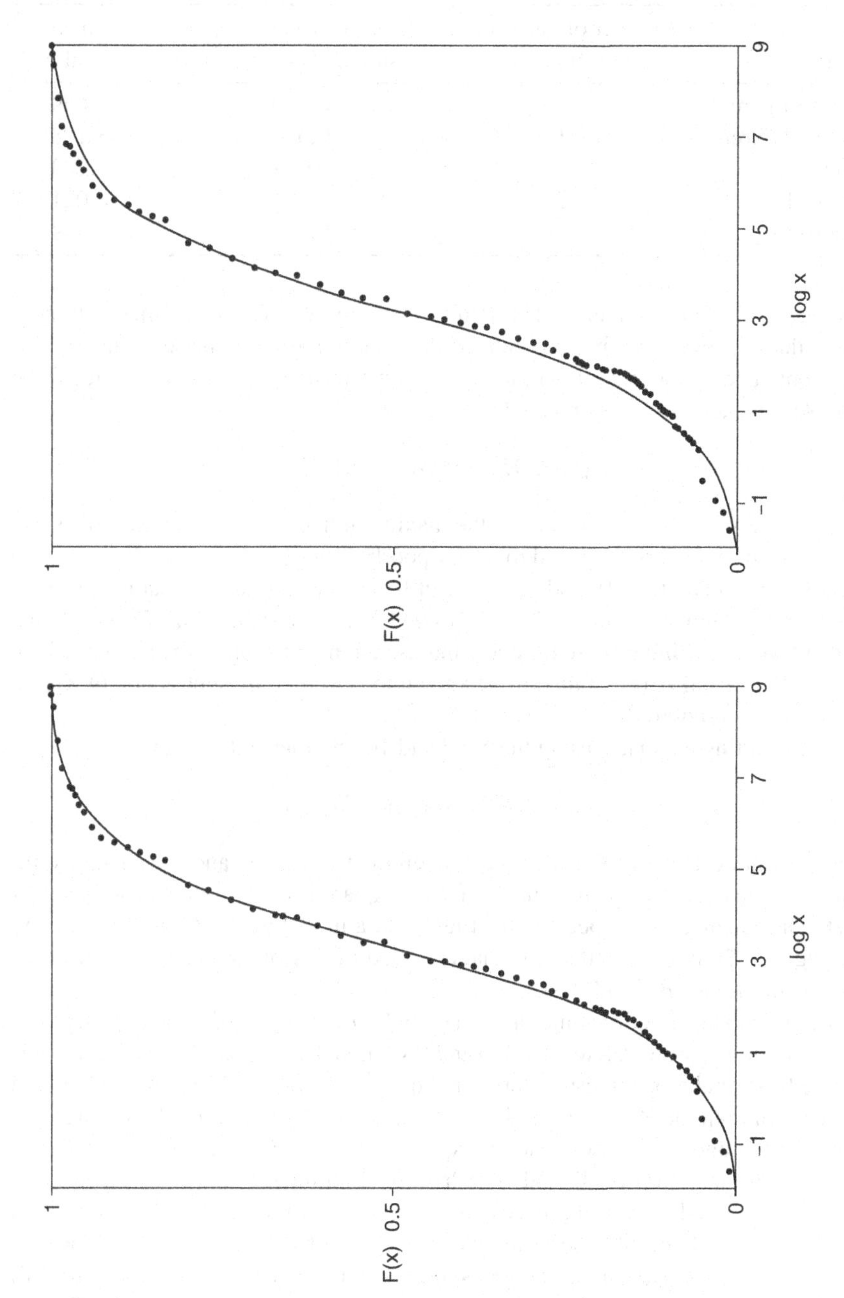

Figure B.2 *Standard Pareto model (left) and classical Pareto model (right), each fitted to Texas counties data.*

Table B.7 *Goodness-of-fit statistics for fitted maximum likelihood models*

Model	Number of parameters estimated	Kolmogorov-Smirnov $(\sup_x \lvert F_n(x) - \hat{F}(x) \rvert)$	Chi-squared (12 cells)
Classical Pareto	1	0.050	9.669
Standard Pareto	1	0.037	8.108
Pareto (II)	2	0.038	8.203
Pareto (III)	2	0.041	8.021
Pareto (IV)	3	0.042	7.809

existence of second moments. The Pareto chart for the Texas counties data suggested that not even the first moment of the underlying distribution is finite. Estimates obtained using (5.2.105–6), under such circumstances, can be expected to be unstable. The actual values obtained were

$$\alpha_M = 2.153 \text{ and } \sigma_M = 201.053.$$

These are very much at variance with the maximum likelihood estimates and, as can be seen from Table B.8, they lead to a very poorly fitting model.

Assuming a Pareto (III) model we may utilize the technique of equating quantiles to their expectations, i.e., use (5.2.116). Instead, since the sample size (157) is large, we can invoke Stirling's approximation and use the much simpler expressions given in (5.2.117) (recall μ is assumed to be known to 0 so that the subtraction of $X_{1:n}$ in (5.2.117) is not required).

The resulting estimates using the 52nd and 105th order statistics are

$$\gamma_{EQ} = 0.969 \text{ and } \sigma_{EQ} = 27.994.$$

These are quite close to the maximum likelihood estimates and only slightly inferior as measured by the likelihood and chi-squared. In fact, though it is not of much consequence, the expected quantiles yield a model whose fit as measured by Kolmogorov-Smirnov is better than that provided by any of the other estimates considered (see Tables B.7 and B.8).

The method of fractional moments (i.e., 5.2.107) using the $\frac{1}{2}$, $\frac{1}{4}$ and $\frac{1}{8}$ moments yields solutions for the Pareto (II), (III) and (IV) models. Perusal of Table B.8 reveals that the fractional moment estimators agree quite well with the maximum likelihood estimates in both the Pareto (II) and Pareto (III) cases. The resulting fit as measured by both Kolmogorov-Smirnov and chi-squared appears to be as good as the fit obtained using maximum likelihood. The fractional moment estimates for the Pareto (IV) differ markedly from the maximum likelihood estimates. It will be observed that the values of the likelihood for the two kinds of estimates are not, however, greatly different. For the Texas counties data, as for the golfer data, the Pareto (IV) likelihood surface is relatively flat in a large region around its maximal point. This is reflected in the fact that two Pareto (IV) models with markedly different parameter values may both fit the data quite well.

Table B.8 *Alternative estimates for the Texas counties data*

Model	Equation used	Estimated parameters (fixed parameters in parentheses) α	γ	σ	$-2\log L$	Kolmogorov-Smirnov	Chi-squared (12 cells)
Pareto II	(5.2.105–6)	2.153	(1)	201.053	1744.628	.288	84.064
Pareto II	(5.2.107)	1.026	(1)	28.954	1676.115	.037	8.063
Pareto III	(5.2.107)	(1)	1.011	27.511	1675.933	.038	8.052
Pareto III	(5.2.117)	(1)	0.969	27.994	1676.583	.036	8.596
Pareto IV	(5.2.107)	1.316	1.142	44.295	1676.572	.054	7.479

In summary, both the golfer and the Texas counties data sets appear to be well fitted by a Pareto (II) model. In the case of the Texas counties, a further reduction to a standard Pareto model appears appropriate. One should not draw the conclusion that all data sets will behave quite this well. Recall that the two data sets were pre-screened by checking for a “roughly linear” Pareto chart. If data come from a P(IV) $(\mu,\sigma,\gamma,\alpha)$ with $\gamma \neq 1$ then its Pareto chart will not be linear. To conclude, we will briefly describe a data set similar to the golfer data for which the Pareto (II) model does not provide a satisfactory fit. If one considers the 1978 tournament earnings of the top 50 professional tennis players (source: World of Tennis, 1979), the Pareto chart of $\log \overline{F}(x)$ vs $\log x$ appears to be distinctly curved. A Pareto (IV) model with $\hat{\gamma} = 1.196$ provides a significant improvement over a Pareto (II) model for this data. Even with this tennis data, the Pareto (II) is the best of the simplified models (i.e., is better than the Pareto (III), the classical and the standard Pareto).

Appendix C

A quarterly household income data set

A more typical example of a data configuration which can be expected to be well described by a Pareto or generalized Pareto model is one involving income of individual households in a population. The National Household Income and Expenditure Survey 2012 (ENIGH-2012) was conducted by the National Institute of Geography and Statistics of Mexico. The ENIGH-2012 objective was to provide a statistical overview of the behavior of the income and expenditure of households in terms of amount, origin and distribution; additionally to provide information on occupational and socio-demographic characteristics of household members, as well as the characteristics of the infrastructure of housing and household equipment in Mexico. The number of sampled households was 46,616. For each household, a quarterly income figure was reported. Rather than analyze the full set of 46,616 data points, a random subset of size 500 was selected for analysis, since this was thought to be a more typical sample size to be encountered in practice. Pareto(I)–(IV) models were fitted to the set of 500 sampled data points using maximum likelihood. The smallest observation among the 500 was used as an estimate of the location parameter μ and was subtracted from each of the other data points prior to estimating the other model parameters, i.e., models with $\mu = 0$ were fitted to the set of data points representing quarterly household income minus minimum quarterly household income. We address the question of whether any or all of the generalized Pareto models (3.2.1), (3.2.2), (3.2.5) and (3.2.8) fit this data set. All four models can be thought of as specializations of the Pareto (IV) model. The results are displayed in Table C.

Table C.1 *Maximum likelihood estimates for generalized Pareto models for the Mexican income data*

	Estimated parameters (fixed parameters in parentheses)			
Model	α	γ	σ	$-2\log L$
Classical Pareto	0.1798	(1)	0.5638	7701.7175
Pareto (II)	1.3558	(1)	233.6233	6886.7850
Pareto (III)	(1)	0.9301	148.1856	6890.4988
Pareto (IV)	2.1195	1.1351	482.6832	6884.6980

Table C.2 *Goodness-of-fit statistics for fitted maximum likelihood models*

Model	Number of parameters estimated	Kolmogorov-Smirnov ($\sup_x \lvert F_n(x) - \hat{F}(x) \rvert$)
Classical Pareto	1	0.3859
Pareto (II)	2	0.0871
Pareto (III)	2	0.0692
Pareto (IV)	3	0.0912

Table C tells us about the relative merits of the competing models. It does not tell us about the goodness-of-fit. The models are all quite good as indicated by the corresponding Kolmogorov-Smirnov statistics which are displayed in Table C. For this data set, based on likelihood statistics, the Pareto (II) model appears to be the most satisfactory choice, although the Pareto (III) appears to be only slightly inferior. The Pareto (IV) model does not represent a significant improvement over the Pareto (II) model.

In Figure C, the sample distribution and the fitted Pareto (II) distribution are compared confirming the observation that the fit is satisfactory. Note that the ranking of the four competing models provided by the Kolmogorov-Smirnov measure is different from the likelihood ranking. Using the Kolmogorov-Smirnov statistics the Pareto (III) model is to be preferred, followed by the Pareto(II) model.

A more extensive study of this data set, jointly authored by Humberto Vaquera, is in preparation. A broader spectrum of popular income models and estimation strategies will be considered in that report.

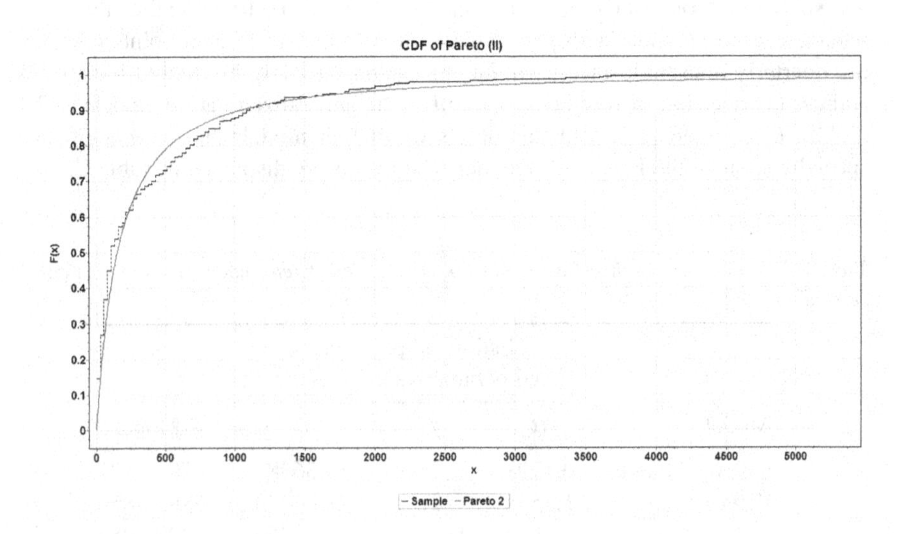

Figure C.1 *Pareto (II) model fitted to Mexican income data.*

References

[1] Aaberge, R. (2001). Axiomatic characterization of the Gini coefficient and Lorenz curve orderings. *J. Econom. Theory*, 101, 115-132. Erratum: 140, e356.

[2] Aaberge, R. (2007). Asymptotic distribution theory of empirical rank-dependent measures of inequality. In *Advances in Statistical Modeling and Inference, Ser. Biostat., 3,*, World Sci. Publ., Hackensack, NJ, 495–512.

[3] Aaberge, R. (2009). Ranking intersecting Lorenz curves. *Soc. Choice Welf.*, 33 , 235-259.

[4] Aban, I. B., Meerschaert, M. M. and Panorska, A. K. (2006). Parameter estimation for the truncated Pareto distribution. *J. Amer. Statist. Assoc.*, 101, 270-277.

[5] Abd-El-Hakim, N. S. and Sultan, K. S. (2004). Maximum likelihood estimation from record-breaking data for the generalized Pareto distribution. *Metron*, 62, 377-389.

[6] Abdell-All, N. H., Mahmoud, M. A. W. and Abd-Ellah, H. N. (2003). Geometrical properties of Pareto distribution. *Appl. Math. Comput.*, 145, 321-339.

[7] Abdel-Ghaly, A. A., Attia, A. F. and Aly, H. M. (1998). Estimation of the parameters of Pareto distribution and the reliability function using accelerated life testing with censoring. *Comm. Statist. Simulation Comput.*, 27, 469-484.

[8] Abdul-Sathar, E. I. and Jeevanand, E. S. (2010). Bayes estimation of Lorenz curve and Gini-index for classical Pareto distribution in some real data situation. *J. Appl. Statist. Sci.*, 17, 315-329.

[9] Abdul-Sathar, E. I., Jeevanand, E. S. and Muraleedharan, K. R. (2006). Bayesian estimation of Lorenz curve, Gini-index and variance of logarithms in a Pareto distribution. *Statistica*, 65, 193-205.

[10] Abusev, R. A. (2004). On statistical estimations for the Pareto distribution. *J. Math. Sci. (N. Y.)*, 123, 3716-3721.

[11] Adams, K. A. H. (1979). Income distribution and remuneration systems in South Africa. *The South African Journal of Economics*, 47, 441.

[12] d'Addario, R. (1930). Un'indagine sulla dinamica distributiva. *Rivista Italiana di Statistica*, 2, 269–304.

[13] d'Addario, R. (1931). La curve dei redditi. *Rivista Italiana di Statistica*, 3, 140–161.

[14] d'Addario, R. (1932). Intorno alla curva dei redditi di Amoroso. *Rivista Italiana di Statistica*, 4, 89–108.

[15] d'Addario, R. (1934a). Intorno alla validita dei due teoremi Paretiani sulla dinamica distributiva. *Atti dell'Istituto Nazionale delle Assicurazioni*, 6, 179–198.

[16] d'Addario, R. (1934b). Sulla Misura della Concentrazione dei Redditi. Instituto Poligrafico dello Stato, Roma.

[17] d'Addario, R. (1936). Sulla curva dei redditi di Amoroso. *Annali dell'Istituto di Statistica (Bari)*, 10, 1–57.

[18] d'Addario, R. (1939). Un metodo per la rappresentazione analitica delle distribuzioni statistiche. *Annali dell'Istituto di Statistica della Reale Universita di Bari*, 16, 3–56.

[19] Adeyemi, S. (2002). Some recurrence relations for single and product moments of order statistics from the generalized Pareto distribution. *J. Statist. Res.*, 36, 169-177.

[20] Adeyemi, S. and Ojo, M. O. (2004). Higher moments of order statistics from the generalized Pareto distribution. *J. Statist. Res.*, 38, 167-181.

[21] Adler, A. (2003). Exact laws for sums of order statistics from the Pareto distribution. *Bull. Inst. Math. Acad. Sinica*, 31, 181-193.

[22] Adler, A. (2004). Exact laws for sums of order statistics from a generalized Pareto distribution. *Stochastic Anal. Appl.*, 22, 459-477.

[23] Adler, A. (2005a). Limit theorems for arrays of ratios of order statistics. *Bull. Inst. Math. Acad. Sinica*, 33, 327-344.

[24] Adler, A. (2005b). Limit theorems for randomly selected adjacent order statistics from a Pareto distribution. *Int. J. Math. Math. Sci.*, 21, 3427-3441.

[25] Adler, A. (2006a). Exact laws for sums of ratios of order statistics from the Pareto distribution. *Cent. Eur. J. Math.*, 4, 1-4.

[26] Adler, A. (2006b). Limit theorems for randomly selected ratios of order statistics from a Pareto distribution. *Stoch. Anal. Appl.*, 24, 339-358.

[27] Afify, W. M. (2010). On estimation of the exponentiated Pareto distribution under different sample schemes. *Stat. Methodol.*, 7, 77-83.

[28] Aggarwal, V. (1984). On optimum aggregation of income distribution data. *Sankhya*, 46B, 343–355.

[29] Aggarwal, V. and Singh, R. (1984). On optimum stratification with proportional allocation for a class of Pareto distributions. *Comm. Statist. Theory Methods*, 13, 3107-3116.

[30] Aghevli, B. B. and Mehran, F. (1981). Optimal grouping of income distribution data. *Journal of the American Statistical Association*, 76, 22–26.

[31] Ahmad, A. E. A. (2001). Moments of order statistics from doubly truncated continuous distributions and characterizations. *Statistics*, 35, 479-494.

[32] Ahmad, A. E. A. and Fawzy, M. A. (2003). Recurrence relations for single moments of generalized order statistics from doubly truncated distributions. *J. Statist. Plann. Inference*, 117, 241-249.

[33] Ahmad, M. (1980). Estimation of the parameters of Burr distribution based on order statistics. Technical Report, University of Petroleum and Minerals, Dhahran, Saudi Arabia.

[34] Ahmad, M. (1985). Estimation of the parameters of Burr distribution based on order statistics. *Studia Sci. Math. Hungar.*, 20, 133-137

[35] Ahn, S., Kim, J. H. T. and Ramaswami, V. (2012). A new class of models for heavy tailed distributions in finance and insurance risk. *Insurance: Mathematics and Economics*,51, 43–52.

[36] Ahsanullah, M. (1976). On a characterization of the exponential distribution by order statistics. *Journal of Applied Probability*, 13, 818–822.

[37] Ahsanullah, M. (1992). Inference and prediction problems of the generalized Pareto distribution based on record values. In *Order Statistics and Nonparametrics: Theory and Applications (Alexandria, 1991)*, North-Holland, Amsterdam, 47–57.

[38] Ahsanullah, M. and Houchens, R. L. (1989). A note on record values from a Pareto distribution. *Pakistan J. Statist.*, 5, 51-57.

[39] Ahsanullah, M. and Kabir, A. B. M. L. (1973). A characterization of the Pareto distribution. *Canadian Journal of Statistics*, 1, 109–112.

[40] Aigner, D. J. (1970). The estimation of moments for a Pareto distribution subject to both sampling and grouping errors. *Review of the International Statistical Institute*, 38, 210–219.

[41] Aigner, D. J. and Goldberger, A. S. (1970). Estimation of Pareto's law from grouped observations. *Journal of the American Statistical Association*, 65, 712–723.

[42] Airth, A. D. (1984). The progression of wage distributions. In *Recent Developments in the Analysis of Large-scale Data Sets: Proceedings of a Seminar Held in Luxembourg, 16-18.11.1983*, Eurostat News, Special number, 1984, 139–164.

[43] Aitchison, J. and Brown, J. A. C. (1957). *The Lognormal Distribution* (with special reference to its uses in economics). Cambridge University Press, Cambridge, England.

[44] Akinsete, A., Famoye, F. and Lee, C. (2008). The beta-Pareto distribution. *Statistics*, 42, 547-563.

[45] Al-Hussaini, E. K. and Jaheen, Z. F. (1992). Bayesian estimation of the parameters, reliability and failure rate functions of the Burr type XII failure model. *J. Statist. Comput. Simulation*, 41, 31-40.

[46] Al-Hussaini, E. K. and Jaheen, Z. F. (1995). Bayesian prediction bounds for the Burr type XII failure model. *Comm. Statist. Theory Methods*, 24, 1829–1842.

[47] Al-Hussaini, E. K., Mousa, M. A. M. A. and Jaheen, Z. F. (1992). Estimation under the Burr type XII failure model based on censored data: a comparative study. *Test*, 1, 47-60.

[48] Al-Marzoug, A. M. and Ahmad, M. (1985). Estimation of parameters of Burr probability model using fractional moments. *Pakistan J. Statisti.*, 1, 67-77.

[49] Al-Ruzaiza, A. S. and El-Gohary, A. (2007). A new class of positively quadrant dependent bivariate distributions with Pareto. *Int. Math. Forum*, 2, 1259-1273.

[50] Ali, M. M. and Khan, A. H. (1987). On order statistics from the log-logistic distribution. *J. Statist. Plann. Inference*, 17, 103-108.

[51] Ali, M. M., Umbach, D. and Hassanein, K. M. (1981). Small sample quantile estimation of Pareto populations using two order statistics. Technical Report, Ball State University, Muncie, Indiana.

[52] Ali, M. M. and Woo, J. (2004). Bayesian analysis of mean parameter in a Pareto distribution. *Calcutta Statist. Assoc. Bull.*, 55, 129-137.

[53] Alker, H. R. (1965). *Mathematics and Politics*. Macmillan, New York.

[54] Alker, H. R. and Russet, B. M. (1966). Indices for comparing inequality. In *Comparing Nations: The Use of Quantitative Data in Cross-National Research*, R. L. Merritt and S. Rokkan, eds. Yale University Press, New Haven, Connecticut, 349–372.

[55] Alzaid, A. A. and Proschan, F. (1994). Max-infinite divisibility and multivariate total positivity. *J. Appl. Probab.*, 31, 721-730.

[56] Amari, S. (1985). *Differential-geometrical Methods in Statistics.*, Lecture Notes in Statistics, 28. Springer-Verlag, New York.

[57] Amati, G. and van Rijsbergen, C. J. (2002). Term frequency normalization via Pareto distributions. In *Advances in Information Retrieval: 24th BCS-IRSG Proceedings*, F. Crestani, M. Girolami and C. J. van Rijsbergen, Eds. Springer–Verlag, Berlin, 183-192.

[58] Amato, V. (1968). *Metodologia Statistica Strutturale, Vol. 1*, Cacucci, Bari.

[59] Amoroso, L. (1925). Ricerche intorno alla curva dei redditi. *Annali di Matematica Pura ed Applicata (Bologna)*, 4, 123–155.

[60] Anderson, D. N. and Arnold, B. C. (1993). Non-parametric estimation of Lorenz curves. *Internat. J. Math. Statist. Sci.*, 2, 57-72.

[61] Arellano-Valle, R. B., Branco, M. D. and Genton, M. G. (2006). A unified view on skewed distributions arising from selections. *Canad. J. Statist.*, 34, 581-601.

[62] Arnold, B. C. (1967). A note on multivariate distributions with specified marginals. *Journal of the American Statistical Association*, 62, 1460–1461.

[63] Arnold, B. C. (1970). Inadmissibility of the usual scale estimate for a shifted exponential distribution. *J. Amer. Statist. Assoc.*, 65, 1260-1264.

[64] Arnold, B. C. (1971). Two characterizations of the exponential distribution. Unpublished manuscript, Iowa State University, Ames, Iowa.

[65] Arnold, B. C. (1975). Multivariate exponential distributions based on hierarchical successive damage. *Journal of Applied Probability*, 12, 142–147.

[66] Arnold, B. C. (1976). On an inequality due to Klamkin and Newman. Unpublished manuscript, Iowa State University, Ames, Iowa.

[67] Arnold, B. C. (1980). Misreporting increases apparent inequality. Technical Report #73, Department of Statistics, University of California, Riverside.

[68] Arnold, B. C. (1981). Transformations which attenuate income inequality. Technical Report #76, Department of Statistics, University of California, Riverside.

[69] Arnold, B. C. (1986). A class of hyperbolic Lorenz curves. *Sankhya Ser. B*, 48 , 427-436

[70] Arnold, B. C. (1987). *Majorization and the Lorenz Order: A Brief Introduction*, Lecture Notes in Statistics, 43. Springer-Verlag, Berlin.

[71] Arnold, B. C. (1988a). A logistic process constructed using geometric minimization. *Statist. Probab. Lett.*, 7, 253-257.

[72] Arnold, B. C. (1988b). Pseudo-likelihood estimation for the Pareto conditionals distribution. Technical Report 170. Statistics Department, University of Claifornia, Riverside.

[73] Arnold, B. C. (1990). A flexible family of multivariate Pareto distributions. *J. Statist. Plann. Inference*, 24, 249-258.

[74] Arnold, B. C. (1992). Logistic and semi-logistic processes. *J. Comput. Appl. Math.*, 40, 139-149.

[75] Arnold, B. C. (1993). Logistic processes involving Markovian minimization. *Comm. Statist. Theory Methods*, 22, 1699-1707.

[76] Arnold, B. C. (2012). On the Amato inequality index. *Statistics and Probability Letters*, 82, 1504–1506.

[77] Arnold, B. C. (2014). On Zenga and Bonferroni curves. *Metron*, to appear.

[78] Arnold, B. C. and Austin, K. (1987). Truncated Pareto distributions: Flexible, tractable and familiar. Technical Report 150, Department of Statistics, University of California, Riverside.

[79] Arnold, B. C., Balakrishnan, N. and Nagaraja, H. N. (1992). *A First Course in Order Statistics.*, Wiley, New York.

[80] Arnold, B. C., Balakrishnan, N. and Nagaraja, H. N. (1998). *Records*. Wiley, New York.

[81] Arnold, B. C. and Beaver, R. J. (2002). Skewed multivariate models related to hidden truncation and/or selective reporting. With discussion and a rejoinder by the authors. *Test*, 11, 7-54.

[82] Arnold, B. C. and Brockett, P. L. (1983). When does the th percentile residual life function determine the distribution? *Oper. Res.*, 31, 391-396.

[83] Arnold, B. C., Castillo, E. and Sarabia, J. M. (1993). Multivariate distributions with generalized Pareto conditionals. *Statist. Probab. Lett.*, 17, 361-368.

[84] Arnold, B. C., Castillo, E. and Sarabia, J. M. (1998). Bayesian analysis for classical distributions using conditionally specified priors. *Sankhya Ser. B*, 60, 228-245.

[85] Arnold, B. C., Castillo, E. and Sarabia, J. M. (1999). *Conditional Specification of Statistical Models.* Springer-Verlag, New York.

[86] Arnold, B. C., Castillo, E. and Sarabia, J. M. (2008). Some characterizations involving uniform and powers of uniform random variables. *Statistics*, 42, 527-534.

[87] Arnold, B. C. and Ganeshalingam, S. (1988). Estimation for multivariate Pareto distributions: A simulation study. Technical Report 162, Statistics Department, University of California, Riverside.

[88] Arnold, B. C. and Ghosh I. (2011). Inference for Pareto data subject to hidden truncation. *Journal of the Indian Society for Probability and Statistics,*, 13, 1–16.

[89] Arnold, B.C. and Ghosh, I. (2013). Hidden truncation in multivariate Pareto data. In *Contemporary Topics in Mathematics and Statistics with Application, Volume 1.*, A. Adhikari, M. R . Adhikari, and Y. P. Chaubey, Eds. Asian Books Pvt. Ltd., New Delhi. 211–226

[90] Arnold, B. C. and Gokhale, D.V. (2014). Parametric constraints for generalized hyperexponential distributions. Technical report. Department of Statistics, University of California, Riverside.

[91] Arnold, B. C. and Hallett, J. T. (1989). A characterization of the Pareto process among stationary stochastic processes of the form $X_n = c\min(X_{n-1}, Y_n)$. *Statist. Probab. Lett.*, 8, 377-380.

[92] Arnold, B. C. and Laguna, L. (1976). A stochastic mechanism leading to asymptotically Paretian distributions. Business and Economic Statistics Section, Proceedings of the American Statistical Association, 208–210.

[93] Arnold, B. C. and Laguna, L. (1977). *On Generalized Pareto Distributions with Applications to Income Data*. International Studies in Economics, Monograph #10, Department of Economics, Iowa State University, Ames, Iowa.

[94] Arnold, B. C., and Nagaraja, H. N. (1991). Lorenz ordering of exponential order statistics. *Statist. Probab. Lett.*, 11, 485–490.

[95] Arnold, B. C. and Ng, H. K. T. (2011). Flexible bivariate beta distributions. *J. Multivariate Anal.*, 102, 1194-1202.

[96] Arnold, B. C. and Pouramahdi, M. (1988). Conditional characterizations of multivariate distributions. *Metrika*, 35, 99–108.

[97] Arnold, B. C. and Press, S. J. (1981). Bayesian inference for Pareto populations. Technical Report #84, Department of Statistics, University of California, Riverside.

[98] Arnold, B. C. and Press, S. J. (1983). Bayesian inference for Pareto populations. *J. Econometrics*, 21, 287-306.

[99] Arnold, B. C. and Press, S. J. (1986). Bayesian analysis of censored or grouped data from Pareto populations. in *Bayesian Inference and Decision Techniques, Stud. Bayesian Econometrics Statist., 6,*, North-Holland, Amsterdam, 157-173.

[100] Arnold, B. C. and Press, S. J. (1989). Bayesian estimation and prediction for Pareto data. *J. Amer. Statist. Assoc.*, 84, 1079-1084.

[101] Arnold, B. C., Robertson, C. A. and Yeh, H-C. (1986). Some properties of a Pareto-type distribution. *Sankhya Ser. A*, 48, 404-408.

[102] Arnold, B. C., Robertson, C. A., Brockett, P. L. and Shu, B-Y. (1987). Generating ordered families of Lorenz curves by strongly unimodal distributions. *Journal of Business and Economic Statistics*, 5, 305–308.

[103] Arnold, B. C. and Strauss, D. J. (1991). Bivariate distributions with conditionals in prescribed exponential families. *J. Roy. Statist. Soc. Ser. B*, 53, 365-375.

[104] Arnold, B. C. and Villaseñor, J. A. (1985). Inequality preserving and inequality attenuating transformations. Technical Report, Department of Statistics, Colegio de Postgraduados, Chapingo, Mexico.

[105] Arnold, B. C. and Villaseñor, J. A. (1986). Lorenz ordering of means and medians. *Statist. Probab. Lett.*, 4, 47–49.

[106] Arnold, B. C. and Villaseñor, J. A. (1991). Lorenz ordering of order statistics. in *Stochastic Orders and Decision under Risk (Hamburg, 1989)*, IMS Lecture Notes Monogr. Ser., 19, Inst. Math. Statist., Hayward, CA, 38–47.

[107] Arnold, B. C. and Villaseñor, J. A. (2012). Generalized order statistic processes and Pfeifer records. *Statistics*, 46, 373-385.

[108] Asadi, M. (2004) Some characterizations on generalized Pareto distributions. *Comm. Statist. Theory Methods*, 33, 2929-2939.

[109] Asadi, M., Rao, C. R. and Shanbhag, D. N. (2001). Some unified characterization results on generalized Pareto distributions. *J. Statist. Plann. Inference*, 93, 29-50.

[110] Asgharzadeh, A., Abdi, M. and Kus, C. (2011). Interval estimation for the two-parameter Pareto distribution based on record values. *Seluk J. Appl. Math.* Special Issue, 149-161.

[111] Ashkar, F. and Ouarda, T. B. M. J. (1996). On some methods of fitting the generalized Pareto distribution. *Journal of Hydrology*, 177, 117-141.

[112] Asimit, A. V., Furman, E. and Vernic, R. (2010). On a multivariate Pareto distribution. *Insurance Math. Econom.*, 46, 308-316.

[113] Asrabadi, B. R. (1990). Estimation in the Pareto distribution. *Metrika*, 37, 199-205.

[114] Athreya, K. B. and Ney, P. E. (1972). *Branching Processes*. Springer, Berlin.

[115] Atkinson, A. B. (1970). On the measurement of inequality. *Journal of Economic Theory*, 2, 244–263.

[116] Awad, A. M. and Marzuq, M. M. (1986). Characterization of distributions through maximization of a-entropies. *IMA J. Math. Control Inform.*, 3, 1-7.

[117] Baklizi, A. (2008). Preliminary test estimation in the Pareto distribution using minimax regret significance levels. *Int. Math. Forum*, 3, 473-478.

[118] Bakouch, H. S., Al-Zahrani, B. M., Al-Shomrani, A. A., Marchi, V. A. A. and Louzada, F. (2012). An extended Lindley distribution. *J. Korean Statist. Soc.*, 41, 75–85.

[119] Balakrishna, N. and Jayakumar, K. (1997). Bivariate semi-Pareto distributions and processes. *Statist. Papers*, 38, 149-165.

[120] Balakrishnan, N. and Ahsanullah, M. (1994). Recurrence relations for single and product moments of record values from generalized Pareto distribution. *Comm. Statist. Theory Methods*, 23, 2841-2852.

[121] Balakrishnan, N. and Joshi, P. C. (1982). Moments of order statistics from doubly truncated Pareto distribution. *J. Indian Statist. Assoc.*, 20, 109-117.

[122] Balakrishnan, N. and Shafay, A. R. (2012). One- and two-sample Bayesian prediction intervals based on type-II hybrid censored data. *Comm. Statist. Theory Methods*, 41, 1511–1531.

[123] Balkema, A. A. and de Haan, L. (1974). Residual life time at great age. *Annals of Probability*, 2, 792–804.

[124] Bancroft, T. A. (1944). On biases in estimation due to the use of preliminary tests of significance. *Ann. Math. Statistics*, 15, 190-204.

[125] Baringhaus, L. (1980). Characterization of distributions by random sums. *SIAM Journal Algebraic Discrete Methods*, 1, 345-347.

[126] Barlow, R. E., Bartholomew, D. J., Bremner, J. M. and Brunk, H. D. (1972). *Statistical Inference Under Order Restrictions: The Theory and Application of Isotonic Regression*. Wiley, New York.

[127] Bartholomew, D. J. (1969). Sufficient conditions for a mixture of exponentials to be a probability density function. *Annals of Math. Stat.*, 40, 2183-2188.

[128] Basmann, R. L., Hayes, K. L., Slottje, D. J. and Johnson, J. D. (1990). A general functional form for approximating the Lorenz curve. *Journal of Econometrics*, 43, 77–90.

[129] Baxter, M. A. (1980). Minimum variance unbiased estimation of the parameters of the Pareto distribution. *Metrika*, 27, 133–138.

[130] Bechofer, R. E., Kiefer, J. and Sobel, M. (1968). *Sequential Identification and Ranking Procedures*. University of Chicago Press, Chicago, Illinois.

[131] Beg, M. I. and Kirmani, S. N. U. A. (1974). On a characterization of exponential and related distributions. *The Australian Journal of Statistics*, 16, 163–6.

[132] Beg, M. A. (1981). Estimation of the tail probability of the truncated Pareto distribution. *J. Inform. Optim. Sci.*, 2, 192-198.

[133] Beg, M. A. (1983). Unbiased estimators and tests for truncation and scale parameters. *Amer. J. Math. Management Sci.*, 3, 251-274.

[134] Behboodian, J. and Tahmasebi, S. (2008). Some properties of entropy for the exponentiated Pareto distribution (EPD) based on order statistics. *J. Math. Ext.*, 3, 43-53.

[135] Benedetti, C. (1961). A proposito dei rapporti tra differenza meaia *e* scostamenti medio quadratico, semplice medio *e* semplice medio della mediana. *Metron*, 21, 181–185.

[136] Benhabib, J., Bisin, A. and Zhu, S. (2011) The distribution of wealth and fiscal policy in economies with finitely lived agents. *Econometrica*, 79, 123–157.

[137] Berg, C. (1981). The Pareto distribution is a generalized G-convolution—a new proof. *Scand. Actuar. J.*, 117-119.

[138] Berkson, J. (1944). Application of the logistic function to bio-assay. *Journal of the American Statistical Association*, 39, 357–364.

[139] Beutner, E. and Kamps, U. (2008). Random contraction and random dilation of generalized order statistics. *Comm. Statist. Theory Methods*, 37, 2185-2201.

[140] Bhandari, S. K. (1986). Characterisation of the parent distribution by inequality measures on its truncations. *Sankhya Ser. B*, 48, 297-300.

[141] Bhandari, S. K. and Mukerjee, R. (1986). Some relations among inequality measures. *Sankhya Ser. B*, 48, 258-261.

[142] Bhattacharjee, M. C. (1993). How rich are the rich? Modeling affluence and inequality via reliability theory. *Sankhya Ser. B*, 55, 1-26.

[143] Bhattacharya, N. (1963). A property of the Pareto distribution. *Sankhya, Series B*, 25, 195–6.

[144] Billingsley, P. (1968). *Convergence of Probability Measures.* Wiley, New York.

[145] Bladt, M. and Nielsen, B. F. (2011). Moment distributions of phase type. *Stoch. Models*, 27, 651-663.

[146] Blitz, R. C. and Brittain, J. A. (1964). An extension of the Lorenz diagram to the correlation of two variables. *Metron*, 23, 137–143.

[147] Block, H. W. (1975). Physical models leading to multivariate exponential and negative binomial distributions. *Modeling and Simulation*, 6, 445–450.

[148] Blum, M. (1970). On the sums of independently distributed Pareto variates. *SIAM Journal of Applied Mathematics*, 19, 191–198.

[149] Bomsdorf, E. (1977). The prize competition distribution – a particular L-distribution as a supplement to the Pareto distribution. *Statistische Hefte*, 18, 254–264.

[150] Bonferroni, C. E. (1930). *Elmenti di Statistica Generale*.Libreria Seber, Firenze.

[151] Bowley, A. L. (1902). *Elements of Statistics* (2nd ed.). Scribners, New York.

[152] Bowley, A. L. (1937). Standard deviation of Gini's mean difference. *Comptes Rendus du Congres International des Mathematiciens*, (Oslo, 1936), Vol. 2, 182–185.

[153] Bowman, M. J. (1945). A graphical analysis of personal income distribution in the United States. *American Economic Review*, 35, 607–628.

[154] Bradford, S. C. (1948). *Documentation.* Crosby Lockwood, London.

[155] Brazauskas, V. (2000). Robust parametric modeling of the proportional reinsurance premium when claims are approximately Pareto-distributed. *Proceedings of the Business and Economic Statistics Section*, American Statistical Association, Alexandria, VA, 144–149.

[156] Brazauskas, V. (2002). Fisher information matrix for the Feller-Pareto distribution. *Statist. Probab. Lett.*, 59, 159-167.

[157] Brazauskas, V. (2003). Information matrix for Pareto(IV), Burr, and related distributions. *Comm. Statist. Theory Methods*, 32, 315-325.

[158] Brazauskas, V. and Serfling, R. (2001). Small sample performance of robust estimators of tail parameters for Pareto and exponential models. *J. Statist. Comput. Simulation*, 70, 1-19.

[159] Brazauskas, V. and Serfling, R. (2003). Favorable estimators for fitting Pareto models: A study using goodness-of-fit measures with actual data. *Astin Bull.*, 33, 365-381.

[160] Brennan, L. E., Reed, I. S. and Sollfrey, W. (1968). A comparison of average likelihood and maximum likelihood tests for detecting radar targets of unknown Doppler frequency. *IEEE Transactions in Information Theory*, IT-4, 104–110.

[161] Bresciani-Turroni, C. (1937). On Pareto's law. *Journal of the Royal Statistical Society*, 100, 421–432.

[162] Bresciani-Turroni, C. (1939). Annual survey of statistical data: Pareto's law and the index of inequality of incomes. *Econometrica*, 7, 107–133.

[163] Brown, G. and Sanders, J. W. (1981). Lognormal genesis. *Journal of Applied Probability*, 18, 542–547.

[164] Bryson, M. C. (1974). Heavy tailed distributions: Properties and tests. *Technometrics*, 16, 61–68.

[165] Burkschat, M. (2010). Linear estimators and predictors based on generalized order statistics from generalized Pareto distributions. *Comm. Statist. Theory Methods*, 39, 311-326.

[166] Burkschat, M., Cramer, E. and Kamps, U. (2007). Linear estimation of location and scale parameters based on generalized order statistics from generalized Pareto distributions. In *Recent Developments in Ordered Random Variables*, Nova Sci. Publ., New York, 253–261.

[167] Burr, I. W. (1942). Cumulative frequency functions. *Annals of Mathematical Statistics*, 13, 215–232.

[168] Burr, I. W. (1968). On a general system of distributions III. The sample range. *Journal of the American Statistical Association*, 63, 636–643.

[169] Burr, I. W. and Cislak, P. J. (1968). On a general system of distributions, (I). Its curve shape characteristics, (II). The sample mean. *Journal of the American Statistical Association*, 63, 627–635.

[170] Cai, Y. (2010) Polynomial power-Pareto quantile function models. *Extremes*, 13, 291-314.

[171] Campbell, G. and Ratnaparkhi, M. V. (1993). An application of Lomax distributions in receiver operating characteristic (ROC) curve analysis. *Communications in Statistics, Theory and Methods*, 22, 1681–1697.

[172] Candel Ato, J. and Bernadic, M. (2011). The characterization of the symmetric Lorenz curves by doubly truncated mean function. *Comm. Statist. Theory Methods*, 40, 3269-3280.

[173] Carreau, J. and Bengio, Y. (2009). A hybrid Pareto model for asymmetric fat-tailed data: The univariate case. *Extremes*, 12, 53-76.

[174] Castellani, M. (1950). On multinomial distributions with limited freedom: A stochastic genesis of Pareto's and Pearson's curves. *Annals of Mathematical Statistics*, 21, 289–293.

[175] Castellano, V. (1933). Sulle relazioni tra curve di frequenza *e* curve di concentrazione *e* sui rapporti di concentrazione corrispondenti *a* determinate distribuzioni. *Metron*, 10, 3–60.

[176] Castellano, V. (1935). Recente letteratura sugli di variabilita. *Metron*, 12, 100–131.

[177] Castillo, E. and Hadi, A. S. (1995). Modeling lifetime data with application to fatigue models. *J. Amer. Statist. Assoc.*, 90, 1041-1054.

[178] Castillo, E. and Hadi, A. S. (1997). Fitting the generalized Pareto distribution to data. *J. Amer. Statist. Assoc.*, 92, 1609-1620.

[179] Castillo, E., Hadi, A. S. and Sarabia, J. M. (1998). A method for estimating Lorenz curves. *Comm. Statist. Theory Methods*, 27, 2037-2063.

[180] Castillo, E., Sarabia, J. M. and Hadi, A. S. (1997). Fitting continuous bivariate distributions to data. *The Statistician*, 46, 355–369.

[181] Cerone, P. (2009a). Bounding the Gini mean difference. *Inequalities and Applications, Internat. Ser. Numer. Math., 157*, Birkhuser, Basel, 77–89.

[182] Cerone, P. (2009b). On Young's inequality and its reverse for bounding the Lorenz curve and Gini mean. *J. Math. Inequal.*, 3, 369-381.

[183] Chacko, M. and Thomas, P. Y. (2007). Estimation of a parameter of bivariate Pareto distribution by ranked set sampling. *J. Appl. Stat.*, 34, 703-714.

[184] Champernowne, D. G. (1937). The theory of income distribution. *Econometrica*, 5, 379–381.

[185] Champernowne, D. G. (1952). The graduation of income distributions. *Econometrica*, 20, 591–615.

[186] Champernowne, D. G. (1953). A model of income distribution. *Economic Journal*, 63, 318–351.

[187] Champernowne, D. G. (1973). *The Distribution of Income*. Cambridge University Press, Cambridge.

[188] Chan, L. K. and Cheng, S. W. (1973). On the optimum spacing for the asymptotically best linear estimate of the scale parameter of the Pareto distribution. *Tamkang Journal of Mathematics*, 4, 9–21.

[189] Chandra, M. and Singpurwalla, N. D. (1978). The Gini index, the Lorenz curve and the total time on test transforms. Technical Report T-368, George Washington University, School of Engineering and Applied Science, Washington, D.C.

[190] Chandra, M. and Singpurwalla, N. D. (1981). Relationships between some notions which are common to reliability theory and economics. *Math. Oper. Res.*, 6, 113-121.

[191] Chaouche, A. and Bacro, J-N. (2004). A statistical test procedure for the shape parameter of a generalized Pareto distribution. *Comput. Statist. Data Anal.*, 45, 787-803.

[192] Chaouche, A. and Bacro, J-N. (2006). Statistical inference for the generalized Pareto distribution: Maximum likelihood revisited. *Comm. Statist. Theory Methods*, 35, 785-802.

[193] Chaturvedi, A. and Alam, M. W. (2010). UMVUE and MLE in a family of lifetimes distributions. *J. Indian Statist. Assoc.*, 48, 189-213.

[194] Chen, W. C. (1979a). On the weak form of Zipf's law. Technical Report #153, Department of Statistics, Carnegie-Mellon University, Pittsburgh, PA.

[195] Chen, W. C. (1979b). Limit theorems for general sizes distributions. Technical Report #166, Department of Statistics, Carnegie-Mellon University, Pittsburgh, PA.

[196] Chen, Z. (1996). Joint confidence region for the parameters of Pareto distribution. *Metrika*, 44, 191-197.

[197] Chen, Z. (2006). Estimating the shape parameter of the log-logistic distribution. *Int. Journal of Reliability, Quality and Safety Engineering*, 13, 257–266.

[198] Cheng, R. C. H. and Amin, N. A. K. (1983). Estimating parameters in continuous univariate distributions with a shifted origin. *J. Roy. Statist. Soc. Ser. B*, 45, 394-403.

[199] Cheng, S. W. and Chou, C. H. (2000a). On the BLUE of the scale parameter of the generalized Pareto distribution. *Tamkang J. Math.*, 31, 165-173.

[200] Cheng, S. W. and Chou, C. H. (2000b). On the ABLUE of the scale parameter of the generalized Pareto distribution. *Tamkang J. Math.*, 31, 317-330.

[201] Chiragiev, A. and Landsman, Z. (2009). Multivariate flexible Pareto model: Dependency structure, properties and characterizations. *Statist. Probab. Lett.*, 79, 1733-1743.

[202] Chotikapanich, D. and Griffiths, W. E. (2002). Estimating Lorenz curves using a Dirichlet distribution. *J. Bus. Econom. Statist.*, 20, 290-295.

[203] Choulakian, V. and Stephens, M. A. (2001). Goodness-of-fit tests for the generalized Pareto distribution. *Technometrics*, 43, 478-484.

[204] Chung, K. L. (2001). *A Course in Probability Theory* (3rd ed.). Academic Press, New York.

[205] Cicchitelli, G. (1976). The sampling distribution of the ratio of the linear combinations of order statistics. *Metron*, 34, 219–228.

[206] Cifarelli, D. M., Gupta, R. P. and Jayakumar, K. (2010). On generalized semi-Pareto and semi-Burr distributions and random coefficient minification processes. *Statist. Papers*, 51, 193-208.

[207] Cirillo, R. (1979). *The Economics of Vilfredo Pareto*. Frank Case, London, England.

[208] Cobham, A. and Sumner, A. (2014). Is inequality all about the tails? The Palma measure of income inequality. *Significance*, 11, 10–13.

[209] Coelho, C. A. and Mexia, J. T. (2007). On the distribution of the product and ratio of independent generalized gamma-ratio random variables. *Sankhya*, 69, 221-255.

[210] Cohen, A. C. (1991). *Truncated and Censored Samples. Theory and Applications.*, Statistics: Textbooks and Monographs, 119. Marcel Dekker, Inc., New York.

[211] Cohen, A. C. and Whitten, B. J. (1988). *Parameter Estimation in Reliability and Life Span Models.* Statistics: Textbooks and Monographs, 96. Marcel Dekker, New York.

[212] Cohen, J. (1966). *A Model of Simple Competition*. Harvard University Press, Cambridge, Massachusetts.

[213] Colombi, R. (1990) A new model of income distribution: The Pareto–lognormal distribution. In *Studies in Contemporary Economics*, C. Dagum and M. Zenga, eds. Springer–Verlag, Heidelberg.

[214] Conlisk, J. (1967). Some cross-state evidence of income inequality. *Review of Economics and Statistics*, 49, 115–118.

[215] Consul, P. C. (1990). Geeta distribution and its properties. *Communications in Statistics, Theory and Methods*, 19, 3051–3068.

[216] Conway, D. A. (1979). Multivariate distributions with specified marginals. Technical Report #145, Department of Statistics, Stanford University, Stanford, California.

[217] Cook, R. D. and Johnson, M. E. (1981). A family of distributions for modelling nonelliptically symmetric multivariate data. *J. Roy. Statist. Soc. Ser. B*, 43 , 210-218.

[218] Cook, R. D. and Johnson, M. E. (1986). Generalized Burr-Pareto-logistic distributions with applications to a uranium exploration data set. *Technometrics*, 28, 123-131.

[219] Cook, W. L. and Mumme, D. C. (1981). Estimation of Pareto parameters by numerical methods. In *Statistical Distributions in Scientific Work, Vol. 5*, C. Taillie, G. P. Patil, and B. Baldessari, eds. Reidel, Dordrecht, Holland, 127–132.

[220] Cormann, U. and Reiss, R-D. (2009). Generalizing the Pareto to the log-Pareto model and statistical inference. *Extremes*, 12, 93-105.

[221] Cramer, E. (2003) Contributions to Generalized Order Statistics, Habilitationsschrift, University of Oldenburg, Germany.

[222] Cramer, E., Kamps, U. and Keseling, C. (2004). Characterizations via linear regression of ordered random variables: A unifying approach. *Comm. Statist. Theory Methods*, 33, 2885-2911.

[223] Cramer, J. S. (1978). A function for size distribution of incomes: Comment. *Econometrica*, 46, 459–460.

[224] Crawford, C. B. (1966). Characterization of geometric and exponential distributions. *Annals of Mathematical Statistics*, 37, 1790–1795.

[225] Cronin, D. C. (1977). *Theory and applications of the log-logistic distribution*. Unpublished M.A. thesis. Polytechnic of Central London.

[226] Cronin, D. C. (1979). A function for size distribution of incomes: A further comment. *Econometrica*, 47, 773–4.

[227] Csorgo, M., Csorgo, S. and Horvath, L. (1987). Estimation of total time on test transforms and Lorenz curves under random censorship. *Statistics*, 18, 77-97.

[228] Csorgo, M., Gastwirth, J. L. and Zitikis, R. (1998). Asymptotic confidence bands for the Lorenz and Bonferroni curves based on the empirical Lorenz curve. *J. Statist. Plann. Inference*, 74, 65-91.

[229] Csorgo, M. and Yu, H. (1999). Weak approximations for empirical Lorenz curves and their Goldie inverses of stationary observations. *Adv. in Appl. Probab.*, 31, 698-719.

[230] Csorgo, M. and Zitikis, R. (1996). Strassen's LIL for the Lorenz curve. *J. Multivariate Anal.*, 59, 1-12.

[231] Csorgo, M. and Zitikis, R. (1997a). On the rate of strong consistency of Lorenz curves. *Statist. Probab. Lett.*, 34, 113-121.

[232] Csorgo, M. and Zitikis, R. (1997b). On confidence bands for the Lorenz and Goldie curves. In *Advances in the Theory and Practice of Statistics*, Wiley, New York, 261–281.

[233] Cuadras, C. M. and Augé, J. (1981). A continuous general multivariate distribution and its properties. *Comm. Statist. Theory Methods*, 10 , 339-353.

[234] Dagum, C. (1977). A new model for personal income distribution: Specification and estimation. *Economie Appliquée*, 30, 413–437

[235] Dagum, C. (1980). Inequality measures between income distributions with applications. *Econometrica*, 48, 1791–1803.

[236] Dallas, A. C. (1976). Characterizing the Pareto and power distributions. *Annals of Institute of Statistical Mathematics*, 28, 491–7.

[237] Dalton, H. (1920).The measurement of the inequality of incomes. *Economic Journal*, 30, 348–361.

[238] Dasgupta, P., Sen, A. and Starrett, D. (1973). Notes on the measurement of inequality. *Journal of Economic Theory*, 6, 180–187.

[239] David, H. A. (1968). Gini's mean difference rediscovered. *Biometrika*, 55, 573–574.

[240] David, H. A. and Nagaraja, H. N. (2003). *Order Statistics*, (3rd ed.). Wiley, New York.

[241] Davis, H. T. and Feldstein, M. L. (1979). The generalized Pareto law as a model for progressively censored survival data. *Biometrika*, 66, 299–306.

[242] Davis, R. L. and Arnold, J. C. (1970). An efficient preliminary test estimator for the variance of a normal population when mean is unknown. *Biometrika*, 56, 674–676.

[243] Davison, A. C. and Smith, R. L. (1990). Models for exceedances over high thresholds. With discussion and a reply by the authors. *J. Roy. Statist. Soc. Ser. B*, 52, 393-442.

[244] De Boeck, B., Thas, O., Rayner, J. C. W. and Best, D. J. (2011). Generalised smooth tests for the generalised Pareto distribution. *J. Stat. Theory Pract.*, 5, 737-750.

[245] De Cani, J. S. and Stine, R. A. (1986). A note on deriving the information matrix for a logistic distribution. *Amer. Statist.*, 40, 220-222.

[246] De Groot, M. H. (1970). *Optimal Statistical Decisions*, McGraw-Hill, New York.

[247] De Luca, G. and Zuccolotto, P. (2006). Regime-switching Pareto distributions for ACD models. *Comput. Statist. Data Anal.*, 51, 2179-2191.

[248] De Zea Bermudez, P. and Kotz, S. (2010a). Parameter estimation of the generalized Pareto distribution. I. *J. Statist. Plann. Inference*, 140, 1353-1373.

[249] De Zea Bermudez, P. and Kotz, S. (2010b). Parameter estimation of the generalized Pareto distribution. II. *J. Statist. Plann. Inference*, 140, 1374-1388.

[250] De Zea Bermudez, P. and Turkman, M. A. A. (2003). Bayesian approach to parameter estimation of the generalized Pareto distribution. *Test*, 12, 259–277.

[251] Del Castillo, J. and Daoudi, J. (2009). Estimation of the generalized Pareto distribution. *Statist. Probab. Lett.*, 79, 684-688.

[252] Dempster, A. P. and Kleyle, R. M. (1968). Distributions determined by cutting a simplex with hyperplanes. *Annals of Mathematical Statistics*, 39, 1473–1478.

[253] Dey, D. K. and Liu, P-S. L. (1992). Improved estimation of the common scale parameters of several Pareto distributions. *Calcutta Statist. Assoc. Bull.*, 42, 177-189.

[254] Dimaki, C. and Xekalaki, E. (1993). Characterizations of the Pareto distribution based on order statistics. In *Stability Problems for Stochastic Models (Suzdal, 1991), Lecture Notes in Math., 1546*, Springer, Berlin, 1–16.

[255] Dimitrov, B. and von Collani, E. (1995). Contorted uniform and Pareto distributions. *Statist. Probab. Lett.*, 23, 157-164.

[256] Dixit, U. J. and Jabbari-Nooghabi, M. (2010). Efficient estimation in the Pareto distribution. *Stat. Methodol.*, 7, 687-691.

[257] Dixit, U. J. and Jabbari-Nooghabi, M. (2011). Efficient estimation in the Pareto distribution with the presence of outliers. *Stat. Methodol.*, 8, 340-355.

[258] Doksum, K. (1969). Starshaped transformations and the power of rank tests. *Ann. Math. Statist.*, 40, 1167-1176.

[259] Donaldson, D. and Weymark, J. A. (1980). A single parameter generalization of the Gini indices of inequality. *Journal of Economic Theory*, 22, 67–86.

[260] Donaldson, D. and Weymark, J. A. (1983). Ethically flexible Gini indices for income distributions in the continuum. *Journal of Economic Theory*, 29, 353–358.

[261] Donath, W. E. (1981). Wire length distribution for placement of computer logic. *IBM Journal Research and Development*, 25, 152–155.

[262] Doostparast, M., Akbari, M. G. and Balakrishnan, N. (2011). Bayesian analysis for the two-parameter Pareto distribution based on record values and times. *J. Stat. Comput. Simul.*, 81, 1393-1403.

[263] Dorfman, R. (1979). A formula for the Gini coefficient. *Review of Economics and Statistics*, 61, 146–149.

[264] Drane, J. W., Owen, D. B. and Seibert, G. B., Jr. (1978). The Burr distribution and quantal responses. *Statistiche Hefte*, 18, 204–210.

[265] Dubey, S. D. (1968). A compound Weibull distribution. *Naval Research Logistics Quarterly*, 15, 179–188.

[266] Dubey, S. D. (1970). Compound gamma, beta and F distributions. *Metrika*, 16, 27–31.

[267] DuMouchel, W. H. and Olshen, R. A. (1975). On the distributions of claims costs. In *Credibility: Theory and Applications*. Academic Press, New York. Pages 23–50 and 409–414.

[268] Dunsmore, I. and Amin, Z. H. (1998). Some prediction problems concerning samples from the Pareto distribution. *Comm. Statist. Theory Methods*, 27, 1221-1238.

[269] Durand, D. (1943). A simple method for estimating the size distribution of a given aggregate income. *Review of Economics and Statistics*, 25, 227–230.

[270] Durling, F. C. (1975). The bivariate Burr distribution. In *Statistical Distributions in Scientific Work, Vol. 1*, G. P. Patil, S. Kotz, and J. K. Ord, eds. Reidel, Dordrecht, Holland, 329–335.

[271] Durling, F. C., Owen, D. B. and Drane, J. W. (1970). A new bivariate Burr distribution (abstract). *Annals of Mathematical Statistics*, 41, 1135.

[272] Dutta, J. and Michel, P. (1998). The distribution of wealth with imperfect altruism. *Journal of Economic Theory*, 82, 379–404.

[273] Dwass, M. (1964). Extremal processes. *Annals of Mathematical Statistics*, 35, 1718–1725.

[274] Dyer, D. (1981). Structural probability bounds for the strong Pareto law. *Canad. J. Statist.*, 9, 71-77.

[275] Dziubdziela, W. (1997). A note on the characterization of some minification processes. *Appl. Math. (Warsaw)*, 24, 425-428.

[276] Dziubdziela, W. and Kopocinski, B. (1976). Limiting properties of the k-th record values. *Zastos. Mat.*, 15, 187-190.

[277] Efron, B. and Thisted, R. (1976). Estimating the number of unseen species: How many words did Shakespeare know? *Biometrika*, 63, 435–447.

[278] Eichhorn, W., Funke, H. and Richter, W. F. (1984). Tax progression and inequality of income distribution. *J. Math. Econom.*, 13, 127-131.

[279] Eichhorn, W. and Gleissner, W. (1985). On a functional differential equation arising in the theory of the distribution of wealth. *Aequationes Mathematicae*, 28, 190–198.

[280] Elfessi, A. and Jin, C. (1996). On robust estimation of the common scale parameter of several Pareto distributions. *Statist. Probab. Lett.*, 29, 345-352.

[281] Elteto, O. and Frigyes, E. (1968). New income inequality measures as efficient tools for causal analysis and planning. *Econometrica*, 36, 383–396.

[282] Emlen, J. M. (1973). *Ecology: An Evolutionary Approach*. Addison-Wesley, Reading, Massachusetts.

[283] Epstein, B. and Sobel, M. (1953). Life testing I. *Journal of the American Statistical Association*, 48, 486–502.

[284] Ericson, O. (1945). Un probleme de la repartition des revenus. *Skand. Aktuarietidskr*, 28, 247–257.

[285] Fakoor, V., Ghalibaf, M. B. and Azarnoosh, H. A. (2011). Asymptotic behaviors of the Lorenz curve and Gini index in sampling from a length-biased distribution. *Statist. Probab. Lett.*, 81, 1425-1435.

[286] Falk, M., Guillou, A. and Toulemonde, G. (2008). A LAN based Neyman smooth test for Pareto distributions. *J. Statist. Plann. Inference*, 138, 2867-2886.

[287] Falk, M. and Michel, R. (2009). Testing for a multivariate generalized Pareto distribution. *Extremes*, 12, 33-51.

[288] Feller, W. (1968). *An Introduction to Probability Theory and its Applications, Vol. 1*, (3rd ed.). Wiley, New York.

[289] Feller, W. (1971). *An Introduction to Probability Theory and its Applications, Vol. 2*, (2nd ed.). Wiley, New York.

[290] Fellman, J. (1976). The effect of transformations on Lorenz curves. *Econometrica*, 44, 823–824.

[291] Fellman, J. (2009). Discontinuous transformations, Lorenz curves and transfer policies. *Soc. Choice Welf.*, 33, 335-342.

[292] Ferguson, T. S. (1962). Location and scale parameters in exponential families of distributions. *Annals of Mathematical Statistics*, 33, 986–1001. Correction 34, 1603.

[293] Ferguson, T. S. (1967). On characterizing distributions by properties of order statistics. *Sankhya, Series A*, 29, 265–278.

[294] Fernández, A. J. (2012). Minimizing the area of a Pareto confidence region. *European J. Oper. Res.*, 221, 205-212.

[295] Ferreira, A. and de Haan, L. (2013). The generalized Pareto process; with a view towards application and simulation. *Bernoulli*, to appear.

[296] Ferreira, M. (2012). On the extremal behavior of a Pareto process: An alternative for ARMAX modeling. *Kybernetika (Prague)*, 48, 31-49.

[297] Ferreri, C. (1964). Una nuova funzione di frequenza per l'analisi delle variabili statistiche semplici. *Statistica*, 24, 225–251.

[298] Ferreri, C. (1967). Su taluni aspetti di una estensione della curva Paretiana di prima approssimazione per l'analisi della distribuzione dei redditi. Atti della XXV Riunione Scientifica, Societa Italiana di Statistica, Bologna, Italy.

[299] Ferreri, C. (1975). Su il campo di variazione dell'indice "alfa" del Pareto e lo schema iperbolico tronco. *Statistica*, 35, 409–423.

[300] Ferreri, C. (1978a). Su un sistema di funzioni-grado di benessere individuale del reddito e relative misure di concentrabilita. Atti della XXIX Riunione Scientifica, Societa Italiana di Statistica, Bologna, Italy.

[301] Ferreri, C. (1978b). Su un sistema di indicatori di concezione Daltoniana del grado di benessere economico individuale del reddito. *Statistica*, 38, 13–42.

[302] Fishlow, A. (1973). Brazilian income size distribution. Unpublished manuscript, University of California, Berkeley.

[303] Fisk, P. R. (1961a). The graduation of income distributions. *Econometrica*, 29, 171–185.

[304] Fisk, P. R. (1961b). Estimation of location and scale parameters in a truncated grouped sech-square distribution. *Journal of the American Statistical Association*, 56, 692–702.

[305] Fisz, M. (1958). Characterization of some probability distributions. *Skand. Aktuarietidskr*, 1–2, 65–70.

[306] Fraser, D. A. S. (1957). *Non-parametric Methods in Statistics*. Wiley, New York.

[307] Frechet, M. (1939). Sur les formules de repartition des revenus. *Revue de l'Institut International de Statistique*, 1–7.

[308] Frechet, M. (1945). Nouveaux essais d'explication de la repartition des revenus. *Revue de l'Institut International de Statistique*, 16–32.

[309] Frechet, M. (1958). Letter to the editor. *Econometrica*, 26, 590–1.

[310] Fu, J., Xu, A. and Tang, Y. (2012). Objective Bayesian analysis of Pareto distribution under progressive Type-II censoring. *Statist. Probab. Lett.*, 82, 1829-1836.

[311] Furlan, V. (1911). Neue literatur zur einkommensverteilung in Italien. *Jahrbucher fur Nationalokonomie und Statistik*, 97, 241–255.

[312] Gail, M. H. (1977). *A scale-free goodness-of-fit test for the exponential distribution based on the sample Lorenz curve*, Ph.D. Thesis, George Washington University, Washington, D.C.

[313] Gail, M. H.and Gastwirth, J. L. (1978a). A scale-free goodness-of-fit test for the exponential distribution based on the Gini statistic. *Journal of the Royal Statistical Society, Series B*, 40, 350–357.

[314] Gail, M. H. and Gastwirth, J. L. (1978b). A scale-free goodness-of-fit test for the exponential distribution based on the Lorenz curve. *Journal of the American Statistical Association*, 73, 787–793.

[315] Galambos, J. and Kotz, S. (1978). *Characterizations of Probability Distributions*. Lecture Notes in Mathematics #675, Springer-Verlag, Berlin.

[316] Garren, S. T., Hume, E. R. and Leppert, G. R. (2007). Improved estimation of the Pareto location parameter under squared error loss. *Comm. Statist. Theory Methods*, 36, 97-102.

[317] Gastwirth, J. L. (1971). A general definition of the Lorenz curve. *Econometrica*, 39, 1037–1039.

[318] Gastwirth, J. L. (1972). The estimation of the Lorenz curve and Gini index. *Review of Economics and Statistics*, 54, 306–316.

[319] Gastwirth, J. L. (1974). Large sample theory of some measures of income inequality. *Econometrica*, 42, 191–196.

[320] Gastwirth, J. L. (1975a). The estimation of a family of measures of economic inequality. *Journal of Econometrics*, 3, 61–70.

[321] Gastwirth, J. L. (1975b). Statistical measures of earnings differentials. *The American Statistician*, 29, 32–35.

[322] Gastwirth, J. L. and Glauberman, M. (1976). The interpolation of the Lorenz curve and Gini index from grouped data. *Econometrica*, 44, 479–483.

[323] Gastwirth, J. L. and Krieger, A. M. (1975). On bounding moments from grouped data. *Journal of the American Statistical Association*, 70, 468–471.

[324] Gastwirth, J. L., Nayak, T. K. and Krieger, A. M. (1986). Large sample theory for the bounds on the Gini and related indices of inequality estimated from grouped data. *Journal of Business and Economic Statistics*, 4, 269–273.

[325] Gastwirth, J. L. and Smith, J. T. (1972). A new goodness of fit test. ASA Proceedings of Business and Economics Section, 320–2.

[326] Gay, R. (2004). Pricing risk when distributions are fat tailed. *Journal of Applied Probability*, 41A, 157-175.

[327] Geisser, S. (1984). Predicting Pareto and exponential observables. *Canad. J. Statist.*, 12, 143-152.

[328] Geisser, S. (1985). Interval prediction for Pareto and exponential observables. *J. Econometrics*, 29, 173-185.

[329] Ghitany, M. E., Atieh, B. and Nadarajah, S. (2008). Lindley distribution and its application. *Mathematics and Computers in Simulation*, 78, 493–506.

[330] Ghosh, M. and Wackerly, D. (1986). Fixed length interval estimation of the location parameter of a Pareto distribution. *Calcutta Statist. Assoc. Bull.*, 35, 67-75.

[331] Gibrat, R. (1931). *Les Inegalites Economiques*. Recueil Sirey, Paris.

[332] Gini, C. (1909). Il diverso accrescimento delle classi sociali e la concentrazione della ricchezza. *Giornale degli Economisti*, 38, 27–83.

[333] Gini, C. (1912) *Variabilità e Mutabilità*, Cuppini, Bologna.

[334] Gini, C. (1914). Sulla misura della concentrazione e della variabilita dei caratteri. *Atti del R. Instituto Veneto di Scienze, Lettere ed Arti*, 73, 1203–1248.

[335] Gini, C. (1936). On the measure of concentration with special reference to income and wealth. Cowles Commission Research Conference on Economics and Statistics. Colorado College Publication, General Series No. 208, 73–80.

[336] Gini, C. (1955). *Variabilita e Concentrazione*, (2nd ed.) Veschi, Roma.

[337] Giorgi, G. M. and Mondani, R. (1995). Sampling distribution of the Bonferroni inequality index from exponential population. *Sankhya Ser. B*, 57, 10-18.

[338] Giorgi, G. M. and Pallini, A. (1987a). On the speed of convergence to the normal distribution of some inequality indices. *Metron*, 45, 137–149.

[339] Giorgi, G. M. and Pallini, A. (1987b). About a general method for the lower and upper distribution-free bounds on Gini's concentration ratio from grouped data. *Statistica*, 47, 171–184.

[340] Giorgi, G. M. and Pallini, A. (1990). Inequality indices: Theoretical and empirical aspects of their asymptotic behaviour. *Statist. Hefte*, 31, 65-76.

[341] Glasser, G. J. (1961). Relationship between the mean difference and other measures of variation. *Metron*, 21, 176–180.

[342] Gnedenko, B. V. (1943). Sur la distribution limite du terme maximum d'une serie aleatoire. *Annals of Mathematics*, 44, 423–453.

[343] Gnedenko, B. V. and Kolmogorov, A. N. (1954). *Limit Distributions for Sums of Independent Random Variables*. Addison-Wesley, Cambridge, Massachusetts.

[344] Godwin, H. J. (1945). On the distribution of the estimate of mean deviation obtained from samples from a normal population. *Biometrika*, 33, 254–256.

[345] Goldie, C. M. (1967). A class of infinitely divisible distributions. *Proceedings of Cambridge Philosophical Society*, 63, 1141–1143.

[346] Goldie, C. M. (1977). Convergence theorems for empirical Lorenz curves and their inverses. *Advances in Applied Probability*, 9, 765–791.

[347] Goodman, I. R. and Kotz, S. (1981). Hazard rates based on isoprobability contours. In *Statistical Distributions in Scientific Work, Vol. 5*, C. Taillie, G. P. Patil and B. Baldessari, eds. Reidel, Dordrecht, Holland, 289–308.

[348] Goodman, L. (1953). A simple method for improving some estimators. *Annals of Mathematical Statistics*, 24, 114–117.

[349] Govindarajulu, Z. (1966). Characterizations of the exponential and power distributions. *Skand. Aktuarietidskr*, 49, 132–136.

[350] Greselin, F. and Pasquazzi, L. (2009). Asymptotic confidence intervals for a new inequality measure. *Comm. Statist. Simulation Comput.*, 38, 1742-1756.

[351] Grimshaw, S. D. (1993). Computing maximum likelihood estimates for the generalized Pareto distribution. *Technometrics*, 35, 185-191.

[352] Gross, A. J. and Lam, C. F. (1980). Paired observations from a survival distribution. Technical Report, Department of Biometry, Medical University of South Carolina, Charleston, South Carolina.

[353] Gulati, S. and Shapiro, S. (2008). Goodness-of-fit tests for Pareto distribution. In *Statistical Models and Methods for Biomedical and Technical Systems*, Birkhuser, Boston, 259274.

[354] Gumbel, E. J. (1928). Das konzentrationsmass. *Allgemeines Statistiches Archiv.*, 18, 279–300.

[355] Gupta, A. K. and Nadarajah, S. (2007). Information matrices for some bivariate Pareto distributions. *Appl. Math. Comput.*, 184, 1069-1079.

[356] Gupta, M. R. (1984). Functional form for estimating the Lorenz curve. *Econometrica*, 52, 1313–1314.

[357] Gupta, R. C. (1979). On the characterization of survival distributions in reliability by properties of their renewal densities. *Comm. Statist. ATheory Methods*, 8, 685-697.

[358] Gupta, R. C. (2001). Reliability studies of bivariate distributions with Pareto conditionals. *J. Multivariate Anal.*, 76, 214-225.

[359] Gupta, R. C. and Keating, J. P. (1986). Relations for reliability measures under length biased sampling. *Scand. J. Statist.*, 13, 49-56.

[360] Gupta, R. C. and Langford, E. S. (1984). On the determination of a distribution by its median residual life function: A functional equation. *J. Appl. Probab.*, 21, 120-128.

[361] Gupta, R. P. and Sankaran, P. G. (1998). Bivariate equilibrium distribution and its applications to reliability. *Comm. Statist. Theory Methods*, 27, 385-394.

[362] Habibullah, M. and Ahsanullah, M. (2000). Estimation of parameters of a Pareto distribution by generalized order statistics. *Comm. Statist. Theory Methods*, 29, 1597-1609.

[363] Hagerbaumer, J. B. (1977). The Gini concentration ratio and the minor concentration ratio: A two parameter index of inequality. *Review of Economics and Statistics*, 59, 377–379.

[364] Hagstroem, K. G. (1925). La loi de Pareto et la reassurance. *Skand. Aktuarietidskr*, 8, 65–88.

[365] Hagstroem, K. G. (1960). Remarks on Pareto distributions. *Skand. Aktuarietidskr*, 43, 59–71.

[366] Hamada, K. and Takayama, N. (1978). Censored income distributions and the measurement of poverty. *Proceedings of the International Statistical Institute*, New Delhi.

[367] Hamdy, H. I. and Pallotta, W. J. (1987). Triple sampling procedure for estimating the scale parameters of Pareto distributions. *Comm. Statist. Theory Methods*, 16, 2155-2164.

[368] Hamedani, G. G. (2002). Characterizations of univariate continuous distributions. II. *Studia Sci. Math. Hungar.*, 39, 407-424.

[369] Hamedani, G. G. (2004). Characterizations of univariate continuous distributions based on hazard function. *J. Appl. Statist. Sci.*, 13, 169-183.

[370] Hanagal, D. D. (1996). A multivariate Pareto distribution. *Comm. Statist. Theory Methods*, 25, 1471-1488.

[371] Hardy, G. H., Littlewood, J. E. and Polya, G. (1929). Some simple inequalities satisfied by convex functions. *Messenger of Mathematics*, 58, 145–152.

[372] Hardy, G. H., Littlewood, J. E. and Polya, G. (1959). *Inequalities*, (2nd ed.). Cambridge University Press, Cambridge.

[373] Harris, C. M. (1968). The Pareto distribution as a queue service discipline. *Operations Research*, 16, 307–313.

[374] Harris, C. M. and Singpurwalla, N. D. (1969). On estimation in Weibull distributions with random scale parameter. *Naval Research Logistics Quarterly*, 16, 405–10.

[375] Hart, P. E. (1975). Moment distributions in economics: An exposition. *Journal of the Royal Statistical Society, Series A*, 138, 423–434.

[376] Hart, P. E. (1976). The comparative statics and dynamics of income distributions. *Journal of the Royal Statistical Society, Series A*, 139, 108–125.

[377] Hart, P. E. (1980). The statics and dynamics of income distributions: A survey. In *The Statics and Dynamics of Income*, N. A. Klevmarken and J. A. Lybeck, eds. Tieto Ltd., Clevedon, Avon.

[378] Harter, H. L. (1961). Estimating the parameters of negative exponential populations from one or two order statistics. *Annals of Mathematical Statistics*, 32, 1078–1090.

[379] Harter, H. L. (1977). *A Chronological Annotated Bibliography on Order Statistics: Vol. 1, pre-1950*. Air Force Flight Dynamics Laboratory, Wright Patterson Air Force Base, Dayton, Ohio.

[380] Hartley, M. J. and Revankar, N. S. (1974). On the estimation of the Pareto law from under-reported data. *Journal of Econometrics*, 2, 327–341.

[381] Hashemi, M. and Asadi, M. (2007). Some characterization results on generalized Pareto distribution based on progressive type-II right censoring. *J. Iran. Stat. Soc.*, 6, 99-110.

[382] Hashemi, M., Tavangar, M. and Asadi, M. (2010). Some properties of the residual lifetime of progressively Type-II right censored order statistics. *Statist. Probab. Lett.*, 80, 848-859.

[383] Hatke, M. A. (1949). A certain cumulative probability function. *Annals of Mathematical Statistics*, 20, 461–3.

[384] Hawkes, A. G. (1972). A bivariate exponential distribution with applications to reliability. *Journal of the Royal Statistical Society, Series B*, 34, 129–131.

[385] Hayakawa, M. (1951). The application of Pareto's law of income to Japanese data. *Econometrica*, 19, 174–183.

[386] He, H., Zhou, N. and Zhang, R. (2014). On estimation for the Pareto distribution. *Statistical Methodology*, 21, 49–58.

[387] Heilmann, W. R. (1980). Basic distribution theory for non-parametric Gini-like measures of location and dispersion. *Biometric Journal*, 22, 51–60.

[388] Helmert, F. R. (1876). Die berechnung des wahrschlein lichen beobachtungsfehlers aus den ersten potenzen der differenzen gleichgenauer directer beobachtungen. *Astrom. Nach.*, 88, 127–132.

[389] Herfindahl, O. C. (1950). *Concentration in the steel industry*. Ph.D. Dissertation, Columbia University, New York.

[390] Herzel, A. and Leti, G. (1977). Italian contributions to statistical inference. *Metron*, 35, 3–48.

[391] Hill, B. M. (1970). Zipf's law and prior distributions for the composition of a population. *Journal of the American Statistical Association*, 65, 1220–1232.

[392] Hill, B. M. (1974). The rank-frequency form of Zipf's law. *Journal of the American Statistical Association*, 69, 1017–1026.

[393] Hill, B. M. (1975). A simple general approach to inference about the tail of a distribution. *Annals of Statistics*, 3, 1163–1174.

[394] Hill, B. M. and Woodroofe, M. (1975). Stronger forms of Zipf's law. *Journal of the American Statistical Association*, 70, 212–219.

[395] Hinkley, D. V. and Revankar, N. S. (1977). Estimation of the Pareto law from under-reported data. *Journal of Econometrics*, 5, 1–11.

[396] Hinz, P. N. and Gurland, J. (1970). Test of fit for the negative binomial and other contagious distributions. *Journal of the American Statistical Association*, 65, 887–903.

[397] Hirschman, A. O. (1945). *National Power and the Structure of Foreign Trade*. University of California Press, Berkeley, California.

[398] Hoeffding, W. (1948). A class of statistics with asymptotically normal distribution. *Annals of Mathematical Statistics*, 19, 293–325.

[399] Holm, J. (1993). Maximum entropy Lorenz curves. *J. Econometrics*, 59, 377-389.

[400] Holmes, G. K. (1905). Measurement of concentration of wealth. *Publications of the American Statistical Association*, 9, 318–319.

[401] Hosking, J. R. M. (1990). L-moments: Analysis and estimation of distributions using linear combinations of order statistics. *J. Roy. Statist. Soc. Ser. B*, 52, 105-124.

[402] Hosking, J. R. M. and Wallis, J. R. (1987). Parameter and quantile estimation for the generalized Pareto distribution. *Technometrics*, 29, 339-349.

[403] Howlader, H. A. and Hossain, A. M. (2002). Bayesian survival estimation of Pareto distribution of the second kind based on failure-censored data. *Comput. Statist. Data Anal.*, 38, 301-314.

[404] Huang, J. S. (1974). On a theorem of Ahsanullah and Rahman. *Journal of Applied Probability*, 11, 216–218.

[405] Huang, J. S. (1975). A note on order statistics from the Pareto distribution. *Scandinavian Actuarial Journal*, 2, 187–190.

[406] Huang, J. S. (1978). On a "lack of memory" property. Technical Report #84, University of Guelph, Statistical Series, Guelph, Ontario, Canada.

[407] Huang, J. S., Arnold, B. C. and Ghosh, M. (1979). On characterizations of the uniform distribution based on identically distributed spacings. *Sankhya, Series B*, 41, 109–115.

[408] Huang, M. L. (2011). Optimal estimation for the Pareto distribution. *J. Stat. Comput. Simul.*, 81, 2059-2076.

[409] Huang, M. L. and Zhao, K. (2010). On estimation of the truncated Pareto distribution. *Adv. Appl. Stat.*, 16, 83-102.

[410] Huang, W-J. and Su, N-C. (2012). Characterizations of distributions based on moments of residual life. *Comm. Statist. Theory Methods*, 41, 2750-2761.

[411] Huang, W-T. (1974). Selection procedures of Pareto populations. *Sankhya, Series B*, 36, 417–26.

[412] Huber, P. J. (1967). The behavior of maximum likelihood estimates under non-standard conditions. *Proceedings of the Fifth Berkeley Symposium, Vol. 1*, 221–233.

[413] Hudson, D. (1976). Statistical distribution of incomes among rural households in Botswana, 1974/75. Central Statistics Office, Gabarone, Botswana.

[414] Hurlimann, W. (2009). From the general affine transform family to a Pareto type IV model. *Journal of Probability and Statistics*, Art. ID 364901.

[415] Husler, J., Li, D. and Raschke, M. (2011). Estimation for the generalized Pareto distribution using maximum likelihood and goodness of fit. *Comm. Statist. Theory Methods*, 40, 2500-2510.

[416] Hutchinson, T. P. (1979). Four applications of a bivariate Pareto distribution. *Biometric Journal*, 21, 553–563.

[417] Iliescu, D. and Viorel, Gh. V. (1974). Some notes on Pareto distribution. Partial Ms.

[418] Irwin, J. O. (1968). The generalized Waring distribution applied to accident theory. *Journal of the Royal Statistical Society, Series A*, 131, 205–225.

[419] Irwin, J. O. (1975a,b,c). The generalized Waring distribution, Parts I, II, III. *Journal of the Royal Statistical Society, Series A*, 138, 18–31, 204–227, 374–384.

[420] Isaacson, D. L. and Madsen, R. W. (1976). *Markov Chains: Theory and Applications*. Wiley, New York.

[421] Ismail, S. (2004). A simple estimator for the shape parameter of the Pareto distribution with economics and medical applications. *Journal of Applied Statistics*, 31, 3–13.

[422] Iwase, K. (1986). UMVU estimation for the cumulative hazard function in a Pareto distribution of the first kind. *Comm. Statist. ATheory Methods*, 15, 3559-3566.

[423] Iwinska, M. (1980). Estimation of the scale parameter of the Pareto distribution. *Fasc. Math.*, No. 12, 113-118.

[424] Iyengar, N. S. (1960). On the standard error of the Lorenz concentration ratio. *Sankhya*, 22, 371–378.

[425] Jain, K. and Arora, S. (2000). Lorenz Gini ratio: some distributional properties. *Proceedings of the IISA 2000-2001 Conference, New Delhi*, 204–207.

[426] Jakobsson, U. (1976). On the measurement of the degree of progression. *Journal of Public Economics*, 5, 161–168.

[427] Jalali, A. and Watkins, A. J. (2009). On maximum likelihood estimation for the two parameter Burr XII distribution. *Comm. Statist. Theory Methods*, 38, 1916-1926

[428] Jambunathan, M. V. and Desai, G. R. (1964). On the efficiency of ban estimates of the Pareto coefficient. *Proceedings of the 4th Indian Econometric Conference*.

[429] James, I. R. (1979). Characterization of a family of distributions by the independence of size and shape variables. *Annals of Statistics*, 7, 869–881.

[430] James, I. R. and Mosimann, J. E. (1980). A new characterization of the Dirichlet distribution through neutrality. *Annals of Statistics*, 8, 183–189.

[431] Janjic, S. (1986). Characterizations of some distributions connected with extremal-type distributions. *Publ. Inst. Math. (Beograd) (N.S.)*, 39(53), 179–186.

[432] Jayakumar, K. (2004). On Pareto process. In *Soft Methodology and Random Information Systems, Adv. Soft Comput.*, Springer, Berlin, 235-242.

[433] Jeevanand, E. S. and Abdul-Sathar, E. I. (2010). Bayes estimation of time to test transform for classical Pareto distribution in the presence of outliers. *J. Indian Statist. Assoc.*, 48, 215-230.

[434] Jeevanand, E. S. and Padamadan, V. (1996). Parameter estimation for a bivariate Pareto distribution. *Statist. Papers*, 37, 153-164.

[435] Jin, C. and Elfessi, A. (2001). On the common scale parameter of several Pareto populations in censored samples. *Comm. Statist. Theory Methods*, 30, 451-462.

[436] Joe, H. (1985). Characterizations of life distributions from percentile residual lifetimes. *Ann. Inst. Statist. Math.*, 37, 165-172.

[437] Johnson, N. L., Kemp, A. W. and Kotz, S. (2005). *Univariate Discrete Distributions*, (3rd ed.). Wiley, New York.

[438] Johnson, N. L., Kotz, S. and Balakrishnan, N. (1994). *Continuous Univariate Distribution - Vol. 1*, (2nd ed.). Wiley, New York.

[439] Johnson, N. L., Kotz, S. and Balakrishnan, N. (2000). *Continuous Multivariate Distributions - Vol. 1: Models and Applications*, (2nd ed.). Wiley, New York.

[440] Johnson, N. O. (1937). The Pareto law. *Review of Economics and Statistics*, 19, 20–26.

[441] Jones, M. C. (2004). Families of distributions arising from distributions of order statistics. *Test*, 13, 143.

[442] Jones, M. C. (2009). Kumaraswamy's distribution: A beta-type distribution with some tractability advantages. *Stat. Methodol.*, 6, 7081.

[443] Jupp, P. E. (1986). Characterization of matrix probability distributions by mean residual lifetime. *Math. Proc. Cambridge Philos. Soc.*, 100, 583-589.

[444] Jupp, P. E. and Mardia, K. V. (1982). A characterization of the multivariate Pareto distribution. *Ann. Statist.*, 10, 1021-1024.

[445] Kabe, D. G. (1972). On moments of order statistics from the Pareto distribution. *Skand. Aktuarietidskr*, 55, 179–181.

[446] Kabe, D. G. (1975). Distributions of Gini's mean difference for exponential and rectangular populations. *Metron*, 33, 94–97.

[447] Kadane, J. B. (1971). A moment problem for order statistics. *Annals of Mathematical Statistics*, 42, 745–751.

[448] Kagan, A. M., Linnik, Y. V. and Rao, C. R. (1973). *Characterization Problems in Mathematical Statistics*. Wiley, New York.

[449] Kagan, Y. Y. and Schoenberg, F. (2001). Estimation of the upper cutoff parameter for the tapered Pareto distribution. *J. Appl. Probab.*, 38A, 158-175.

[450] Kakwani, N. C. (1974). A note on the efficient estimation of the new measures of income inequality. *Econometrica*, 42, 597–600.

[451] Kakwani, N. C. (1980a). On a class of poverty measures. *Econometrica*, 48, 437–446.

[452] Kakwani, N. C. (1980b). Functional forms for estimating the Lorenz curve: A reply. *Econometrica*, 48, 1063–1064.

[453] Kakwani, N. C. (1984). Welfare ranking of income distributions. *Advances in Econometrics*, 3, 191–213.

[454] Kakwani, N. C. and Podder, N. (1973). On the estimation of Lorenz curves from grouped observations. *Economic Review*, 14, 278–292.

[455] Kakwani, N. C. and Podder, N. (1976). Efficient estimation of the Lorenz curve and associated inequality measures from grouped observations. *Econometrica*, 44, 137–148.

[456] Kalbfleisch, J. D. and Prentice, R. L. (1980). *The Statistical Analysis of Failure Time Data*. Wiley, New York.

[457] Kalecki, M. (1945). On the Gibrat distribution. *Econometrica*, 13, 161–170.

[458] Kamat, A. R. (1953). The third moment of Gini's mean difference. *Biometrika*, 40, 451–452.

[459] Kaminsky, K. S. (1973). Best linear unbiased estimation and prediction in censored samples from Pareto populations. Unpublished manuscript, Bucknell University, Lewisburg, PA.

[460] Kaminsky, K. S. and Nelson, P. I. (1975). Best linear unbiased prediction of order statistics in location and scale families. *Journal of the American Statistical Association*, 70, 145–150.

[461] Kaminsky, K. S. and Rhodin, L. S. (1978). The prediction information in the latest failure. *Journal of the American Statistical Association*, 73, 863–866.

[462] Kamps, U. (1991). Inequalities for moments of order statistics and characterizations of distributions. *J. Statist. Plann. Inference*, 27, 397-404.

[463] Kamps, U. (1995). *A Concept of Generalized Order Statistics*. Teubner Texts on Mathematical Stochastics, B. G. Teubner, Stuttgart.

[464] Kang, S. G., Kim, D. H. and Lee, W. D. (2012). Noninformative priors for common shape parameter in the Pareto distributions. *Adv. Stud. Contemp. Math. (Kyungshang)*, 22, 173-182.

[465] Karlin, S. and Rinott, Y. (1988) A generalized Cauchy-Binet formula and applications to total positivity and majorization. *J. Multivariate Anal.*, 27, 284-299.

[466] Katz, L. (1963). Unified treatment of a broad class of discrete probability distributions. *Proceedings of International Symposium on Discrete Distributions*, 175–182. Pergammon Press, Montreal.

[467] Keiding, N., Hansen, O. K. H., Sorensen, D. N. and Slama, R. (2012). The current duration approach to estimating time to pregnancy. *Scandinavian Journal of Statistics*, 39, 185–204.

[468] Keiding, N., Kvist, K., Hartvig, H., Tvede, M. and Juul, S. (2002). Estimating time to pregnancy from current durations in a cross-sectional sample. *Biostatistics*, 3, 565–578.

[469] Kendall, M.G. and Gibbons, J. D. (1990) *Rank Correlation Methods*, (5th ed.). Edward Arnold, London.

[470] Kendall, M. G., Stuart, A. and Ord, J. K. (1994) *Kendall's advanced theory of statistics. Vol. 1. Distribution theory*, (6th ed.). Edward Arnold, London

[471] Kern, D. M. (1983). Minimum variance unbiased estimation in the Pareto distribution. *Metrika*, 30, 15-19.

[472] Khan, R. U. and Kumar, D. (2010). On moments of lower generalized order statistics from exponentiated Pareto distribution and its characterization. *Appl. Math. Sci. (Ruse)*, 4, 2711-2722.

[473] Khurana, A. P. and Jha, V. D. (1991). Recurrence relations between moments of order statistics from a doubly truncated Pareto distribution. *Sankhya Ser. B*, 53, 11–16.

[474] Khurana, A. P. and Jha, V. D. (1995). Some recurrence relations between product moments of order statistics of a doubly truncated Pareto distribution. *Sankhya Ser. B*, 57, 1-9.

[475] Kibble, W. F. (1941). A two variable gamma type distribution. *Sankhya*, 5, 137–150.

[476] Kibria, B. M. G. and Saleh, A. K. M. E. (2010). Preliminary test estimation of the parameters of exponential and Pareto distributions for censored samples. *Statist. Papers*, 51, 757-773.

[477] Kimball, A. W. (1960). Estimation of mortality intensities in animal experiments. *Biometrics*, 16, 505.

[478] Klamkin, M. S. and Newman, D. J. (1976). Inequalities and identities for sums and integrals. *American Mathematical Monthly*, 83, 26–30.

[479] Klefsjö, B. (1982). On aging properties and total time on test transforms. *Scand. J. Statist.*, 9, 37-41.

[480] Klefsjö, B. (1984). Reliability interpretations of some concepts from Economics. *Naval Research Logistics*, 31. 301–308.

[481] Kleiber, C. (1999). On the Lorenz order within parametric families of income distributions. *Sankhya Ser. B*, 61, 514-517.

[482] Kleiber, C. (2002). Variability ordering of heavy-tailed distributions with applications to order statistics. *Statist. Probab. Lett.*, 58, 381-388.

[483] Kleiber, C. and Kotz, S. (2003). *Statistical Size Distributions in Economics and Actuarial Sciences.* Wiley, Hoboken, New Jersey.

[484] Kloek, T. and van Dijk, H. K. (1978). Efficient estimation of income distribution parameters. *Journal of Econometrics*, 8, 61–74.

[485] Kondor, Y. (1971). An old-new measure of income inequality. *Econometrica*, 39, 1041–1042.

[486] Korwar, R. M. (1985). On the joint characterization of the distributions of the true income and the proportion of the reported income to the true income through a model of under-reported incomes. *Statistics and Probability Letters*, 3, 201–203.

[487] Koshevoy, G. (1995). Multivariate Lorenz majorization. *Soc. Choice Welf.*, 12, 93–102.

[488] Koshevoy, G. and Mosler, K. (1996). The Lorenz zonoid of a multivariate distribution. *J. Amer. Statist. Assoc.*, 91, 873-882.

[489] Koshevoy, G. and Mosler, K. (1997). Multivariate Gini indices. *J. Multivariate Anal.*, 60, 252-276.

[490] Koutrouvelis, I. A. (1981). Large sample quantile estimation in Pareto laws. *Communications in Statistics, Theory and Methods*, A10, 189–201.

[491] Krause, M. (2013). Corrigendum to "Elliptical Lorenz curves" [J. Econom. 40 (1989) 327–338]. *J. Econometrics*, 174, 44.

[492] Krieger, A. M. (1979). Bounding moments, the Gini index and Lorenz curve from grouped data for unimodal density functions. *Journal of the American Statistical Association*, 74, 375–378.

[493] Krishna, H. and Singh Pundir, P. (2009). Discrete Burr and discrete Pareto distributions. *Stat. Methodol.*, 6, 177-188.

[494] Krishnaji, N. (1970). Characterization of the Pareto distribution through a model of under-reported incomes. *Econometrica*, 38, 251–5.

[495] Krishnaji, N. (1971). Note on a characterizing property of the exponential distribution. *Annals of Mathematical Statistics*, 42, 361–2.

[496] Krishnan, P., Ng, E. and Shihadeh, E. (1991) Some generalized forms of the Pareto curve to approximate income distributions. Technical report, Department of Sociology, University of Alberta, Canada.

[497] Krysicki, W. (1999). On some new properties of the beta distribution. *Statist. Probab. Lett.*, 42, 131-137.

[498] Kulldorff, G. and Vannman, K. (1973). Estimation of the location and scale parameters of a Pareto distribution by linear functions of order statistics. *Journal of the American Statistical Association*, 68, 218–227. Corrigenda, 70, 494.

[499] Kumar, S. and Gangopadhyay, A. K. (2005). Estimating parameters of a selected Pareto population. *Stat. Methodol.*, 2, 121-130.

[500] Kurabayashi, Y. and Yatsuka, A. (1977). Redistribution of income and measures of income inequality. *Lecture Notes in Economics and Mathematical Systems, Vol. 147*. Springer-Verlag, Berlin. Pages 47–59.

[501] Kus, C. and Kaya, M. F. (2007). Estimation for the parameters of the Pareto distribution under progressive censoring. *Comm. Statist. Theory Methods*, 368, 1359-1365.

[502] Lam, D. (1986). The dynamics of population growth, differential fertility, and inequality. *Am. Econ. Rev.*, 76, 1103–1116.

[503] Lambert, P. J. (1993). Inequality reduction through the income tax. *Economica*, 60, 357–365.

[504] Lambert, P. J. and Lanza, G. (2006) The effect on inequality of changing one or two incomes. *Journal of Economic Inequality*, 4, 253–277.

[505] Lamperti, J. (1966). *Probability*. W. A. Benjamin, New York.

[506] Laurent, A. G. (1974). On characterization of some distributions by truncation properties. *Journal of the American Statistical Association*, 69, 823–827.

[507] Lee, M-Y. (2003). Characterizations of the Pareto distribution by conditional expectations of record values. *Commun. Korean Math. Soc.*, 18, 127-131.

[508] Lee, M-Y. and Kim, Y. I. (2011). On characterizations of Pareto distribution by hazard rate of record value. *Ann. Univ. Sci. Budapest. Sect. Comput.*, 34, 195-199.

[509] Leimkuhler, F. F. (1967). The Bradford distribution. *Journal of Documentation*, 23, 197–207.

[510] Levine, D. B. and Singer, N. M. (1970). The mathematical relation between the income density function and the measurement of income inequality. *Econometrica*, 38, 324–330.

[511] Levy, M. (2003). Are rich people smarter? *Journal of Economic Theory*, 110, 42–64.

[512] Liang, T. (1993). Convergence rates for empirical Bayes estimation of the scale parameter in a Pareto distribution. *Comput. Statist. Data Anal.*, 16, 35-45.

[513] Likes, J. (1969). Minimum variance unbiased estimates of the parameters of power function and Pareto's distribution. *Statistische Heft*, 10, 104–110.

[514] Likes, J. (1985). Estimation of the quantiles of Pareto's distribution. *Statistics*, 16, 541-547.

[515] Lin, C-T. and Wang, W-Y. (2000). Estimation for the generalized Pareto distribution with censored data. *Comm. Statist. Simulation*, 29, 1183–1213.

[516] Lin, G. D. (1988). Characterizations of distributions via relationships between two moments of order statistics. *J. Statist. Plann. Inference*, 19, 73–80.

[517] Lindley, D. V. (1958) Fiducial distributions and Bayes' theorem. *J. Roy. Statist. Soc. Ser. B*, 20, 102-107.

[518] Lindley, D. V. (1980). Approximate Bayesian methods. With discussion. in *Bayesian Statistics (Valencia, 1979)*, Univ. Press, Valencia, 223-245.

[519] Lindley, D. V. and Singpurwalla, N. D. (1986). Multivariate distributions for the life lengths of components of a system sharing a common environment. *J. Appl. Probab.*, 23, 418-431.

[520] Loeve, M. (1977). *Probability Theory 1*, (4th ed.). Springer-Verlag, New York.

[521] Lomax, K. S. (1954). Business failures. Another example of the analysis of failure data. *Journal of the American Statistical Association*, 49, 847–852.

[522] Lomnicki, Z. A. (1952). The standard error of Gini's mean difference. *Annals of Mathematical Statistics*, 23, 635–637.

[523] Long, J. E., Rasmussen, D. W. and Haworth, C. T. (1977). Income inequality and city size. *Review of Economics and Statistics*, 59, 244–6.

[524] Lorah, D. J. and Stark, R. M. (1978). On elementary functions of Pareto variables. *Metrika*, 25, 59-63.

[525] Lorenz, M. O. (1905). Methods of measuring the concentration of wealth. *Publication of the American Statistical Association*, 9, 209–219.

[526] Lucke, J., Myhre, J. and Williams, P. (1977). Comparison of parameter estimation for the Pareto distribution. Manuscript.

[527] Luceño, A. (2006). Fitting the generalized Pareto distribution to data using maximum goodness-of-fit estimators. *Comput. Statist. Data Anal.*, 51, 904-917.

[528] Lunetta, G. (1972). Di un indice di cocentrazione per variabili statistische doppie. *Annali della Facoltá di Economia e Commercio dell Universitá di Catania*, A 18.

[529] Lwin, T. (1972). Estimation of the tail of the Paretian law. *Skand. Aktuarietidskr*, 55, 170–178.

[530] Lwin, T. (1974). Empirical Bayes approach to statistical estimation in the Paretian law. *Scand. Actuar. J.*, 221-236.

[531] Lydall, H. F. (1959). The distribution of employment income. *Econometrica*, 27, 110–115.

[532] Macgregor, D. H. (1936). Pareto's law. *Economic Journal*, 46, 80–87.

[533] Madi, M. T. and Raqab, M. Z. (2008). Analysis of record data from generalized Pareto distribution. *J. Appl. Statist. Sci.*, 16, 211-221.

[534] Maguire, B. A., Pearson, E. S. and Wynn, A. H. A. (1952). The time intervals between industrial accidents. *Biometrika*, 39, 168–180.

[535] Mahmoud, M. A. W., Sultan, K. S. and Moshref, M. E. (2005). Inference based on order statistics from the generalized Pareto distribution and application. *Comm. Statist. Simulation Comput.*, 34, 267-282.

[536] Mahmoud, M. A. W., Rostom, N. K. A. and Yhiea, N. M. (2008). Recurrence relations for the single and product moments of progressively type-II right censored order statistics from the generalized Pareto distribution. *J. Egyptian Math. Soc.*, 16, 61-74.

[537] Mahmoud, M. R. and Abd El-Ghafour, A. (2014) Fisher information matrix for the generalized Feller-Pareto distribution. *Communications in Statistics, Theory and Methods*, to appear.

[538] Mahmoudi, E. (2011). The beta generalized Pareto distribution with application to lifetime data. *Math. Comput. Simulation*, 81, 2414-2430.

[539] Malik, H. J. (1966). Exact moments of order statistics from the Pareto distribution. *Skand. Aktuarietidskr*, 49, 144–157.

[540] Malik, H. J. (1970a). Distribution of product statistics from a Pareto population. *Metrika*, 15, 19–22.

[541] Malik, H. J. (1970b). Estimation of the parameters of the Pareto distribution. *Metrika*, 15, 126–132.

[542] Malik, H. J. (1970c). Bayesian estimation of the Paretian index. *Skand. Aktuarietidskr*, 53, 6–9.

[543] Malik, H. J. (1970d). A characterization of the Pareto distribution. *Skand. Aktuarietidskr*, 53, 115–7.

[544] Malik, H. J. (1980). Exact formula for the cumulative distribution function of the quasi-rank from the logistic distribution. *Communications in Statistics, Theory and Methods*, A9(14), 1527–1534.

[545] Malik, H. J. and Abraham, B. (1973). Multivariate logistic distributions. *Ann. Statist.*, 1, 588-590.

[546] Malik, H. J. and Trudel, R. (1985). Distributions of the product and the quotient from bivariates t,F and Pareto distribution. *Comm. Statist. Theory Methods*, 14, 2951-2962.

[547] Mallows, C. L. (1973). Bounds on distribution functions in terms of expectations of order statistics. *Annals of Probability*, 1, 297–303.

[548] Manas, G. J. (1997). A new estimation procedure based on frequency moments. *South African Statist. J.*, 31, 91-106.

[549] Manas, G. J. and Boyd, A. V. (1997). New estimators for Pareto parameters based on frequency moments. *South African Statist. J.*, 31, 107-124.

[550] Manas, A. (2011). The Paretian Ratio Distribution: An application to the volatility of GDP. *Econom. Lett.*, 111, 180–183.

[551] Mandelbrot, B. (1953). An informational theory of the statistical structure of language. In *Communication Theory*, W. Jackson, ed. Butterworths, London, 486–502.

[552] Mandelbrot, B. (1959). A note on a class of skew distribution functions: Analysis and critique of a paper by H. A. Simon. *Information and Control*, 2, 90–99.

[553] Mandelbrot, B. (1960). The Pareto-Levy law and the distribution of income. *International Economic Review*, 1, 79–106.

[554] Mandelbrot, B. (1961). Stable Paretian random functions and the multiplicative variation of income. *Econometrica*, 29, 517–543.

[555] Mandelbrot, B. (1963). New methods in statistical economics. *Journal of Political Economy*, 71, 421–440.

[556] Mardia, K. V. (1962). Multivariate Pareto distributions. *Annals of Mathematical Statistics*, 33, 1008–1015.

[557] Mardia, K. V. (1964). Some results on the order statistics of the multivariate normal and Pareto type 1 populations. *Annals of Mathematical Statistics*, 35, 1815–1818.

[558] Mardia, K. V. (1970). *Families of Bivariate Distributions*. Hafner, Darien, Connecticut.

[559] Marohn, F. (2002). A characterization of generalized Pareto distributions by progressive censoring schemes and goodness-of-fit tests. *Comm. Statist. Theory Methods*, 31, 1055–1065.

[560] Marsaglia, G. and Tubilla, A. (1975). A note on the "lack of memory" property of the exponential distribution. *Annals of Probability*, 3, 353–4.

[561] Marshall, A. W. and Olkin, I. (1967). A multivariate exponential distribution. *J. Amer. Statist. Assoc.*, 62, 30-44.

[562] Marshall, A. W., Olkin, I. and Arnold, B. C. (2011). *Inequalities: Theory of Majorization and Its Applications*, (2nd ed.). Springer, New York.

[563] Marshall, A. W., Olkin, I. and Proschan, F. (1967). Monotonicity of ratios of means and other applications of majorization, In *Inequalities*, O. Shisha, ed. Academic Press, New York. Pages 177–190.

[564] Marshall, A. W. and Proschan, F. (1965). Maximum likelihood estimation for distributions with monotone failure rate. *Annals of Mathematical Statistics*, 36, 69–77.

[565] Mathai, A. M. and Moschopoulos, P. G. (1991). On a multivariate gamma. *J. Multivariate Anal.*, 39, 135-153.

[566] Mauldon, J. G. (1951). Random division of an interval. *Proceedings of the Cambridge Philosophical Society*, 47, 331–336.

[567] McCune, E. D. and McCune, S. L. (2000). Estimation of the Pareto shape parameter. *Comm. Statist. Simulation*, 29, 1317–1324.

[568] McDonald, J. B. (1981). Some issues associated with the measurement of income inequality. In *Statistical Distributions in Scientific Work, Vol. 6*, C. Taillie, G. P. Patil and B. Baldessari, eds. Reidel, Dordrecht, 161–180.

[569] McDonald, J. B. and Jensen, B. C. (1979a). An analysis of some properties of alternative measures of income inequality based on the gamma distribution function. *Journal of the American Statistical Association*, 74, 856–860.

[570] McDonald, J. B. and Jensen, B. C. (1979b). An analysis of some properties of alternative measures of income inequality based upon the gamma distribution function. Technical Report, March 1979, Department of Economics, Brigham Young University, Provo, Utah.

[571] McDonald, J. B. and Ransom, M. R. (1979). Functional forms, estimation techniques and the distribution of incomes. *Econometrica*, 47, 1513–1526.

[572] McDonald, J. B. and Ransom, M. R. (1981). An analysis of the bounds for the Gini coefficient. *Journal of Econometrics*, 17, 177–188.

[573] Mehran, F. (1975). Bounds on the Gini index based on observed points of the Lorenz curve. *Journal of the American Statistical Association*, 10, 64–66.

[574] Mehran, F. (1976). Linear measures of income inequality. *Econometrica*, 44, 805–809.

[575] Meintanis, S. G. (2009). A unified approach of testing for discrete and continuous Pareto laws. *Statist. Papers*, 50, 569-580.

[576] Mert, M. and Saykan, Y. (2005). On a bonus-malus system where the claim frequency distribution is geometric and the claim severity distribution is Pareto. *Hacet. J. Math. Stat.*, 34, 75-81.

[577] Metcalf, C. E. (1969). The size distribution of personal income during the business cycle. *American Economic Review*, 59, 657–668.

[578] Meyer, P. A. (1966). *Probability and Potentials*. Blaisdell, Waltham, Mass.

[579] Michel, R. (2009). Parametric estimation procedures in multivariate generalized Pareto models. *Scand. J. Stat.*, 36, 60-75.

[580] Midlarsky, M. I. (1989). A distribution of extreme inequality with applications to conflict behavior: A geometric derivation of the Pareto distribution. *Mathematical and Computer Modelling*, 12, 577–587.

[581] Mitzenmacher, M. (2004) Dynamic models for file sizes and double Pareto distributions. *Internet Math.*, 1, 305–333.

[582] Mohie El-Din, M. M., Abdel-Aty, Y. and Shafay, A. R. (2012). Two-sample Bayesian prediction intervals of generalized order statistics based on multiply Type II censored data. *Comm. Statist. Theory Methods*, 41, 381-392.

[583] Moore, A. H. and Harter, H. L. (1967). One order statistic conditional estimators of shape parameters of limited and Pareto distributions and scale parameters of type II asymptotic distributions of smallest and largest values. *IEEE Transactions on Reliability*, R16, 100–103.

[584] Moore, A. H. and Harter, H. L. (1969). Conditional maximum-likelihood estimation, from single censored samples, of the shape parameters of Pareto and limited distributions. *IEEE Transactions on Reliability*, R18, 76–78.

[585] Moothathu, T. S. K. (1983). Properties of Gastwirth's Lorenz curve bounds for general Gini index. *Journal of the Indian Statistical Association*, 21, 149–154.

[586] Moothathu, T. S. K. (1985). Sampling distributions of Lorenz curve and Gini index of the Pareto distribution. *Sankhya Ser. B*, 47, 247-258.

[587] Moothathu, T. S. K. (1986). The best estimators of quantiles and the three means of the Pareto distribution. *Calcutta Statist. Assoc. Bull.*, 35, 111-121.

[588] Moothathu, T. S. K. (1988). On the best estimators of coefficients of variation, skewness and kurtosis of Pareto distribution and their variances. *Calcutta Statist. Assoc. Bull.*, 37, 29-39.

[589] Moothathu, T. S. K. (1990). A characterization property of Weibull, exponential and Pareto distributions. *J. Indian Statist. Assoc.*, 28, 69-74.

[590] Moothathu, T. S. K. (1990). The best estimator and a strongly consistent asymptotically normal unbiased estimator of Lorenz curve Gini index and Theil entropy index of Pareto distribution. *Sankhya Ser. B*, 52, 115-127.

[591] Moothathu, T. S. K. (1991). On a sufficient condition for two non-intersecting Lorenz curves. *Sankhya Ser. B*, 53, 268-274.

[592] Moothathu, T. S. K. (1993). A characterization of the a-mixture Pareto distribution through a property of Lorenz curve. *Sankhya Ser. B*, 55, 130-134.

[593] Morgan, J. (1962). The anatomy of income distribution. *Review of Economics and Statistics*, 44, 270–283.

[594] Morrison, D. G. (1978). On linearly increasing mean residual lifetimes. *Journal of Applied Probability*, 15, 617–620.

[595] Mosimann, J. E. (1970). Size allometry: Size and shape variables with characterizations of the lognormal and generalized gamma distributions. *Journal of the American Statistical Association*, 65, 930–945.

[596] Mosler, K. (2002). *Multivariate Dispersion, Central Regions and Depth: The Lift Zonoid Approach.* Lecture Notes in Statistics, 165. Springer-Verlag, Berlin.

[597] Mosler, K. and Muliere, P. (1996). Inequality indices and the starshaped principle of transfers. *Statist. Papers*, 37, 343-364.

[598] Muirhead, R. F. (1903). Some methods applicable to identities and inequalities of symmetric algebraic functions of *n* letters. *Proceedings of Edinburgh Mathematical Society*, 21, 144–157.

[599] Mukhopadhyay, N. and Ekwo, M. E. (1987). Sequential estimation problems for the scale parameter of a Pareto distribution. *Scand. Actuar. J.*, 83-103.

[600] Muliere, P. and Scarsini, M. (1987). Characterization of a Marshall-Olkin type class of distributions. *Ann. Inst. Statist. Math.*, 39, 429-441.

[601] Muliere, P. and Scarsini, M. (1989) A note on stochastic dominance and inequality measures. *Journal of Economic Theory*, 49, 314–323.

[602] Muniruzzaman, A. N. M. (1957). On measures of location and dispersion and tests of hypotheses in a Pareto population. *Bulletin of Calcutta Statistical Association*, 7, 115–123.

[603] Muniruzzaman, A. N. M. (1968). Estimation of the parameter of Pareto distribution – A Bayesian approach. Statistics for Development, Proceedings, Institute of Statistical Research and Training, University of Dacca.

[604] Muralidharan, K. and Khabia, A. (2011). A modified Pareto distribution. *Journal of the Indian Statistical Association*, 49, 73–89.

[605] Murray, D. (1978). Extreme values for Gini coefficients calculated from grouped data. *Economic Letters*, 1, 389–392.

[606] Myhre, J. and Saunders, S. C. (1976). Problems of estimation for decreasing failure rate distributions applied to reliability. Manuscript.

[607] Nadarajah, S. (2005). Exponentiated Pareto distributions. *Statistics*, 39, 255-260.

[608] Nadarajah, S. (2007). Colombi-type Pareto models for income. *Nuovo Cimento Soc. Ital. Fis. B*, 122, 425–446.

[609] Nadarajah, S. (2009). A bivariate Pareto model for drought. *Stoch. Environ. Res. Risk Assess.*, 23, 811-822.

[610] Nadarajah, S. (2012). The gamma beta ratio distribution. *Brazilian Journal of Probability and Statistics*, 26, 178–207.

[611] Nadarajah, S., Cordeiro, G. M. and Ortega, E. M. M. (2012). General results for the Kumaraswamy-G distribution. *J. Stat. Comput. Simul.*, 82, 951–979.

[612] Nadarajah, S. and Espejo, M. R. (2006). Sums, products, and ratios for the generalized bivariate Pareto distribution. *Kodai Math. Journal*, 29, 72–83.

[613] Nadarajah, S. and Gupta, A. K. (2008). A product Pareto distribution. *Metrika*, 68, 199-208.

[614] Nadarajah, S. and Kotz, S. (2005a). Information matrix for a mixture of two Pareto distributions. *Iran. J. Sci. Technol. Trans. A Sci.*, 29, 377-385.

[615] Nadarajah, S. and Kotz, S. (2005b)). Muliere and Scarsini's bivariate Pareto distribution: Sums, products, and ratios. *SORT*, 29, 183-199.

[616] Nadarajah, S. and Kotz, S. (2006a). On the Laplace transform of the Pareto distribution. *Queueing Syst.*, 54, 243-244.

[617] Nadarajah, S. and Kotz, S. (2006b). Performance measures for some bivariate Pareto distributions. *Int. J. Gen. Syst.*, 35, 387-393.

[618] Nagata, Y. (1983). Estimation of the Pareto parameter and its admissibility. *Math. Japon.*, 28, 149-155.

[619] Nair, U. S. (1936). The standard error of Gini's mean difference. *Biometrika*, 28, 428–436.

[620] Nair, N. U. and Hitha, N. (1990). Characterizations of Pareto and related distributions. *J. Indian Statist. Assoc.*, 28 , 75-79.

[621] Nasri-Roudsari, D. and Cramer, E. (1999). On the convergence rates of extreme generalized order statistics. *Extremes*, 2, 421-447

[622] Needleman, L. (1978). On the approximation of the Gini coefficient of concentration. *The Manchester School*, 46, 105–122.

[623] Neuts, M. F. (1975). Computational uses of the method of phases in the theory of queues. *Comput. Math. Appl.*, 1, 151166.

[624] Newbery, D. (1970). A theorem on the measurement of inequality. *Journal of Economic Theory*, 2, 264–6.

[625] Niehans, J. (1958). An index of the size of industrial establishments. *International Economic Papers*, 8, 122–132.

[626] Nigm, A. M., Al-Hussaini, E. K. and Jaheen, Z. F. (2003). Bayesian one-sample prediction of future observations under Pareto distribution. *Statistics*, 37, 527-536.

[627] Nigm, A. M. and Hamdy, H. I. (1987). Bayesian prediction bounds for the Pareto lifetime model. *Comm. Statist. Theory Methods*, 16, 1761-1772.

[628] Oakes, D. and Dasu, T. (1990). A note on residual life. *Biometrika*, 77, 409-410.

[629] Ogwang, T. and Rao, U. L. G. (1996). A new functional form for approximating the Lorenz curve. *Economic letters*, 52, 21–29.

[630] Olkin, I. and Liu, R. (2003). A bivariate beta distribution. *Statist. Probab. Lett.*, 62, 407-412.

[631] Omar, A., Ibrahim, K. and Razali, A. M. (2012). Confidence interval estimation of the shape parameter of Pareto distribution using extreme order statistics. *Appl. Math. Sci. (Ruse)*, 6, 4627-4640.

[632] Oncel, S. Y., Ahsanullah, M., Aliev, F. A. and Aygun, F. (2005). Switching record and order statistics via random contractions. *Statist. Probab. Lett.*, 73, 207-217.

[633] Ord, J. K. (1972). *Families of Frequency Distributions*. Griffin, London.

[634] Ord, J. K. (1975). Statistical models for personal income distributions. In *Statistical Distributions in Scientific Work, Vol. 2*, G. P. Patil, S. Kotz, and J. K. Ord, eds. Reidel, Dordrecht, Holland, 151–158.

[635] Ord, J. K., Patil, G. P. and Taillie, C. (1978). The choice of a distribution to describe personal incomes. Unpublished manuscript.

[636] Ord, J. K., Patil, G. P. and Taillie, C. (1981a). Relationships between income distributions for individuals and for house-holds. In *Statistical Distributions in Scientific Work, Vol. 6*, C. Taillie, G. P. Patil and B. Baldessari, eds. Reidel, Dordrecht, Holland, 203–210.

[637] Ord, J. K., Patil, G. P. and Taillie, C. (1981b). The choice of a distribution to describe personal incomes. In *Statistical Distributions in Scientific Work, Vol. 6*, C. Taillie, G. P. Patil and B. Baldessari, eds. Reidel, Dordrecht, Holland, 193–202.

[638] Osborne, J. A. and Severini, T. A. (2002) The Lorenz curve for model assessment in exponential order statistic models. *J. Stat. Comput. Simul.*, 72, 87-97.

[639] Padamadan, V. and Muraleedharan Nair, K. R. (1994). Characterization of a bivariate Pareto distribution. *J. Indian Statist. Assoc.*, 32, 15-20.

[640] Pakes, A. G. (1981). On income distributions and their Lorenz Curves. Technical Report. Department of Mathematics, The University of Western Australia, Nedlands.

[641] Pakes, A. G. (1983). Remarks on a model of competitive bidding for employment. *Journal of Applied Probability*, 20, 349-357.

[642] Palma, J. G. (2011). Homogenous middles vs. heterogenous tails, and the end of the "inverted-U": It's all about the share of the rich. *Development and Change*, 42, 87–153.

[643] Pandey, B. N., Singh, B. P. and Mishra, C. S. (1996). Bayes estimation of shape parameter of classical Pareto distribution under LINEX loss function. *Comm. Statist. Theory Methods*, 25, 3125-3145.

[644] Papadakis, E. N. and Tsionas, E. G. (2010). Multivariate Pareto distributions: Inference and financial applications. *Comm. Statist. Theory Methods*, 39, 1013-1025.

[645] Papathanasiou, V. (1990). Some characterizations of distributions based on order statistics. *Statist. Probab. Lett.*, 9, 145-147.

[646] Paranaiba, P. F., Ortega, E. M. M., Cordeiro, G. M. and de Pascoa, M. A. R. (2013). The Kumaraswamy Burr XII distribution: Theory and practice. *J. Stat. Comput. Simul.*, 83, 2117-2143.

[647] Pareto, V. (1897).*Cours d'economie Politique, Vol. II*. F. Rouge, Lausanne.

[648] Parsi, S., Ganjali, M. and Farsipour, N. S. (2010). Simultaneous confidence intervals for the parameters of Pareto distribution under progressive censoring. *Comm. Statist. Theory Methods*, 39, 94-106.

[649] Parsian, A. and Farsipour, N. S. (1997) Estimation of the Pareto parameter under entropy loss. *Calcutta Statistical Association Bulletin*, 47, 153–165.

[650] Patil, G. P. and Ratnaparkhi, M. V. (1979). On additive and multiplicative damage models and the characterizations of linear and logarithmic exponential families. *Canad. J. Statist.*, 7, 61-64.

[651] Paulson, A. S. (1973). A characterization of the exponential distribution and a bivariate exponential distribution. *Sankhya, Series A*, 35, 69–78.

[652] Pearson, E. S., Johnson, N. L. and Burr, I. W. (1978). Comparisons of the percentage points of distributions with the same first four moments, chosen from eight different systems of frequency curves. Institute of Statistics Mimeo Series 1181, Chapel Hill, North Carolina.

[653] Pederzoli, G. and Rathie, P. N. (1980). Distribution of product and quotient of Pareto variates. *Metrika*, 27, 165–169.

[654] Peng, L., Sun, H. and Jiu, L. (2007). The geometric structure of the Pareto distribution. *Bol. Asoc. Mat. Venez.*, 14, 5-13.

[655] Pestieau, P. and Possen, U. M. (1979). A model of wealth distribution. *Econometrica*, 47, 761–772.

[656] Petersen, H. G. (1979). Effects of growing incomes on classified income distributions, the derived Lorenz curves, and Gini indices. *Econometrica*, 47, 183–198.

[657] Pfeifer, D. (1984). Limit laws for inter-record times from nonhomogeneous record values. *J. Organ. Behav. Stat.*, 1, 69-74.

[658] Picard, D. and Tribouley, K. (2002). Evolutionary Pareto distributions. *Ann. Inst. H. Poincaré Probab. Statist.*, 38, 1023-1037.

[659] Pickands, J. III. (1975). Statistical inference using extreme order statistics. *Annals of Statistics*, 3, 119-131.

[660] Pickands, J. III. (1993). Bayes quantile estimation and threshhold selection for the generalized Pareto family. Technical Report. Department of Statistics, University of Pennsylvania, Philadelphia.

[661] Pietra, G. (1915). Delle relazioni tra gli indici di variabilita. Note I, II. *Atti del Reale Istituto Veneto di Scienze, Lettere ed Arti*, 74, 775–792, 793–804.

[662] Pietra, G. (1932). Nuovi contributi alia metodologia degli indici di variabilita e di concentrazione. *Atti del Reale Istituto Veneto di Scienze, Lettere ed Arti*, 91, 989–1008.

[663] Pietra, G. (1935a). Intorno alla discordanza fra gli indici di variabilita di concentrazione. *Bulletin de l'Institut International de Statistique*, 28, 171–191.

[664] Pietra, G. (1935b). A proposito della misura della variabilita e della concentrazione dei caratteri. *Supplemento Statistico ai Nuovi Problemi di Politica, Storia ed Eoonomia (Ferrara)*, 1, 3–12.

[665] Pillai, R. N. (1991). Semi-Pareto processes. *Journal of Applied Probability*, 28, 461-465.

[666] Pillai, R. N., Jose, K. K. and Jayakumar, K. (1995). Autoregressive minification processes and the class of distributions of universal geometric minima. *J. Indian Statist. Assoc.*, 33, 53-61.

[667] Pisarewska, H. (1982). Conjugate classes for Pareto distributions. *Zeszyty Nauk. Politech. Ldz. Mat.*, No. 15, 49-58.

[668] Pizzetti, E. (1955). Osservazioni sul calcolo aritmetico del rapporto di concentrazione. *Studi in Onore di Gaetano Pietra, Biblioteca di Statistica, n. 1.*, Cappelli, Bologna, 580–593.

[669] Podder, C. K., Roy, M. K., Bhuiyan, K. J. and Karim, A. (2004). Minimax estimation of the parameter of the Pareto distribution under quadratic and MLINEX loss functions. *Pakistan J. Statist.*, 20, 137-149.

[670] Porru, E. (1912). La concentrazione della ricchezza nelle diverse regioni d'Italia. *Studi Ecanomico-Giurdici della Reale Universita di Cagliari*, 4, 79–115.

[671] Prakash, G., Singh, D. C. and Singh, R. D. (2006). Some test estimators for the scale parameter of classical Pareto distribution. *J. Statist. Res.*, 40, 41-54.

[672] Preda, V. and Ciumara, R. (2007). Convergence rates in empirical Bayes problems with a weighted squared-error loss. The Pareto distribution case. *Rev. Roumaine Math. Pures Appl.*, 52, 673-682.

[673] Proschan, F. and Sullo, P. (1976). Estimating the parameters of a multivariate exponential distribution. *J. Amer. Statist. Assoc.*, 71, 465-472.

[674] Pusz, J. (1989). On characterization of generalized logistic and Pareto distributions. *Stability problems for stochastic models, Lecture Notes in Math.*, 1412, Springer, Berlin, 288-295.

[675] Pyke, R. (1965). Spacings. *Journal of the Royal Statistical Society, Series B*, 27, 395–436.

[676] Quandt, R. E. (1966). Old and new methods of estimation and the Pareto distribution. *Metrika*, 10, 55–82.

[677] Rachev, S. T. and Resnick, S. (1991). Max-geometric infinite divisibility and stability. *Comm. Statist. Stochastic Models*, 7, 191-218.

[678] Radouane, O. and Cretois, E. (2002). Neyman smooth tests for the generalized Pareto distribution. *Comm. Statist. Theory Methods*, 31, 1067-1078.

[679] Rahman, M. and Pearson, L. M. (2003). A note on estimating parameters in two-parameter Pareto distributions. *Internat. J. Math. Ed. Sci. Tech.*, 34, 298-306.

[680] Rakonczai, P. and Zempléni, A. (2011). Bivariate generalized Pareto distribution in practice: Models and estimation. *Environmetrics*, 23, 219–227.

[681] Ramachandran, B. (1977). On the strong Markov property of the exponential laws. *Proceedings of Colloquium on Methods of Complex Analysis in the Theory of Probability and Statistics*, Debrecen, Hungary.

[682] Ramberg, J. S. and Schmeiser, B. W. (1974) An approximate method for generating asymmetric random variables. *Commun. ACM*, 17, 78–82.

[683] Ramsay, C. M. (2003). A solution to the ruin problem for Pareto distributions. *Insurance Math. Econom.*, 33, 109-116.

[684] Ranadive, K. R. (1965). The “equality” of incomes in India. *Bulletin of the Oxford Institute of Statistics*, 27, 119–134.

[685] Rao, C. R. (1985). Weighted distributions arising out of methods of ascertainment: What population does a sample represent? In *A Celebration of Statistics*, Springer, New York, 543-569.

[686] Rao, C. R. and Shanbhag, D. N. (1986). Recent results on characterization of probability distributions: A unified approach through extensions of Denys theorem. *Advances in Applied Probability*, 18, 660–678.

[687] Rao, C. R. and Zhao, L. C. (1995). Strassen’s law of the iterated logarithm for the Lorenz curves. *J. Multivariate Anal.*, 54, 239–252.

[688] Rao Tummala, V. M. (1977). Minimum expected loss estimators of parameters of Pareto distribution. Proceedings of the Business and Economics Section of the American Statistical Association, 357–358.

[689] Raqab, M. Z., Ahmadi, J. and Doostparast, M. (2007). Statistical inference based on record data from Pareto model. *Statistics*, 41, 105-118.

[690] Raqab, M. Z., Asgharzadeh, A. and Valiollahi, R. (2010). Prediction for Pareto distribution based on progressively type-II censored samples. *Comput. Statist. Data Anal.*, 54, 1732-1743.

[691] Raqab, M. Z. and Awad, A. M. (2000). Characterizations of the Pareto and related distributions. *Metrika*, 52, 63-67.

[692] Raqab, M. Z. and Awad, A. M. (2001). A note on characterization based on Shannon entropy of record statistics. *Statistics*, 35, 411-413.

[693] Rasche, R. H., Gaffney, J., Koo, A. Y. C. and Obst, N. (1980). Functional forms for estimating the Lorenz curve. *Econometrica*, 48, 1061–1062.

[694] Ratnaparkhi, M. V. (1981). Some bivariate distributions of (X,Y) where the conditional distribution of Y, given X, is either beta or unit-gamma. In *Statistical Distributions in Scientific Work, Vol. 4*, C. Taillie, G. P. Patil and B. Baldessari, eds. Reidel, Dordrecht, 389–400.

[695] Ravi, S. (2010). On a characteristic property of generalized Pareto distributions, extreme value distributions and their max domains of attraction. *Statist. Papers*, 51, 455-463.

[696] Reed, W. J. (2003). The Pareto law of incomes–an explanation and an extension. *Physica A*, 319, 469–486.

[697] Reed, W. J. and Jorgensen, M. (2004). The double Pareto-lognormal distribution: A new parametric model for size distributions. *Communications in Statistics: Theory and Methods*, 33, 1733-1753.

[698] Renyi, A. (1953). On the theory of order statistics. *Acta Mathematica Acad. Sci. Hung.*, 4, 191–231.

[699] Republic of Botswana (1976). The rural income distribution survey in Botswana 1974-5. Central Statistics Office, Ministry of Finance and Planning, Republic of Botswana.

[700] Resnick, S. (1973). Limit laws for record values. *Stochastic Processes Appl.*, 1, 67–82.

[701] Revankar, N. S., Hartley, M. J. and Pagano, M. (1974). A characterization of the Pareto distribution. *Annals of Statistics*, 2, 599–601.

[702] Rezaei, S., Tahmasbi, R. and Mahmoodi, M. (2010). Estimation of P[Y¡X] for generalized Pareto distribution. *J. Statist. Plann. Inference*, 140, 480-494.

[703] Rider, P. R. (1964). Distribution of product and of quotient of maximum values in samples from a power function population. *Journal of the American Statistical Association*, 59, 877–880.

[704] Rizzo, M. L. (2009). New goodness-of-fit tests for Pareto distributions. *Astin Bull.*, 39, 691-715.

[705] Robertson, C. A. (1972). Analyses of forest fire data in California, Technical Report 11, Department of Statistics, University of California, Riverside, USA.

[706] Rodriguez, R. N. (1977). A guide to the Burr type XII distributions. *Biometrika*, 64, 129–134.

[707] Roehner, B. and Winiwarter, P. (1985) Aggregation of independent Paretian random variables. *Adv. in Appl. Probab.*, 17, 465469.

[708] Rogers, G. S. (1959). A note on stochastic independence of functions of order statistics. *Annals of Mathematical Statistics*, 30, 1263–1264.

[709] Rogers, G. S. (1963). An alternative proof of the characterization of the density Ax^B. *American Mathematical Monthly*, 70, 857–8.

[710] Rohatgi, V. K. and Saleh, A. K. M. E. (1987). Estimation of the common scale parameter of two Pareto distributions in censored samples. *Naval Res. Logist.*, 34, 235-238.

[711] Rohde, N. (2009). An alternative functional form for estimating the Lorenz curve. *Econom. Lett.*, 105, 61-63.

[712] Rootzen, H. and Tajvidi, N. (2006). Multivariate generalized Pareto distributions. *Bernoulli*, 12, 917-930.

[713] Rosenblatt, M. (1952). Remarks on a multivariate transformation. *Ann. Math. Statistics*, 23, 470-472.

[714] Rossberg, H-J. (1972a). Characterization of the exponential and the Pareto distributions by means of some properties of the distributions which the differences and quotients of order statistics are subject to. *Math. Operationsforsch. u. Statist.*, 3, 207–216.

[715] Rossberg, H-J. (1972b). Characterization of distribution functions by the independence of certain functions of order statistics. *Sankhya, Series A*, 34, 111–120.

[716] Rothschild, M. and Stiglitz, J. E. (1973). Some further results on the measurement of inequality. *Journal of Economic Theory*, 6, 188–204.

[717] Ruiz, J. M. and Navarro, J. (1996). Characterizations based on conditional expectations of the doubled truncated distribution. *Ann. Inst. Statist. Math.*, 48, 563-572.

[718] Rutherford, R. S. G. (1955). Income distributions: A new model. *Econometrica*, 23, 277–294.

[719] Sahin, I. and Hendrick, D. J. (1978). On strike durations and a measure of termination. *Applied Statistics*, 27, 319–24.

[720] Saksena, S. K. (1978). *Estimation of parameters in a Pareto distribution and simultaneous comparison of estimators*. Ph.D. Thesis, Louisiana Tech. University, Ruston, Louisiana.

[721] Saksena, S. K. and Johnson, A. M. (1984). Best unbiased estimators for the parameters of a two-parameter Pareto distribution. *Metrika*, 31, 77-83.

[722] Saldaña-Zepeda, D. P., Vaquera-Huerta, H. and Arnold, B. C. (2010). A goodness of fit test for the Pareto distribution in the presence of Type II censoring, based on the cumulative hazard function. *Comput. Statist. Data Anal.*, 54, 833-842.

[723] Saleh, A. K. M. E., Ali, M. M. and Umbach, D. (1985). Large sample estimation of Pareto quantiles using selected order statistics. *Metrika*, 32, 49-56.

[724] Salem, A. B. Z. and Mount, T. D. (1974). A convenient descriptive model of income distribution: The gamma density. *Econometrica*, 42, 1115–1127.

[725] Samanta, M. (1972). Characterization of the Pareto distribution. *Skand. Aktuarietidskr*, 55, 191–2.

[726] Samuel, P. (2003). Some characterizations of a Pareto distribution by generalized order statistics. *Calcutta Statist. Assoc. Bull.*, 54, 269-274.

[727] Sandhya, E. and Satheesh, S. (1996). On distribution functions with completely monotone derivative. *Statist. Probab. Lett.*, 27, 127-129.

[728] Sankaran, P. G. and Unnikrishnan Nair, N. (1993a). A bivariate Pareto model and its applications to reliability. *Naval Res. Logist.*, 40, 1013-1020.

[729] Sankaran, P. G. and Unnikrishnan Nair, N. (1993b). Characterizations by properties of residual life distributions. *Statistics*, 24, 245-251.

[730] Sarabia, J. M. (1997). A hierarchy of Lorenz curves based on the generalized Tukey's lambda distribution. *Econometric Reviews*, 16, 305–320.

[731] Sarabia, J. M. (2008a). Parametric Lorenz curves: Models and applications. In *Modeling Income Distributions and Lorenz Curves*, D. Chotikapanich, ed. Springer, New York, Pages 167–190.

[732] Sarabia, J. M. (2008b). A general definition of the Leimkuhler curve. *Journal of Infometrics*, 2, 156–163.

[733] Sarabia, J. M. (2013) Building new Lorenz cuves from old ones. Personal communication.

[734] Sarabia, J. M., Castillo, E., Pascual, M. and Sarabia, M. (2005). Mixture Lorenz curves. *Economics Letters*, 89, 89–94.

[735] Sarabia, J. M., Castillo, E. and Slottje, D. J. (1999). An ordered family of Lorenz curves. *Journal of Econometrics*, 91, 43–60.

[736] Sarabia, J. M., Castillo, E. and Slottje, D. J. (2001). An exponential family of Lorenz curves. *Southern Economic Journal*, 67, 748–756.

[737] Sarabia, J. M., Gómez-Déniz, E., Sarabia, M. and Prieto, F. (2010). A general method for generating parametric Lorenz and Leimkuhler curves. *Journal of Infometrics*, 4, 524–539.

[738] Sarabia, J. M. and Jordá, V. (2013). Modeling Bivariate Lorenz Curves with Applications to Multidimensional Inequality in Well-Being. *Fifth ECINEQ Meeting, Bari, Italy*, 201. Document available at: http://www.ecineq.org/ecineq_bari13/documents/booklet05.pdf.

[739] Sarabia, J. M., Jordá, V. and Trueba, C. (2013). The Lamé class of Lorenz curves. *Comm. Statist. Theory Methods*, to appear.

[740] Sarabia, J. M. and Pascual, M. (2002). A class of Lorenz curves based on linear exponential loss functions. *Comm. Statist. Theory Methods*, 31, 925–942.

[741] Sarabia, J. M., Prieto, F. and Sarabia, M. (2010). Revisiting a functional form for the Lorenz curve. *Economics Letters*, 107, 249–252.

[742] Saran, J. and Pandey, A. (2004). Recurrence relations for marginal and joint moment generating functions of generalized order statistics from Pareto distribution. *Stat. Methods*, 6, 12-21.

[743] Sargan, J. D. (1957). The distribution of wealth. *Econometrica*, 25, 568–590.

[744] Sarhan, A. E., Greenberg, B. G. and Ogawa, J. (1960). Simplified estimates for the exponential distribution. Interim Technical Report #1, grant #DA-ORD-7, University of North Carolina, Chapel Hill, North Carolina.

[745] Schur, I. (1923). Uber eine klasse von mittelbildungen mit anwendungen die determinaten. *Theorie Sitzungsber Berlin Math. Gesellschaft*, 22, 9–20.

[746] Schutz, R. R. (1951). On the measurement of income inequality. *American Economic Review*, 41, 107–122.

[747] Schmittlein, D. C. (1983). Some sampling properties of a model for income distribution. *Journal of Business and Economic Statistics*, 1, 147–153.

[748] Schmittlein, D. C. and Morrison, D. G. (1981). The median residual lifetime: A characterization theorem and an application. *Oper. Res.*, 29, 392-399.

[749] Schwartz, J. E. (1985). The utility of the cube root of income. *Journal of Official Statistics*, 1, 5–19

[750] Seailles, J. (1910). *La Repartition des Fortunes en France*. F. Alcan, Paris.

[751] Seal, H. L. (1952). The maximum likelihood fitting of the discrete Pareto law. *Journal of the Institute of Actuaries*, 78, 115–121.

[752] Seal, H. L. (1980). Survival probabilities based on Pareto claim distributions. *Astin Bulletin*, 11, 61–71.

[753] Sen, A. K. (1976). Poverty: An ordinal approach to measurement. *Econometrica*, 44, 219–231.

[754] Sendler, W. (1979). On statistical inference in concentration measurement. *Metrika*, 26, 109-122.

[755] Seto, N. and Iwase, K. (1982). Uniformly minimum variance unbiased estimation of quantiles and probabilities for the Pareto distribution of the first kind. *J. Japan Statist. Soc.*, 12, 105-112.

[756] Shaked, M. (1980). On mixtures from exponential families. *J. Roy. Statist. Soc. Ser. B*, 42, 192-198.

[757] Shaked, M. (1982). Dispersive ordering of distributions. *J. Appl. Probab.*, 19, 310-320.

[758] Shaked, M. and Shanthikumar, J. G. (1998). Two variability orders. *Probab. Engrg. Inform. Sci.*, 12, 1–23.

[759] Shaked, M. and Shanthikumar, J. G. (2007). *Stochastic Orders*. Springer, New York.

[760] Shanmugam, R. (1987). Estimating the fraction of population in an income bracket using Pareto distribution. *Rebrape*, 1, 139-156.

[761] Shawky, A. I. and Abu-Zinadah, H. H. (2008). Characterizations of the exponentiated Pareto distribution based on record values. *Appl. Math. Sci. (Ruse)*, 2, 1283-1290.

[762] Shawky, A. I.; Abu-Zinadah, H. H. (2009). Exponentiated Pareto distribution: Different method of estimations. *Int. J. Contemp. Math. Sci.*, 4, 677-693.

[763] Sheshinski, E. (1972). Relation between a social welfare function and the Gini index of income inequality. *Journal of Economic Theory*, 4, 98–100.

[764] Shimizu, R. (1979). On a lack of memory property of the exponential distribution. *Annals of Institute of Statistical Mathematics*, 31, 309–313.

[765] Shirras, G. F. (1935). The Pareto law and the distribution of income. *Economic Journal*, 45, 663–681.

[766] Shorrocks, A. F. (1975). On stochastic models of size distributions. *Review of Economic Studies*, 43, 631–641.

[767] Shorrocks, A. F. (1980). The class of additively decomposable inequality measures. *Econometrica*, 48, 613–625.

[768] Shorrocks, A. F. (1983). Ranking income distributions. *Economica*, 50, 3–17.

[769] Shorrocks, A. F. and Foster, J. E. (1987). Transfer sensitive inequality measures. *Rev. Econom. Stud.*, 54, 485-497.

[770] Shortle, J., Gross, D., Fischer, M. and Masi, D. (2006). Using the Pareto distribution in queueing modeling. *Journal Probab. Stat. Sci.*, 4, 85–98.

[771] Silcock, H. (1954). The phenomenon of labour turnover. *Journal of the Royal Statistical Society, Series A*, 117, 429–440.

[772] Simon, H. A. (1955). On a class of skew distribution functions. *Biometrika*, 52, 425–440.

[773] Simon, H. A. (1960). Some further notes on a class of skew distribution functions. *Information and Control*, 3, 80–88.

[774] Simon, H. A. and Bonini, C. P. (1958). The size distribution of business firms. *American Economic Review*, 48, 607–617.

[775] Simpson, E. H. (1949). Measurement of diversity. *Nature*, 163, 688.

[776] Singh, D. C., Prakash, G. and Singh, P. (2007). Shrinkage testimators for the shape parameter of Pareto distribution using LINEX loss function. *Comm. Statist. Theory Methods*, 36, 741-753.

[777] Singh, S. K. and Maddala, G. S. (1975). A stochastic process for income distribution and tests of income distribution functions. ASA Proceedings of Business and Economics Section, 551–3.

[778] Singh, S. K. and Maddala, G. S. (1976). A function for size distribution of incomes. *Econometrica*, 44, 963–970.

[779] Singh, S. K. and Maddala, G. S. (1978). A function for size distribution of incomes: Reply. *Econometrica*, 46, 461.

[780] Singh, V. P. and Ahmad, M. (2004). A comparative evaluation of the estimators of the three-parameter generalized Pareto distribution. *J. Stat. Comput. Simul.*, 74, 91-106.

[781] Sinha, S. K. and Howlader, H. A. (1980). On the sampling distributions of the Bayesian estimators of the Pareto parameter with proper and improper priors and associated goodness of fit. Technical Report #103, Department of Statistics, University of Manitoba, Winnipeg, Canada.

[782] Sordo, M. A., Navarro, J. and Sarabia, J. M. (2013). Distorted Lorenz curves: Models and comparisons. *Social Choice and Welfare*, 42, 761–780.

[783] Srivastava, M. S. (1965). A characterization of Pareto's distribution and $(k+1)x^k/\theta^{k+1}$ (Abstract). *Annals of Mathematical Statistics*, 36, 361–2.

[784] Stacy, E. W. (1962). A generalization of the gamma distribution. *Annals of Mathematical Statistics*, 33, 1187–1192.

[785] Steindl, J. (1965). *Random Processes and the Growth of Firms: A Study of the Pareto Law*. Hafner, New York.

[786] Steutel, F. W. (1969). Note on completely monotone densities. *Annals of Mathematical Statistics*, 40, 1130–1131.

[787] Stigler, S. M. (1974). Linear functions of order statistics with smooth weight functions. *Annals of Statistics*, 2, 676–693.

[788] Stiglitz, J. E. (1969). Distribution of income and wealth among individuals. *Econometrica*, 37, 382–397.

[789] Strassen, V. (1965). The existence of probability measures with given marginals. *Annals of Mathematical Statistics*, 36, 423–439.

[790] Sukhatme, P. V. (1937). Tests of significance for samples of the χ^2 population with two degrees of freedom. *Annals of Eugenics*, 8, 52–56.

[791] Sullo, P. and Rutherford, D. (1977). Characterizations of the power distribution by conditional exceedence. Research Report 37-77-P6, Renesselaer Poly. Inst., Troy, New York.

[792] Tadikamalla, P. R. (1980). A look at the Burr and related distributions. *Internat. Statist. Rev.*, 48, 337–344.

[793] Taguchi, T. (1964). On Pareto's distribution and curves. *Proceedings of the Institute of Statistical Mathematics*, 12, 293–313.

[794] Taguchi, T. (1967). On some properties of concentration curve and its applications. *Metron*, 26, 381–395.

[795] Taguchi, T.(1968). Concentration-curve methods and structures of skew populations. *Annals of Institute of Statistical Mathematics*, 20, 107–141.

[796] Taguchi, T. (1972a). On the two-dimensional concentration surface and extensions of concentration coefficient and Pareto distribution to the two dimensional case-I. *Annals of Institute of Statistical Mathematics*, 24, 355–382.

[797] Taguchi, T. (1972b). On the two-dimensional concentration surface and extensions of concentration coefficient and Pareto distribution to the two dimensional case-II. *Annals of Institute of Statistical Mathematics*, 24, 599–619.

[798] Taguchi, T. (1973). On the two-dimensional concentration surface and extensions of concentration coefficient and Pareto distribution to the two dimensional case-III. *Annals of Institute of Statistical Mathematics*, 25, 215–237.

[799] Taguchi, T. (1978). On an unbiased, consistent and asymptotically efficient estimate of Gini's concentration coefficient. *Metron*, 36, 57–72.

[800] Taguchi, T. (1980). On a multi-dimensional mean difference and the multi-dimensional regression and the multiple correlation coefficient on it. A seminar at the University of Rome, May 28, 1980.

[801] Taguchi, T. (1987). On the structure of multivariate concentration - Some relationships among the concentration surface and two variate mean difference and regressions. *Computational Statistics and Data Analysis*, 6, 307–334.

[802] Taillie, C. (1979). Species equitability: A comparative approach. In *Ecological Diversity in Theory and Practice*, J. F. Grassle, G. P. Patil, W. Smith and C. Taillie, eds. International Co-operative Publishing House, Fairland, Maryland. Pages 51–62.

[803] Taillie, C. (1981). Lorenz ordering within the generalized gamma family of income distributions. In *Statistical Distributions in Scientific Work, Vol. 6*, C. Taillie, G. P. Patil and B. Baldessari, eds. Reidel, Dordrecht, Holland. 181–192.

[804] Takahasi, K. (1965). Note on the Multivariate Burr's Distribution. *Annals of the Institute of Statistical Mathematics*, 17, 257–260.

[805] Takano, K. (1992). On relations between the Pareto distribution and a differential equation. *Bull. Fac. Sci. Ibaraki Univ. Ser. A*, 24, 15-30.

[806] Takano, K. (2007). On the confluent hypergeometric function coming from the Pareto distribution. *General Mathematics*, 15 (2007), no. 2, 6476.

[807] Takayama, N. (1979). Poverty, income inequality, and their measures: Professor Sen's axiomatic approach reconsidered. *Econometrica*, 47, 747–759.

[808] Talwalker, S. (1980). On the property of dullness of Pareto distribution. *Metrika*, 27, 115–119.

[809] Targhetta, M. L. (1979). Confidence intervals for a one parameter family of multivariate distributions, *Biometrika*, 66, 687–688.

[810] Tarsitano, A. (1990). The Bonferroni index of income inequality. In *Income and Wealth Distribution, Inequality, and Poverty*, C. Dagum and M. Zenga, eds. Springer-Verlag, Berlin, 228–242.

[811] Tavangar, M. and Asadi, M. (2007). Generalized Pareto distributions characterized by generalized order statistics. *Comm. Statist. Theory Methods*, 36, 1333-1341.

[812] Tavangar, M. and Asadi, M. (2008). On a characterization of generalized Pareto distribution based on generalized order statistics. *Comm. Statist. Theory Methods*, 37, 1347-1352.

[813] Teodorescu, S. (2010). On the truncated composite lognormal-Pareto model. *Math. Rep. (Bucur.)*, 12(62), 71-84.

[814] Teodorescu, S. and Vernic, R. (2006). A composite exponential-Pareto distribution. *An. Stiint. Univ. "Ovidius" Constanta Ser. Mat.*, 14, 99-108.

[815] Thirugnanasambanthan, K. (1980). *Functional equations associated with Markov models for labor mobility and with distribution theory*. Ph. D. Dissertation, Department of Statistics, University of California, Riverside, California.

[816] Thomas, A. and Jose, K. K. (2003). Marshall-Olkin Pareto processes. *Far East J. Theor. Stat.*, 9, 117-132.

[817] Thomas, A. and Jose, K. K. (2004). Bivariate semi-Pareto minification processes. *Metrika*, 59, 305-313.

[818] Thomas, D. L. (1976). Reciprocal moments of linear combinations of exponential variates. *Journal of the American Statistical Association*, 71, 506–512.

[819] Thompson, W. A., Jr. (1976). Fisherman's luck. *Biometrics*, 32, 265–271.

[820] Thorin, O. (1977). On the infinite divisibility of the Pareto distribution. *Scandinavian Actuarial Journal*, 31–40.

[821] Tierney, L. and Kadane, J. B. (1986). Accurate approximations for posterior moments and marginal densities. *J. Amer. Statist. Assoc.*, 81, 82-86.

[822] Tiwari, R. C. and Zalkikar, J. N. (1990). Empirical Bayes estimation of the scale parameter in a Pareto distribution. *Comput. Statist. Data Anal.*, 10, 261-270.

[823] Tong, H. (1974). A note on the estimation of Pr(X¡Y) in the exponential case. *Technometrics*, 16, 625.

[824] Trader, R. L. (1985). Bayesian inference for truncated exponential distributions. *Comm. Statist. Theory Methods*, 14, 585-592.

[825] Tripathi, R. C. and Gurland, J. (1976). A general family of discrete distributions with hypergeometric probabilities. Technical Report #459, Department of Statistics, University of Wisconsin, Madison, Wisconsin.

[826] Tsokos, C. and Nadarajah, S. (2003) Extreme value models for software reliability. *Stochastic Anal. Appl.*, 21, 719-735.

[827] Turnbull, B. W., Brown, B. W., Jr. and Hu, M. (1974). Survivorship analysis of heart transplant data. *Journal of the American Statistical Association*, 69, 74–80.

[828] Uppuluri, V. R. R. (1981). Some properties of the log-Laplace distribution. In *Statistical Distributions in Scientific Work, Vol. 4*, C. Taillie, G. P. Patil and B. Baldessari, eds. Reidel, Dordrecht, Holland, 105–110.

[829] Vannman, K. (1976). Estimators based on order statistics from a Pareto distribution. *Journal of the American Statistical Association*, 71, 704–8.

[830] van Zyl, J. M. (2012). A median regression model to estimate the parameters of the three-parameter generalized Pareto distribution. *Comm. Statist. Simulation Comput.*, 41, 544-553.

[831] Vartak, M. N. (1974). Characterization of certain classes of probability distributions. *Journal of the Indian Statistical Association*, 12, 67–74.

[832] Vartia, P. L. I. and Vartia, Y. O. (1980). Description of the income distribution by the scaled *F* distribution model. In *The Statics and Dynamics of Income*, N. A. Klevmarken and J. A. Lybeck, eds. Tieto Ltd., Clevedon, Avon.

[833] Vernic, R. (2011). Tail conditional expectation for the multivariate Pareto distribution of the second kind: Another approach. *Methodol. Comput. Appl. Probab.*, 13, 121-137.

[834] Verway, D. I. (1966). A ranking of states by inequality using census and tax data. *Review of Economics and Statistics*, 48, 314–321.

[835] Villaseñor, J. A. and Arnold, B. C. (1984a). The general quadratic Lorenz curve. Technical Report, Colegio de Postgraduados, Chapingo, Mexico.

[836] Villaseñor, J. A. and Arnold, B. C. (1984b). Some examples of fitted general quadratic Lorenz curves. Technical Report 130, Statistics Department, University of California, Riverside.

[837] Villaseñor, J. A. and Arnold, B. C. (1989). Elliptical Lorenz curves. *J. Econometrics*, 40, 327-338.

[838] Villaseñor, J. A. and Gonzalez, E. (2009). A bootstrap goodness of fit test for the generalized Pareto distribution. *Comput. Statist. Data Anal.*, 53, 3835-3841.

[839] Voinov, V. G. and Nikulin, M. S. (1993) *Unbiased estimators and their applications. Vol. 1. Univariate case.* Translated from the 1989 Russian original by L. E. Strautman and revised by the authors. *Mathematics and its Applications*, 263. Kluwer, Dordrecht.

[840] von Bortkiewicz, L. (1931). Die disparitatsmasse der einkommenstatistik. *Bulletin de l'Institut International de Statistique*, 25, 189–298.

[841] Voorn, W. J. (1987). Characterization of the logistic and loglogistic distributions by extreme value related stability with random sample size. *J. Appl. Probab.*, 24, 838-851.

[842] Walther, A. (1926). Anschauliches zur Riemannschen zetafunktion. *Acta. Math.*, 48, 393–400.

[843] Wang, Y. H. (1971). Sequential estimation of the scale parameter of the Pareto distribution. Technical Report 71-6, Ohio State University, Columbus, Ohio.

[844] Wang, Y. H. and Srivastava, R. C. (1980). A characterization of the exponential and related distributions by linear regression. *Annals of Statistics*, 8, 217–220.

[845] Wani, J. K. and Kabe, D. G. (1972). Distributions of linear functions of ordered rectangular variates. *Skand. Aktuarietidskr*, 71, 58–60.

[846] Watkins, G. P. (1905). Comment on the method of measuring concentration of wealth. *Publications of the American Statistical Association*, 9, 349–355.

[847] Watkins, G. P. (1908). An interpretation of certain statistical evidence of concentration of wealth. *Publications of the American Statistical Association*, 11, 27–55.

[848] Wei, L. and Yang, H-L. (2004) Explicit expressions for the ruin probabilities of Erlang risk processes with Pareto individual claim distributions. *Acta Mathematicae Applicatae Sinica. English Series*, 20, 495–506.

[849] Weissman, I. (1975). Multivariate extremal processes generated by independent non-identically distributed random variables. *Journal of Applied Probability*, 12, 477–487.

[850] Wesolowski, J. (1995). Bivariate distributions via a Pareto conditional distribution and a regression function. *Ann. Inst. Statist. Math.*, 47, 177-183.

[851] Wesolowski, J. and Ahsanullah, M. (1995). Conditional specifications of multivariate Pareto and Student distributions. *Comm. Statist. Theory Methods*, 24, 1023-1031.

[852] Weymark, J. A. (1979). Generalized Gini inequality indices. Discussion paper #79-12, Department of Economics, University of British Columbia, Vancouver, British Columbia.

[853] Whitt, W. (1976). Bivariate distributions with given marginals. *Annals of Statistics*, 4, 1280–1289.

[854] Whitt, W. (1980). The effect of variability in the GI/G/s queue. *Journal of Applied Probability*, 17, 1062–1071.

[855] Wicksell, S. D. (1933). On correlation functions of type III. *Biometrika*, 25, 121–133.

[856] Wilfling, B. (1996a). A sufficient condition for Lorenz ordering. *Sankhya*, B58, 62-69.

[857] Wilfling, B. (1996b) Lorenz ordering of power-function order statistics. *Statist. Probab. Lett.*, 30, 313-319.

[858] Wilks, S. S. (1959). Recurrence of extreme observations. *Journal of Australian Mathematical Society*, 1, 106–112.

[859] Wilks, S. S. (1962). *Mathematical Statistics*. Wiley, New York.

[860] Willis, J. C. (1922). *Age and Area*. Cambridge University Press, Cambridge.

[861] Winkler, W. (1924). Einkommen. *Handworterbuch der Staatswissenschaften*, 4, 367–401.

[862] Wise, M. (1980). Income distributions: Gamma, lognormal, inverse Gaussian or Pareto? Personal communication.

[863] Wold, H. (1935). A study on the mean difference, concentration curves and concentration ratio. *Metron*, 12, 39–58.

[864] Wold, H. O. A. and Whittle, P. (1957). A model explaining the Pareto distribution of wealth. *Econometrica*, 25, 591–5.

[865] Wong, A. (2012). Interval estimation of P(Y¡X) for generalized Pareto distribution. *J. Statist. Plann. Inference*, 142, 601-607.

[866] Wu, J-W. and Lee, W-C. (2001). On the characterization of generalized extreme value, power function, generalized Pareto and classical Pareto distributions by conditional expectation of record values. *Statist. Papers*, 42, 225-242.

[867] Wu, J-W., Lee, W-C. and Chen, S-C. (2006). Computational comparison for weighted moments estimators and BLUE of the scale parameter of a Pareto distribution with known shape parameter under type II multiply censored sample. *Appl. Math. Comput.*, 181, 1462-1470.

[868] Wu, J-W., Lee, W-C. and Chen, S-C. (2007a). Computational comparison of the prediction intervals of future observation for three-parameter Pareto distribution with known shape parameter. *Appl. Math. Comput.*, 190, 150-178.

[869] Wu, J-W., Lee, W-C. and Chen, S-C. (2007b). Computational comparison of prediction future lifetime of electronic components with Pareto distribution based on multiply type II censored samples. *Appl. Math. Comput.*, 184, 374-406.

[870] Wu, S-F. (2008). Interval estimation for a Pareto distribution based on a doubly type II censored sample. *Comput. Statist. Data Anal.*, 52, 3779-3788.

[871] Wu, S-F. (2010). Interval estimation for the Pareto distribution based on the progressive type II censored sample. *J. Stat. Comput. Simul.*, 80, 463-474.

[872] Wu, S-J. (2003). Estimation for the two-parameter Pareto distribution under progressive censoring with uniform removals. *J. Stat. Comput. Simul.*, 73, 125-134.

[873] Wu, S-J. and Chang, C-T. (2003). Inference in the Pareto distribution based on progressive type II censoring with random removals. *J. Appl. Stat.*, 30, 163-172.

[874] Xekalaki, E. and Dimaki, C. (2005). Identifying the Pareto and Yule distributions by properties of their reliability measures. *J. Statist. Plann. Inference*, 131, 231-252.

[875] Xekalaki, E. and Panaretos, J. (1988). On the association of the Pareto and the Yule distribution. (Russian) *Teor. Veroyatnost. i Primenen.*, 33, 206–210; translation in *Theory Probab. Appl.*, 33, 191195.

[876] Yeh, H-C. (1983). Pareto Processes. Unpublished Ph.D. dissertation, University of California, Riverside.

[877] Yeh, H-C. (1990). Some bivariate Pareto distributions generated by geometric maximization procedures. *Bull. Inst. Math. Acad. Sinica*, 18, 213-232.

[878] Yeh, H-C. (1994). Some properties of the homogeneous multivariate Pareto (IV) distribution. *J. Multivariate Anal.*, 51, 46-53.

[879] Yeh, H-C. (2002). Six multivariate Zipf distributions and their related properties. *Statist. Probab. Lett.*, 56, 131-141.

[880] Yeh, H-C. (2004a). Some properties and characterizations for generalized multivariate Pareto distributions. *J. Multivariate Anal.*, 88, 47-60.

[881] Yeh, H-C. (2004b). The generalized Marshall-Olkin type multivariate Pareto distributions. *Comm. Statist. Theory Methods*, 33, 1053-1068.

[882] Yeh, H-C. (2007). Three general multivariate semi-Pareto distributions and their characterizations. *J. Multivariate Anal.*, 98, 1305-1319.

[883] Yeh, H-C. (2009). Some related minima stability and minima infinite divisibility of the general multivariate Pareto distributions. *Comm. Statist. Theory Methods*, 38, 497-510.

[884] Yeh, H-C., Arnold, B. C. and Robertson, C. A. (1988). Pareto processes. *J. Appl. Probab.*, 25, 291-301.

[885] Yitzhaki, S. (1983). On an extension of the Gini inequality index. *International Economic Review*, 24, 617–628.

[886] Yntema, D. B. (1933). Measures of the inequality in the personal distribution of wealth or income. *Journal of the American Statistical Association*, 28, 423–433.

[887] Yule, G. U. (1924). A mathematical theory of evolution, based on the conclusions of Dr. J. C. Willis, F. R. S. *Philosophical Transactions B*, 213, 21.

[888] Zahedi, H. (1985). Some new classes of multivariate survival distribution functions. *J. Statist. Plann. Inference*, 11, 171-188.

[889] Zandonatti, A. (2001). Distribuzioni de Pareto generalizzate. Tesi di Laurea. Department of Economics, University of Trento, Italy.

[890] Zellner, A. (1971). *An Introduction to Bayesian Inference in Econometrics*. Wiley, New York.

[891] Zellner, A. (1986). On assessing prior distributions and Bayesian regression analysis with g-prior distributions. in *Bayesian Inference and Decision Techniques, Stud. Bayesian Econometrics Statist., 6.*, North-Holland, Amsterdam, 233-243.

[892] Zenga, M. (1984). Proposta per un indice di concentrazione basato sui rapporti tra quantili di popolazione e quantili reddito. *Giornale degli Economisti e Annali di Economia*, 48, 301–326.

[893] Zenga, M. (2007) Inequality curve and inequality index based on the ratios between lower and upper arithmetic means. *Statistica e Applicazioni*, 4, 3–27.

[894] Zhang, J. (2007). Likelihood moment estimation for the generalized Pareto distribution. *Aust. N. Z. J. Stat.*, 49, 69-77.

[895] Zhang, J. (2010). Improving on estimation for the generalized Pareto distribution. *Technometrics*, 52, 335-339.

[896] Zhang, J. and Stephens, M. A. (2009). A new and efficient estimation method for the generalized Pareto distribution. *Technometrics*, 51, 316-325.

[897] Zidek, J. V. (1973). Estimating the scale parameter of the exponential distribution with unknown location. *Ann. Statist.*, 1, 264-278.

[898] Zipf, G. (1949). *Human Behavior and the Principle of Least Effort*. Addison-Wesley, Cambridge, Massachusetts.

[899] Zisheng, O. and Chi, X. (2006) Generalized Pareto distribution fit to medical insurance claims data. *Appl. Math. Journal Chinese Univ. Ser. B*, 21, 21–29.

[900] Zografos, K. and Nadarajah, S. (2005). Expressions for Renyi and Shannon entropies for multivariate distributions. *Statist. Probab. Lett.*, 71, 71-84.

Subject Index

Aaberge indices, 125
Absolute inequality measures, 5
Absolute mean deviation, 118, 175
Accelerated failure, 83
Additive utility, 200, 201
Additive welfare function, 14, 200
Admissibility, 227
Aging concepts, 149
Alpha-mixture Pareto distribution, 84
Amato index, 133, 134, 200, 202, 220
Amato index, sample, 174, 178
Amato index, series representation, 134
Anderson-Darling, 263, 269
Arc-sine distribution, 156
Association measure, 142
Asymmetrical Laplace distribution, 115
Asymptotic behavior of extreme order statistics, 72
Asymptotic behavior of non-extreme order statistics, 71, 72
Asymptotic distribution of S_n, 53, 73
Asymptotic normality, 15, 16, 137, 176–179, 182, 184–191, 227, 251, 252, 266, 267
Asymptotically best linear unbiased estimates (ABLUE's), 261
Asymptotically best linear unbiased predictors (ABLUP's), 276
Averaging, 206, 207

Balls-in-boxes model, 33
Bartlett's test, 268
Bayes estimates, 236–239, 249, 263, 264, 347
Bayes estimates, generalized order statistics, 246
Bayes estimates, grouped data, 245, 287
Bayes estimates, various data configurations, 243
Bayes prediction, 246, 247
Bayesian analysis, 46
Benedetti's bound, 194
Best linear unbiased estimates (BLUE's), 11, 254, 261, 262
Best linear unbiased prediction, 276
Best unbiased estimates, 227–230, 232, 233, 254, 260
Beta distribution, 11, 12, 29, 36, 41, 42
Beta-Pareto distribution, 39, 102, 296, 334
Bhandari and Mukerjee bounds, 194
Bilateral bivariate Pareto distribution, 323
Birth and death process, 25, 27, 28
Bivariate Beta distribution, 319, 321
Bivariate Beta distribution, Olkin-Liu, 319, 320
Bivariate Feller-Pareto distribution, 320
Bivariate geometric distribution, 316
Bivariate Pareto distribution, 349
Bivariate Pareto I distribution, Marshall-Olkin type, 347
Bivariate Pareto II distribution, 343
Bivariate Pareto III distribution, 46, 317, 318
Bivariate uniform distribution, 322
Bonferroni curve, 167
Bonferroni inequality index, 168

Bonferroni inequality index, sample, 178, 188
Bonferroni partial order, 168
Bose-Einstein statistics, 9, 24, 33, 34
Bowley's interquartile measure, 3, 120, 190
Bradford distribution, 12, 36
Branching process, 32
Broken stick model, 33
Building Lorenz curves, 158–161, 165
Burr distribution, 8, 9, 37, 41, 49, 57, 261
Burrits, 279

Castillo-Hadi estimates, 259
Castillo-Hadi-Sarabia estimates, 281
Castillo-Sarabia-Hadi estimates, 345
Cauchy functional equation, 99
Censored data, 254, 260, 266, 274, 275, 277
Change point, 17, 297
Characteristic function, 17
Characterizations, 78, 89, 90, 95, 99, 100, 114, 115, 271, 309, 327, 332
Chi-squared test, 269
Chotikapanich-Griffiths estimate, 282
Circular Lorenz curves, 164
Claim premiums, 17
Classical Pareto distribution, 2, 3, 11, 12, 25, 34, 35, 41, 56, 57, 70, 82–84, 92, 93, 115, 146, 147, 177, 223, 224, 227, 233, 235, 236, 260, 264, 268, 272, 277, 284, 287
Coefficient of variation, 5, 16, 119, 146, 149–151, 153–157, 226
Coefficient of variation, sample, 179, 187
Coin shower, 9, 24, 25, 33, 36
Competitive bidding for employment, 31, 33, 55, 76, 98
Compound interest, 30
Concentration measure, 16, 146
Conditional likelihood function, 344
Conditionally likelihood estimates, 344
Conditionally specified prior, 241, 243, 244, 247, 263, 288
Confidence interval, 263–267, 269
Confidence interval for α in multivariate Pareto I distribution, 347
Confidence interval, sequential, 268
confidence region, 265, 266, 269
conjugate prior, 237, 239, 241–244, 247, 249, 263, 347
Contorted Pareto distribution, 109
Convex combinations of densities, 38
Convex ordering, 218, 219
Convolution, 51, 55
Convolutions of Pareto variables, 12, 51–53
Copula, 321
Covariables, 254
Cramer-von Mises, 263, 269
Cube root model, 17, 37

Dagum (I) distribution, 152
Dalton's inequality index, 190
Decreasing failure rate, 12
Decreasing hazard rate, 16
Density crossing condition, 212
Density crossing ordering, 215
Differencing measures, 147
Diffusion model, 9, 10, 36
Dirichlet distribution, 58, 59, 183, 188, 282, 306
Discrete Pareto distribution, 9, 10, 31, 33, 110, 111, 288
Dispersion ordering, 216
Double Pareto model, 17, 21
Doubly stochastic matrix, 198, 199
Dullness property, 115

Effect of grouping, 135, 136
Effect of taxation, 204
Effects of transformations, 203
Egalitarian distribution, 3, 15, 118, 123, 150
Elemental quantile estimates, 236
Elliptical distribution, 156
Elliptical Lorenz curves, 164

Elteto and Frigyes inequality measures, 15, 129, 130, 132, 150, 151, 153, 154, 156, 157
Elteto and Frigyes inequality measures, sample, 188, 262
Elteto and Frigyes vector, 130
Emlen's measure, 192
Empirical Bayes estimation, 248, 254, 264
Empirical Bayes predictor, 248, 249
Entropy maximization, 97
Entropy measures, 121, 191
Estimating equations, 234, 254, 257
Estimation after selection, 278
Evolutionary Pareto distribution, 297
Excess wealth transform, 169
Exchange rate Lorenz ordering, 218
Exponential distribution, 9–11, 15, 16, 24, 29, 33, 34, 41, 42, 44, 56, 64, 68, 75, 78, 82, 153, 183, 187, 188, 271, 272
Exponential order statistics, 56
Exponential order statistics, Lorenz ordering, 213, 214
Exponentiated Pareto distribution, 37, 102, 296
Extended neighboring order statistics, 96
Extreme inequality, 39

Failure rate function, 71, 80, 81, 226, 332
Farley-Gumbel-Morgenstern family, 325
Feller-Pareto density, 45, 252
Feller-Pareto distribution, 9, 17, 36, 41, 43–47, 50, 55, 57, 59, 211, 250, 252, 255, 256, 270, 319
Feller-Pareto mode, 47
Feller-Pareto moments, 48
Feller-Zipf distribution, 111
File sizes, 17
Financial durations, 17
Firm sizes, 33
First moment distribution, 81
Fishlow's poverty measure, 192
Flexible multivariate Pareto model, 325, 343
Forest fires, 17
Forward recurrence time, 81
Frechet family, 324
Frechet-Hoeffding bounds, 323
Frechet-Hoeffding distribution, 324
Functional equation, 32, 79, 103, 317, 329, 331, 335

g-prior, 249
Gamma distribution, 11, 12, 24, 36, 45, 46, 154, 211, 237, 239–241, 277
Gamma distribution, k-variate, 312
Gamma-power, 239
Geeta distribution, 37
General quadratic Lorenz curves, 162
Generalized Γ-transform, 50
Generalized beta distribution, 212
Generalized F distribution, 45
Generalized Feller-Pareto distribution, 45, 253
Generalized gamma distribution, 211
Generalized gamma-ratio variables, 115
Generalized likelihood ratio test, 268–270
Generalized Lwin priors, 238, 240–242, 244, 247, 248, 263, 288
Generalized median estimate, 235
Generalized order statistics, 67, 94, 96, 226, 245, 248, 251, 266, 279, 280
Generalized order statistics, sequence of, 68
Generalized order statistics, uniform, 67, 68
Generalized Pareto distribution, 5, 17, 37, 44–46, 250–252, 254, 259, 262, 263, 268, 269, 288
  Pickands type, 41, 42, 69, 83, 92, 94, 96, 97, 251, 254, 255, 262, 263, 269
  Pickands type, generalized order statistics, 95
Generalized quantile estimates, 235

Generalized semi-Pareto distribution, 104
Generalized Tukey lambda distribution, 162
Geometric distribution, 9, 25–27, 29, 31, 98, 154
Geometric ergodicity, 24
Geometric maximization, 55, 75, 77, 99, 318, 327, 351
Geometric minimization, 11, 32, 54, 55, 75, 77, 98, 103, 111, 271, 308, 309, 315, 325, 326, 338, 351
Geometric spacings, 86, 88, 89, 115
Geometric-Pareto III process, 339
Gestation, current duration, 17
Gibbs sampler, 243, 247, 347
Gini chart, 4, 6, 173
Gini index, 4, 5, 13, 14, 16, 41–43, 80, 83, 119, 120, 123–125, 133, 142, 147–157, 164, 173, 200–202, 205, 220, 226, 249, 272
Gini index bounds, 15, 119, 126–128, 137, 270, 287
Gini index, k-dimensional, 141
Gini index, bivariate, 141
Gini index, confidence interval for, 267
Gini index, generalized, 124, 191
Gini index, higher order, 124, 126
Gini index, sample, 15, 174, 183, 191–193, 272
Gini index, sample, distribution of, 183
Gini index, sample, variant definitions, 183, 187
Gini index, unbiased estimation, 184
Gini indices, extended family of, 124
Gini mean difference, 7, 15, 119, 180, 181
Gini mean difference, sample, 179, 182
Gini's δ, 4, 84, 173, 174
Glasser's bound, 193
Glasser's inequality, 193
Golfer data, 359
Goodness of fit, 262, 270, 273, 274
Graphical techniques, 6, 171, 236
Grouped data, 12, 15, 16, 135, 138, 224, 253, 270, 283, 284, 286, 287
Grouped data and inflation, 138
Grouping corrections, 136, 137
Grouping, optimal, 138, 139

Hazard function, 115, 274, 284
Herfindahl index, 121, 146
Hidden truncation, 40, 105, 108
Hidden truncation Pareto distribution, 109, 296
Hidden truncation, multivariate, 332, 333
Hierarchical successive damage, 313
Hill's estimate, 298
Historical income data sources, 353
Husler, Li and Raschke estimate, 257
Hybrid censoring, 248
Hybrid estimation, 257, 258
Hyperbolic Lorenz curves, 163

Imperfect altruism, 35
Improper prior, 29
Income and wealth distributions, 19, 27, 30
Income inequality, 13
Income inequality measures, 13
Income inequality principles, 13
Income model, 19
Income power, 8–10, 23, 25, 31
Income power shower, 25
Income share ratios, 134
Increasing failure rate, 38, 148
Independent priors, 240, 242, 263, 347
Indifference zone, 277
Inequality accentuating transformations, 204
Inequality attenuating transformations, 203, 204, 207
Inequality measures, 3, 6, 15, 83, 117, 118, 149, 262
Inequality measures, confidence intervals for, 266
Inequality measures, multivariate, 219, 220
Inequality measures, sample, 15, 267

Inequality measures, truncation invariance, 83
Inequality principles, 194–196, 199, 201
Inequality trend, 4
Infinite divisibility, 12, 50
Infinitesimal transfers, 201
Information matrix, 109, 252, 253, 266, 295
Information theory, 29
Inherited wealth, 30
Insurance risk pricing, 17
Inter-quartile range, 120
Interval estimates, 223, 264
Investment talent, 35
Isotonic regression, 279

Jackknife, 234
Jain-Arora indices, 125
Johnson's estimate, 228

k-dimensional Beta distribution, 319
k-dimensional Feller-Pareto distribution, 303, 305
k-dimensional Gamma distribution, 311
k-dimensional Kumaraswamy distribution, 334
k-dimensional normal distribution, 322
k-dimensional Pareto distribution, 299
k-dimensional semi-Pareto distribution, 336
k-dimensional standard Pareto distribution, 325, 340
k-dimensional Zipf distribution, 350
k-records, 67, 97
Kakwani and Podder coordinates, 286
Klefsjö indices, 125, 133
Kolmogorov-Smirnov, 262, 269, 272
Kumaraswamy distribution, 102, 103
Kumaraswamy-Pareto distribution, 103, 334

L-moments, 258
Lack of memory, 82
Lamé Lorenz curve, 161
Laplace distribution, 9
Laplace transform, 17, 49
Latest failure, 277
Law of proportional effect, 5, 7, 19, 20, 35
Least squares estimates, 236, 271, 285
Leimkuhler curve, 167
Leimkuhler partial order, 167
Library data, 17
Lindley distribution, 38
Lindley distribution, generalized, 38
Linear combinations of densities, 40, 108, 109, 333
Linear combinations of Lorenz curves, 162
Linear combinations ordering, 218
Linear distribution, 157
Linear measures of inequality, 15, 133, 191
Linear systematic estimates, 12, 260
Linear systematic statistics, 261
Linear tax, 205
LINEX loss, 246, 249, 275, 296
Locally equality preferring, 201
Locally inequality preferring, 201
Locally most powerful, 273
Location invariance, 169, 170
Log-geometric distribution, 284
Log-logistic distribution, 8, 12, 36, 47, 50, 78, 98, 252
Log-normal distribution, 5, 7, 9, 10, 13, 15, 16, 20, 21, 25, 27, 132, 145, 147, 153, 166, 189, 211, 273, 274
Log-Pareto distribution, 39
Log-phase-type distribution, 40
Logistic distribution, 9, 12, 46, 78, 252
Logistic process, 338
Lomax distribution, 24, 36
Lomax distribution, second, 36
Long run distribution, 23, 25, 26
Lorenz curve, 3, 4, 6, 13–17, 81–84, 121–123, 128, 134, 140, 143, 144, 146, 148, 150–157, 169, 174, 197, 226, 249, 281
Lorenz curve of truncated distribution, 162

Lorenz curve, bounds for, 138
Lorenz curve, discrete distribution, 123
Lorenz curve, effect of transformations, 16
Lorenz curve, empirical, 15, 16
Lorenz curve, general quadratic, 37
Lorenz curve, generalized, 167, 219
Lorenz curve, hyperbolic, 37
Lorenz curve, kernel based estimate, 187
Lorenz curve, length of, 133
Lorenz curve, maximal entropy, 164
Lorenz curve, properties, 158
Lorenz curve, sample, 184–187, 286
Lorenz curve, sample, variant definitions, 185, 187
Lorenz curve, symmetric, 16, 131, 132, 164
Lorenz curves, convergence of, 7
Lorenz curves, corresponding density, 158
Lorenz curves, intersecting, 200, 203
Lorenz curves, mutually symmetric, 130–132
Lorenz curves, reverse, 143
Lorenz order and order statistics, 212
Lorenz order, preservation of, 208
Lorenz ordering, 16, 194, 197, 198, 200–202
Lorenz ordering, generalized, 218
Lorenz ordering, in parametric families, 211
Lorenz process, 185, 186
Lorenz process, confidence bounds, 186
Lorenz process, convergence, 186
Lorenz process, functionals of, 182, 187
Lorenz process, law of iterated logarithm, 186
Lorenz surface, 140–142, 144, 145
Lorenz zonoid, 16, 17, 139, 143, 144, 167, 217
Lower records, 244

Majorization, 14, 195, 197–200
Majorization, preservation of, 208
Mardia's first multivariate Pareto distribution, 299, 313, 330
Mardia's second multivariate Pareto distribution, 300, 313
Mardia's second multivariate Pareto distribution, estimation for, 340, 341
Marginal Lorenz ordering, 218
Markov chain model, 9, 21–24, 39, 283, 338, 340
Marshall-Olkin bivariate exponential distribution, 310, 313, 314
Marshall-Olkin multivariate exponential distribution, 322
Marshall-Olkin multivariate exponential distribution, transformed, 344
Matrix-variate exponential distribution, 166, 327
Matrix-variate Feller-Pareto distribution, 327
Matrix-variate Pareto distribution, 327, 340
Maximal attraction, 75
Maximum entropy, 12
Maximum likelihood, 11, 224–226, 231, 233–235, 237, 250–254, 256, 257, 260, 262, 266, 268, 269, 271, 275, 277, 282–284, 289, 292, 294–298, 342–344, 349
Maximum likelihood, various data configurations, 251
Maximum product of spacings, 236
Maxwell-Boltzmann, 34
Mean deviation, 5, 118
Mean residual life, 5, 10, 70, 71, 79–81, 115, 149
Mean residual life, linear, 71, 79, 82, 86, 87, 115, 271, 309
Median regression estimate, 281
Median residual life, 80
Medical insurance claims, 17
Meintanis' test, 274
Mellin transform, 49, 83
Method of (fractional) moments, 234, 255, 256

Method of moments, 13, 233, 254–259, 264, 266, 282, 289, 294, 298
Min-geometric infinite divisibility, 308
Min-geometric stable, 308
Minification process, 336–338
Minima of Pareto (IV) variables, 54, 57
Minimal majority, 134, 192
Minimax estimate, 249
Minimum chi-square, 291
Minimum discrepancy estimates, 284
Minimum income level, 9, 10, 22, 23, 30, 31, 36
Minimum mean squared error estimate, 227, 229
Minimum of a sample of random size, 31
Minimum risk estimate, 233
Minimum variance unbiased estimates, 11
Minor concentration ratio, 16, 132, 174
Misreported income, 206, 207
Mixture of exponential distributions, 11, 24, 29, 34, 46
Mixture of geometric distributions, 29, 34, 112
Mixture of normal distributions, 26, 27
Mixture of Pareto distributions, 9, 54, 253
Mixture of Weibull distributions, 32, 36, 47, 112, 302, 307, 310
Mixtures, inequality in, 209, 210
Model selection, 223
Modified Pareto distribution, 37
Moment distribution, 16, 145–147
Moment distribution, closure, 147, 166
Moment estimates, 233
Moral fortune, 9
Mortality rate, 25
Multiplicative central limit theorem, 7, 20
Multiplicative lack of memory, 82, 83
Multivariate Burr distribution, 10, 302
Multivariate concentration curves, 10
Multivariate extremes, 350
Multivariate Feller-Pareto distribution, 343
Multivariate logistic distribution, 301
Multivariate Lorenz curves, 139
Multivariate Lorenz orderings, 216–218
Multivariate Lorenz surface, 144, 147
Multivariate majorization, 217, 219
Multivariate Pareto distribution, 10, 17, 309, 322
    Pickands type, 322, 349
Multivariate Zipf distribution, 350
Multivariate Zipf III distribution, 351
Multivariate Zipf IV distribution, 350

Negative binomial distribution, 77
Negative income, 42
Neyman smooth test, 269
Non-additive utility functions, 14
Normal distribution, 9, 24, 25, 37

Open population model, 25
Order statistic estimation, 259, 260
Order statistics, 7, 10, 12, 55, 56, 91, 92, 179
Order statistics, covariances, 61–63
Order statistics, expected values, 60, 258
Order statistics, moments of, 60, 62
Order statistics, ratios of, 57, 58
Outliers, 282
Overlapping generations, 35

$P(X < Y)$, 254, 282, 283
Palma index, 134
Parabolic Lorenz curves, 163
Parametric families of Lorenz curves, 157, 158
Parametric hypotheses, 268
Pareto characterizations, 5, 10, 90–92
Pareto chart, 6, 171, 173
Pareto conditionals, 328
Pareto conditionals distribution, 329–332, 344
Pareto distribution, 9, 10, 12, 17, 24, 30, 31, 33, 35, 41, 145
Pareto distribution, generalized order statistics, 74
Pareto distribution, records, 65, 66, 74
Pareto I distribution, 41, 43, 46, 78–80,

82, 83, 85, 90, 91, 97, 100, 110, 150, 213, 242, 243, 247–249, 254, 259, 270, 271, 273, 274, 282
Pareto I distribution, generalized order statistics, 68, 69, 95
Pareto I distribution, maximum likelihood estimates, 225, 226
Pareto I distribution, moments of generalized order statistics, 69
Pareto I distribution, moments of records, 66, 93
Pareto I distribution, quotients of records, 65, 66
Pareto I distribution, records, 93, 94
Pareto I distribution, truncated, 63, 139, 151, 165
Pareto I–II distributions, variance, 49
Pareto I–IV distributions, modes, 47
Pareto I–IV distributions, moments, 48, 255
Pareto II distribution, 11, 12, 17, 24, 29, 36, 42–44, 46, 50, 75, 77–81, 85, 86, 92, 97, 151, 250–253, 256, 259, 260, 263, 264, 269–271, 274, 279, 280, 296
Pareto II distribution, truncated, 42, 70, 152
Pareto III distribution, 9, 11, 12, 32, 42, 43, 45, 50, 55, 60, 78, 91, 98, 104, 115, 152, 250, 252, 259, 264, 271, 272, 283, 338
Pareto III distribution, characterization, 98, 99
Pareto III distribution, Gini index, 126
Pareto III distribution, residual life, 70
Pareto III process, 337, 338
Pareto index, 3, 172
Pareto IV distribution, 8, 9, 11, 12, 17, 33, 36, 37, 39, 41, 43–45, 47, 57, 60, 70, 77, 91, 106, 152, 250, 252, 254, 259, 261, 262, 264, 270, 283
Pareto IV distribution, Lorenz curve, 122
Pareto IV distribution, sample median, 57
Pareto IV distribution, skewness and kurtosis, 49
Pareto process, 336, 338, 340
Pareto variables, random sums of, 55
Pareto's α, 41, 84, 173, 174
Pareto's law, 1–3, 10, 19, 27, 37
Pareto's second model, 298
Pearson IV distribution, 9
Percentile residual life function, 80
Permutation matrices, 199
Pfeifer records, 68
Phase-type distribution, 39, 166
Pickands generalized Pareto distribution, 11, 223
Pietra index, 118, 129–131, 133, 147, 149–151, 153–157, 200, 202, 220
Pietra index of order k, 130
Pietra index, sample, 174, 177, 187
Pigou-Dalton transfer principle, 194
Pitman closeness, 278
Pivotal quantity, 264–267
Plackett family, 324
Platoon formation in traffic, 112
Poisson distribution, 26, 27, 34, 55, 115
Pooled populations, 209, 210
Positive combinations ordering, 218
Posterior median, 240
Poverty indices, 133, 203, 205
Poverty line, 4, 133, 196, 205
Power function distribution, 63, 213, 238–241
Powered Cauchy distribution, 36
Precision parameter, 241, 243
Prediction interval, 267
Preliminary test estimate, 275
Price Lorenz order, 218
Principle of transfers, 195
Prize competition distribution, 12, 13, 36
Probability weighted moments, 256, 257, 259
Probit diagram, 26
Product models, 38

Product Pareto model, 17, 21, 39
Products of Pareto variables, 17, 49, 53, 54, 115
Progressive censoring, 226, 246, 266
Progressive tax, 9, 205, 210
Progressive transfer, 197
Pyramid model, 10, 30, 31

Quantal response, 279
Quantile estimates, 234, 235
Quantile function, 15, 39, 52, 101, 121, 162, 202, 261
Quantile ratios, 192
Quasi-concavity, 15, 201
Quasi-convex, 201
Quasi-range, 115
Queueing models, 17

Random contraction, 96
Random dilation, 91, 96
Random taxation, 207
Randomly selected order statistics, 72
Ranked set sample, 266, 342
Ratio models, 38
Ratio of concentration, 4, 119
Ratio of Pareto variables, 115
Receiver operating characteristics, 17
Record values, 64, 65, 68, 93, 226, 243, 254
Recurrence relation, 60–64, 66, 67, 69, 102
Regressive transfer, 197
Regular variation, 51
Relative deprivation curve, 170
Relative inequality aversion, 14
Relative inequality measures, 5
Relative mean deviation, 5, 6, 13, 15, 16, 118, 129, 177
Relative mean deviation, sample, 15, 177
Relative mean difference, 6
Reliability, 11, 16, 46, 50, 69, 148, 219
Renyi entropy, 97
Residual life at great age, 75
Residual life distribution, 11, 69, 75, 79, 80
Reversed Gini chart, 6, 174
Rizzo tests, 274
Robin Hood axiom, 5, 13, 14, 194–196, 202
Robin Hood axiom, modified, 202
Robin Hood, modified activity, 203
Rosenblatt construction, 348

Sample distribution function, 171, 175, 196
Sample first moment distribution function, 171
Sample income share ratios, 192
Sample means, Lorenz ordering, 214
Sample medians, Lorenz ordering, 213
Scale invariance, 3, 118, 169, 170, 195, 196, 201, 202, 215, 269
Scale mixture, 206
Scaled F distribution, 36, 44
Schur convex function, 199–201
Schur convexity, 200
Sech^2 distribution, 8, 9, 15, 36, 46, 283
Selection mechanism, 106
Selection problems, 12, 277
Semi-log graph, 174
Semi-Pareto (IV) distribution, 105
Semi-Pareto distribution, 43, 103, 335–337
Semi-Pareto process, 336, 337, 340
Separable convex function, 200, 201
Sequential classification, 21
Sequential estimation, 275
Shannon entropy, 97, 121
Sheppard's corrections, 136
Shifted exponential distribution, 165
Sign change ordering, 215
Significance level, 269
Simon distribution, 29, 33, 112, 113, 290
Simpson's index, 191
Size and shape, 88, 89
Size biased distribution, 80
Software reliability, 17
Spacings, 12, 180
Spacings test, 270
Species diversity, 10, 33

Species equitability, 202
Spliced models, 37, 38
Stable law, 12, 20, 73
Standard deviation, 119
Standard deviation, sample, 178
Standard Pareto distribution, 10, 12, 50, 59, 76
Standard Zipf distribution, 34, 112, 113
Star ordering, 16, 215, 216
Statistical manifold, 115
Stochastic dominance, 124, 167
Strassen's theorem, 208, 211, 212, 218
Stream of dollars model, 10, 28, 29
Strongly unimodal, 166
Structural probability, 247
Subset selection, 277, 278
Sum of a sample of random size, 55
Survival analysis, 67
Survival probability, 229, 230, 239
Symmetric mean, 199

Taguchi's inequality, 149, 193
Tapered Pareto distribution, 37, 298
Tax evaders, 6
Texas counties data, 359
Theil index, 121
Three stage estimation, 275
Time-on-test transform, 16, 148, 169, 187, 248, 249
Top 100α percent, 192
Trader's prior, 249
Transfer ordering, $\geq_T$, 197, 198
Transfers, 14, 15, 197
Transformed data estimates, 292
Transformed exponential variables, 310, 312
Transformed Pareto variables, 101
Transforms, 49
Trivariate reduction, 310, 313
Truncated Cauchy, 17
Truncation, 16, 63, 82, 84, 205, 226, 244, 251, 284, 297
Truncation equivalent to rescaling, 82, 83, 85
Type I multivariate Pareto distribution, 299
Type I multivariate Pareto distribution, estimation for, 342
Type II multivariate Pareto distribution, 300, 303, 304
Type II multivariate Pareto distribution, Bayes estimates, 346
Type II multivariate Pareto distribution, ordered coordinates, 306
Type II multivariate Pareto distribution, ratio of coordinates, 304
Type III multivariate Pareto distribution, 301, 308, 309
Type IV multivariate Pareto distribution, 10, 302–304, 313, 345
Type IV multivariate Pareto distribution, Bayes estimates, 345, 346
Type IV multivariate Pareto distribution, min-infinite-divisibility, 308
Type IV multivariate Pareto distribution, sample minima, 308, 314

U statistic, 180, 182
Unbiased test, 11
Under-reported income, 5, 10, 12, 84–86, 115, 205, 293, 296
Uniform distribution, 9, 15, 33, 36, 58, 90, 156
Uniform order statistics, 212
Uniform representation, 51
Uniformly most powerful tests, 11
Universal geometric minima, 105
Universal pivotal technique, 347
Upper records, 244
Utility, 14, 15, 191, 194, 198, 200
Utility theory, 14

Vague prior information, 237, 263
Vitality function, 115

Wage structure, 30
Waiting for large observations, 77
Waiting for outstanding observations, 77
Waring distribution, 29, 34, 110, 113–115, 290, 292

Waring distribution, factorial moments, 113
Waring distribution, generalized, 114
Weak Pareto law, 10
Weibull distribution, 11, 12, 74, 155
Weighted distributions, 81, 209
Weighted least squares, 16, 285
Weighted minima, 309
Weighted quadratic loss, 237, 249
Weighting, inequality attenuating, 209
Weighting, inequality preserving, 209
Welfare, 200
Wicksell-Kibble bivariate exponential, 313
Wiener process, 23
Wire lengths, 17

Yntema indices, sample, 189
Yule distribution, 10, 27–29, 34, 110, 113, 115
Yule distribution, factorial moments, 113

Zenga-I curve, 168
Zenga-I curve, constant, 169
Zenga-I partial order, 169
Zenga-II curve, 168
Zenga-II inequality index, 169
Zenga-II inequality index, sample, 178
Zenga-II partial order, 169
Zeta distribution, 110, 289
Zipf (II) distribution, 111, 114
Zipf (III) distribution, 111
Zipf (IV) distribution, 111
Zipf distribution, 27, 33, 110, 111, 288
Zonoid, 142, 221
Zonoid ordering, 218–220
Zonoid volume, 220

Author Index

Aaberge, R., 125, 182, 203, 371
Aban, I. B., 297, 371
Abd El-Ghafour, A., 253, 393
Abd-El-Hakim, N. S., 280, 371
Abd-Ellah, H. N., 115, 371
Abdel-Aty, Y., 248, 395
Abdel-Ghaly, A. A., 277, 371
Abdell-All, N. H., 115, 371
Abdi, M., 280, 375
Abdul-Sathar, E. I., 240, 249, 371
Abraham, B., 301, 393
Abu-Zinadah, H. H., 102, 296, 402
Abusev, R. A., 232, 371
Adams, K. A. H., 221, 371
Adeyemi, S., 60, 61, 371, 372
Adler, A., 72, 372
Afify, W. M., 102, 296, 372
Aggarwal, V., 139, 151, 163, 372
Aghevli, B. B., 139, 372
Ahmad, A. E. A., 61, 64, 69, 372
Ahmad, M., 256, 257, 259, 261, 372, 373, 403
Ahmadi, J., 280, 399
Ahn, S., 40, 372
Ahsanullah, M., 67, 87, 90, 91, 93, 254, 280, 332, 372, 376, 386, 397, 406
Aigner, D. J., 12, 224, 284–286, 291, 356, 372
Airth, A. D., 17, 372
Aitchison, J., 153, 373
Akbari, M. G., 243, 382
Akinsete, A., 39, 102, 373
Al-Hussaini, E. K., 247, 254, 264, 373, 396
Al-Marzoug, A. M., 256, 373
Al-Ruzaiza, A. S., 313, 373
Al-Shomrani, A. A., 38, 376
Al-Zahrani, B. M., 38, 376
Alam, M. W., 232, 379
Ali, M. M., 61, 232, 249, 261, 373, 401
Aliev, F. A., 91, 93, 397
Alker, H. R., 134, 192, 373
Aly, H. M., 277, 371
Alzaid, A. A., 308, 373
Amari, S., 115, 373
Amati, G., 17, 373
Amato, V., 133, 373
Amin, N. A. K., 236, 379
Amin, Z. H., 247, 382
Amoroso, L., 5, 373
Anderson, D. N., 187, 373
Arellano-Valle, R. B., 106, 373
Arnold, B. C., 11, 31, 32, 36, 37, 40, 42, 47, 60, 64, 68, 74, 79, 80, 91, 93, 94, 98, 106, 109, 119, 127, 134, 151, 152, 158, 161, 162, 164, 166, 169, 187, 192, 197–199, 204, 206–208, 212–214, 217, 219, 224, 227, 234, 238, 240–243, 245–247, 250, 255, 258, 263, 264, 267, 271, 272, 274, 287, 288, 297, 310, 313, 317, 320, 325, 328, 330–332, 337, 338, 340, 344, 346, 373–375, 387, 394, 400, 405, 407
Arnold, J. C., 275, 381
Arora, S., 125, 388
Asadi, M., 82, 83, 92, 96, 375, 386
Asgharzadeh, A., 280, 375, 399

Ashkar, F., 255, 375
Asimit, A. V., 315, 375
Asrabadi, B. R., 231, 375
Athreya, K. B., 32, 376
Atieh, B., 38, 384
Atkinson, A. B., 14, 191, 198, 200, 201, 376
Attia, A. F., 277, 371
Augé, J., 322, 380
Austin, K., 42, 151, 161, 374
Awad, A. M., 97, 376
Aygun, F., 91, 93, 397
Azarnoosh, H. A., 186, 383

Bacro, J-N., 253, 269, 379
Baklizi, A., 275, 376
Bakouch, H. S., 38, 376
Balakrishna, N., 335–337, 376
Balakrishnan, N., 8, 47, 49, 63, 64, 67, 93, 94, 243, 246, 248, 302, 303, 341, 374, 376, 382, 389
Balkema, A. A., 11, 75, 376
Bancroft, T. A., 275, 376
Baringhaus, L., 33, 98, 99, 376
Barlow, R. E., 184, 376
Bartholomew, D. J., 109, 184, 376
Basmann, R. L., 161, 376
Baxter, M. A., 225, 227, 229, 376
Beaver, R. J., 106, 374
Bechofer, R. E., 277, 376
Beg, M. A., 297, 376
Beg, M. I., 89, 376
Behboodian, J., 102, 376
Benedetti, C., 194, 376
Bengio, Y., 38, 378
Benhabib, J., 35, 376
Berg, C., 50, 376
Berkson, J., 279, 377
Bernadic, M., 131, 378
Best, D. J., 269, 381
Beutner, E., 97, 377
Bhandari, S. K., 83, 194, 377
Bhattacharjee, M. C., 149, 377
Bhattacharya, N., 5, 82, 83, 85, 377
Bhuiyan, K. J., 249, 398
Billingsley, P., 175, 377
Bisin, A., 35, 376
Bladt, M., 166, 377
Blitz, R. C., 10, 142, 355, 377
Block, H. W., 316, 317, 377
Blum, M., 12, 52, 53, 73, 377
Bomsdorf, E., 13, 36, 377
Bonferroni, C. E., 167, 377
Bonini, C. P., 10, 28, 402
Bowley, A. L., 3, 6, 120, 377
Bowman, M. J., 6, 7, 171, 173, 174, 193, 355, 377
Boyd, A. V., 234, 393
Bradford, S. C., 12, 36, 377
Branco, M. D., 106, 373
Brazauskas, V., 17, 235, 252, 253, 377
Bremner, J. M., 184, 376
Brennan, L. E., 53, 377
Bresciani-Turroni, C., 6, 377
Brittain, J. A., 10, 142, 355, 377
Brockett, P. L., 80, 166, 374, 375
Brown, G., 21, 378
Brown, J. A. C., 153, 373
Brunk, H. D., 184, 376
Bryson, M. C., 11, 79, 174, 224, 271, 378
Burkschat, M., 262, 280, 378
Burr, I. W., 8, 9, 48, 57, 378

Cai, Y., 39, 378
Campbell, G., 17, 378
Candel Ato, J., 131, 378
Carreau, J., 38, 378
Castellani, M., 10, 35, 378
Castellano, V., 5, 154, 156, 157, 378
Castillo, E., 91, 159–161, 236, 241, 243, 259, 281, 328, 330, 331, 344, 345, 374, 378, 401
Cerone, P., 128, 378
Chacko, M., 342, 378
Champernowne, D. G., 8–10, 21, 23, 25, 28, 36, 101, 228, 283, 353–357, 378
Chan, L. K., 12, 261, 379
Chandra, M., 16, 148, 149, 169, 175, 184, 185, 187, 379

Chang, C-T., 280, 407
Chaouche, A., 253, 269, 379
Chaturvedi, A., 232, 379
Chen, S-C., 267, 277, 406, 407
Chen, W. C., 292, 379
Chen, Z., 263, 266, 379
Cheng, R. C. H., 236, 379
Cheng, S. W., 12, 261, 262, 379
Chi, X., 17, 408
Chiragiev, A., 311, 379
Chotikapanich, D., 281, 282, 379
Chou, C. H., 262, 379
Choulakian, V., 269, 379
Chung, K. L., 73, 379
Cicchitelli, G., 183, 379
Cifarelli, D. M., 99, 104, 379
Cirillo, R., 3, 379
Cislak, P. J., 9, 57, 378
Ciumara, R., 248, 398
Cobham, A., 134, 379
Coelho, C. A., 115, 379
Cohen, A. C., 48, 297, 379
Cohen, J., 33, 380
Colombi, R., 21, 38, 380
Conlisk, J., 13, 380
Consul, P. C., 37, 380
Conway, D. A., 323–325, 380
Cook, R. D., 301, 380
Cook, W. L., 227–229, 234, 380
Cordeiro, G. M., 103, 334, 396, 397
Cormann, U., 39, 380
Cramer, E., 74, 96, 262, 378, 380, 396
Cramer, J. S., 12, 380
Crawford, C. B., 88, 380
Cretois, E., 269, 399
Cronin, D. C., 36, 47, 126, 152, 380
Csorgo, M., 186, 380
Csorgo, S., 186, 380
Cuadras, C. M., 322, 380

d'Addario, R., 5, 79, 371
Dagum, C., 36, 153, 380, 381
Dallas, A. C., 82, 87, 88, 381
Dalton, H., 5, 13, 174, 190, 191, 194–196, 198, 381
Daoudi, J., 251, 381
Dasgupta, P., 14, 15, 198, 381
Dasu, T., 83, 396
David, H. A., 56, 60, 119, 179, 381
Davis, H. T., 298, 381
Davis, R. L., 275, 381
Davison, A. C., 254, 381
De Boeck, B., 269, 381
De Cani, J. S., 252, 381
De Groot, M. H., 323, 381
de Haan, L., 11, 75, 323, 376, 383
De Luca, G., 17, 381
de Pascoa, M. A. R., 103, 397
De Zea Bermudez, P., 255, 259, 263, 381
Del Castillo, J., 251, 381
Dempster, A. P., 183, 381
Dey, D. K., 278, 381
Dimaki, C., 79, 90, 92, 114, 381, 407
Dimitrov, B., 109, 381
Dixit, U. J., 231, 282, 381
Doksum, K., 216, 381
Donaldson, D., 124, 192, 381, 382
Donath, W. E., 17, 382
Doostparast, M., 243, 280, 382
Dorfman, R., 119, 154, 382
Drane, J. W., 279, 316, 382
Dubey, S. D., 11, 47, 48, 382
DuMouchel, W. H., 224, 273, 382
Dunsmore, I., 247, 382
Durand, D., 174, 382
Durling, F. C., 316, 326, 382
Dutta, J., 35, 382
Dwass, M., 72, 382
Dyer, D., 247, 382
Dziubdziela, W., 67, 338, 382

Efron, B., 29, 382
Eichhorn, W., 30, 204, 382
Ekwo, M. E., 275, 395
El-Gohary, A., 313, 373
Elfessi, A., 278, 382
Elteto, O., 6, 15, 129, 150, 153, 155, 156, 188, 382
Emlen, J. M., 192, 382
Epstein, B., 260, 382

Ericson, O., 9, 24, 25, 33, 355, 382
Espejo, M. R., 326, 396

Fakoor, V., 186, 383
Falk, M., 269, 350, 383
Famoye, F., 39, 102, 373
Farsipour, N. S., 233, 266, 397
Fawzy, M. A., 69, 372
Feldstein, M. L., 298, 381
Feller, W., 24, 34, 43, 51, 54, 58, 73, 75, 112, 182, 307, 383
Fellman, J., 16, 203, 204, 383
Ferguson, T. S., 87, 89, 383
Fernández, A. J., 266, 383
Ferreira, A., 323, 383
Ferreira, M., 338, 383
Ferreri, C., 35, 383
Fischer, M., 17, 402
Fishlow, A., 192, 383
Fisk, P. R., 8, 9, 46, 126, 152, 224, 283, 284, 353–355, 383
Fisz, M., 86, 383
Foster, J. E., 167, 203, 402
Fraser, D. A. S., 181, 182, 383
Frechet, M., 9, 383
Frigyes, E., 6, 15, 129, 150, 153, 155, 156, 188, 382
Fu, J., 246, 384
Funke, H., 204, 382
Furlan, V., 4, 384
Furman, E., 315, 375

Gómez-Déniz, E., 165, 401
Gaffney, J., 286, 399
Gail, M. H., 16, 175, 183–188, 384
Galambos, J., 78, 89, 90, 99, 271, 384
Ganeshalingam, S., 344, 374
Gangopadhyay, A. K., 278, 391
Ganjali, M., 266, 397
Garren, S. T., 279, 384
Gastwirth, J. L., 15, 16, 121, 126, 127, 136–139, 150, 154, 156, 158, 175–177, 183–188, 224, 270, 271, 286, 356, 357, 380, 384
Gay, R., 17, 384
Geisser, S., 240, 246, 247, 384
Genton, M. G., 106, 373
Ghalibaf, M. B., 186, 383
Ghitany, M. E., 38, 384
Ghosh, I., 40, 109, 297, 374
Ghosh, M., 91, 268, 384, 387
Gibbons, J. D., 120, 390
Gibrat, R., 5, 7–10, 19–21, 25, 384
Gini, C., 3, 4, 6, 7, 173, 384, 385
Giorgi, G. M., 137, 182, 188, 385
Glasser, G. J., 149, 193, 385
Glauberman, M., 138, 384
Gleissner, W., 30, 382
Gnedenko, B. V., 72, 73, 75, 385
Godwin, H. J., 176, 385
Gokhale, D.V., 109, 374
Goldberger, A. S., 12, 224, 284–286, 291, 356, 372
Goldie, C. M., 12, 15, 175, 184, 185, 187, 385
Gonzalez, E., 274, 405
Goodman, I. R., 307, 385
Goodman, L., 227, 385
Govindarajulu, Z., 87, 385
Green, R. F., 34
Greenberg, B. G., 260, 401
Greselin, F., 178, 385
Griffiths, W. E., 281, 282, 379
Grimshaw, S. D., 253, 385
Gross, D., 17, 402
Guillou, A., 269, 383
Gulati, S., 270, 385
Gumbel, E. J., 5, 385
Gupta, A. K., 17, 38, 343, 344, 385, 396
Gupta, M. R., 166, 385
Gupta, R. C., 80, 81, 332, 385, 386
Gupta, R. P., 99, 104, 326, 379, 386
Gurland, J., 289–291, 387, 405

Habibullah, M., 280, 386
Hadi, A. S., 236, 259, 281, 344, 345, 378
Hagerbaumer, J. B., 16, 132, 174, 357, 386
Hagstroem, K. G., 4, 5, 9, 12, 24, 52, 55, 78, 79, 353–355, 357, 386

Hallett, J. T., 338, 374
Hamada, K., 133, 386
Hamdy, H. I., 247, 275, 386, 396
Hamedani, G. G., 80, 81, 386
Hanagal, D. D., 315, 344, 386
Hansen, O. K. H., 17, 390
Hardy, G. H., 14, 127, 197–199, 386
Harris, C. M., 11, 12, 250, 254, 255, 386
Hart, P. E., 16, 145–147, 192, 356, 386
Harter, H. L., 4, 7, 12, 173, 259, 260, 265, 266, 386, 395
Hartley, M. J., 10–12, 86, 293–295, 386, 400
Hartvig, H., 17, 46, 390
Hashemi, M., 96, 386
Hassanein, K. M., 261, 373
Hatke, M. A., 9, 387
Hawkes, A. G., 317, 387
Haworth, C. T., 13, 392
Hayakawa, M., 6, 354, 387
Hayes, K. L., 161, 376
He, H., 233, 387
Helmert, F. R., 176, 387
Hendrick, D. J., 24, 400
Herfindahl, O. C., 146, 387
Hill, B. M., 33, 34, 297, 298, 387
Hinkley, D. V., 12, 293, 295, 387
Hinz, P. N., 291, 387
Hirschman, A. O., 146, 387
Hitha, N., 81, 396
Hoeffding, W., 180, 182, 387
Holm, J., 164, 387
Holmes, G. K., 3, 6, 387
Horvath, L., 186, 380
Hosking, J. R. M., 257, 258, 387
Hossain, A. M., 263, 387
Houchens, R. L., 280, 372
Howlader, H. A., 236, 263, 387, 403
Huang, J. S., 12, 56, 62, 79, 85, 90, 387
Huang, M. L., 260, 297, 388
Huang, W-J., 79, 388
Huang, W-T., 12, 277, 278, 388
Huber, P. J., 295, 388
Hume, E. R., 279, 384
Hurlimann, W., 44, 388
Husler, J., 257, 258, 388
Hutchinson, T. P., 349, 388

Ibrahim, K., 266, 397
Iliescu, D., 265, 268, 388
Irwin, J. O., 114, 388
Isaacson, D. L., 22, 388
Ismail, S., 284, 388
Iwase, K., 233, 388, 402
Iwinska, M., 229, 388

Jabbari-Nooghabi, M., 231, 282, 381
Jaheen, Z. F., 247, 254, 264, 373, 396
Jain, K., 125, 388
Jakobsson, U., 204, 388
Jalali, A., 254, 388
James, I. R., 88, 89, 388
Janjic, S., 77, 99, 388
Jayakumar, K., 99, 104, 105, 335–338, 376, 379, 398
Jeevanand, E. S., 240, 249, 347, 371, 389
Jensen, B. C., 155, 394
Jha, V. D., 63, 64, 390
Jin, C., 278, 382
Jiu, L., 115, 398
Joe, H., 80, 389
Johnson, A. M., 227–229, 400
Johnson, J. D., 161, 376
Johnson, M. E., 301, 380
Johnson, N. L., 8, 9, 47, 48, 113, 114, 302, 303, 341, 389
Johnson, N. O., 6, 355, 389
Jones, M. C., 101, 103, 389
Jordá, V., 159, 161, 217, 401
Jorgensen, M., 21, 399
Jose, K. K., 105, 337, 338, 398, 404
Joshi, P. C., 63, 376
Jupp, P. E., 309, 327, 389
Juul, S., 17, 46, 390

Kabe, D. G., 12, 15, 62, 181, 182, 389
Kabir, A. B. M. L., 87, 372
Kadane, J. B., 147, 263, 389, 405
Kagan, Y. Y., 37, 298, 389
Kakwani, N. C., 16, 124, 133, 161, 170, 262, 286, 353, 389

Kalbfleisch, J. D., 45, 270, 389
Kalecki, M., 7, 20, 25, 36, 355, 389
Kamat, A. R., 181, 390
Kaminsky, K. S., 12, 261, 276, 277, 390
Kamps, U., 67, 69, 74, 92, 96, 97, 245, 262, 279, 377, 378, 380, 390
Kang, S. G., 249, 390
Karim, A., 249, 398
Karlin, S., 214, 390
Katz, L., 290, 390
Kaya, M. F., 280, 391
Keating, J. P., 81, 386
Keiding, N., 17, 46, 390
Kemp, A. W., 113, 114, 389
Kendall, M.G., 120, 136, 176, 177, 390
Kern, D. M., 230, 390
Keseling, C., 96, 380
Khabia, A., 38, 396
Khan, A. H., 61, 373
Khan, R. U., 102, 390
Khurana, A. P., 63, 64, 390
Kibble, W. F., 313, 390
Kibria, B. M. G., 275, 390
Kiefer, J., 277, 376
Kim, D. H., 249, 390
Kim, J. H. T., 40, 372
Kim, Y. I., 94, 392
Kimball, A. W., 284, 390
Kirmani, S. N. U. A., 89, 376
Klamkin, M. S., 127, 128, 390
Klefsjö, B., 125, 148, 390
Kleiber, C., 45, 211–213, 390, 391
Kleyle, R. M., 183, 381
Kloek, T., 13, 391
Kolmogorov, A. N., 73, 385
Kondor, Y., 6, 129, 391
Koo, A. Y. C., 286, 399
Kopocinski, B., 67, 382
Korwar, R. M., 85, 391
Koshevoy, G., 16, 139, 142, 145, 218, 220, 391
Kotz, S., 8, 45, 47, 49, 78, 89, 90, 99, 109, 113, 114, 253, 255, 259, 271, 302, 303, 307, 313, 326, 341, 381, 384, 385, 389, 391, 396
Koutrouvelis, I. A., 235, 391
Krause, M., 164, 391
Krieger, A. M., 137, 138, 384, 391
Krishna, H., 288, 391
Krishnaji, N., 5, 10, 84, 85, 293, 391
Krishnan, P., 37, 391
Krysicki, W., 321, 391
Kulldorff, G., 12, 251, 259, 261, 391
Kumar, D., 102, 390
Kumar, S., 278, 391
Kus, C., 280, 375, 391
Kvist, K., 17, 46, 390

Laguna, L., 11, 31, 32, 36, 47, 60, 119, 152, 224, 234, 250, 255, 258, 267, 271, 272, 375
Lam, D., 210, 391
Lam, D.., 210
Lambert, P. J., 203, 210, 211, 391
Lamperti, J., 72, 391
Landsman, Z., 311, 379
Langford, E. S., 80, 386
Lanza, G., 203, 391
Laurent, A. G., 70, 79, 391
Lee, C., 39, 102, 373
Lee, M-Y., 94, 392
Lee, W-C., 94, 267, 277, 406, 407
Lee, W. D., 249, 390
Leimkuhler, F. F., 12, 392
Leppert, G. R., 279, 384
Levine, D. B., 16, 205, 392
Levy, M., 35, 392
Li, D., 257, 258, 388
Liang, T., 248, 392
Likes, J., 229, 232, 233, 392
Lin, C-T., 254, 392
Lin, G. D., 92, 93, 392
Lindley, D. V., 38, 263, 303, 392
Littlewood, J. E., 14, 127, 197–199, 386
Liu, P-S. L., 278, 381
Liu, R., 319, 397
Lomax, K. S., 11, 36, 392
Lomnicki, Z. A., 180, 392

Long, J. E., 13, 392
Lorah, D. J., 49, 392
Lorenz, M. O., 3, 121–123, 174, 355, 356, 392
Louzada, F., 38, 376
Luceño, A., 262, 263, 392
Lucke, J., 12, 251, 262, 392
Lunetta, G., 140, 392
Lwin, T., 12, 229, 231, 232, 236, 238–240, 249, 392
Lydall, H. F., 9, 10, 30, 31, 392

Macgregor, D. H., 6, 355, 392
Maddala, G. S., 12, 36, 152, 285, 286, 356, 403
Madi, M. T., 246, 392
Madsen, R. W., 22, 388
Maguire, B. A., 11, 46, 296, 393
Mahmoodi, M., 254, 400
Mahmoud, M. A. W., 69, 115, 262, 371, 393
Mahmoud, M. R., 253, 393
Mahmoudi, E., 102, 296, 393
Malik, H. J., 12, 54, 58, 60, 62, 87, 115, 225, 236, 301, 305, 393
Mallows, C. L., 126, 393
Manas, A., 17, 38, 393
Manas, G. J., 234, 393
Mandelbrot, B., 9, 10, 12, 13, 18, 20, 29, 393
Marchi, V. A. A., 38, 376
Mardia, K. V., 10, 299, 300, 308–310, 313, 340–342, 389, 393
Marohn, F., 96, 394
Marsaglia, G., 85, 394
Marshall, A. W., 134, 148, 192, 197–199, 204, 214, 217, 219, 313, 394
Marzuq, M. M., 97, 376
Masi, D., 17, 402
Mathai, A. M., 312, 394
Mauldon, J. G., 188, 394
McCune, E. D., 234, 394
McCune, S. L., 234, 394
McDonald, J. B., 12, 36, 126, 137, 152, 155, 156, 205, 206, 211, 285, 286, 294, 296, 394
Meerschaert, M. M., 297, 371
Mehran, F., 15, 16, 133, 135, 136, 138, 139, 191, 202, 372, 394
Meintanis, S. G., 273, 274, 394
Mert, M., 17, 394
Metcalf, C. E., 13, 36, 394
Mexia, J. T., 115, 379
Meyer, P. A., 219, 394
Michel, P., 35, 382
Michel, R., 350, 383, 394
Midlarsky, M. I., 35, 394
Mishra, C. S., 249, 397
Mitzenmacher, M., 17, 38, 394
Mohie El-Din, M. M., 248, 395
Mondani, R., 188, 385
Moore, A. H., 12, 259, 260, 265, 395
Moothathu, T. S. K., 84, 99, 100, 149, 202, 227, 230, 395
Morgan, J., 13, 193, 395
Morrison, D. G., 70, 79, 80, 395, 402
Moschopoulos, P. G., 312, 394
Moshref, M. E., 262, 393
Mosimann, J. E., 88, 388, 395
Mosler, K., 142, 145, 202, 218, 220, 391, 395
Mount, T. D., 12, 36, 356, 401
Mousa, M. A. M. A., 254, 373
Muirhead, R. F., 199, 395
Mukerjee, R., 194, 377
Mukhopadhyay, N., 275, 395
Muliere, P., 124, 202, 313, 315, 395
Mumme, D. C., 227–229, 234, 380
Muniruzzaman, A. N. M., 11, 225, 230, 268, 395
Muraleedharan Nair, K. R., 315, 397
Muraleedharan, K. R., 240, 371
Muralidharan, K., 38, 396
Myhre, J., 12, 251, 260, 262, 392, 396

Nadarajah, S., 17, 37, 38, 49, 50, 102, 109, 253, 304, 308, 313, 326, 334, 343, 344, 384, 385, 396, 405, 408

Nagaraja, H. N., 56, 60, 64, 93, 94, 214, 243, 246, 374, 375, 381
Nagata, Y., 227, 396
Nair, N. U., 81, 396
Nair, U. S., 180–182, 396
Nasri-Roudsari, D., 74, 396
Navarro, J., 80, 166, 400, 403
Nayak, T. K., 137, 384
Nelson, P. I., 12, 276, 390
Neuts, M. F., 39, 396
Newbery, D., 14, 15, 198, 200, 396
Newman, D. J., 127, 128, 390
Ney, P. E., 32, 376
Ng, E., 37, 391
Ng, H. K. T., 320, 375
Niehans, J., 146, 396
Nielsen, B. F., 166, 377
Nigm, A. M., 247, 396
Nikulin, M. S., 232, 405

Oakes, D., 83, 396
Obst, N., 286, 399
Ogawa, J., 260, 401
Ogwang, T., 164, 397
Ojo, M. O., 61, 372
Olkin, I., 134, 192, 197–199, 204, 214, 217, 219, 313, 319, 394, 397
Olshen, R. A., 224, 273, 382
Omar, A., 266, 397
Oncel, S. Y., 91, 93, 397
Ord, J. K., 10, 12, 16, 19, 20, 23, 25, 36, 47, 83, 97, 120, 121, 127, 136, 176, 177, 191, 204–207, 292, 390, 397
Ortega, E. M. M., 103, 334, 396, 397
Osborne, J. A., 16, 397
Ouarda, T. B. M. J., 255, 375
Owen, D. B., 279, 316, 382

Padamadan, V., 315, 347, 389, 397
Pagano, M., 10–12, 86, 400
Pakes, A. G., 24, 32, 33, 77, 123, 158, 397
Pallini, A., 137, 182, 385
Pallotta, W. J., 275, 386
Palma, J. G., 134, 397
Panaretos, J., 115, 407
Pandey, A., 69, 401
Pandey, B. N., 249, 397
Panorska, A. K., 297, 371
Papadakis, E. N., 346, 397
Papathanasiou, V., 92, 397
Paranaiba, P. F., 103, 397
Pareto, V., 1, 2, 19, 35, 46, 172, 173, 221, 298, 353–355, 397
Parsi, S., 266, 397
Parsian, A., 233, 397
Pascual, M., 160, 161, 401
Pasquazzi, L., 178, 385
Patil, G. P., 12, 16, 79, 83, 85, 97, 120, 121, 191, 204–207, 397
Paulson, A. S., 317, 397
Pearson, E. S., 9, 11, 46, 296, 393, 398
Pearson, L. M., 236, 399
Pederzoli, G., 115, 398
Peng, L., 115, 398
Pestieau, P., 21, 398
Petersen, H. G., 16, 138, 139, 354, 398
Pfeifer, D., 68, 398
Picard, D., 17, 38, 297, 398
Pickands, J. III., 11, 223, 263, 398
Pietra, G., 5, 6, 15, 84, 121, 398
Pillai, R. N., 43, 103–105, 337, 398
Pisarewska, H., 238, 398
Pizzetti, E., 137, 398
Podder, C. K., 249, 398
Podder, N., 16, 161, 286, 353, 389
Polya, G., 14, 127, 197–199, 386
Porru, E., 4, 398
Possen, U. M., 21, 398
Pouramahdi, M., 332, 375
Prakash, G., 275, 398, 403
Preda, V., 248, 398
Prentice, R. L., 45, 270, 389
Press, S. J., 238, 240, 242, 245, 247, 263, 264, 287, 288, 346, 375
Prieto, F., 163, 165, 401
Proschan, F., 148, 204, 308, 344, 373, 394, 398
Pusz, J., 309, 398

Pyke, R., 270, 398

Quandt, R. E., 11, 224, 225, 233–235, 270, 399

Rachev, S. T., 308, 399
Radouane, O., 269, 399
Rahman, M., 236, 399
Rakonczai, P., 323, 350, 399
Ramachandran, B., 85, 399
Ramaswami, V., 40, 372
Ramberg, J. S., 162, 399
Ramsay, C. M., 17, 53, 399
Ranadive, K. R., 13, 354, 356, 357, 399
Ransom, M. R., 12, 126, 137, 152, 156, 285, 286, 394
Rao Tummala, V. M., 236, 237, 399
Rao, C. R., 81–83, 85, 96, 186, 209, 375, 399
Rao, U. L. G., 164, 397
Raqab, M. Z., 97, 246, 280, 392, 399
Rasche, R. H., 286, 399
Raschke, M., 257, 258, 388
Rasmussen, D. W., 13, 392
Rathie, P. N., 115, 398
Ratnaparkhi, M. V., 17, 79, 85, 293, 378, 397, 399
Ravi, S., 100, 399
Rayner, J. C. W., 269, 381
Razali, A. M., 266, 397
Reed, I. S., 53, 377
Reed, W. J., 17, 21, 399
Reiss, R-D., 39, 380
Republic of Botswana, 354, 400
Resnick, S., 74, 308, 399, 400
Resnick, S. I., 74
Revankar, N. S., 10–12, 86, 293–295, 386, 387, 400
Rezaei, S., 254, 400
Rhodin, L. S., 276, 277, 390
Richter, W. F., 204, 382
Rider, P. R., 54, 63, 400
Rinott, Y., 214, 390
Rizzo, M. L., 274, 400
Robertson, C. A., 17, 98, 130, 166, 337, 375, 400, 407
Rodriguez, R. N., 9, 49, 400
Roehner, B., 51, 400
Rogers, G. S., 86, 87, 400
Rohatgi, V. K., 283, 400
Rohde, N., 163, 400
Rootzen, H., 323, 400
Rosenblatt, M., 348, 400
Rossberg, H-J., 86, 87, 90, 400
Rostom, N. K. A., 69, 393
Rothschild, M., 14, 15, 197–199, 201, 400
Roy, M. K., 249, 398
Ruiz, J. M., 80, 400
Russet, B. M., 134, 192, 373
Rutherford, D., 70, 79, 80, 403
Rutherford, R. S. G., 10, 25–27, 36, 353–355, 400

Sahin, I., 24, 400
Saksena, S. K., 12, 227–229, 234, 261, 400
Saldaña-Zepeda, D. P., 274, 400
Saleh, A. K. M. E., 261, 275, 283, 390, 400, 401
Salem, A. B. Z., 12, 36, 356, 401
Samanta, M., 88, 401
Samuel, P., 96, 401
Sanders, J. W., 21, 378
Sandhya, E., 105, 401
Sankaran, P. G., 326, 386, 401
Sarabia, J. M., 91, 153, 155, 158–163, 165–167, 217, 241, 243, 281, 328, 330, 331, 344, 345, 374, 378, 401, 403
Sarabia, M., 160, 163, 165, 401
Saran, J., 69, 401
Sargan, J. D., 9, 20, 401
Sarhan, A. E., 260, 401
Satheesh, S., 105, 401
Saunders, S. C., 260, 396
Saykan, Y., 17, 394
Scarsini, M., 124, 313, 315, 395
Schmeiser, B. W., 162, 399
Schmittlein, D. C., 36, 80, 253, 402
Schoenberg, F., 37, 298, 389

Schur, I., 199, 401
Schutz, R. R., 6, 42, 129, 147, 355, 401
Schwartz, J. E., 17, 37, 402
Seailles, J., 3, 402
Seal, H. L., 17, 49, 289, 402
Seibert, G. B., Jr., 279, 382
Sen, A., 14, 15, 198, 381
Sen, A. K., 205, 402
Sendler, W., 187, 402
Serfling, R., 235, 377
Seto, N., 233, 402
Severini, T. A., 16, 397
Shafay, A. R., 248, 376, 395
Shaked, M., 169, 212, 214, 216, 402
Shanbhag, D. N., 82, 83, 85, 96, 375, 399
Shanmugam, R., 231, 232, 402
Shanthikumar, J. G., 169, 214, 402
Shapiro, S., 270, 385
Shawky, A. I., 102, 296, 402
Sheshinski, E., 14, 198, 200, 402
Shihadeh, E., 37, 391
Shimizu, R., 85, 402
Shirras, G. F., 6, 173, 354, 402
Shorrocks, A. F., 24, 121, 167, 203, 402
Shortle, J., 17, 402
Shu, B-Y., 166, 375
Silcock, H., 11, 250, 402
Simon, H. A., 10, 27–29, 402
Simpson, E. H., 191, 403
Singer, N. M., 16, 205, 392
Singh Pundir, P., 288, 391
Singh, B. P., 249, 397
Singh, D. C., 275, 398, 403
Singh, P., 275, 403
Singh, R., 151, 163, 372
Singh, R. D., 275, 398
Singh, S. K., 12, 36, 152, 285, 286, 356, 403
Singh, V. P., 257, 403
Singpurwalla, N. D., 11, 12, 16, 148, 149, 169, 175, 184, 185, 187, 250, 303, 386, 392
Sinha, S. K., 236, 403
Slama, R., 17, 390
Slngpurwalla, N. D., 16, 379
Slottje, D. J., 159, 161, 376, 401
Smith, J. T., 15, 270, 271, 356, 384
Smith, R. L., 254, 381
Sobel, M., 260, 277, 376, 382
Sollfrey, W., 53, 377
Sordo, M. A., 166, 403
Sorensen, D. N., 17, 390
Srivastava, M. S., 88, 403
Srivastava, R. C., 92, 406
Stark, R. M., 49, 392
Starrett, D., 14, 15, 198, 381
Stephens, M. A., 258, 269, 379, 408
Steutel, F. W., 12, 403
Stigler, S. M., 191, 403
Stiglitz, J. E., 9, 14, 15, 36, 197–199, 201, 400, 403
Stine, R. A., 252, 381
Strassen, V., 208, 219, 403
Strauss, D. J., 328, 375
Stuart, A., 136, 176, 177, 390
Su, N-C., 79, 388
Sukhatme, P. V., 56, 57, 59, 403
Sullo, P., 70, 79, 80, 344, 398, 403
Sultan, K. S., 262, 280, 371, 393
Sumner, A., 134, 379
Sun, H., 115, 398

Tadikamalla, P. R., 49, 403
Taguchi, T., 10, 12, 16, 130–132, 140, 142, 149, 157, 184, 185, 193, 403, 404
Tahmasbi, R., 254, 400
Tahmasebi, S., 102, 376
Taillie, C., 12, 16, 83, 97, 120, 121, 191, 202, 204–207, 211, 397, 404
Tajvidi, N., 323, 400
Takahasi, K., 10, 302, 303, 310, 404
Takano, K., 17, 49, 404
Takayama, N., 133, 205, 386, 404
Talwalker, S., 115, 404
Tang, Y., 246, 384
Targhetta, M. L., 347, 348, 404
Tarsitano, A., 168, 404
Tavangar, M., 96, 386
Teodorescu, S., 38, 404

Thas, O., 269, 381
Thirugnanasambanthan, K., 90
Thisted, R., 29, 382
Thomas, A., 337, 338, 404
Thomas, D. L., 235, 260, 404
Thomas, P. Y., 342, 378
Thompson, W. A., Jr., 123, 404
Thorin, O., 12, 50, 404
Tierney, L., 263, 405
Tiwari, R. C., 248, 405
Tong, H., 283, 405
Toulemonde, G., 269, 383
Trader, R. L., 249, 405
Tribouley, K., 17, 38, 297, 398
Tripathi, R. C., 289–291, 405
Trudel, R., 58, 305, 393
Trueba, C., 159, 161, 401
Tsionas, E. G., 346, 397
Tsokos, C., 17, 405
Tubilla, A., 85, 394
Turkman, M. A. A., 263, 381
Tvede, M., 17, 46, 390

Umbach, D., 261, 373, 401
Unnikrishnan Nair, N., 326, 401

Valiollahi, R., 280, 399
van Dijk, H. K., 13, 391
van Rijsbergen, C. J., 17, 373
van Zyl, J. M., 280, 281, 405
Vannman, K., 12, 251, 259, 261, 391, 405
Vaquera-Huerta, H., 274, 400
Vartak, M. N., 70, 79, 405
Vartia, P. L. I., 36, 354, 405
Vartia, Y. O., 36, 354, 405
Vernic, R., 38, 307, 315, 375, 404, 405
Verway, D. I., 13, 356, 405
Villaseñor, J. A., 37, 68, 74, 162, 164, 204, 207, 208, 212–214, 274, 375, 405
Viorel, Gh. V., 265, 268, 388
Voinov, V. G., 232, 405
von Bortkiewicz, L., 5, 6, 405
von Collani, E., 109, 381
Voorn, W. J., 99, 405

Wackerly, D., 268, 384
Wallis, J. R., 257, 387
Walther, A., 289, 405
Wang, W-Y., 254, 392
Wang, Y. H., 92, 406
Wani, J. K., 182, 406
Watkins, A. J., 254, 388
Watkins, G. P., 4, 406
Wei, L., 17, 406
Weissman, I., 72, 406
Wesolowski, J., 91, 332, 406
Weymark, J. A., 124, 191, 192, 381, 382, 406
Whitt, W., 219, 323, 406
Whitten, B. J., 48, 379
Whittle, P., 10, 30, 406
Wicksell, S. D., 313, 406
Wilfling, B., 211–213, 406
Wilks, S. S., 71, 72, 77, 406
Williams, P., 12, 251, 262, 392
Willis, J. C., 292, 406
Winiwarter, P., 51, 400
Winkler, W., 5, 406
Wold, H., 7, 122, 155, 179, 406
Wold, H. O. A., 10, 30, 406
Wong, A., 267, 406
Woo, J., 232, 249, 373
Woodroofe, M., 33, 34, 387
Wu, J-W., 94, 267, 277, 406, 407
Wu, S-F., 266, 407
Wu, S-J., 280, 407
Wynn, A. H. A., 11, 46, 296, 393

Xekalaki, E., 79, 90, 92, 114, 115, 381, 407
Xu, A., 246, 384

Yang, H-L., 17, 406
Yeh, H-C., 98, 307–309, 314, 315, 318, 335–338, 350, 351, 375, 407
Yhiea, N. M., 69, 393
Yitzhaki, S., 124, 407
Yntema, D. B., 5, 6, 129, 189, 193, 355–357, 407

Yu, H., 186, 380
Yule, G. U., 27, 407

Zahedi, H., 326, 407
Zalkikar, J. N., 248, 405
Zandonatti, A., 45, 253, 407
Zellner, A., 224, 236, 249, 287, 407
Zempléni, A., 323, 350, 399
Zenga, M., 168, 169, 408
Zhang, J., 257, 258, 408
Zhang, R., 233, 387
Zhao, K., 297, 388
Zhao, L. C., 186, 399
Zhou, N., 233, 387
Zhu, S., 35, 376
Zidek, J. V., 227, 408
Zipf, G., 3, 18, 27, 110, 408
Zisheng, O., 17, 408
Zitikis, R., 186, 380
Zografos, K., 308, 408
Zuccolotto, P., 17, 381

Printed in the United States
by Baker & Taylor Publisher Services